Real Analysis

Real Analysis: An Undergraduate Problem Book for Mathematicians, Applied Scientists, and Engineers is a classical Real Analysis/Calculus problem book. This topic has been a compulsory subject for every undergraduate studying mathematics or engineering for a very long time. This volume contains a huge number of engaging problems and solutions, as well as detailed explanations of how to achieve these solutions. This latter quality is something that many problem books lack, and it is hoped that this feature will be useful to students and instructors alike.

Features
- Hundreds of problems and solutions
- Can be used as a stand-alone problem book, or in conjunction with the author's textbook, **Real Analysis: An Undergraduate Textbook for Mathematicians, Applied Scientists, and Engineers,** ISBN 9781032481487
- Perfect resource for undergraduate students studying a first course in Calculus or Real Analysis
- Contains explanatory figures, detailed techniques, tricks, hints, and "recipes" on how to proceed once we have a calculus problem in front of us.

Real Analysis
An Undergraduate Problem Book for Mathematicians, Applied Scientists, and Engineers

Gustavo da Silva Araújo
State University of Paraíba, Brazil

Luis Bernal González
Universidad de Sevilla, Spain

José L. Gámez Merino
Universidad Complutense de Madrid, Spain

María E. Martínez Gómez
Rey Juan Carlos University, Spain

Gustavo A. Muñoz Fernández
Universidad Complutense de Madrid, Spain

Daniel L. Rodríguez Vidanes
Universidad Politécnica de Madrid, Spain

Juan B. Seoane Sepúlveda
Universidad Complutense de Madrid, Spain

CRC Press
Taylor & Francis Group
Boca Raton London New York

CRC Press is an imprint of the
Taylor & Francis Group, an **informa** business

A CHAPMAN & HALL BOOK

Designed cover image: ShutterStock Images

First edition published 2024
by CRC Press
2385 NW Executive Center Drive, Suite 320, Boca Raton FL 33431

and by CRC Press
4 Park Square, Milton Park, Abingdon, Oxon, OX14 4RN

CRC Press is an imprint of Taylor & Francis Group, LLC

© 2024 Gustavo da Silva Araújo, Gustavo A. Muñoz Fernández, María E. Martínez Gómez, Luis Bernal González, José L. Gámez Merino,Juan B. Seoane Sepúlveda, Daniel L. Rodríguez Vidanes

ISBN: 978-1-032-50461-2 (hbk)
ISBN: 978-1-032-51026-2 (pbk)
ISBN: 978-1-003-40074-5 (ebk)

DOI: 10.1201/9781003400745

Typeset in CMR10 font
by KnowledgeWorks Global Ltd.

Publisher's note: This book has been prepared from camera-ready copy provided by the authors.

Contents

Preface

This text does not intend to be a theoretical guide with exercises, but a complement and help to all the classical undergraduate Calculus textbooks by providing a carefully chosen collection of exercises that, with detailed worked out solutions, would certainly enlighten the student in their understanding of the subject of one variable Calculus.

Certainly the topic of Real Analysis is a very vast and wide one. The study of Calculus dates back, at least, to the time of the Greek mathematician and philosopher Eudoxus of Cnidus (IV century BC), who rigorously developed Antiphon's method of exhaustion (a precursor to the integral calculus) using (without developing it rigorously) the idea of limit in order to approximate areas of irregular geometric figures. However, the modern formalization of Mathematical Analysis did not come until two millennia later, when Descartes and Fermat developed Analytic Geometry in the middle of the XVII century, the subsequent development of Infinitesimal Calculus by Leibniz and Newton in the XVIII century and (already in the XIX century) the work of Cauchy in which he established formalism and rigor in Calculus as we know it, essentially, today and through the sequences bearing his name.

More authors should be named in these aforementioned time periods together with their respective seminal contributions to the area. We cannot forget the Bernoulli family (XVIII century) to whom, among many other results, we owe the famous L'Hôpital's rule. Nor can we overlook the enormous contributions of Euler and his magical numerical series, in addition to, among many other contributions, having established the current notation we nowadays use to represent a function, or the number e. Furthermore, it is a must to remember Bolzano and Weierstrass (XIX century) and, in particular, the famous theorem that bears their names. This theory could have not been developed without the help of Cantor (XIX–XX centuries), his theory of the infinity and his famous ternary set. Darboux and Riemann (XIX–XX centuries) also deserve a special place, since the version of the integral that we study here is due to them (although there are other, more advanced versions of the integral, but they are not studied in an initial course in real analysis, such as, for example, the Lebesgue integral). In the XX century, Mathematical Analysis diversified into more varied and essential topics, such as the study of function spaces (functional analysis), complex analysis, measure theory or operator theory, among others.

This problem book is divided into eight main chapters or topics, that are those typically included in any first year Calculus textbook, namely:

(1.) The Field of Real Numbers.

(2.) The Field of Complex Numbers.

(3.) Sequences of Real Numbers. Convergence.

(4.) Continuous Functions.

(5.) Differentiability.

(6.) Riemann Integrability.

(7.) Numerical Series.

(8.) Power Series. Function Sequences and Series.

Of course, this book serves as the perfect guide to students who wish to have a thorough understanding of the theory involved in Calculus by means of a very rich and extended selection of problems (easy ones and "less easy" ones!) including the typical tricks or recipes that usually appear in Calculus problems and less conventional techniques that are useful in order to tackle them successfully. At the same time it is also an excellent text for instructors to find ways to illustrate the theory they deliver in class. Most of the problems appearing in this text are extracted from the proposed problem list of the book entitled "Real Analysis: An Undergraduate Textbook for Mathematicians, Applied Scientists, and Engineers" [6] currently being revised by the time this present text is in press. Although, by now, there are quite an amount of Real Analysis (or Calculus) textbooks published around the World, we believe that a problem book with detailed worked out solutions (and not just a mere list of exercises with the corresponding list of answers to them) is something that, in our experience as instructors, a student will always need. It would neither be the first time nor the last that a mathematics instructor, during class, is asked the question "where can I find the solutions to all the exercises?" coming from a student. It does not matter the college we are in or the country we live in, students always want to have the answers to the proposed exercises and we believe that this book covers the most important type of questions that any Calculus student could encounter while pursuing a degree involving mathematics.

Gustavo da Silva Araújo
Luis Bernal González
José Luis Gámez Merino
María Elena Martínez Gómez
Gustavo Adolfo Muñoz Fernández
Daniel Luis Rodríguez Vidanes,
Juan Benigno Seoane Sepúlveda

Author Bios

Gustavo da Silva Araújo is an Associate Professor at the State University of Paraíba, Brazil. His primary research interests encompass real and complex analysis, the geometry of Banach spaces, operator theory, series and summability, mathematical inequalities, and lineability. He has authored several papers in these areas, and also serves as reviewer for many mathematics journals. He earned his Ph.D. in mathematics from the Federal University of Paraíba in 2016.

Luis Bernal González graduated in 1980 from the Universidad de Sevilla, Spain. He obtained his Ph.D. in Mathematics from the same university in 1984. Bernal has been a permanent faculty member at Sevilla since 1980 and was promoted to Associate Professor in 1987, and to Full Professor in 2010. He was an invited speaker at the International Congress on Hypercyclicity and Chaos for Linear Operators and Semigroups in Valencia (Spain) in 2009. His main interests are Complex Analysis, Operator Theory and, lately, the interdisciplinary subject of Lineability. Bernal has authored or co-authored more than 130 papers in these areas, many of them concerning the structure of the sets of the mathematical objects discovered. He has been plenary lecturer at many international conferences.

José L. Gámez Merino graduated from Universidad Complutense de Madrid (Spain) in 1989 and obtained his Ph.D. degree in Mathematics from the same university in 1997. He is an expert in Real Analysis. Gámez is, currently, an Associate Professor at the Department of Mathematical Analysis and Applied Mathematics at the Universidad Complutense de Madrid.

María E. Martínez Gómez is currently an Assistant Professor at the Department of Applied Mathematics, Materials Science and Engineering and Electronic Technology, Rey Juan Carlos University (Spain). In 2017, she graduated in Mathematics from the Complutense University of Madrid (UCM) and defended her Ph.D. Thesis in 2021. Her areas of expertise include Real and Convex Analysis and Set Theory.

Gustavo A. Muñoz Fernández graduated in Mathematics from Universidad Complutense in 1994 and in Physics from UNED in 2001. He earned his Ph.D. in Mathematics from Universidad Complutense in 1999. Muñoz is currently the Academic Secretary of the Interdisciplinary Mathematics Institute (IMI) and a Full Professor at the Department of Mathematical Analysis and Applied Mathematics at Universidad Complutense. Muñoz has co-authored more than 70 publications including a research book and several textbooks. The scientific interests of Muñoz are related, mainly, to geometry of Banach spaces, polynomials in normed spaces and algebraic genericity (lineability).

Daniel L. Rodríguez Vidanes is currently an Assistant Professor within the Department of Applied Mathematics to Industrial Engineering at Universidad Politécnica de Madrid (UPM). He defended his Ph.D. thesis on 2023 under the supervision of professors Juan B. Seoane Sepúlveda and Gustavo A. Muñoz Fernández from UCM (Spain), alongside Krzysztof C. Ciesielski from West Virginia University (WVU, USA). His academic journey includes over 20 scientific international publications. Additionally, he also co-authored a research book on the geometry of spaces of polynomials. Daniel's scholarly pursuits are deeply rooted in various domains within mathematics. His research interests span the analysis of real functions, functional analysis, geometry of Banach spaces, spaces of polynomials, and lineability.

Juan B. Seoane Sepúlveda earned his first Ph.D. at the Universidad de Cádiz (Spain) jointly with Universität Karlsruhe (Germany) in 2005. He earned his second Ph.D. at Kent State University (Kent, Ohio, USA) in 2006. His main interests include Real and Complex Analysis, Operator Theory, Number Theory, Banach Space Geometry, and Lineability. He has co-authored about 200 papers up to this day, together with several books. Seoane is currently a Full Professor at Universidad Complutense de Madrid (Spain), where he also holds the position of Director of the Master's Studies in Advanced Mathematics. He has delivered invited lectures at many international conferences and research institutes around the world.

Chapter 1

The Field of Real Numbers

Exercise 1.1. *Decide if each one of the following statements is true or false:*

(I) $\displaystyle\sum_{j=1}^{30} j^4 = \sum_{j=0}^{30} j^4,$ (III) $\displaystyle\sum_{j=1}^{20} \left(2 + j^2\right) = 2 + \sum_{j=1}^{20} j^2,$

(II) $\displaystyle\sum_{j=0}^{100} 2 = 200,$ (IV) $\displaystyle\sum_{k=1}^{100} k^2 = \left(\sum_{k=1}^{100} k\right)^2.$

Solution to Exercise 1.1.
- (I). True.
- (II). False.
- (III). False.
- (IV). False. $\square$

Exercise 1.2. *Express with summation notation:*

(I) $\dfrac{1}{1\cdot 2} + \dfrac{1}{2\cdot 3} + \dfrac{1}{3\cdot 4} + \cdots + \dfrac{1}{10\cdot 11},$

(II) $1 + 40 + 900 + 16\,000 + 250\,000 + 3\,600\,000,$

(III) $1 - 2x + 3x^2 - 4x^3 + 5x^4,$

(IV) $a_0 x^4 + a_1 x^3 + a_2 x^2 + a_3 x + a_4,$

(V) $a^5 - a^4 b + a^3 b^2 - a^2 b^3 + ab^4 - b^5,$

(VI) $a^5 + a^4 b + a^3 b^2 + a^2 b^3 + ab^4 + b^5.$

DOI: 10.1201/9781003400745-1

1

Solution to Exercise 1.2.

If the expression of any of your summations does not match the ones below, do not panic!, there are many ways to do it. The important thing is that the expression is equivalent to that of the statement.

(I). $\dfrac{1}{1 \cdot 2} + \dfrac{1}{2 \cdot 3} + \dfrac{1}{3 \cdot 4} + \cdots + \dfrac{1}{10 \cdot 11} = \displaystyle\sum_{j=1}^{10} \dfrac{1}{j(j+1)}.$

(II). $1 + 40 + 900 + 16\,000 + 250\,000 + 3\,600\,000 = \displaystyle\sum_{k=0}^{5}(k+1)^2 10^k.$

(III). $1 - 2x + 3x^2 - 4x^3 + 5x^4 = \displaystyle\sum_{i=1}^{5}(-1)^{i+1} i x^{i-1}.$

(IV). $a_0 x^4 + a_1 x^3 + a_2 x^2 + a_3 x + a_4 = \displaystyle\sum_{n=0}^{4} a_n x^{4-n}.$

(V). $a^5 - a^4 b + a^3 b^2 - a^2 b^3 + ab^4 - b^5 = \displaystyle\sum_{m=0}^{5}(-1)^m a^{5-m} b^m.$

(VI). $a^5 + a^4 b + a^3 b^2 + a^2 b^3 + ab^4 + b^5 = \displaystyle\sum_{l=0}^{5} a^{5-l} b^l.$

$\square$

Exercise 1.3. *Knowing that* $\dfrac{1}{j(j+1)} = \dfrac{1}{j} - \dfrac{1}{j+1}$ *for each* $j \in \mathbb{N}$, *find the sum of* $\displaystyle\sum_{j=1}^{n} \dfrac{1}{j(j+1)}$ *where* $n \in \mathbb{N}$.

Solution to Exercise 1.3.

We know $\frac{1}{j(j+1)} = \frac{1}{j} - \frac{1}{j+1}$ for each $j \in \mathbb{N}$, hence

$$\sum_{j=1}^{n} \frac{1}{j(j+1)} = \sum_{j=1}^{n}\left(\frac{1}{j} - \frac{1}{j+1}\right)$$

$$= \left(1 - \frac{1}{2}\right) + \left(\frac{1}{2} - \frac{1}{3}\right) + \cdots + \left(\frac{1}{n-1} - \frac{1}{n}\right) + \left(\frac{1}{n} - \frac{1}{n+1}\right),$$

where in the last summation observe that the terms are canceling consecutively (this is known as a *telescopic sum*), and the only terms that do not cancel are

1 and $-\frac{1}{n+1}$. That is,

$$\sum_{j=1}^{n} \frac{1}{j(j+1)} = 1 - \frac{1}{n+1}.$$

Another, although a little bit more elaborate, way to do it is as follows: For $n = 1$, we have $\sum_{j=1}^{1} \frac{1}{j(j+1)} = \frac{1}{2}$. Therefore, assume that $n \in \mathbb{N}$ is such that $n \geq 2$, thus

$$\sum_{j=1}^{n} \frac{1}{j(j+1)} = \sum_{j=1}^{n} \left(\frac{1}{j} - \frac{1}{j+1} \right)$$

$$= \sum_{j=1}^{n} \frac{1}{j} - \sum_{j=1}^{n} \frac{1}{j+1}$$

$$= 1 + \sum_{j=2}^{n} \frac{1}{j} - \left(\sum_{j=1}^{n-1} \frac{1}{j+1} + \frac{1}{n+1} \right). \tag{1.1}$$

In the first summation of (1.1) we apply the change of variable $k = j - 1$ (we leave the second summation as it is, but we simply write k instead of j since the value of the sum does not change), and we obtain

$$\sum_{j=1}^{n} \frac{1}{j(j+1)} = 1 + \sum_{k=1}^{n-1} \frac{1}{k+1} - \sum_{k=1}^{n-1} \frac{1}{k+1} - \frac{1}{n+1} = 1 - \frac{1}{n+1}.$$

$\square$

Exercise 1.4. *Find an explicit expression for the following sums ($n \in \mathbb{N}$):*

(I) $\sum_{j=1}^{n} (2j - 1)$. *Hint: Use the equality $j^2 - (j-1)^2 = 2j - 1$ where $j \in \mathbb{N}$.*

(II) $\sum_{j=1}^{n} j$. *Hint: Recall (I).*

(III) *Use a similar strategy to calculate $\sum_{j=1}^{n} j^2$.*

Solution to Exercise 1.4.

(I). First, observe that for each $j \in \mathbb{N}$,

$$j^2 - (j-1)^2 = j^2 - j^2 + 2j - 1 = 2j - 1.$$

Then, it is enough to notice that the summation

$$\sum_{j=1}^{n} \left(j^2 - (j-1)^2 \right) =$$

$$= 1 + (4-1) + (9-4) + \cdots + \left((n-1)^2 - (n-2)^2 \right) + \left(n^2 - (n-1)^2 \right)$$

is a telescopic sum and only the term n^2 does not cancel. Hence,

$$\sum_{j=1}^{n} (2j - 1) = n^2.$$

Another more elaborate way to do it is the following: On the one hand, $\sum_{j=1}^{1} j^2 = 1$. Therefore, assume that $n \geq 2$. Recall that $j^2 - (j-1)^2 = j^2 - j^2 + 2j - 1 = 2j - 1$ for all $j \in \mathbb{N}$, thus

$$\sum_{j=1}^{n} (2j-1) = \sum_{j=1}^{n} \left(j^2 - (j-1)^2 \right)$$

$$= \sum_{j=1}^{n} j^2 - \sum_{j=1}^{n} (j-1)^2$$

$$= n^2 + \sum_{j=1}^{n-1} j^2 - \sum_{j=2}^{n} (j-1)^2. \tag{1.2}$$

In the last equality we have used for the second summation the fact that when $j = 1$, we have $(j-1)^2 = 0$. Now, in the first summation of the third line of (1.2) we make the change of variable $k = j-1$ (we leave the second summation as it is, but we write k instead j since the value does not change), and we obtain

$$\sum_{j=1}^{n} (2j-1) = n^2 + \sum_{k=2}^{n} k^2 - \sum_{k=2}^{n} (k-1)^2 = n^2.$$

(II). We are using part (I) above:

$$n^2 = \sum_{j=1}^{n} (2j-1) = 2\sum_{j=1}^{n} j - \sum_{j=1}^{n} 1 = 2\sum_{j=1}^{n} j - n.$$

Leaving only the summation $\sum_{j=1}^{n} j$ on one side yields

$$\sum_{j=1}^{n} j = \frac{1}{2} \left(n^2 + n \right) = \frac{n(n+1)}{2}.$$

(III). First, we are going to apply a similar method to the one used in part (I). Observe that $j^3 - (j-1)^3 = 3j^2 - 3j + 1$ with $j \in \mathbb{N}$. Indeed,

$$j^3 - (j-1)^3 = j^3 - j^3 + 3j^2 - 3j + 1 = 3j^2 - 3j + 1.$$

Then,

$$\sum_{j=1}^{n} \left(j^3 - (j-1)^3 \right) = \sum_{j=1}^{n} \left(3j^2 - 3j + 1 \right)$$

$$= 3\sum_{j=1}^{n} j^2 - 3\sum_{j=1}^{n} j + \sum_{j=1}^{n} 1$$

$$= 3\sum_{j=1}^{n} j^2 - 3\sum_{j=1}^{n} j + n. \qquad (1.3)$$

Observe that the first summation of (1.3) is a telescopic sum:

$$\sum_{j=1}^{n} \left(j^3 - (j-1)^3 \right) = 1 + (8-1) + (27-8) + \cdots$$

$$+ \left((n-1)^3 - (n-2)^3 \right) + \left(n^3 - (n-1)^3 \right),$$

and the only term that does not cancel is n^3. Now, applying part (II) on the right-hand side of (1.3), we arrive at

$$n^3 = 3\sum_{j=1}^{n} j^2 - \frac{3n(n+1)}{2} + n.$$

Leaving only the summation $\sum_{j=1}^{n} j^2$ on one side, we obtain

$$\sum_{j=1}^{n} j^2 = \frac{1}{3} \left(n^3 + \frac{3n(n+1)}{2} - n \right)$$

$$= \frac{2n^3 + 3n(n+1) - 2n}{6}$$

$$= \frac{n\left(2n^2 + 3n + 1\right)}{6}$$

$$= \frac{n(n+1)(2n+1)}{6},$$

where in the last equality we have factorized $2n^2 + 3n + 1$ by solving the quadratic equation.

$\square$

Exercise 1.5. *Some of these statements about natural numbers n and p are true and others false. Decide which one is which and justify your answer.*

(I) n^2 *is even if, and only if, n is even.*

(II) $(n+p)^2$ *is even if, and only if, $(n-p)^2$ is even.*

(III) *If np is odd, then $n+p$ is even.*

(IV) *If $n^2 + np + p^2$ is even, then np is even.*

(V) *If $n^2 + np + p^2$ is even, then n and p are even.*

Solution to Exercise 1.5.
(I). True. Observe that it is an equivalence, so we have to prove two things. The left to right implication and the right to left one.

($\Leftarrow$) First, we are going to prove the right to left one; that is, we have to show that if n is even, then n^2 is even. To do so, we will use the direct method. Assume that n is even, then there is a $k \in \mathbb{Z}$ such that $n = 2k$. Now, squaring the equality $n = 2k$, we have

$$n^2 = (2k)^2 = 4k^2 = 2(2k^2) = 2k',$$

where $k' = 2k^2 \in \mathbb{Z}$, that is, n^2 is even.

($\Rightarrow$) Second, we are proving the left to right implication; that is, we are showing that if n^2 is even, then n is even. Once again, we are using the direct method. Assume that n^2 is even. We know that n is a natural number, so it is either even or odd. We can express the latter claim in the following way: $n = 2k + r$ with $k \in \mathbb{Z}$ and $r \in \{0, 1\}$. Squaring $n = 2k + r$ gives us:

$$n^2 = (2k+r)^2 = 4k^2 + 4k + r^2 = 2(2k^2 + 2k) + r^2 = 2k' + r^2,$$

where $k' = 2k^2 + 2k \in \mathbb{Z}$. Since we are assuming that n^2 is even, it is clear that r^2 must be 0. So, as $r^2 = 0$, we have $r = 0$. Thus, $n = 2k + r = 2k$ with $k \in \mathbb{Z}$, that is, n is even.

Since we have verified both implications, the statement is true.

(II). True.

($\Leftarrow$) First, we are proving the right to left implication using the direct method. Assume that $(n - p)^2$ is even. If $n - p \in \mathbb{N}$, then by (I) we have that $n - p$ is even. If $n - p \in \mathbb{Z} \setminus \mathbb{N}$, then $n - p = 0$ provided that $n = p$, or $n - p = (-1) \cdot (p - n)$ if $n < p$. Now, if $n < p$, then $p - n \in \mathbb{N}$; hence, by (I) we have that $p - n$ is even. Therefore, in any case we have that $n - p$ is even, which implies that n and p have the same parity. That is, $n = 2k' + r$ for some $k' \in \mathbb{Z}$ and $r \in \{0, 1\}$, and $p = 2k'' + r$ with $k'' \in \mathbb{Z}$. Thus,

$$(n + p)^2 = (2k' + r + 2k'' + r)^2 = 2 \cdot 2(k' + r + k'')^2 = 2k''',$$

where $k''' = 2(k' + r + k'')^2 \in \mathbb{Z}$, that is, $(n + p)^2$ is even.

($\Rightarrow$) Finally, we are proving the left to right implication using the direct method. Assume that $(n + p)^2$ is even, then by (I) we have that $n + p$ is even. The latter implies that n and p have the same parity; that is, there exist $k, k' \in \mathbb{Z}$ and $r \in \{0, 1\}$ such that $n = 2k + r$ y $p = 2k' + r$. Hence,

$$(n - p)^2 = [2k + r - (2k' + r)]^2 = 2 \cdot 2(k - k')^2 = 2k''$$

where $k' = 2(k - k')^2 \in \mathbb{Z}$, that is, $(n - p)^2$ is even.

(III). True. We will prove it using the direct method.

If np is odd, then n and p are odd. That is, there are $k, k' \in \mathbb{Z}$ such that $n = 2k - 1$ y $p = 2k' - 1$. Therefore,

$$n + p = 2k - 1 + 2k' - 1 = 2(k + k' - 1) = 2k'',$$

where $k'' = k + k' - 1 \in \mathbb{Z}$, that is, $n + p$ is even.

(IV). True. We are going to prove it by contrapositive. Assume that np is odd and let us see show that then $n^2 + np + p^2$ is odd.

Since np is odd, we have that n and p are odd; that is, there exist $k, k', k'' \in \mathbb{Z}$ such that $n = 2k - 1$, $p = 2k' - 1$, and $np = 2k'' - 1$. Hence,

$$\begin{aligned}
n^2 + np + p^2 &= (2k - 1)^2 + (2k'' - 1) + (2k' - 1)^2 \\
&= 4k^2 - 4k + 1 + 2k'' - 1 + 4(k')^2 - 4k' + 1 \\
&= 2(2k^2 - 2k + k'' + 2(k')^2 - 2k') + 1 \\
&= 2k''' + 1,
\end{aligned}$$

where $k''' = 2k^2 - 2k + k'' + 2(k')^2 - 2k' \in \mathbb{Z}$, which means that $n^2 + np + p^2$ is odd.

(V). True. We will prove it by contrapositive, that is, assume that one of the numbers n, p is odd, and let us see that $n^2 + np + p^2$ is odd. If n and p are odd, then we have that np is odd. Assume that n is odd and p is even (analogously when n is even and p is odd). Then there are $k, k' \in \mathbb{Z}$ such that $n = 2k - 1$ and $p = 2k'$. Hence,

$$
\begin{aligned}
n^2 + np + p^2 &= (2k - 1)^2 + 2k'(2k - 1) + 4(k')^2 \\
&= 4k^2 + 1 - 4k + 2k'(2k - 1) + 4(k')^2 \\
&= 2(2k^2 - 2k + k'(2k - 1) + 2(k')^2) + 1 \\
&= 2k'' + 1,
\end{aligned}
$$

where $k'' = 2k^2 - 2k + k'(2k - 1) + 2(k')^2 \in \mathbb{Z}$, which implies that $n^2 + np + p^2$ is odd.

$\square$

Exercise 1.6. *Prove that*

$$
x^n - y^n = (x - y)(x^{n-1} + x^{n-2}y + \cdots + xy^{n-2} + y^{n-1})
$$

for each $n \in \mathbb{N} \setminus \{1\}$, $x, y \in \mathbb{R}$. Write the right-hand side of the equality using summation notation. This formula is known as Cyclotomic Equation.

Solution to Exercise 1.6.
We shall prove the Cyclotomic Equation using the direct method, and in order to do so, we begin by writing the term $x^{n-1} + x^{n-2}y + \cdots + xy^{n-2} + y^{n-1}$ using the summation notation:

$$
x^{n-1} + x^{n-2}y + \cdots + xy^{n-2} + y^{n-1} = \sum_{k=0}^{n-1} x^{n-1-k}y^k. \tag{1.4}
$$

Multiplying the summation of (1.4) by $x - y$, we have

$$
(x - y)\sum_{k=0}^{n-1} x^{n-1-k}y^k = \sum_{k=0}^{n-1} x^{n-k}y^k - \sum_{k=0}^{n-1} x^{n-1-k}y^{k+1}
$$

$$
= x^n + \sum_{k=1}^{n-1} x^{n-k}y^k - \sum_{k=0}^{n-2} x^{n-1-k}y^{k+1} - y^n. \tag{1.5}
$$

Now, rewriting the second summation of the second line of (1.5) as $\sum_{k=1}^{n-1} x^{n-k}y^k$, we have that

$$
(x - y)\left(x^{n-1} + x^{n-2}y + \cdots + xy^{n-2} + y^{n-1}\right)
$$

$$
= x^n + \sum_{k=1}^{n-1} x^{n-k}y^k - \sum_{k=1}^{n-1} x^{n-k}y^k - y^n
$$

$$
= x^n - y^n.
$$

$\square$

Exercise 1.7. *Infer from the Cyclotomic Equation the sum $\sum_{j=0}^{n} x^j$ where $x \neq 1$. Do calculations in the expression $(1-x)\sum_{j=1}^{n} jx^j$ to deduce the sum $\sum_{j=1}^{n} jx^j$ where $x \neq 1$. Analogously in $(1-x)\sum_{j=1}^{n} j^2 x^j$ to deduce the sum $\sum_{j=1}^{n} j^2 x^j$ where $x \neq 1$.*

Solution to Exercise 1.7.

First, we will calculate $\sum_{j=0}^{n} x^j$ with $x \neq 1$. To do so, we are using the Cyclotomic Equation of Exercise 1.6 for $n+1$:

$$x^{n+1} - y^{n+1} = (x-y)\left(x^n + x^{n-1}y + \cdots + xy^{n-1} + y^n\right).$$

Taking $y = 1$ we have that

$$x^{n+1} - 1 = (x-1)\left(x^{n-1} + x^{n-2} + \cdots + x + 1\right) = (x-1)\sum_{j=0}^{n} x^j.$$

Thus,

$$\sum_{j=0}^{n} x^j = \frac{x^{n+1} - 1}{x - 1}.$$

Second, we will calculate $\sum_{j=1}^{n} jx^j$ with $x \neq 1$. Observe that the value of the sum is trivial if $n = 1$. Hence, assume that $n \geq 2$. Doing calculations in $(1-x)\sum_{j=1}^{n} jx^j$, we have

$$(1-x)\sum_{j=1}^{n} jx^j = \sum_{j=1}^{n} jx^j - \sum_{j=1}^{n} jx^{j+1}$$

$$= x + \sum_{j=2}^{n} jx^j - \sum_{j=1}^{n-1} jx^{j+1} - nx^{n+1}$$

$$= x + \sum_{j=2}^{n} jx^j - \sum_{j=2}^{n} (j-1)x^j - nx^{n+1}$$

$$= x + \sum_{j=2}^{n} jx^j - \sum_{j=2}^{n} jx^j + \sum_{j=2}^{n} x^j - nx^{n+1}$$

$$= x + \sum_{j=2}^{n} x^j - nx^{n+1}$$

$$= \sum_{j=0}^{n} x^j - nx^{n+1} - 1$$

$$= \frac{x^{n+1} - 1}{x - 1} - nx^{n+1} - 1$$

$$= \frac{(1+n)x^{n+1} - nx^{n+2} - x}{x - 1}.$$

Therefore,

$$\sum_{j=1}^{n} jx^j = \frac{nx^{n+2} - (1+n)x^{n+1} + x}{(x-1)^2}.$$

Finally, let us calculate $\sum_{j=1}^{n} j^2 x^j$ with $x \neq 1$. It is straightforward if $n = 1$, so assume that $n \geq 2$. We begin by doing calculations in $(1-x)\sum_{j=1}^{n} j^2 x^j$:

$$(1-x)\sum_{j=1}^{n} j^2 x^j = \sum_{j=1}^{n} j^2 x^j - \sum_{j=1}^{n} j^2 x^{j+1}$$

$$= x + \sum_{j=2}^{n} j^2 x^j - \sum_{j=1}^{n-1} j^2 x^{j+1} - n^2 x^{n+1}$$

$$= x + \sum_{j=2}^{n} j^2 x^j - \sum_{j=2}^{n} (j-1)^2 x^j - n^2 x^{n+1}$$

$$= x + \sum_{j=2}^{n} j^2 x^j - \sum_{j=2}^{n} j^2 x^j + 2\sum_{j=2}^{n} j x^j - \sum_{j=2}^{n} x^j - n^2 x^{n+1}$$

$$= x + 2\sum_{j=2}^{n} j x^j - \sum_{j=2}^{n} x^j - n^2 x^{n+1}$$

$$= 2x - x + 1 - 1 + 2\sum_{j=2}^{n} j x^j - \sum_{j=2}^{n} x^j - n^2 x^{n+1}$$

$$= 2\sum_{j=1}^{n} j x^j - \sum_{j=0}^{n} x^j - n^2 x^{n+1} + 1$$

$$= 2\frac{n x^{n+2} - (1+n) x^{n+1} + x}{(x-1)^2} - \frac{x^{n+1} - 1}{x - 1} - n^2 x^{n+1} + 1$$

$$= \frac{-n^2 x^{n+3} + (2n^2 + 2n - 1) x^{n+2} - (n+1)^2 x^{n+1} + x^2 + x}{(x-1)^2}.$$

Hence,

$$\sum_{j=1}^{n} j^2 x^j = \frac{n^2 x^{n+3} - (2n^2 + 2n - 1) x^{n+2} + (n+1)^2 x^{n+1} - x^2 - x}{(x-1)^3}.$$

$\square$

Exercise 1.8. *Prove the following properties by induction on $n \in \mathbb{N}$.*

(I) $\displaystyle\sum_{k=1}^{n} k^2 = \frac{n(n+1)(2n+1)}{6},$

(II) $\displaystyle\sum_{k=1}^{n} \frac{k+4}{k(k+1)(k+2)} = \frac{n(3n+7)}{2(n+1)(n+2)},$

(III) $\displaystyle\sum_{k=1}^{n} k^3 = \left(\frac{n(n+1)}{2}\right)^2,$

(IV) $\displaystyle\sum_{k=1}^{n} k(k+1) = \frac{n(n+1)(n+2)}{3}$,

(V) $\displaystyle\sum_{k=1}^{n} \frac{1}{\sqrt{k}} \geq \sqrt{n}$,

(VI) $\displaystyle\sum_{k=n+1}^{2n} \frac{1}{k} = \sum_{k=1}^{2n} \frac{(-1)^{k+1}}{k}$,

(VII) $\displaystyle\sum_{j=1}^{n} ar^{j-1} = \frac{a(r^n - 1)}{r - 1}$ *where* $a \in \mathbb{R}$ *and* $r \neq 1$.

Solution to Exercise 1.8.

(I).

- Base case ($n = 1$): $\displaystyle\sum_{k=1}^{1} k^2 = 1^2 = 1$ and $\frac{1(1+1)(2\cdot1+1)}{6} = \frac{1\cdot2\cdot3}{6} = 1$, so the base case is true.

- Induction step. Assume that $\displaystyle\sum_{k=1}^{n} k^2 = \frac{n(n+1)(2n+1)}{6}$ and let us see

  that $\displaystyle\sum_{k=1}^{n+1} k^2 = \frac{(n+1)(n+2)(2n+3)}{6}$. Indeed,

$$
\begin{aligned}
\sum_{k=1}^{n+1} k^2 &= \sum_{k=1}^{n} k^2 + (n+1)^2 \\
&= \frac{n(n+1)(2n+1)}{6} + (n+1)^2 \\
&= \frac{(n+1)(n(2n+1) + 6(n+1))}{6} \\
&= \frac{(n+1)(2n^2 + 7n + 6)}{6} \\
&= \frac{(n+1)(n+2)(2n+3)}{6},
\end{aligned}
$$

where in the last equality we have used the second degree equation formula to factorize the numerator.

(II).

- Base case ($n = 1$):

$$\sum_{k=1}^{1} \frac{k+4}{k(k+1)(k+2)} = \frac{1+4}{1(2)(3)} = \frac{5}{6} \text{ and } \frac{1(3+7)}{2(2)(3)} = \frac{5}{6}.$$

- Induction step. Assume that $\displaystyle\sum_{k=1}^{n} \frac{k+4}{k(k+1)(k+2)} = \frac{n(3n+7)}{2(n+1)(n+2)}$ and

let us show that $\displaystyle\sum_{k=1}^{n+1} \frac{k+4}{k(k+1)(k+2)} = \frac{(n+1)(3(n+1)+7)}{2(n+2)(n+3)}$. Indeed,

$$\sum_{k=1}^{n+1} \frac{k+4}{k(k+1)(k+2)} = \sum_{k=1}^{n} \frac{k+4}{k(k+1)(k+2)} + \frac{n+5}{(n+1)(n+2)(n+3)}$$

$$= \frac{n(3n+7)}{2(n+1)(n+2)} + \frac{n+5}{(n+1)(n+2)(n+3)}$$

$$= \frac{n(3n+7)(n+3) + 2(n+5)}{2(n+1)(n+2)(n+3)}$$

$$= \frac{3n^3 + 16n^2 + 23n + 10}{2(n+1)(n+2)(n+3)}$$

$$= \frac{(n+1)^2(3n+10)}{2(n+1)(n+2)(n+3)}$$

$$= \frac{(n+1)(3n+10)}{2(n+2)(n+3)}$$

$$= \frac{(n+1)(3(n+1)+7)}{2(n+2)(n+3)},$$

where we have used Ruffini's Rule to factorize the polynomial of degree three in the numerator.

(III).

- Base case ($n = 1$): $\displaystyle\sum_{k=1}^{1} k^3 = 1^3 = 1$ and $\frac{1^2(2^2)}{4} = 1$.

- Induction step. Assume that $\displaystyle\sum_{k=1}^{n} k^3 = \frac{n^2(n+1)^2}{4}$ and let us see that

$$\sum_{k=1}^{n+1} k^3 = \frac{(n+1)^2(n+2)^2}{4}. \text{ Indeed,}$$

$$\begin{aligned} \sum_{k=1}^{n+1} k^3 &= \sum_{k=1}^{n} k^3 + (n+1)^3 \\ &= \frac{n^2(n+1)^2}{4} + (n+1)^3 \\ &= \frac{(n+1)^2(n^2+4n+4)}{4} \\ &= \frac{(n+1)^2(n+2)^2}{4}. \end{aligned}$$

(IV).

- Base case ($n=1$): $\displaystyle\sum_{k=1}^{1} k(k+1) = 1 \cdot 2 = 2$ and $\frac{1(1+1)(1+2)}{3} = \frac{1 \cdot 2 \cdot 3}{3} = 2.$

- Induction step. Assume that $\displaystyle\sum_{k=1}^{n} k(k+1) = \frac{n(n+1)(n+2)}{3}$ and let us show that $\displaystyle\sum_{k=1}^{n+1} k(k+1) = \frac{(n+1)(n+2)(n+3)}{3}$. Indeed,

$$\begin{aligned} \sum_{k=1}^{n+1} k(k+1) &= \sum_{k=1}^{n} k(k+1) + (n+1)(n+2) \\ & \frac{n(n+1)(n+2)}{3} + (n+1)(n+2) \\ &= \frac{(n+1)(n+2)(n+3)}{3}. \end{aligned}$$

(V).

- Base case ($n=1$): $\displaystyle\sum_{k=1}^{1} \frac{1}{\sqrt{k}} = \frac{1}{\sqrt{1}} = 1$ and $\sqrt{1} = 1.$

- Induction step. Assume that $\displaystyle\sum_{k=1}^{n} \frac{1}{\sqrt{k}} \geq \sqrt{n}$ and let us see that $\displaystyle\sum_{k=1}^{n+1} \frac{1}{\sqrt{k}} \geq \sqrt{n+1}$. Indeed,

$$\sum_{k=1}^{n+1} \frac{1}{\sqrt{k}} = \sum_{k=1}^{n} \frac{1}{\sqrt{k}} + \frac{1}{\sqrt{n+1}} \geq \sqrt{n} + \frac{1}{\sqrt{n+1}}.$$

Observe that it is enough to prove that $\sqrt{n} + \frac{1}{\sqrt{n+1}} \geq \sqrt{n+1}$ for each $n \in \mathbb{N}$. To do so, we are going to prove it backwards, that is, we are going to use equivalent statements until we reach one that is clearly true.

$$\sqrt{n} + \frac{1}{\sqrt{n+1}} \geq \sqrt{n+1}$$

$$\Longleftrightarrow \frac{1}{\sqrt{n+1}} \geq \sqrt{n+1} - \sqrt{n}$$

$$\Longleftrightarrow \frac{1}{\sqrt{n+1}} \geq \frac{(\sqrt{n+1} - \sqrt{n})(\sqrt{n+1} + \sqrt{n})}{\sqrt{n+1} + \sqrt{n}}$$

$$\Longleftrightarrow \frac{1}{\sqrt{n+1}} \geq \frac{1}{\sqrt{n+1} + \sqrt{n}}$$

$$\Longleftrightarrow \sqrt{n+1} + \sqrt{n} \geq \sqrt{n+1}$$

$$\Longleftrightarrow \sqrt{n} \geq 1$$

$$\Longleftrightarrow n \geq 1.$$

The last inequality is always true for all $n \in \mathbb{N}$ and we are done.

(VI).

- Base case ($n = 1$): $\displaystyle\sum_{k=2}^{2} \frac{1}{k} = \frac{1}{2}$ and $\displaystyle\sum_{k=1}^{2} \frac{(-1)^{k+1}}{k} = \frac{(-1)^2}{1} + \frac{(-1)^3}{2} = 1 - \frac{1}{2} = \frac{1}{2}$.

- Induction step. Assume that $\displaystyle\sum_{k=n+1}^{2n} \frac{1}{k} = \sum_{k=1}^{2n} \frac{(-1)^{k+1}}{k}$ (Induction Hypothesis) and let us show that $\displaystyle\sum_{k=n+2}^{2n+2} \frac{1}{k} = \sum_{k=1}^{2n+2} \frac{(-1)^{k+1}}{k}$.

Indeed,

$$\sum_{k=n+2}^{2n+2}\frac{1}{k} = -\frac{1}{n+1} + \frac{1}{n+1} + \sum_{k=n+2}^{2n}\frac{1}{k} + \frac{1}{2n+1} + \frac{1}{2n+2}$$

$$= -\frac{1}{n+1} + \sum_{k=n+1}^{2n}\frac{1}{k} + \frac{1}{2n+1} + \frac{1}{2n+2}$$

$$= -\frac{1}{n+1} + \sum_{k=1}^{2n}\frac{(-1)^{k+1}}{k} + \frac{1}{2n+1} + \frac{1}{2(n+1)}$$

$$= \sum_{k=1}^{2n}\frac{(-1)^{k+1}}{k} + \frac{1}{2n+1} - \frac{1}{2(n+1)}$$

$$= \sum_{k=1}^{2n}\frac{(-1)^{k+1}}{k} + \frac{(-1)^{2n+2}}{2n+1} + \frac{(-1)^{2n+3}}{2n+2}$$

$$= \sum_{k=1}^{2n+2}\frac{(-1)^{k+1}}{k}.$$

(VII).

- Base case $(n=1)$: $\sum_{j=1}^{1} ar^{j-1} = ar^0 = a$ and $\frac{a(r^1-1)}{r-1} = a$.

- Induction step. Assume that $\sum_{j=1}^{n} ar^{j-1} = \frac{a(r^n-1)}{r-1}$ (Induction Hypothesis) and let us show that $\sum_{j=1}^{n+1} ar^{j-1} = \frac{a(r^{n+1}-1)}{r-1}$. Indeed,

$$\sum_{j=1}^{n+1} ar^{j-1} = \sum_{j=1}^{n} ar^{j-1} + ar^n = \frac{a(r^n-1)}{r-1} + ar^n$$

$$= \frac{a(r^n-1) + (r-1)ar^n}{r-1} = \frac{ar^n - a + ar^n(r-1)}{r-1}$$

$$= \frac{ar^n(1+r-1) - a}{r-1} = \frac{ar^{n+1} - a}{r-1}$$

$$= \frac{a(r^{n+1}-1)}{r-1}.$$

$\square$

Exercise 1.9. *Infer, from the equations*

$$1 = 1,$$
$$1 - 4 = -(1 + 2),$$
$$1 - 4 + 9 = 1 + 2 + 3,$$
$$1 - 4 + 9 - 16 = -(1 + 2 + 3 + 4),$$

a general and simple formula that includes the previous ones as particular cases, and prove it using Mathematical Induction.

Solution to Exercise 1.9.
Observe that the following equation is satisfied for all $n \in \{1, 2, 3, 4\}$:

$$\sum_{j=1}^{n} (-1)^{j+1} j^2 = (-1)^{n+1} \sum_{j=1}^{n} j.$$

Now, by Exercise 1.4 part (II) we know that $\sum_{j=1}^{n} j = \dfrac{n(n+1)}{2}$. Thus, the following equality is also a general formula:

$$\sum_{j=1}^{n} (-1)^{j+1} j^2 = (-1)^{n+1} \cdot \frac{n(n+1)}{2}. \tag{1.6}$$

Let us prove by induction (1.6).

- Base case ($n = 1$): $\sum_{j=1}^{1} (-1)^{j+1} j^2 = (-1)^2 = 1$ and $(-1)^{1+1} \cdot \frac{1(1+1)}{2} = 1$.

- Induction step. Assume that $\sum_{j=1}^{n} (-1)^{j+1} j^2 = (-1)^{n+1} \cdot \dfrac{n(n+1)}{2}$ (Induction Hypothesis) and let us show that

$$\sum_{j=1}^{n+1} (-1)^{j+1} j^2 = (-1)^{n+2} \cdot \frac{(n+1)(n+2)}{2}.$$

Indeed,

$$\sum_{j=1}^{n+1}(-1)^{j+1}j^2 = \sum_{j=1}^{n}(-1)^{j+1}j^2 + (-1)^{n+2}(n+1)^2$$

$$= (-1)^{n+1}\cdot\frac{n(n+1)}{2} + (-1)^{n+2}(n+1)^2$$

$$= (-1)^{n+2}(n+1)\left(n+1-\frac{n}{2}\right)$$

$$= (-1)^{n+2}\cdot\frac{(n+1)(n+2)}{2}.$$

$\square$

Exercise 1.10. *Given* $n \in \mathbb{N}$*, we define its* factorial *by*

$$n! = n(n-1)(n-2)\cdots 2\cdot 1,$$

and we also define $0! = 1$*. Given two numbers* $m,n \in \mathbb{N}\cup\{0\}$ *with* $m \geq n$*, we define the binomial coefficient "m over n" as*

$$\binom{m}{n} = \frac{m!}{n!(m-n)!}.$$

Prove the following properties:

(I) $\quad\dbinom{m}{0} = \dbinom{m}{m} = 1, \qquad \dbinom{m}{1} = \dbinom{m}{m-1} = m,$

(II) $\quad\dbinom{m}{n} = \dbinom{m}{m-n},$

(III) $\quad\dbinom{m}{n} + \dbinom{m}{n+1} = \dbinom{m+1}{n+1}.$

Solution to Exercise 1.10.

(I).

- First, let us see that $\binom{m}{0} = \binom{m}{m} = 1$:

$$\binom{m}{0} = \frac{m!}{0!(m-0)!} = \frac{m!}{m!} = 1,$$

$$\binom{m}{m} = \frac{m!}{m!(m-m)!} = \frac{m!}{m!0!} = \frac{m!}{m!} = 1.$$

- Second, let us show that $\binom{m}{1} = \binom{m}{m-1} = m$:

$$\binom{m}{1} = \frac{m!}{1!(m-1)!} = \frac{m \cdot (m-1)!}{(m-1)!} = m,$$

$$\binom{m}{m-1} = \frac{m!}{(m-1)![m-(m-1)]!} = \frac{m \cdot (m-1)!}{(m-1)!1!} = m.$$

(II).

$$\binom{m}{m-n} = \frac{m!}{(m-n)![m-(m-n)]!} = \frac{m!}{(m-n)!n!} = \binom{m}{n}.$$

(III).

$$\binom{m}{n} + \binom{m}{n+1} = \frac{m!}{n!(m-n)!} + \frac{m!}{(n+1)![m-(n+1)]!}$$

$$= \frac{m!}{n!(m-n)\cdot(m-n-1)!} + \frac{m!}{(n+1)\cdot n!(m-n-1)!}$$

$$= \frac{m!(n+1) + m!(m-n)}{(n+1)\cdot n!(m-n)\cdot(m-n-1)!}$$

$$= \frac{(n+1+m-n)\cdot m!}{(n+1)!(m-n)!} = \frac{(m+1)\cdot m!}{(n+1)![m+1-(n+1)]!}$$

$$= \frac{(m+1)!}{(n+1)![m+1-(n+1)]!} = \binom{m+1}{n+1}.$$

$\square$

Exercise 1.11. *Prove* Newton's Binomial Theorem: *for every $x, y \in \mathbb{R}$ and each $n \in \mathbb{N}$,*

$$(x+y)^n = \sum_{j=0}^{n} \binom{n}{j} x^j y^{n-j}.$$

Deduce that:

(I) $1 + n + \binom{n}{2} + \cdots + \binom{n}{n-1} + 1 = 2^n,$

(II) $1 - n + \binom{n}{2} + \cdots + (-1)^{n-1}\binom{n}{n-1} + (-1)^n = 0.$

Solution to Exercise 1.11.

Let us prove Newton's Binomial Theorem, $(x+y)^n = \sum_{j=0}^{n} \binom{n}{j} x^j y^{n-j}$, by induction.

- Base case $(n=1)$: $\sum_{j=0}^{1} \binom{1}{j} x^j y^{n-j} = \binom{1}{0} y + \binom{1}{1} x = x + y$, where we have used Exercise 1.10 part (I).

- Induction step. Assume that Newton's Binomial Theorem is satisfied for some $n \in \mathbb{N}$ (Induction Hypothesis) and let us show that it is true for $n+1$, that is, $(x+y)^{n+1} = \sum_{j=0}^{n+1} \binom{n+1}{j} x^j y^{n+1-j}$. Indeed,

$$
\begin{aligned}
(x+y)^{n+1} &= (x+y)^n (x+y) \\
&= \left(\sum_{j=0}^{n} \binom{n}{j} x^j y^{n-j} \right) (x+y) \\
&= \sum_{j=0}^{n} \binom{n}{j} x^{j+1} y^{n-j} + \sum_{j=0}^{n} \binom{n}{j} x^j y^{n-j+1}.
\end{aligned} \tag{1.7}
$$

Notice that the first summation of the third line of (1.7) can be written as

$$
\sum_{j=0}^{n} \binom{n}{j} x^{j+1} y^{n-j} = \sum_{j=1}^{n+1} \binom{n}{j-1} x^{(j-1)+1} y^{n-(j-1)} = \sum_{j=1}^{n+1} \binom{n}{j-1} x^j y^{n+1-j}.
$$

Thus,

$$
\begin{aligned}
(x+y)^{n+1} \\
&= \sum_{j=0}^{n} \binom{n}{j} x^j y^{n-j+1} + \sum_{j=1}^{n+1} \binom{n}{j-1} x^j y^{n-j+1} \\
&= \binom{n}{0} y^{n+1} + \sum_{j=1}^{n} \binom{n}{j} x^j y^{n+1-j} + \sum_{j=1}^{n} \binom{n}{j-1} x^j y^{n-j+1} + \binom{n}{n} x^{n+1} \\
&= \binom{n+1}{0} y^{n+1} + \sum_{j=1}^{n} \left[\binom{n}{j} + \binom{n}{j-1} \right] x^j y^{n+1-j} + \binom{n+1}{n+1} x^{n+1} \\
&= \binom{n+1}{0} y^{n+1} + \sum_{j=1}^{n} \binom{n+1}{j} x^j y^{n+1-j} + \binom{n+1}{n+1} x^{n+1} \\
&= \sum_{j=0}^{n+1} \binom{n+1}{j} x^j y^{n+1-j},
\end{aligned}
$$

where we have used Exercise 1.10 parts (I) and (III).

(I). To prove the equality, we will use Newton's Binomial Theorem with $x = y = 1$:

$$2^n = (1+1)^n = \sum_{j=0}^{n} \binom{n}{j} 1^j 1^{n-j}$$

$$= \binom{n}{0} + \binom{n}{1} + \binom{n}{2} + \cdots + \binom{n}{n-1} + \binom{n}{n}$$

$$= 1 + n + \binom{n}{2} + \cdots + \binom{n}{n-1} + 1,$$

where we have used Exercise 1.10 part (I).

(II). To prove the equality, we will use Newton's Binomial Theorem with $x = 1$ and $y = -1$:

$$0 = [1 + (-1)]^n = \sum_{j=0}^{n} \binom{n}{j} 1^j (-1)^{n-j}$$

$$= \binom{n}{0} + (-1) \cdot \binom{n}{1} + (-1)^2 \binom{n}{2} + \cdots + (-1)^{n-1} \binom{n}{n-1} + (-1)^n \binom{n}{n}$$

$$= 1 - n + \binom{n}{2} + \cdots + (-1)^{n-1} \binom{n}{n-1} + (-1)^n,$$

where we have used Exercise 1.10 part (I). $\qquad\square$

Exercise 1.12. *Prove by complete induction over k that for all $k \in \mathbb{N} \cup \{0\}$ there is a polynomial $P_k(x)$ of degree $k+1$ such that, for any $n \in \mathbb{N}$ we have*

$$\sum_{j=1}^{n} j^k = P_k(n).$$

Hint: For the induction step, follow the technique used in Exercise 1.4.

Solution to Exercise 1.12.

- Base case ($k = 0$): $\sum_{j=1}^{n} j^0 = \sum_{j=1}^{n} 1 = n$, for all $n \in \mathbb{N}$. Notice that if we take the polynomial $P(x) = x$, which is of degree 1, we have that $P(n) = n$.

- Induction step. Assume that for some $k \in \mathbb{N} \cup \{0\}$ we have that for each $i \in \{0, \ldots, k\}$, there is a polynomial $P_i(x)$ of degree $i + 1$ such that $\sum_{j=1}^{n} j^i = P_i(n)$, for all $n \in \mathbb{N}$ (Induction Hypothesis). Let us see that there is a polynomial $P_{k+1}(x)$ of degree $k+2$ such that $\sum\limits_{j=1}^{n} j^{k+1} = P_{k+1}(n)$ for all $n \in \mathbb{N}$. Fix $n \in \mathbb{N}$ arbitrary. Recall that in Exercise 1.4 part (III), we studied the difference $j^2 - (j-1)^2$ in order to calculate $\sum\limits_{j=1}^{n} j$. In our case, we begin by analyzing the difference $j^{k+2} - (j-1)^{k+2}$ to study $\sum\limits_{j=1}^{n} j^{k+1}$. By Newton's Binomial Theorem, we have that

$$
\begin{aligned}
j^{k+2} &- (j-1)^{k+2} \\
&= j^{k+2} - \sum_{i=0}^{k+2} \binom{k+2}{i} j^i (-1)^{k+2-i} \\
&= j^{k+2} - \binom{k+2}{k+2} j^{k+2} - \sum_{i=0}^{k+1} \binom{k+2}{i} j^i (-1)^{k+2-i} \\
&= - \sum_{i=0}^{k+1} \binom{k+2}{i} j^i (-1)^{k+2-i},
\end{aligned}
$$

where we have used Exercise 1.10 part (I). Hence, summing in terms of j from 1 until n, we obtain

$$
\begin{aligned}
\sum_{j=1}^{n} \left(j^{k+2} - (j-1)^{k+2} \right) &= \sum_{j=1}^{n} \left[- \sum_{i=0}^{k+1} \binom{k+2}{i} j^i (-1)^{k+2-i} \right] \\
&= - \sum_{i=0}^{k+1} \sum_{j=1}^{n} \binom{k+2}{i} j^i (-1)^{k+2-i} \\
&= - \sum_{i=0}^{k+1} \left[\binom{k+2}{i} (-1)^{k+2-i} \left(\sum_{j=1}^{n} j^i \right) \right] \\
&= \binom{k+2}{k+1} \sum_{j=1}^{n} j^{k+1} - \sum_{i=0}^{k} \left[\binom{k+2}{i} (-1)^{k+2-i} \left(\sum_{j=1}^{n} j^i \right) \right] \\
&= \binom{k+2}{k+1} \sum_{j=1}^{n} j^{k+1} - \sum_{i=0}^{k} \binom{k+2}{i} (-1)^{k+2-i} P_i(n) \\
&= (k+2) \sum_{j=1}^{n} j^{k+1} - \sum_{i=0}^{k} \binom{k+2}{i} (-1)^{k+2-i} P_i(n),
\end{aligned}
$$

where we have used Exercise 1.10 part (I). Observe that

$$\sum_{j=1}^{n} \left(j^{k+2} - (j-1)^{k+2} \right)$$

is a telescopic sum and it is equal to n^{k+2}. Thus,

$$n^{k+2} = (k+2) \sum_{j=1}^{n} j^{k+1} - \sum_{i=0}^{k} \binom{k+2}{i} (-1)^{k+2-i} P_i(n). \qquad (1.8)$$

Now, in (1.8) we have the summation that we wanted to study: $\sum_{j=1}^{n} j^{k+1}$.

Then, what we do is to leave on one side of the equality $\sum_{j=1}^{n} j^{k+1}$ and

the rest on the other, which gives us

$$\sum_{j=1}^{n} j^{k+1} = \frac{1}{k+2} \left(n^{k+2} + \sum_{i=0}^{k} \binom{k+2}{i} (-1)^{k+2-i} P_i(n) \right). \qquad (1.9)$$

Since x^{k+2} is a polynomial of degree $k+2$ and the polynomials $P_i(x)$ have degree $i+1$ for all $i \in \{0, \dots, k\}$, we have that there is

no monomial in the polynomial $\sum_{i=0}^{k} \binom{k+2}{i} (-1)^{k+2-i} P_i(x)$ that can-

cels with the monomial x^{k+2}. Therefore, the polynomial $P_{k+2}(x) =$

$\frac{1}{k+2} \left(x^{k+2} + \sum_{i=0}^{k} \binom{k+2}{i} (-1)^{k+2-i} P_i(x) \right)$ has degree $k+2$ and it

is satisfied that $\sum_{j=1}^{n} j^{k+1} = P_{k+2}(n)$ with $n \in \mathbb{N}$ arbitrary.

$\square$

Exercise 1.13. *Prove that for all $n \in \mathbb{N}$:*

(I) $5^n - 1$ *is a multiple of* 4.

(II) $7^{2n+1} - 48n - 7$ *is divisible by* 48.

(III) $2^{2n} + 15n - 1$ *is a multiple of* 9.

(IV) $11^{n+2} + 12^{2n+1}$ *is a multiple of* 133.

Solution to Exercise 1.13.

(I). Let us see that for all $n \in \mathbb{N}$, there is a $k \in \mathbb{Z}$ such that $5^n - 1 = 4k$. We will prove it by induction over $n \in \mathbb{N}$.

- Base case ($n = 1$) is trivial: $5^1 - 1 = 4 = 4 \cdot 1$, where $1 \in \mathbb{Z}$.

- Induction step. Assume that for some $n \in \mathbb{N}$, there is a $k \in \mathbb{Z}$ such that $5^n - 1 = 4k$ (Induction Hypothesis), and let us show that there exists $k' \in \mathbb{Z}$ such that $5^{n+1} - 1 = 4k'$. Indeed,

$$\begin{aligned}
5^{n+1} - 1 &= 5^n \cdot 5 - 1 - 4 + 4 \\
&= 5^n \cdot 5 - 5 + 4 \\
&= 5(5^n - 1) + 4 \\
&= 5 \cdot 4k + 4 \\
&= 4(5k + 1) \\
&= 4k',
\end{aligned}$$

where $k' = 5k + 1 \in \mathbb{Z}$.

(II). It is enough to prove that for each $n \in \mathbb{N}$, there is a $k \in \mathbb{Z}$ such that $7^{2n+1} - 48^n - 7 = 48k$. We will prove it by induction over $n \in \mathbb{N}$.

- Base case ($n = 1$): $7^{2 \cdot 1 + 1} - 48^1 - 7 = 7^3 - 48 - 7 = 288 = 48 \cdot 6$, where $6 \in \mathbb{Z}$.

- Induction step. Assume that for some $n \in \mathbb{N}$, there exists $k \in \mathbb{Z}$ such that $7^{2n+1} - 48^n - 7 = 48k$ or equivalently $7^{2n+1} = 48k + 48^n + 7$ (Induction Hypothesis), and let us show that there is a $k' \in \mathbb{Z}$ such that $7^{2(n+1)+3} - 48^{n+1} - 7 = 48k'$. Indeed,

$$\begin{aligned}
7^{2(n+1)+1} - 48^{n+1} - 7 &= 7^{2n+3} - 48^{n+1} - 7 \\
&= 7^{2n+1} \cdot 7^2 - 48^n 48 - 7 \\
&= 7^2(48k + 48^n + 7) - 48^n \cdot 48 - 7 \\
&= 48(7^2 k + 7^2 \cdot 48^{n-1} - 48^n) + 7^3 - 7 \\
&= 48(7^2 k + 7^2 \cdot 48^{n-1} - 48^n) + 7(7^2 - 1) \\
&= 48(7^2 k + 7^2 \cdot 48^{n-1} - 48^n) + 7 \cdot 48 \\
&= 48(7^2 k + 7^2 \cdot 48^{n-1} - 48^n + 7) \\
&= 48k',
\end{aligned}$$

where $k' = 7^2 k + 7^2 \cdot 48^{n-1} - 48^n + 7 \in \mathbb{Z}$.

(III). Let us show that for all $n \in \mathbb{N}$, there is a $k \in \mathbb{Z}$ such that $2^{2n} + 15n - 1 = 9k$. We will prove it by induction over $n \in \mathbb{N}$.

- Base case ($n = 1$): $2^{2 \cdot 1} + 15 \cdot 1 - 1 = 18 = 9 \cdot 2$, where $2 \in \mathbb{Z}$.

- Induction step. Assume that for some $n \in \mathbb{N}$, there exists $k \in \mathbb{Z}$ such that $2^{2n} + 15n - 1 = 9k$ or equivalently $2^{2n} = 9k - 15n + 1$ (Induction Hypothesis), and let us see that there is a $k' \in \mathbb{Z}$ such that $2^{2(n+1)} + 15(n+1) - 1 = 9k'$. Indeed,

$$
\begin{aligned}
2^{2(n+1)} + 15(n+1) - 1 &= 2^2 \cdot 2^{2n} + 15n + 14 \\
&= 2^2(9k - 15n + 1) + 15n + 14 \\
&= 36k - 60n + 4 + 15n + 14 \\
&= 36k - 45n + 18 \\
&= 9(4k - 5n + 2) \\
&= 9k',
\end{aligned}
$$

where $k' = 4k - 5n + 2 \in \mathbb{Z}$.

(IV). Let us show that for all $n \in \mathbb{N}$, there is a $k \in \mathbb{Z}$ such that $11^{n+2} + 12^{2n+1} = 133k$. We will prove it by induction over $n \in \mathbb{N}$.

- Base case ($n = 1$): $11^{1+2} + 12^{2 \cdot 1 + 1} = 11^3 + 12^3 = 3059 = 133 \cdot 23$, where $23 \in \mathbb{Z}$.

- Induction step. Assume that for some $n \in \mathbb{N}$, there exists $k \in \mathbb{Z}$ such that $11^{n+2} + 12^{2n+1} = 133k$ or equivalently $11^{n+2} = 133k - 12^{2n+1}$ (Induction Hypothesis), and let us show that there is a $k' \in \mathbb{Z}$ such that $11^{(n+1)+2} + 12^{2(n+1)+1} = 133k'$. Indeed,

$$
\begin{aligned}
11^{(n+1)+2} + 12^{2(n+1)+1} &= 11^{n+2} \cdot 11 + 12^{2n+1} \cdot 12^2 \\
&= 11(133k - 12^{2n+1}) + 12^{2n+1} \cdot 12^2 \\
&= 133k \cdot 11 - 12^{2n+1} \cdot 11 + 12^{2n+1} \cdot 12^2 \\
&= 133k \cdot 11 + 12^{2n+1}(12^2 - 11) \\
&= 133(11k - 12^{2n+1}) \\
&= 133k',
\end{aligned}
$$

where $k' = 11k - 12^{2n+1} \in \mathbb{Z}$.

$\square$

Exercise 1.14. *Let f_n be the* Fibonacci *sequence, that is,*

$$f_1 = f_2 = 1, \qquad f_n = f_{n-2} + f_{n-1} \quad if\ n \geq 3.$$

Prove that, for all $n \in \mathbb{N}$, the number

$$f_n + f_{n+1} + f_{n+2} + \cdots + f_{n+9}$$

is a multiple of 11.

Solution to Exercise 1.14.

We have to prove that for any $n \in \mathbb{N}$, there exists $k \in \mathbb{Z}$ such that $f_n + f_{n+1} + \cdots + f_{n+9} = 11k$. To do so, we will write recursively for each $n \in \mathbb{N}$, the terms $f_n, f_{n+1}, \ldots, f_{n+9}$ depending on f_n and f_{n+1}:

$$f_n = f_n,$$
$$f_{n+1} = f_{n+1},$$
$$f_{n+2} = f_n + f_{n+1},$$
$$f_{n+3} = f_{n+1} + f_{n+2} = f_n + 2f_{n+1},$$
$$f_{n+4} = f_{n+2} + f_{n+3} = 2f_n + 3f_{n+1},$$
$$f_{n+5} = f_{n+3} + f_{n+4} = 3f_n + 5f_{n+1},$$
$$f_{n+6} = f_{n+4} + f_{n+5} = 5f_n + 8f_{n+1},$$
$$f_{n+7} = f_{n+5} + f_{n+6} = 8f_n + 13f_{n+1},$$
$$f_{n+8} = f_{n+6} + f_{n+7} = 13f_n + 21f_{n+1},$$
$$f_{n+9} = f_{n+7} + f_{n+8} = 21f_n + 34f_{n+1}.$$

Hence,

$$\sum_{i=0}^{9} f_{n+i} = 55f_n + 88f_{n+1} = 11(5f_n + 8f_{n+1}) = 11k,$$

where $k = 5f_n + 8f_{n+1} \in \mathbb{Z}$.

$\square$

Exercise 1.15 (Bernoulli's Inequality). *Prove that for all $x > -1$ and every $n \in \mathbb{N}$ it is satisfied that*

$$(1+x)^n \geq 1 + nx,$$

fulfilling equality if, and only if, $n = 1$ or $x = 0$.
Hint: Prove directly in the last two cases the equality, and when $n \geq 2$ and $x \neq 0$ use induction to show the strict inequality.

Solution to Exercise 1.15.

We begin by proving that $(1+x)^n = 1 + nx$ if, and only if, $n = 1$ or $x = 0$.

($\Leftarrow$) First, assume that $n = 1$, and $x > -1$ is arbitrary; then

$$(1+x)^1 = 1 + x = 1 + 1 \cdot x.$$

Second, assume that $x = 0$, and $n \in \mathbb{N}$ is arbitrary; then

$$(1+0)^n = 1^n = 1 = 1 + 0 = 1 + n \cdot 0.$$

That is, in any of the two cases we have $(1+x)^n = 1 + nx$.

($\Rightarrow$) We are going to prove this implication by contrapositive; that is, we have to show that if $n \in \mathbb{N} \setminus \{1\}$ and $x > -1$ with $x \neq 0$, then $(1+x)^n \neq 1+nx$. To be more precise, we are proving that $(1 + x)^n > 1 + nx$ for all $n \in \mathbb{N} \setminus \{1\}$ and $x > -1$ with $x \neq 0$. To do so, let $x > -1$ with $x \neq 0$ and let us see by induction over $n \in \mathbb{N} \setminus \{1\}$ that $(1+x)^n > 1 + nx$.

– Base case ($n = 2$): $(1 + x)^2 = x^2 + 2x + 1$. Since $x \neq 0$, we have that $(1 + x)^2 = 1 + 2x + x^2 > 1 + 2x$.

– Induction step. Assume that for some $n \in \mathbb{N} \setminus \{1\}$ we have that $(1 + x)^n > 1 + nx$ (Induction Hypothesis) and let us see that $(1 + x)^{n+1} > 1 + (n + 1)x$:

$$\begin{aligned}
(1+x)^{n+1} &= (1+x)^n(1+x) \\
&> (1+nx)(1+x) \\
&= 1 + nx + x + nx^2 \\
&= 1 + x(n+1) + nx^2.
\end{aligned}$$

As $x^2 > 0$ since $x \neq 0$, we have that $nx^2 > 0$. Thus,

$$(1+x)^{n+1} > 1 + (n+1)x.$$

Finally, it remains to prove that $(1 + x)^n \geq 1 + nx$ for every $x > -1$ and $n \in \mathbb{N}$. However, the latter claim is an immediate consequence of the implications ($\Leftarrow$) and ($\Rightarrow$). Indeed, using $x > -1$ with $x \neq 0$ and $n \in \mathbb{N} \setminus \{1\}$ we have seen that $(1+x)^n > 1+nx$, while in the remaining cases (when $x = 0$ or $n = 1$) we have that $(1 + x)^n = 1 + nx$. Combining these two results, we are done. $\qquad \square$

Exercise 1.16. *Prove the following inequalities for $n \in \mathbb{N}$:*

(I) $n! > 2^{n-1} \quad (n \geq 3)$,

(II) $(2n)! < 2^{2n}(n!)^2$,

(III) $\sqrt{n + \sqrt{(n-1) + \cdots + \sqrt{2 + \sqrt{1}}}} < \sqrt{n} + 1.$

Solution to Exercise 1.16.

(I). We will prove the inequality by induction over $n \in \mathbb{N} \setminus \{1, 2\}$.

• Base case ($n = 3$): $3! = 2 \cdot 3 = 6$ and $2^{3-1} = 4$, where $6 > 4$.

- Induction step. Assume that for some $n \in \mathbb{N}$ we have that $n! > 2^{n-1}$ (Induction Hypothesis) and let us show that $(n+1)! > 2^n$. Indeed,

$$(n+1)! = n!(n+1) > 2^{n-1}(n+1) = 2^n \cdot 2^{-1}(n+1) = 2^n \frac{n+1}{2}.$$

Since $n \geq 3$, we arrive at $\frac{n+1}{2} \geq 1$. Thus,

$$(n+1)! > 2^n \frac{n+1}{2} \geq 2^n.$$

(II). We will prove the inequality by induction over $n \in \mathbb{N}$.

- Base case ($n = 1$): $(2 \cdot 1)! = 2! = 2$ and $2^{2 \cdot 1} \cdot (1!)^2 = 4 \cdot 1 = 4$, where $2 < 4$.

- Induction step. Assume that $(2n)! < 2^{2n} \cdot (n!)^2$ for some $n \in \mathbb{N}$ (Induction Hypothesis) and let us see that $(2(n+1))! < 2^{2(n+1)} \cdot [(n+1)!]^2$. Indeed,

$$
\begin{aligned}
[2(n+1)]! &= (2n+2)! \\
&= (2n)!(2n+1)(2n+2) \\
&< 2^{2n} \cdot (n!)^2(2n+1)(2n+2) \\
&< 2^{2n} \cdot (n!)^2(2n+2)(2n+2) \\
&= 2^{2n} \cdot (n!)^2 \cdot 2^2 \cdot (n+1)^2 \\
&= 2^{2n} \cdot 2^2 \cdot [n!(n+1)]^2 \\
&= 2^{2n+2} \cdot [(n+1)!]^2.
\end{aligned}
$$

(III). We will prove the inequality by induction over $n \in \mathbb{N}$.

- Base case ($n = 1$): $\sqrt{1} = 1$ and $\sqrt{1} + 1 = 2$, where $1 < 2$.

- Induction step. Assume that $\sqrt{n + \sqrt{(n-1) + \cdots + \sqrt{2 + \sqrt{1}}}} < \sqrt{n} + 1$ for some $n \in \mathbb{N}$ (Induction Hypothesis). It is satisfied that

$$\sqrt{n+1 + \sqrt{n + \sqrt{(n-1) + \cdots + \sqrt{2 + \sqrt{1}}}}} < \sqrt{n+1} + 1.$$

Indeed, first observe that

$$
\begin{aligned}
n+1 + &\sqrt{n + \sqrt{(n-1) + \cdots + \sqrt{2 + \sqrt{1}}}} \\
&< n+1 + \sqrt{n} + 1 = n + 2 + \sqrt{n}. \quad (1.10)
\end{aligned}
$$

Since the square root $\sqrt{\cdot}$ is an increasing function, inequality (1.10) is still satisfied after taking square roots, that is,

$$\sqrt{n+1+\sqrt{n+\sqrt{(n-1)+\cdots+\sqrt{2+\sqrt{1}}}}} < \sqrt{n+1+\sqrt{n}+1}$$

$$= \sqrt{n+2+\sqrt{n}}.$$

Now let us see that $\sqrt{n+2+\sqrt{n}} < \sqrt{n+1}+1$. We will prove it backwards:

$$\sqrt{n+2+\sqrt{n}} < \sqrt{n+1}+1$$

$$\Longleftrightarrow \sqrt{n+2+\sqrt{n}} - \sqrt{n+1} < 1$$

$$\Longleftrightarrow \frac{(\sqrt{n+2+\sqrt{n}} - \sqrt{n+1})(\sqrt{n+2+\sqrt{n}} + \sqrt{n+1})}{\sqrt{n+2+\sqrt{n}} + \sqrt{n+1}} < 1$$

$$\Longleftrightarrow \frac{n+2+\sqrt{n}-(n+1)}{\sqrt{n+2+\sqrt{n}} + \sqrt{n+1}} < 1$$

$$\Longleftrightarrow \frac{1+\sqrt{n}}{\sqrt{n+2+\sqrt{n}} + \sqrt{n+1}} < 1$$

$$\Longleftrightarrow 1+\sqrt{n} < \sqrt{n+2+\sqrt{n}} + \sqrt{n+1}. \qquad (1.11)$$

As $1 < \sqrt{n+2+\sqrt{n}}$ and $\sqrt{n} < \sqrt{n+1}$ for every $n \in \mathbb{N}$, inequality (1.11) is true, and therefore, we have that $\sqrt{n+2+\sqrt{n}} < \sqrt{n+1}+1$. This means that

$$\sqrt{n+1+\sqrt{n+\sqrt{(n-1)+\cdots+\sqrt{2+\sqrt{1}}}}} < \sqrt{n+1}+1,$$

as desired.

$\square$

Exercise 1.17. *For which natural numbers n it is true that $\dfrac{1}{n!} > \dfrac{8^n}{(2n)!}$?*
Prove it by induction.

Solution to Exercise 1.17.

Taking $n \in \{1, 2, 3, 4, 5\}$ it is easy to check that the inequality is not satisfied. However, for $n = 6$ the inequality is true. Thus, we claim that the inequality is satisfied for $n \geq 6$ and we proceed to prove it. To be more precise, we will prove by induction that $\frac{1}{n!} > \frac{8^n}{(2n)!}$ for every $n \in \mathbb{N}$ with $n \geq 6$.

- Base case ($n = 6$): $\frac{1}{6!} = \frac{1}{720} \approx 0,001388$ and $\frac{8^6}{(2 \cdot 6)!} = \frac{256}{467775} \approx 0,0005472$.

- Induction step. Assume that $\frac{1}{n!} > \frac{8^n}{(2n)!}$ for some $n \in \mathbb{N}$ (Induction Hypothesis) and let us show that $\frac{1}{(n+1)!} > \frac{8^{n+1}}{[2(n+1)]!}$. Indeed,

$$
\begin{aligned}
\frac{8^{n+1}}{[2(n+1)]!} &= \frac{8^n}{(2n+2)!} \\
&= \frac{8^n}{(2n)!} \cdot \frac{8}{2(n+1)(2n+1)} \\
&< \frac{1}{n!} \cdot \frac{8}{2(n+1)(2n+1)} \\
&= \frac{1}{(n+1)!} \cdot \frac{4}{2n+1}.
\end{aligned}
$$

Since $2n + 1 > 4$ for any $n \geq 6$, we have that

$$
\frac{8^{n+1}}{[2(n+1)]!} < \frac{1}{(n+1)!} \cdot \frac{4}{2n+1} < \frac{1}{(n+1)!}.
$$

$\square$

Exercise 1.18. *Find and prove by induction an explicit formula of a_n with $a_1 = 1$ and, for every $n \in \mathbb{N}$,*

(I) $a_{n+1} = \frac{a_n}{(n+1)(2n+1)}$,

(II) $a_{n+1} = \frac{3a_n}{(2n+2)(2n+3)}$,

(III) $a_{n+1} = \frac{2n+1}{n+1} \cdot a_n$,

(IV) $a_{n+1} = \left(1 + \frac{1}{n}\right)^n a_n$.

Solution to Exercise 1.18.

We are going to find an explicit formula of a_n with $n \in \mathbb{N}$ in each one of the following parts. To do so, we will be calculating first a possible candidate for an explicit formula of a_n, and finally we will prove, by induction over $n \in \mathbb{N}$, that this explicit formula is the one that we are looking for.

(I).

- For $n = 1$: $a_2 = \dfrac{1}{2 \cdot 3} = \dfrac{1}{2! \cdot 3}$.

- For $n = 2$: $a_3 = \dfrac{\frac{1}{2! \cdot 3}}{3 \cdot 5} = \dfrac{1}{2! \cdot 3 \cdot 3 \cdot 5} = \dfrac{1}{3! \cdot 3 \cdot 5}$.

- For $n = 3$: $a_4 = \dfrac{\frac{1}{3! \cdot 3 \cdot 5}}{4 \cdot 7} = \dfrac{1}{3! \cdot 4 \cdot 3 \cdot 5 \cdot 7} = \dfrac{1}{4! \cdot 3 \cdot 5 \cdot 7}$.

- For $n = 4$: $a_5 = \dfrac{\frac{1}{4! \cdot 3 \cdot 5 \cdot 7}}{5 \cdot 9} = \dfrac{1}{4! \cdot 5 \cdot 3 \cdot 5 \cdot 7 \cdot 9} = \dfrac{1}{5! \cdot 3 \cdot 5 \cdot 7 \cdot 9}$.

Given the cases $n \in \{1, 2, 3, 4\}$, we claim that the formula

$$a_n = \frac{1}{n! \displaystyle\prod_{i=1}^{n}(2i - 1)}$$

is the desired one for every $n \in \mathbb{N}$. Indeed:

- Base case ($n = 1$): It is enough to see that $a_1 = 1$: $a_1 = \dfrac{1}{1! \displaystyle\prod_{i=1}^{1}(2i - 1)} =$

$\frac{1}{1 \cdot 1} = 1$.

- Induction step. Assume that for some $n \in \mathbb{N}$ the formula of a_n is satisfied (Induction Hypothesis) and let us see that $a_{n+1} = \dfrac{1}{(n+1)! \cdot \displaystyle\prod_{i=1}^{n+1}(2i - 1)}$.

Indeed,

$$
\begin{aligned}
a_{n+1} &= \frac{a_n}{(n + 1)(2n + 1)} \\[2mm]
&= \frac{\dfrac{1}{n! \displaystyle\prod_{i=1}^{n}(2i - 1)}}{(n + 1)(2n + 1)} \\[2mm]
&= \frac{1}{n!(n + 1) \displaystyle\prod_{i=1}^{n}(2i - 1) \cdot [2(n + 1) - 1]} \\[2mm]
&= \frac{1}{(n + 1)! \displaystyle\prod_{i=1}^{n+1}(2i - 1)}.
\end{aligned}
$$

(II).

- For $n = 1$: $a_2 = \frac{3}{4 \cdot 5}$.

- For $n = 2$: $a_3 = \dfrac{3 \cdot \frac{3}{4 \cdot 5}}{6 \cdot 7} = \dfrac{3^2}{4 \cdot 5 \cdot 6 \cdot 7}$.

- For $n = 3$: $a_4 = \dfrac{3 \cdot \frac{3^2}{4 \cdot 5 \cdot 6 \cdot 7}}{8 \cdot 9} = \dfrac{3^3}{4 \cdot 5 \cdot 6 \cdot 7 \cdot 8 \cdot 9}$.

- For $n = 4$: $a_5 = \dfrac{3 \cdot \frac{3^3}{4 \cdot 5 \cdot 6 \cdot 7 \cdot 8 \cdot 9}}{10 \cdot 11} = \dfrac{3^4}{4 \cdot 5 \cdot 6 \cdot 7 \cdot 8 \cdot 9 \cdot 10 \cdot 11}$.

Given the cases $n \in \{1, 2, 3, 4\}$, we claim that the formula

$$a_n = \frac{2 \cdot 3^n}{(2n + 1)!}$$

is the desired one for every $n \in \mathbb{N}$. Indeed:

- Base case ($n = 1$): It is enough to notice that $a_1 = 1$: $a_1 = \frac{2 \cdot 3^1}{3!} = 1$.

- Induction step. Assume that for some $n \in \mathbb{N}$ we have that $a_n = \frac{2 \cdot 3^n}{(2n+1)!}$ (Induction Hypothesis) and let us show that the formula is satisfied for $n + 1$, that is, $a_{n+1} = \frac{2 \cdot 3^{n+1}}{(2n+3)!}$. Indeed,

$$
\begin{aligned}
a_{n+1} &= \frac{3 a_n}{(2n + 2)(2n + 3)} \\
&= \frac{3 \cdot \frac{2 \cdot 3^n}{(2n+1)!}}{(2n + 2)(2n + 3)} \\
&= \frac{2 \cdot 3^{n+1}}{(2n + 1)!(2n + 2)(2n + 3)} \\
&= \frac{2 \cdot 3^{n+1}}{(2n + 3)!}.
\end{aligned}
$$

(III).

- For $n = 1$: $a_2 = \frac{3}{2} = \frac{3}{2!}$.

- For $n = 2$: $a_3 = \dfrac{5}{3} \cdot \dfrac{3}{2} = \dfrac{3 \cdot 5}{3!}$.

- For $n = 3$: $a_4 = \dfrac{7}{4} \cdot \dfrac{3 \cdot 5}{3!} = \dfrac{3 \cdot 5 \cdot 7}{4!}$.

- For $n = 4$: $a_5 = \dfrac{9}{5} \cdot \dfrac{3 \cdot 5 \cdot 7}{4!} = \dfrac{3 \cdot 5 \cdot 7 \cdot 9}{5!}$.

Given the cases $n \in \{1, 2, 3, 4\}$, we claim that the formula

$$a_n = \frac{\prod_{i=1}^{n}(2i-1)}{n!}$$

is the desired one for every $n \in \mathbb{N}$. Indeed:

- Base case ($n = 1$): It is enough to see that $a_1 = 1$: $a_1 = \frac{\prod_{i=1}^{1}(2i-1)}{1!} = 1$.

- Induction step. Assume that for some $n \in \mathbb{N}$ we have that $a_n = \frac{\prod_{i=1}^{n}(2i-1)}{n!}$ (Induction Hypothesis) and let us show that the formula is satisfied for $n + 1$, that is, $a_{n+1} = \frac{\prod_{i=1}^{n+1}(2i-1)}{(n+1)!}$. Indeed,

$$a_{n+1} = \frac{2n+1}{n+1} \cdot a_n$$

$$= \frac{2n+1}{n+1} \cdot \frac{\prod_{i=1}^{n}(2i-1)}{n!}$$

$$= \frac{(2n+1) \cdot \prod_{i=1}^{n}(2i-1)}{(n+1) \cdot n!}$$

$$= \frac{\prod_{i=1}^{n+1}(2i-1)}{(n+1)!}.$$

(IV).

- For $n = 1$: $a_2 = \left(1 + \frac{1}{1}\right)^1 a_1 = 2 = \left(\frac{2}{1}\right)^1$.

- For $n = 2$: $a_3 = \left(1 + \frac{1}{2}\right)^2 \left(\frac{2}{1}\right)^1 = \left(\frac{2}{1}\right)^1 \cdot \left(\frac{3}{2}\right)^2$.

- For $n = 3$: $a_4 = \left(1 + \frac{1}{3}\right)^3 \cdot \left(\frac{2}{1}\right)^1 \cdot \left(\frac{3}{2}\right)^2 = \left(\frac{2}{1}\right)^1 \cdot \left(\frac{3}{2}\right)^2 \cdot \left(\frac{4}{3}\right)^3$.

- For $n = 4$: $a_5 = \left(1 + \frac{1}{4}\right)^4 \cdot \left(\frac{2}{1}\right)^1 \cdot \left(\frac{3}{2}\right)^2 \cdot \left(\frac{4}{3}\right)^3 = \left(\frac{2}{1}\right)^1 \cdot \left(\frac{3}{2}\right)^2 \cdot \left(\frac{4}{3}\right)^3 \cdot \left(\frac{5}{4}\right)^4$.

Given the cases $n \in \{1, 2, 3, 4\}$, we claim that the formula

$$a_n = \begin{cases} 1 & \text{if } n = 1, \\ \prod_{i=2}^{n}\left(\frac{i}{i-1}\right)^{i-1} & \text{if } n > 1, \end{cases} \tag{1.12}$$

is the needed one for every $n \in \mathbb{N}$.

It is important to mention that there are many ways to write a general formula in this case.

Let us show that the general formula (1.12) is the right one by induction:

- Base case ($n = 1$) is trivial.

- Induction step. Assume that for some $n \in \mathbb{N}$ we have that a_n satisfies the formula (1.12) (Induction Hypothesis) and let us show that the formula is also true for $n+1$. Indeed, first, if $n = 1$, then $a_1 = 1$ by the Induction Hypothesis, and therefore, we have to see that $a_2 = 2$ in the general formula:

$$a_2 = \prod_{i=2}^{2} \left(\frac{i}{i-1} \right)^{i-1} = \left(\frac{2}{1} \right)^1 = 2.$$

Finally, assume that $n > 1$ and let us show that $a_{n+1} = \prod_{i=2}^{n+1} \left(\frac{i}{i-1} \right)^{i-1}$.

Indeed,

$$a_{n+1} = \left(1 + \frac{1}{n} \right)^n a_n$$

$$= \left(\frac{n+1}{n} \right)^n \prod_{i=2}^{n} \left(\frac{i}{i-1} \right)^{i-1}$$

$$= \prod_{i=2}^{n+1} \left(\frac{i}{i-1} \right)^{i-1}.$$

$\square$

Exercise 1.19. *Assume that $x_0 > 0$ and $x_n = 1 - e^{-x_{n-1}}$ for every $n \in \mathbb{N}$. Show that $0 < x_n < 1$ for all $n \in \mathbb{N}$.*

Solution to Exercise 1.19.

We will prove it by induction over $n \in \mathbb{N}$.

- Base case ($n = 1$): By hypothesis, we know that $x_0 > 0$; hence, $e^{x_0} > e^0 = 1$ since the exponential function is an increasing function, which implies that $e^{-x_0} < 1$. Moreover, since the exponential is positive we have that $e^{-x_0} > 0$, that is, $0 < e^{-x_0} < 1$ or, equivalently, $-1 < -e^{x_0} < 0$ from which, $0 < 1 - e^{-x_0} < 1$. But also we know by construction that $x_1 = 1 - e^{-x_0}$, that is, $0 < x_1 < 1$.

- Induction step. Assume that $0 < x_n < 1$ for some $n \in \mathbb{N}$ (Induction Hypothesis) and let us show that $0 < x_{n+1} < 1$. From the Induction Hypothesis and recalling that the exponential function is increasing, we have that the following chain of implications:

$$0 < x_n < 1 \Rightarrow 0 > -x_n > -1$$
$$\Rightarrow e^0 = 1 > e^{-x_n} > e^{-1}$$
$$\Rightarrow -1 < -e^{-x_n} < -e^{-1}$$
$$\Rightarrow 0 < 1 - e^{-x_n} < 1 - e^{-1} < 1.$$

As $x_{n+1} = 1 - e^{-x_n}$, we are done.

$\square$

Exercise 1.20. *Assume that $R > 0$, $x_0 > 0$ and*

$$x_{n+1} = \frac{1}{2}\left(\frac{R}{x_n} + x_n\right), \qquad if\ n \in \mathbb{N} \cup \{0\}.$$

Prove that for every $n \in \mathbb{N}$ we have that $x_n \geq x_{n+1} \geq \sqrt{R}$ and

$$x_n - \sqrt{R} \leq \frac{1}{2^n} \cdot \frac{(x_0 - \sqrt{R})^2}{x_0}.$$

Is there a $c > 0$ such that $x_n = c$ for every $n \in \mathbb{N} \cup \{0\}$?

Solution to Exercise 1.20.

Assume that $R > 0$, $x_0 > 0$ and $x_{n+1} = \frac{1}{2}\left(\frac{R}{x_n} + x_n\right)$ provided that $n \in \mathbb{N} \cup \{0\}$. Let us prove that for every $n \in \mathbb{N}$, we have

- $x_n \geq x_{n+1} \geq \sqrt{R}$.

- $x_n - \sqrt{R} \leq \frac{1}{2^n} \frac{(x_0 - \sqrt{R})^2}{x_0}$.

Moreover, we will see that if $x_0 = \sqrt{R}$, then $x_n = \sqrt{R}$ for all $n \in \mathbb{N} \cup \{0\}$.

First, let us show that for every $n \in \mathbb{N} \cup \{0\}$, we have that $x_{n+1} \geq \sqrt{R}$ or equivalently, that for every $n \in \mathbb{N}$, we have $x_n \geq \sqrt{R}$:

$$0 \leq (x_n - \sqrt{R})^2 = x_n^2 - 2\sqrt{R}x_n + R \iff x_n^2 + R \geq 2\sqrt{R}x_n.$$

If we prove that $x_n > 0$ for every $n \in \mathbb{N}$ we have that

$$\sqrt{R} \leq \frac{x_n^2 + R}{2x_n} = \frac{1}{2}\left(\frac{R}{x_n} + x_n\right) = x_{n+1}.$$

Let us prove then, by induction over $n \in \mathbb{N}$, that $x_n > 0$.

- Base case ($n = 1$): $x_1 = \frac{1}{2}\left(\frac{R}{x_0} + x_0\right) > 0$ since $x_0 > 0$ and $R > 0$.

- Induction step. Assume that $x_n > 0$ for some $n \in \mathbb{N}$ (Induction Hypothesis) and let us see that $x_{n+1} > 0$. Indeed, $x_{n+1} = \frac{1}{2}\left(\frac{R}{x_n} + x_n\right) > 0$ as $R > 0$ and $x_n > 0$.

Hence, we have proven that $x_n \geq \sqrt{R}$ for every $n \in \mathbb{N}$.

Second, let us see that $x_n \geq x_{n+1}$ for every $n \in \mathbb{N}$ or equivalently, that $x_{n+1} - x_n \leq 0$ for every $n \in \mathbb{N}$. By definition, we have that

$$
\begin{aligned}
x_{n+1} - x_n &= \frac{1}{2}\left(\frac{R}{x_n} + x_n\right) - x_n \\
&= \frac{1}{2} \cdot \frac{R}{x_n} - \frac{x_n}{2} \\
&= \frac{1}{2}\left(\frac{R - x_n^2}{x_n}\right) \\
&= \frac{1}{2} \cdot \frac{(\sqrt{R} - x_n)(\sqrt{R} + x_n)}{x_n}.
\end{aligned}
$$

Since $x_n \geq \sqrt{R}$ for every $n \in \mathbb{N}$, we have that $\sqrt{R} - x_n \leq 0$ and $\sqrt{R} + x_n \geq 0$, that is,

$$
x_{n+1} - x_n = \frac{1}{2} \cdot \frac{(\sqrt{R} - x_n)(\sqrt{R} + x_n)}{x_n} \leq 0.
$$

Third, let us prove by induction that $x_n - \sqrt{R} \leq \frac{1}{2^n}\frac{\left(x_0 - \sqrt{R}\right)^2}{x_0}$ for every $n \in \mathbb{N}$.

- Base case ($n = 1$):

$$
\begin{aligned}
x_1 - \sqrt{R} &= \frac{1}{2}\left(\frac{R}{x_0} + x_0\right) - \sqrt{R} \\
&= \frac{R + x_0^2 - 2\sqrt{R}x_0}{2x_0} \\
&= \frac{\left(x_0 - \sqrt{R}\right)^2}{2x_0} \\
&= \frac{1}{2^1} \cdot \frac{\left(x_0 - \sqrt{R}\right)^2}{x_0}.
\end{aligned}
$$

In particular, we have an equality.

- Induction step. Assume that $x_n - \sqrt{R} \leq \frac{1}{2^n} \frac{(x_0 - \sqrt{R})^2}{x_0}$ for some $n \in \mathbb{N}$ (Induction Hypothesis) and let us show that $x_{n+1} - \sqrt{R} \leq \frac{1}{2^{n+1}} \frac{(x_0 - \sqrt{R})^2}{x_0}$. Indeed,

$$
\begin{aligned}
x_{n+1} - \sqrt{R} &= \frac{1}{2}\left(\frac{R}{x_n} + x_n\right) - \sqrt{R} \\
&= \frac{x_n^2 + R - 2\sqrt{R}x_n}{2x_n} \\
&= \frac{(x_n - \sqrt{R})^2}{2x_n} \\
&= \left(x_n - \sqrt{R}\right)\frac{\left(x_n - \sqrt{R}\right)}{2x_n} \\
&\leq \frac{1}{2^n} \cdot \frac{\left(x_0 - \sqrt{R}\right)^2}{x_0} \cdot \frac{\left(x_n - \sqrt{R}\right)}{2x_n} \\
&= \frac{1}{2^{n+1}} \cdot \frac{\left(x_0 - \sqrt{R}\right)^2}{x_0}\left(1 - \frac{\sqrt{R}}{x_n}\right).
\end{aligned}
$$

Since $x_n \geq \sqrt{R} > 0$, we obtain $1 - \frac{\sqrt{R}}{x_n} < 1$. Thus,

$$
x_{n+1} - \sqrt{R} \leq \frac{1}{2^{n+1}}\frac{\left(x_0 - \sqrt{R}\right)^2}{x_0} \cdot \left(1 - \frac{\sqrt{R}}{x_n}\right) < \frac{1}{2^{n+1}}\frac{\left(x_0 - \sqrt{R}\right)^2}{x_0}.
$$

Therefore, we have proven that $x_n - \sqrt{R} \leq \frac{1}{2^n}\frac{(x_0 - \sqrt{R})^2}{x_0}$ for every $n \in \mathbb{N}$.

Notice that we have also proven that $x_n - \sqrt{R} < \frac{1}{2^n}\frac{(x_0 - \sqrt{R})^2}{x_0}$ for every $n \in \mathbb{N}\setminus\{1\}$ and $x_1 - \sqrt{R} = \frac{1}{2^1}\frac{(x_0 - \sqrt{R})^2}{x_0}$.

Finally, take $x_0 = \sqrt{R}$. Let us prove by induction that $x_n = \sqrt{R}$ for every $n \in \mathbb{N}$.

- Base case ($n = 1$): $x_1 = \frac{1}{2}\left(\frac{R}{\sqrt{R}} + \sqrt{R}\right) = \frac{1}{2}\left(\sqrt{R} + \sqrt{R}\right) = \sqrt{R}$.

- Induction step. Assume that for some $n \in \mathbb{N}$ we have that $x_n = \sqrt{R}$ (Induction Hypothesis) and let us show that $x_{n+1} = \sqrt{R}$. Indeed,

$$
x_{n+1} = \frac{1}{2}\left(\frac{R}{x_n} + x_n\right) = \frac{1}{2}\left(\frac{R}{\sqrt{R}} + \sqrt{R}\right) = \sqrt{R}.
$$

$\square$

Exercise 1.21. *Prove by induction that for every* $n \in \mathbb{N}$

(I) $\sin(x)(\sin(x) + \sin(3x) + \cdots + \sin((2n-1)x)) = \sin^2(nx)$.

(II) $2^{n+1} \sin(x) \cos(x) \cos(2x) \cos(4x) \cdots \cos(2^n x) = \sin(2^{n+1} x)$.

(III) $\sin\left(\dfrac{x}{2}\right) (\sin(x) + \sin(2x) + \cdots + \sin(nx)) = \sin\left(\dfrac{nx}{2}\right) \sin\left(\dfrac{(n+1)x}{2}\right)$.

Solution to Exercise 1.21.

(I). To solve this part of the exercise, recall the following trigonometric formulas:

$$\cos(A+B) = \cos A \cos B - \sin A \sin B,$$
$$\sin(A+B) = \sin A \cos B + \sin B \cos A,$$
$$\cos(2A) = \cos^2 A - \sin^2 A,$$
$$\sin(2A) = 2 \sin A \cos A.$$

Let us prove by induction that $\sin x \cdot \displaystyle\sum_{i=1}^{n} \sin((2i-1)x) = \dfrac{1 - \cos(2nx)}{2}$ for every $n \in \mathbb{N}$.

- Base case ($n = 1$): Since $\sin x \cdot \displaystyle\sum_{i=1}^{1} \sin[(2i-1)x] = \sin^2 x$ and

$$\frac{1 - \cos(2nx)}{2} = \frac{1 - \cos(2x)}{2}$$
$$= \frac{1 - (\cos^2 x - \sin^2 x)}{2}$$
$$= \frac{1 - \cos^2 x + \sin^2 x}{2}$$
$$= \frac{\sin^2 x + \sin^2 x}{2}$$
$$= \sin^2 x,$$

where we have used that $\cos^2 x + \sin^2 x = 1$, so the base case is true.

- **Induction step.** Assume that $\sin x \cdot \sum_{i=1}^{n} \sin[(2i-1)x] = \dfrac{1 - \cos(2nx)}{2}$ for

some $n \in \mathbb{N}$ (Induction Hypothesis) and let us see that

$$\sin x \cdot \sum_{i=1}^{n+1} \sin[(2i-1)x] = \frac{1 - \cos[2(n+1)x]}{2}.$$

Indeed,

$$\sin x \cdot \sum_{i=1}^{n+1} \sin[(2i-1)x]$$

$$= \sin x \cdot \sum_{i=1}^{n} \sin[(2i-1)x] + \sin x \cdot \sin[(2n+1)x]$$

$$= \frac{1 - \cos(2nx)}{2} + \sin x \cdot \sin((2n+1)x).$$

If we prove that $\frac{1-\cos(2nx)}{2} + \sin x \cdot \sin[(2n+1)x] = \frac{1-\cos[2(n+1)x]}{2}$, we
are done. Let us prove the latter:

$$\frac{1 - \cos[2(n+1)x]}{2}$$

$$= \frac{1 - \cos(2nx + 2x)}{2}$$

$$= \frac{1}{2} \cdot [1 - \cos(2nx)\cos(2x) + \sin(2nx)\sin(2x)]$$

$$= \frac{1}{2} \cdot [1 - \cos(2nx)(\cos^2 x - \sin^2 x) + \sin(2nx)2\sin x \cos x]$$

$$= \frac{1}{2} \cdot [1 - \cos(2nx)(1 - 2\sin^2 x) + 2\sin x \cos x \sin(2nx)]$$

$$= \frac{1}{2} - \frac{1}{2}\cos(2nx) + \frac{1}{2} \cdot 2\sin^2 x \cos(2nx) + \frac{1}{2} \cdot 2\sin x \cos x \sin(2nx)$$

$$= \frac{1 - \cos(2nx)}{2} + \sin x[\sin x \cos(2nx) + \cos x \sin(2nx)]$$

$$= \frac{1 - \cos(2nx)}{2} + \sin x \cdot \sin(x + 2nx)$$

$$= \frac{1 - \cos(2nx)}{2} + \sin x \cdot \sin[(2n+1)x].$$

(II). For this part, we will only make use of

$$\sin(2A) = 2\sin A \cos A.$$

Let us prove the formula by induction on $n \in \mathbb{N}$.

- Base case ($n = 1$):

$$\sin(2^{1+1}x) = \sin(4x) = \sin(2 \cdot 2x) = 2\sin(2x)\cos(2x)$$
$$= 4\sin(x)\cos(x)\cos(2x).$$

- Induction step. Assume that $2^{n+1}\sin(x)\cos(x)\cos(2x)\cos(4x)\cdots$ $\cos(2^n x) = \sin(2^{n+1}x)$ for some $n \in \mathbb{N}$ (Induction Hypothesis), and let us prove the formula for $n + 1$; that is, we have to prove that

$$2^{n+2}\sin(x)\cos(x)\cos(2x)\cos(4x)\cdots\cos(2^n x)\cos(2^{n+1}x) = \sin(2^{n+2}x).$$

Indeed,

$$\sin(2^{n+2}x) = \sin(2 \cdot 2^{n+1}x) = 2\sin(2^{n+1}x)\cos(2^{n+1}x)$$
$$= 2 \cdot 2^{n+1}\sin(x)\cos(x)\cos(2x)\cos(4x)\cdots\cos(2^n x)\cos(2^{n+1}x)$$
$$= 2^{n+2}\sin(x)\cos(x)\cos(2x)\cos(4x)\cdots\cos(2^n x)\cos(2^{n+1}x).$$

(III). Let us prove by induction on $n \in \mathbb{N}$ that $\sin\frac{x}{2}(\sin x + \sin 2x + \cdots + \sin nx) = \sin\frac{nx}{2}\sin\frac{(n+1)x}{2}$.

- Base case ($n = 1$) is trivial: $\sin\frac{x}{2}\sin x = \sin\frac{x}{2}\sin\frac{(1+1)x}{2}$.

- Induction step. Assume that

$$\sin\frac{x}{2}(\sin x + \sin 2x + \cdots + \sin nx) = \sin\frac{nx}{2}\sin\frac{(n+1)x}{2}$$

and let us see that

$$\sin\frac{x}{2}(\sin x + \sin 2x + \cdots + \sin nx + \sin(n+1)x) = \sin\frac{(n+1)x}{2}\sin\frac{(n+2)x}{2}.$$

First,

$$\sin\frac{x}{2}(\sin x + \sin 2x + \cdots + \sin nx + \sin(n+1)x)$$
$$= \sin\frac{x}{2}(\sin x + \sin 2x + \cdots + \sin nx) + \sin\frac{x}{2}\sin(n+1)x$$
$$= \sin\frac{nx}{2}\sin\frac{(n+1)x}{2} + \sin\frac{x}{2}\sin(n+1)x. \tag{1.13}$$

Now let us apply the formula $\sin(2A) = 2\sin A\cos A$ to $\sin(n+1)x$ in the following way:

$$\sin(n+1)x = \sin\left(2\frac{(n+1)x}{2}\right) = 2\sin\frac{(n+1)x}{2}\cos\frac{(n+1)x}{2}.$$

Hence, we have by (1.13)

$$\sin\frac{x}{2}(\sin x + \sin 2x + \cdots + \sin nx + \sin(n+1)x)$$

$$= \sin\frac{nx}{2}\sin\frac{(n+1)x}{2} + 2\sin\frac{x}{2}\sin\frac{(n+1)x}{2}\cos\frac{(n+1)x}{2}$$

$$= \sin\frac{(n+1)x}{2}\left(\sin\frac{nx}{2} + 2\sin\frac{x}{2}\cos\frac{(n+1)x}{2}\right). \qquad (1.14)$$

Apply now the formula $\sin(A-B) = \sin A \cos B - \cos A \sin B$ to $\sin\frac{nx}{2}$ as follows:

$$\sin\frac{nx}{2} = \sin\left(\frac{(n+1)x}{2} - \frac{x}{2}\right) = \sin\frac{(n+1)x}{2}\cos\frac{x}{2} - \cos\frac{(n+1)x}{2}\sin\frac{x}{2}.$$

Thus, by (1.14) we have

$$\sin\frac{x}{2}(\sin x + \sin 2x + \cdots + \sin nx + \sin(n+1)x)$$

$$= \sin\frac{(n+1)x}{2}\left(\sin\frac{(n+1)x}{2}\cos\frac{x}{2} - \cos\frac{(n+1)x}{2}\sin\frac{x}{2}\right.$$

$$\left. +2\sin\frac{x}{2}\cos\frac{(n+1)x}{2}\right)$$

$$= \sin\frac{(n+1)x}{2}\left(\sin\frac{(n+1)x}{2}\cos\frac{x}{2} + \sin\frac{x}{2}\cos\frac{(n+1)x}{2}\right). \qquad (1.15)$$

Finally, using the formula $\sin(A+B) = \sin A \cos B + \sin B \cos A$ inside the last parenthesis of (1.15):

$$\sin\frac{(n+1)x}{2}\cos\frac{x}{2} + \sin\frac{x}{2}\cos\frac{(n+1)x}{2} = \sin\left(\frac{(n+1)x}{2} + \frac{x}{2}\right)$$

$$= \sin\frac{(n+2)x}{2},$$

it yields

$$\sin\frac{x}{2}(\sin x + \sin 2x + \cdots + \sin nx + \sin(n+1)x) = \sin\frac{(n+1)x}{2}\sin\frac{(n+2)x}{2}.$$

$\square$

Exercise 1.22. *Let* $a_1 = a_2 = 5$ *and*

$$a_{n+1} = a_n + 6a_{n-1}, \qquad \text{if } n \geq 2.$$

Prove by induction that $a_n = 3^n - (-2)^n$ *for every* $n \in \mathbb{N}$.

Solution to Exercise 1.22.

- Base case ($n = 1$): since by hypothesis $a_1 = 5$, and $3^1 - (-2)^1 = 3 - (-2) = 5$, the base case is true.

- Induction step. Assume that for some $n \in \mathbb{N}$ we have $a_k = 3^k - (-2)^k$ for all $k \in \{1, \ldots, n\}$ (complete Induction Hypothesis) and let us show that $a_{n+1} = 3^{n+1} - (-2)^{n+1}$. Indeed,

$$
\begin{aligned}
a_{n+1} &= a_n + 6a_{n-1} \\
&= 3^n - (-2)^n + 6[3^{n-1} - (-2)^{n-1}] \\
&= 3^n - (-2)^n + 6 \cdot 3^{n-1} - 6(-2)^{n-1} \\
&= 3^n(1 + 6 \cdot 3^{-1}) - (-2)^n(1 + 6(-2)^{-1}) \\
&= 3^n \cdot 3 - (-2)^n(-2) \\
&= 3^{n+1} - (-2)^{n+1}.
\end{aligned}
$$

$\square$

Exercise 1.23. *Let $a_1 = 2$, $a_2 = 0$, $a_3 = -14$, and*

$$a_{n+1} = 9a_n - 23a_{n-1} + 15a_{n-2}, \qquad \text{if } n \geq 3.$$

Prove by induction that $a_n = 3^{n-1} - 5^{n-1} + 2$ for every $n \in \mathbb{N}$.

Solution to Exercise 1.23.

- Base case ($n = 1$): since by hypothesis $a_1 = 2$, and $3^{1-1} - 5^{1-1} + 2 = 1 - 1 + 2 = 2$, the base case is true.

- Induction step. Assume that for some $n \in \mathbb{N}$ we have that $a_k = 3^{k-1} - 5^{k-1} + 2$ for all $k \in \{1, \ldots, n\}$ (complete Induction Hypothesis) and let us show that $a_{n+1} = 3^n - 5^n + 2$.

  If $n = 1$ we have by hypothesis that $a_2 = 0$ and $3^1 - 5^1 + 2 = 3 - 5 + 2 = 0$, that is, $a_{n+1} = 3^n - 5^n + 2$.

  If $n = 2$ we have by hypothesis that $a_3 = -14$ y $3^2 - 5^2 + 2 = 9 - 25 + 2 = -14$, that is, $a_{n+1} = 3^n - 5^n + 2$.

Finally, assume that $n \geq 3$ is an integer, then

$$
\begin{aligned}
a_{n+1} &= 9a_n - 23a_{n-1} + 15a_{n-2} \\
&= 9\left(3^{n-1} - 5^{n-1} + 2\right) - 23\left(3^{n-2} - 5^{n-2} + 2\right) \\
&\quad + 15\left(3^{n-3} - 5^{n-3} + 2\right) \\
&= 3^n \left(9 \cdot 3^{-1} - 23 \cdot 3^{-2} + 15 \cdot 3^{-3}\right) \\
&\quad - 5^n \left(9 \cdot 5^{-1} - 23 \cdot 5^{-2} + 15 \cdot 5^{-3}\right) + 18 - 46 + 30 \\
&= 3^n \cdot \frac{9 \cdot 3^2 - 23 \cdot 3 + 15}{3^3} - 5^n \cdot \frac{9 \cdot 5^2 - 23 \cdot 5 + 15}{5^3} + 2 \\
&= 3^n \cdot \frac{81 - 69 + 15}{27} - 5^n \cdot \frac{225 - 115 + 15}{125} + 2 \\
&= 3^n \cdot \frac{27}{27} - 5^n \cdot \frac{125}{125} + 2 \\
&= 3^n - 5^n + 2.
\end{aligned}
$$

$\square$

Exercise 1.24 (Binet's Formula). *Let (f_n) be the Fibonacci sequence defined as in Exercise 1.14. Prove by complete induction that*

$$
f_n = \frac{(1 + \sqrt{5})^n - (1 - \sqrt{5})^n}{2^n \sqrt{5}}, \qquad n \in \mathbb{N}.
$$

Solution to Exercise 1.24.

- Base case ($n = 1$): since by hypothesis $f_1 = 1$, and

$$
\frac{\left(1 + \sqrt{5}\right)^1 - \left(1 - \sqrt{5}\right)^1}{2^1 \cdot \sqrt{5}} = \frac{2\sqrt{5}}{2\sqrt{5}} = 1,
$$

the base case is true.

- Induction step. Assume that for some $n \in \mathbb{N}$ it is satisfied that

$$
f_k = \frac{\left(1 + \sqrt{5}\right)^k - \left(1 - \sqrt{5}\right)^k}{2^k \cdot \sqrt{5}}
$$

for all $k \in \{1, \ldots, n\}$ (complete Induction Hypothesis or strong induction method) and let us show that

$$
f_{n+1} = \frac{\left(1 + \sqrt{5}\right)^{n+1} - \left(1 - \sqrt{5}\right)^{n+1}}{2^{n+1} \cdot \sqrt{5}}.
$$

Indeed,

$$f_{n+1} = f_n + f_{n-1} =$$

$$= \frac{\left(1+\sqrt{5}\right)^n - \left(1-\sqrt{5}\right)^n}{2^n \cdot \sqrt{5}} + \frac{\left(1+\sqrt{5}\right)^{n-1} - \left(1-\sqrt{5}\right)^{n-1}}{2^{n-1} \cdot \sqrt{5}}$$

$$= \frac{2\left[\left(1+\sqrt{5}\right)^n - \left(1-\sqrt{5}\right)^n\right]}{2^{n+1} \cdot \sqrt{5}} + \frac{2^2\left[\left(1+\sqrt{5}\right)^{n-1} - \left(1-\sqrt{5}\right)^{n-1}\right]}{2^{n+1} \cdot \sqrt{5}}$$

$$= \frac{\left(1+\sqrt{5}\right)^n \left[2 + 4\left(1+\sqrt{5}\right)^{-1}\right] - \left(1-\sqrt{5}\right)^n \left[2 + 4\left(1-\sqrt{5}\right)^{-1}\right]}{2^{n+1} \cdot \sqrt{5}}$$

$$= \frac{\left(1+\sqrt{5}\right)^n \cdot \frac{2\left(1+\sqrt{5}\right)+4}{1+\sqrt{5}} - \left(1-\sqrt{5}\right)^n \cdot \frac{2\left(1-\sqrt{5}\right)+4}{1-\sqrt{5}}}{2^{n+1} \cdot \sqrt{5}}$$

$$= \frac{\left(1+\sqrt{5}\right)^n \cdot \frac{1+5+2\sqrt{5}}{1+\sqrt{5}} - \left(1-\sqrt{5}\right)^n \cdot \frac{1+5-2\sqrt{5}}{1-\sqrt{5}}}{2^{n+1} \cdot \sqrt{5}}$$

$$= \frac{\left(1+\sqrt{5}\right)^n \cdot \frac{\left(1+\sqrt{5}\right)^2}{1+\sqrt{5}} - \left(1-\sqrt{5}\right)^n \cdot \frac{\left(1-\sqrt{5}\right)^2}{1-\sqrt{5}}}{2^{n+1} \cdot \sqrt{5}}$$

$$= \frac{\left(1+\sqrt{5}\right)^n \left(1+\sqrt{5}\right) - \left(1-\sqrt{5}\right)^n \left(1-\sqrt{5}\right)}{2^{n+1} \cdot \sqrt{5}}$$

$$= \frac{\left(1+\sqrt{5}\right)^{n+1} - \left(1-\sqrt{5}\right)^{n+1}}{2^{n+1} \cdot \sqrt{5}}.$$

□

Exercise 1.25. *Let us consider the property* P_n *given by* $\sum_{k=1}^{n} k = \frac{(2n+1)^2}{8}$.

(I) *Prove that if* P_n *is true, then* P_{n+1} *is true.*

(II) *Discuss the claim: "by induction* P_n *is true for every* $n \in \mathbb{N}$*".*

Solution to Exercise 1.25.

(I). Assuming that P_n is satisfied, we have to prove that $\displaystyle\sum_{k=1}^{n+1} k = \frac{(2n+3)^2}{8}$.

$$\sum_{k=1}^{n+1} k = \sum_{k=1}^{n} k + (n+1)$$
$$= \frac{(2n+1)^2}{8} + n + 1$$
$$= \frac{4n^2 + 1 + 4n + 8n + 8}{8}$$
$$= \frac{4n^2 + 9 + 12n}{8}$$
$$= \frac{(2n+3)^2}{8}.$$

(II). In order to assure that property P_n is satisfied for every $n \in \mathbb{N}$ by induction, it is necessary that the base case is fulfilled, that is, that property P_1 is true. However, property P_1 is not satisfied: $\displaystyle\sum_{k=1}^{1} k = 1 \neq \frac{9}{8} = \frac{(2 \cdot 1 + 1)^2}{8}$.

In fact, it is easy to verify that $\displaystyle\sum_{k=1}^{n} k \neq \frac{(2n+1)^2}{8}$ for all $n \in \mathbb{N}$.

□

Exercise 1.26. *Decide for which natural numbers n is the following inequality true $2^n > n^2$. Prove it by induction.*

Solution to Exercise 1.26.

- For $n = 1$: $2^1 = 2 > 1 = 1^2$.

- For $n = 2$: $2^2 = 4$.

- For $n = 3$: $2^3 = 8 < 9 = 3^2$.

- For $n = 4$: $2^4 = 16 = 4^2$.

- For $n = 5$: $2^5 = 32 > 25 = 5^2$.

We will prove by induction that $2^n > n^2$ is fulfilled for every integer $n \geq 5$.

- Base case ($n = 5$) has already been studied.

- Induction step. Assume that $2^n > n^2$ for some integer $n \geq 5$ (Induction Hypothesis) and let us show that $2^{n+1} > (n+1)^2$. Indeed,

$$2^{n+1} = 2 \cdot 2^n > 2 \cdot n^2 = n^2 + n^2.$$

Since $n^2 > 2n + 1 = n \left(2 + \frac{1}{n}\right)$ for every integer $n > 5$, we have that

$$2^{n+1} > n^2 + n^2 > n^2 + n\left(2 + \frac{1}{n}\right) = n^2 + 2n + 1 = (n+1)^2.$$

If the reader is not convinced that $n^2 > 2n + 1$ for every integer $n \geq 6$, let us prove it by induction.

- Base case ($n = 6$): $6^2 = 36 > 13 = 2 \cdot 6 + 1$.

- Induction step. Assume that $n^2 > 2n + 1$ for some integer $n \geq 6$ (Induction Hypothesis) and let us show that $(n+1)^2 > 2n + 3$. Indeed,

$$(n+1)^2 = n^2 + 1 + 2n > 2n + 1 + 1 + 2n = 2n + 2 + 2n > 2n + 2 + 1 = 2n + 3,$$

where we have used that $2n > 1$ for every $n \in \mathbb{N}$.

$\square$

Exercise 1.27. *Prove that for every natural number n we have*

$$\left(1 + \frac{1}{n}\right)^n < 3.$$

Solution to Exercise 1.27.

To prove the inequality, we will use Newton's Binomial Theorem (Exercise 1.11). For $n = 1$ is obvious that $\left(1 + \frac{1}{1}\right)^1 = 2 < 3$. Therefore, assume for the rest of the proof that $n \in \mathbb{N} \setminus \{1\}$. Taking $x = 1$ and $y = \frac{1}{n}$, we have the following:

$$\left(1+\frac{1}{n}\right)^n = \sum_{k=0}^{n}\binom{n}{k}1^{n-k}\frac{1}{n^k}$$

$$= \binom{n}{0} + \binom{n}{1}\frac{1}{n} + \sum_{k=2}^{n}\binom{n}{k}1^{n-k}\frac{1}{n^k}$$

$$= 1 + n\cdot\frac{1}{n} + \sum_{k=2}^{n}\binom{n}{k}1^{n-k}\frac{1}{n^k}$$

$$= 1 + 1 + \sum_{k=2}^{n}\binom{n}{k}1^{n-k}\frac{1}{n^k}$$

$$= 2 + \sum_{k=2}^{n}\frac{n!}{(n-k)!k!}\cdot\frac{1}{n^k}$$

$$= 2 + \sum_{k=2}^{n}\frac{(n-k)![n-(k-1)]\cdots(n-1)n}{(n-k)!k!}\cdot\frac{1}{n^k}$$

$$= 2 + \sum_{k=2}^{n}\frac{[n-(k-1)]\cdots(n-1)}{k!}\cdot\frac{1}{n^{k-1}}$$

$$= 2 + \sum_{k=2}^{n}\left[\frac{1}{k!}\cdot\frac{n-1}{n}\cdot\frac{n-2}{n}\cdots\frac{n-(k-1)}{n}\right].$$

Since $1-\frac{i}{n} < 1$ for every $i \in \{1,\ldots,n-1\}$, we have

$$\left(1+\frac{1}{n}\right)^n = 2 + \sum_{k=2}^{n}\left[\frac{1}{k!}\cdot\frac{n-1}{n}\cdot\frac{n-2}{n}\cdots\frac{n-(k-1)}{n}\right] < 2 + \sum_{k=2}^{n}\frac{1}{k!}.$$

Moreover, by Exercise 1.16 part (I) and knowing that $2! = 2^{2-1}$, we arrive at

$$\left(1+\frac{1}{n}\right)^n < 2 + \sum_{k=2}^{n}\frac{1}{k!} \leq 2 + \sum_{k=2}^{n}\frac{1}{2^{n-1}}.$$

Finally, by Exercise 1.8 part (VII) with $a = 1$ and $r = \frac{1}{2}$, we obtain

$$\left(1+\frac{1}{n}\right)^n < 2 + \sum_{k=2}^{n}\frac{1}{2^{n-1}}$$

$$= 2 + \frac{\frac{1}{2^n}-1}{\frac{1}{2}-1} - \frac{1}{2^{1-1}}$$

$$= 1 - \frac{1}{2^{n-1}} + 2$$

$$= 3 - \frac{1}{2^{n-1}}$$

$$< 3.$$

$\square$

> **Exercise 1.28.** *Compare* n^{n+1} *and* $(n+1)^n$ *with* $n \in \mathbb{N}$, *and state and prove which inequality is fulfilled between these two numbers.*

Solution to Exercise 1.28.

- For $n = 1$: $1^{1+1} = 1 < 2 = (1+1)^1$.

- For $n = 2$: $2^{2+1} = 8 < 9 = (2+1)^2$.

- For $n = 3$: $3^{3+1} = 81 > 64 = (3+1)^3$.

Let us show that $n^{n+1} > (n+1)^n$ for every integer $n \geq 3$. To do so, we will use Exercise 1.27. Let $n \geq 3$ be an integer, then

$$\left(1 + \frac{1}{n}\right)^n < 3 \implies \frac{(n+1)^n}{n^n} < n \iff (n+1)^n < n^{n+1}.$$

$\square$

> **Exercise 1.29.** *Prove that the cardinal number of the power set of a set that has* $n \in \mathbb{N} \cup \{0\}$ *elements is* 2^n.

Solution to Exercise 1.29.

Notice that if A is the empty set, then the power set of A is formed only by the empty set. So the power set of A has $1 = 2^0$ elements. This can also be written as: $\binom{n}{0}$, which means that we have taken n elements in groups of 0 elements.

If we have a set B which has only one element x, then the power set of B is formed by 2 subsets: the empty set and $\{x\}$. Hence, the power set of B has $2 = 2^1$ elements. The latter can also be written as: $\binom{n}{0} + \binom{n}{1}$, which means that we have grouped n elements in groups of 1 element and in groups of 0 elements.

Therefore, in general, if we have a set C with n elements, we can group the n elements in groups of 0, 1, 2, $\ldots$, $n-1$ and n elements. This can be written as:

$$\binom{n}{0} + \binom{n}{1} + \binom{n}{2} + \cdots + \binom{n}{n-1} + \binom{n}{n}.$$

Using Exercise (I) we have that:

$$2^n = (1+1)^n = \sum_{k=0}^{n} 1^{n-k} 1^k \binom{n}{k} = \sum_{k=0}^{n} \binom{n}{k}.$$

It is important to mention that this exercise can be solved in many different ways. So if your solution does not follow the above reasoning, do not worry. It is possible that the proof that you have is also correct.

□

Exercise 1.30. *Let $m, n \in \mathbb{Z}$. We say that m is a divisor of n, and we denote it by $m|n$, if there exists a $k \in \mathbb{Z}$ such that $n = mk$.*

(I) *Prove that the relation $m|n$ satisfies the* reflexive *and* transitive *properties.*

(II) *Prove that the relation $m|n$ does not satisfy the* antisymmetric *property.*

Solution to Exercise 1.30.

(I).

- Reflexive ($m\mathcal{R}m$): for each $m \in \mathbb{Z}$ we have $m = mk$, where $k = 1 \in \mathbb{Z}$, that is, the reflexive property is fulfilled.

- Transitive (If $m\mathcal{R}n$ and $n\mathcal{R}p$, then $m\mathcal{R}p$): Assume that $m\mathcal{R}n$ and $n\mathcal{R}p$, then there are $k, k' \in \mathbb{Z}$ such that $n = mk$ and $p = nk'$. Thus, $p = nk' = mkk' = m(kk') = mk''$, where $k'' = kk' \in \mathbb{Z}$, that is, $m\mathcal{R}p$.

(II). To prove that the antisymmetric property is not satisfied (If $m\mathcal{R}n$ and $n\mathcal{R}m$, then $m = n$), it is enough to find a counterexample: If $m = -2$ and $n = 2$ then we have that mRn and nRm, but $2 \neq -2$.

□

Exercise 1.31. *Consider the relation from Exercise 1.30, but now defined over $\mathbb{N}$.*

(I) *Prove that it is an order relation, that is, it satisfies the reflexive, antisymmetric and transitive properties.*

(II) *Prove that it does not fulfill the* total order *property.*

Solution to Exercise 1.31.

(I).

- Reflexive: given any $m \in \mathbb{N}$, we have that $m = mk$, where $k = 1 \in \mathbb{Z}$, that is, $m\mathcal{R}m$.

- Antisymmetric (If $m\mathcal{R}n$ and $n\mathcal{R}m$, then $m = n$): Assume that $m\mathcal{R}n$ and $n\mathcal{R}m$, then there exist $k, k' \in \mathbb{Z}$ such that $n = mk$ and $m = nk'$. Hence, $m = nk' = mkk' = m(kk')$ or, equivalently, $kk' = 1$ with $k, k' \in \mathbb{Z}$. Therefore, $k = k' = 1$ (or $k = k' = -1$, but since $n, m \in \mathbb{N}$ we cannot have $n = -m$ so the case $k = k' = -1$ does not occur). In conclusion, $m = n$.

- Transitive: Assume that $m\mathcal{R}n$ and $n\mathcal{R}p$, then there are $k, k' \in \mathbb{Z}$ such that $n = mk$ and $p = nk'$. Thus, $p = nk' = mkk' = m(kk') = mk''$, where $k'' = kk' \in \mathbb{Z}$, that is, $m\mathcal{R}p$.

(II). Let us show that it is not a total order relation with a counterexample. Let $n = 3$ and $m = 5$, since 5 does not divide 3 and 3 does not divide 5, it is not a total order relation.

$\square$

Exercise 1.32. *A relation over a set X that satisfies the reflexive and transitive properties is called a* preorder. *Let $\mathcal{R}$ be a preorder.*

(I) *Prove that the relation $\equiv$, defined over X by*

$$x \equiv y \qquad \text{if, and only if,} \qquad x\mathcal{R}y \quad \text{and} \quad y\mathcal{R}x,$$

is an equivalence relation.

(II) *Denote by $[x]$ the equivalence class of the quotient set $X/\equiv$ corresponding to the element $x \in X$, and define the relation $\leq$ in $X/\equiv$ by*

$$[x] \leq [y] \qquad \text{if, and only if,} \qquad x\mathcal{R}y.$$

Prove that this relation is well defined (that is, it does not depend on the x and y chosen) and it is an order relation.

Solution to Exercise 1.32.

We have to prove that $\equiv$ is a reflexive, symmetric and transitive relation.

- Reflexive: Let $x \in X$. As $\mathcal{R}$ is a preorder, in particular, it is a reflexive relation, that is, $x\mathcal{R}x$; but this is $x \equiv x$.

- Symmetric: Let $x, y \in X$ such that $x \equiv y$, that is, $x\mathcal{R}y$ and $y\mathcal{R}x$. Thus, by definition of $\equiv$, it is clear that $y \equiv x$, so the relation $\equiv$ is symmetric.

- Transitive: Let $x, y, z \in X$ such that $x \equiv y$ and $y \equiv z$. Hence, we have that $x\mathcal{R}y$ and $y\mathcal{R}x$, and $y\mathcal{R}z$ and $z\mathcal{R}y$. Since $\mathcal{R}$ is a transitive relation as a consequence of the fact that it is a preorder, it follows that $x\mathcal{R}z$ and $z\mathcal{R}x$. Therefore $x \equiv z$ and the relation $\equiv$ is transitive.

(II) Let us see that the relation $\leq$ is well defined and it is an order relation.

- Well defined: Let $[x], [y] \in X/\equiv$ such that $[x] \leq [y]$. Let us show that given any $x' \in [x]$ and $y' \in [y]$ we have $[x'] \leq [y']$. Take $x' \in [x]$ and $y' \in [y]$. By definition of equivalence class, it follows that $x \equiv x'$ and $y \equiv y'$. Equivalently, $x\mathcal{R}x'$ and $x'\mathcal{R}x$, and $y\mathcal{R}y'$ and $y'\mathcal{R}y$. Moreover, since $[x] \leq [y]$, we have that $x\mathcal{R}y$. Thus, we obtain $x'\mathcal{R}x$, $x\mathcal{R}y$ and $y\mathcal{R}y'$. Hence, since $\mathcal{R}$ is a transitive relation, it yields $x'\mathcal{R}y'$ or $[x'] \leq [y']$.

- Reflexive: Let $[x] \in X/\equiv$. Since $x \in X$ and $x\mathcal{R}x$ as a consequence of $\mathcal{R}$ being a reflexive relation, it follows that $[x] \leq [x]$.

- Antisymmetric: Let $[x], [y] \in X/\equiv$ such that $[x] \leq [y]$ and $[y] \leq [x]$. Then $x\mathcal{R}y$ and $y\mathcal{R}x$. By definition of the relation $\equiv$ we have that $x \equiv y$, and since the definition of equivalence class does not depend on the chosen representative in the set of the equivalence class it yields $[x] = [y]$.

- Transitive: Let $[x], [y], [z] \in X/\equiv$ such that $[x] \leq [y]$ and $[y] \leq [z]$. Then $x\mathcal{R}y$ and $y\mathcal{R}z$. As $\mathcal{R}$ is a transitive relation we obtain that $x\mathcal{R}z$, that is, $[x] \leq [z]$.

$\square$

Exercise 1.33 (Cauchy-Schwarz Inequality). *If* $x_1, \ldots, x_n, y_1, \ldots, y_n \in \mathbb{R}$ *prove that*

$$\left(\sum_{j=1}^{n} x_j y_j \right)^2 \leq \left(\sum_{j=1}^{n} x_j^2 \right) \left(\sum_{j=1}^{n} y_j^2 \right).$$

Deduce that if $a^2 + b^2 = c^2 + d^2 = 1$, *then* $|ac + bd| \leq 1$.

Solution to Exercise 1.33.

Let $n \in \mathbb{N}$ and $x_1, \ldots, x_n, y_1, \ldots, y_n \in \mathbb{R}$. Observe that

$$0 \le \sum_{i=1}^{n} \sum_{j=1}^{n} (x_i y_j - x_j y_i)^2 = \sum_{i=1}^{n} \sum_{j=1}^{n} (x_i^2 y_j^2 + x_j^2 y_i^2 - 2x_i y_j x_j y_i)$$

$$= \sum_{i=1}^{n} \sum_{j=1}^{n} x_i^2 y_j^2 + \sum_{i=1}^{n} \sum_{j=1}^{n} x_j^2 y_i^2 - 2 \sum_{i=1}^{n} \sum_{j=1}^{n} x_i y_i x_j y_j$$

$$= \sum_{i=1}^{n} x_i^2 \sum_{j=1}^{n} y_j^2 + \sum_{i=1}^{n} y_i^2 \sum_{j=1}^{n} x_j^2 - 2 \sum_{i=1}^{n} x_i y_i \sum_{j=1}^{n} x_j y_j$$

$$= 2 \left(\sum_{i=1}^{n} x_i^2 \right) \left(\sum_{i=1}^{n} y_i^2 \right) - 2 \left(\sum_{i=1}^{n} x_i y_i \right)^2$$

$$= 2 \left[\left(\sum_{i=1}^{n} x_i^2 \right) \left(\sum_{i=1}^{n} y_i^2 \right) - \left(\sum_{i=1}^{n} x_i y_i \right)^2 \right].$$

Thus,

$$0 \le \left(\sum_{i=1}^{n} x_i^2 \right) \left(\sum_{i=1}^{n} y_i^2 \right) - \left(\sum_{i=1}^{n} x_i y_i \right)^2,$$

or equivalently

$$\left(\sum_{i=1}^{n} x_i y_i \right)^2 \le \left(\sum_{i=1}^{n} x_i^2 \right) \left(\sum_{i=1}^{n} y_i^2 \right).$$

Using Cauchy-Schwarz inequality, we observe that, if $a, b, c, d \in \mathbb{R}$ are such that $a^2 + b^2 = c^2 + d^2 = 1$, then $|ac + bd| \le 1$. Indeed, take $x_1 = a$, $x_2 = b$, $y_1 = c$ and $y_2 = d$. Then

$$(ac + bd)^2 \le (a^2 + b^2)(c^2 + d^2) = 1. \tag{1.16}$$

Since the square root function is an increasing function, applying the square root in (1.16) preserves the inequality $\le$, that is,

$$|ac + bd| = \sqrt{(ac + bd)^2} \le \sqrt{1} = 1.$$

$\square$

Exercise 1.34. *Find the solutions of the following inequalities:*

(I) $2x^2 + 9x + 6 \geq x + 2,$

(V) $\dfrac{2x-1}{3x+2} \leq 1,$

(II) $x + \dfrac{1}{x} < 1,$

(VI) $\dfrac{2x^2 + 9x + 6}{x+2} \geq 1,$

(III) $\dfrac{x}{x+5} < 0,$

(VII) $\dfrac{x^2 - 4x + 4}{1 + x^3} > 0,$

(IV) $\dfrac{3x^2 - 1}{1 + x^2} > 0,$

(VIII) $\dfrac{x-1}{3x+4} \leq \dfrac{3x+2}{x}.$

Solution to Exercise 1.34.

(I). Simplify the inequality $2x^2 + 9x + 6 \geq x + 2$ in the following way:

$$2x^2 + 9x + 6 \geq x + 2 \iff 2x^2 + 8x + 4 \geq 0 \iff x^2 + 4x + 2 \geq 0$$
$$\iff \left(x + 2 + \sqrt{2}\right)\left(x + 2 - \sqrt{2}\right) \geq 0,$$

where in the last equivalence we have used the second degree equation formula to obtain the roots of $x^2 + 4x + 2$.

Study the sign of $x^2 + 4x + 2$ depending on the value of its roots:

	$-\infty$	$-2 - \sqrt{2}$	$-2 + \sqrt{2}$	$+\infty$
$x + 2 + \sqrt{2}$	$-$	$+$	$+$	
$x + 2 - \sqrt{2}$	$-$	$-$	$+$	
$x^2 + 4x + 2$	$+$	$-$	$+$	

Hence, the solution is $x \in \left(-\infty, -2 - \sqrt{2}\right] \cup \left[-2 + \sqrt{2}, +\infty\right)$.

(II). Looking at the inequality, it is clear that $x \neq 0$. Now simplify the inequality $x + \dfrac{1}{x} < 1$ as follows:

$$x + \dfrac{1}{x} < 1 \iff \dfrac{x^2 + 1}{x} - 1 < 0 \iff \dfrac{x^2 - x + 1}{x} < 0.$$

Solving the second degree equation $x^2 - x + 1 = 0$, we note that it does not have real roots and it is always positive for every $x \in \mathbb{R}$. Hence, in order to study the sign of $\frac{x^2 - x + 1}{x}$ it is enough to analyze the sign of the denominator, that is, when $x > 0$ or $x < 0$:

	$-\infty$	0	$+\infty$
$x^2 - x + 1$	$+$	$+$	
x	$-$	$+$	
$\frac{x^2 - x + 1}{x}$	$-$	$+$	

Therefore, the solution is $x \in (-\infty, 0)$.

(III). Observe that $x \notin \{-5, 0\}$ and let us study the sign of $\dfrac{x}{x+5}$ depending on the values of x and $x + 5$:

	$-\infty$	-5	0	$+\infty$
x	$-$	$-$	$+$	
$x + 5$	$-$	$+$	$+$	
$\dfrac{x}{x+5}$	$+$	$-$	$+$	

Thus, the solution is $x \in (-5, 0)$.

(IV). Observe that the denominator $1 + x^2$ does not have real roots and it is positive for each $x \in \mathbb{R}$. Let us calculate the roots of the numerator: $3x^2 - 1 = 0$ if, and only if, $x = \pm\frac{\sqrt{3}}{3}$.

Now we will study the sign of $\frac{3x^2-1}{1+x^2}$:

	$-\infty$	$-\frac{\sqrt{3}}{3}$	$\frac{\sqrt{3}}{3}$	$+\infty$
$x + \frac{\sqrt{3}}{3}$	$-$	$+$	$+$	
$x - \frac{\sqrt{3}}{3}$	$-$	$-$	$+$	
$x^2 + 1$	$+$	$+$	$+$	
$\frac{3x^2-1}{x^2+1}$	$+$	$-$	$+$	

Thus, the solution is $x \in \mathbb{R} \setminus \left[-\frac{\sqrt{3}}{3}, \frac{\sqrt{3}}{3}\right]$ or equivalently $x \in \left(-\infty, -\frac{\sqrt{3}}{3}\right) \cup \left(\frac{\sqrt{3}}{3}, +\infty\right)$.

(v). First, observe that the inequality $\frac{2x-1}{3x+2} \leq 1$ is equivalent to $\frac{-x-3}{3x+2} \leq 0$, and moreover $x \neq -\frac{2}{3}$.

We proceed to study the sign:

	$-\infty$	-3	$-\frac{2}{3}$	$+\infty$
$-x - 3$	$+$	$-$	$-$	
$3x + 2$	$-$	$-$	$+$	
$\frac{-x-3}{3x+2}$	$-$	$+$	$-$	

Hence, the solution is $x \in \mathbb{R} \setminus (-3, -\frac{2}{3}]$ or equivalently $x \in (-\infty, -3] \cup (-\frac{2}{3}, +\infty)$.

(VI). Observe that the inequality $\frac{2x^2+9x+6}{x+2} \geq 1$ is equivalent to $\frac{2(x^2+4x+2)}{x+2} \geq 0$ and $x \neq -2$. Calculating the roots of $x^2 + 4x + 2$ using the second degree equation formula we observe: $x = \frac{-4\pm\sqrt{16-8}}{2} = -2 \pm \sqrt{2}$.

Let us now study the sign:

	$-\infty$		$-2-\sqrt{2}$		-2		$-2+\sqrt{2}$		$+\infty$
$x+2+\sqrt{2}$		$-$		$+$		$+$		$+$	
$x+2-\sqrt{2}$		$-$		$-$		$-$		$+$	
$x+2$		$-$		$-$		$+$		$+$	
$\frac{2(x^2+4x+2)}{x+2}$		$-$		$+$		$-$		$+$	

Thus, the solution is $x \in \left[-2-\sqrt{2}, -2\right) \cup \left[-2+\sqrt{2}, +\infty\right)$.

(VII). We observe that $x \neq -1$ and the numerator can be simplified as $(x-2)^2$. So the numerator is always strictly positive except when $x \in \{-2, 2\}$. However, when $x = -2$ or $x = 2$ we obtain that $\frac{x^2-4x+4}{1+x^3} = 0$.

We proceed to study the sign:

	$-\infty$		-1		$+\infty$
$x^2 - 4x + 4$		$+$		$+$	
$x^3 + 1$		$-$		$+$	
$\frac{x^2-4x+4}{x^3+1}$		$-$		$+$	

Hence, the solution is $x \in (-1, 2) \cup (2, +\infty)$ or equivalently $x \in (-1, +\infty) \setminus \{2\}$.

(VIII). We begin by studying the inequality $\frac{x-1}{3x+4} \leq \frac{3x+2}{x}$:

$$\frac{x-1}{3x+4} \leq \frac{3x+2}{x} \iff \frac{x^2-x}{x(3x+4)} - \frac{(3x+2)(3x+4)}{x(3x+4)} \leq 0$$

$$\iff \frac{x^2-x-9x^2-12x-6x-8}{x(3x+4)} \leq 0$$

$$\iff \frac{-8x^2-19x-8}{x(3x+4)} \leq 0.$$

Calculating the roots of $-8x^2 - 19x - 8$ using the second degree equation formula it yields that $x = -\frac{19 \pm \sqrt{105}}{16}$.

Let us now study the sign of $\frac{-8x^2-19x-8}{x(3x+4)}$:

	$-\infty$		$-\frac{19+\sqrt{105}}{16}$		$-\frac{4}{3}$		$-\frac{19-\sqrt{105}}{16}$		0		$+\infty$
$x + \frac{19+\sqrt{105}}{16}$		$-$		$+$		$+$		$+$		$+$	
$x + \frac{19-\sqrt{105}}{16}$		$-$		$-$		$-$		$+$		$+$	
x		$-$		$-$		$-$		$-$		$+$	
$3x + 4$		$-$		$-$		$+$		$+$		$+$	
$\frac{8x^2+19x+8}{x(3x+4)}$		$+$		$-$		$+$		$-$		$+$	

Hence, the solution is $x \in \left[-\frac{19+\sqrt{105}}{16}, -\frac{4}{3}\right) \cup \left[-\frac{19-\sqrt{105}}{16}, 0\right)$.

$\square$

Exercise 1.35. *Solve the equations:*

(I) $|x^2 - 5x + 6| = -(x^2 - 5x + 6)$,

(II) $\left|\frac{x-1}{x+1}\right| = \frac{x-1}{x+1}$,

(III) $|(x^2 + 4x + 9) + (2x - 3)| = |x^2 + 4x + 9| + |2x - 3|$,

(IV) $|x - 1||x + 1| = 0$,

(V) $|x - 1||x + 2| = 3$.

Solution to Exercise 1.35.

(I). We begin by studying the sign of $x^2 - 5x + 6$ depending on the values of $x \in \mathbb{R}$. To do so, we calculate the roots of $x^2 - 5x + 6 = 0$ using the second degree equation formula and we obtain $x = 2$ and $x = 3$.

We now proceed to study the sign of $x^2 - 5x + 6$:

	$-\infty$		2		3		$+\infty$
$x - 2$		$-$		$+$		$+$	
$x - 3$		$-$		$-$		$+$	
$x^2 - 5x + 6$		$+$		$-$		$+$	

Assume that $x \in [2, 3]$, then we have that $x^2 - 5x + 6 \leq 0$, that is, the equation $|x^2 - 5x + 6| = -(x^2 - 5x + 6)$ is satisfied.

Now, if $x \in (-\infty, 2) \cup (3, \infty)$ we have that $x^2 - 5x + 6 > 0$, that is, $|x^2 - 5x + 6| = x^2 - 5x + 6$. Hence, when $x \in (-\infty, 2) \cup (3, \infty)$ it yields the equations $x^2 - 5x + 6 = -(x^2 - 5x + 6)$ if, and only if, $x^2 - 5x + 6 = 0$, but the latter equation is only satisfied when $x = 2$ or 3, which contradicts that $x \in (-\infty, 2) \cup (3, \infty)$.

Therefore, the equation $|x^2 - 5x + 6| = -(x^2 - 5x + 6)$ is satisfied when $x \in [2, 3]$.

(II). First, observe that $x \neq -1$. Second, let us study the sign of $\frac{x-1}{x+1}$. The equality is satisfied when $x = 1$ and moreover we have the following table:

	$-\infty$		-1		1		$+\infty$
$x - 1$		$-$		$-$		$+$	
$x + 1$		$-$		$+$		$+$	
$\frac{x-1}{x+1}$		$+$		$-$		$+$	

Hence, when $x \in (-\infty, -1) \cup [1, +\infty)$ we have that $\frac{x-1}{x+1} \geq 0$, that is, the equation $\left|\frac{x-1}{x+1}\right| = \frac{x-1}{x+1}$ is satisfied.

Finally, if $x \in (-1, 1)$ we have that $\left|\frac{x-1}{x+1}\right| = -\frac{x-1}{x+1}$. So, we obtain the equation $-\frac{x-1}{x+1} = \frac{x-1}{x+1}$ or, equivalently, $\frac{x-1}{x+1} = 0$, but this equation is only satisfied when $x = 1$ which contradicts the fact that $x \in (-1, 1)$.

(III). Let us study the sign of $x^2 + 4x + 9$. To do so, we can check using the second degree equation formula that the roots of $x^2 + 4x + 9$ are complex and that $x^2 + 4x + 9 > 0$ for each $x \in \mathbb{R}$. Observe that $2x - 3 \geq 0$ if, and only if, $x \geq \frac{3}{2}$. Finally, we will study the sign of $(x^2 + 4x + 9) + (2x - 3) = x^2 + 6x + 6$. Using the second degree equation formula, we note that the roots of $x^2 + 6x + 6$ are $-3 \pm \sqrt{3}$. Hence, we have the following table:

	$-\infty$		$-3 - \sqrt{3}$		$-3 + \sqrt{3}$		$+\infty$
$x + 3 + \sqrt{3}$		$-$		$+$		$+$	
$x + 3 - \sqrt{3}$		$-$		$-$		$+$	
$x^2 + 6x + 6$		$+$		$-$		$+$	

Knowing that $-3 + \sqrt{3} < \frac{3}{2}$, we have the following four cases in order to solve the equation $|(x^2 + 4x + 9) + (2x - 3)| = |x^2 + 4x + 9| + |2x - 3|$.

- Case 1: If $x \in \left(-\infty, -3 - \sqrt{3}\right]$, we have $x^2 + 6x + 6 = x^2 + 4x + 9 - 2x + 3$ if, and only if, $4x - 6 = 0$ if, and only if, $x = \frac{3}{2}$. But $\frac{3}{2} \notin \left(-\infty, -3 - \sqrt{3}\right]$.

- Case 2: If $x \in \left(-3 - \sqrt{3}, -3 + \sqrt{3}\right]$, we have $-x^2 - 6x - 6 = x^2 + 4x + 9 - 2x + 3$ if, and only if, $x^2 + 4x + 9 = 0$. But $x^2 + 4x + 9 > 0$ for every $x \in \mathbb{R}$.

- Case 3: If $x \in \left(-3 + \sqrt{3}, \frac{3}{2}\right]$, we have $x^2 + 6x + 6 = x^2 + 4x + 9 - 2x + 3$ if, and only if, $4x - 6 = 0$ if, and only if, $x = \frac{3}{2}$.

- Case 4: If $x \in \left(\frac{3}{2}, +\infty\right)$, we have $x^2 + 6x + 6 = (x^2 + 4x + 9) + (2x - 3)$.

Hence, the equation is satisfied when $x \in \left[\frac{3}{2}, +\infty\right)$.

(IV). The equation $|x - 1||x + 1| = 0$ is satisfied if, and only if, $|x - 1| = 0$ or $|x + 1| =$ if, and only if, $x = 1$ or $x = -1$.

(V). The equation $|x - 1||x + 2| = 3$ is not satisfied when $x = 1$ or $x = -2$, so we can assume that $x \notin \{-2, 1\}$ for the rest of this part. To study the equation $|x - 1||x + 2| = 3$, we have to study the sign of $(x - 1)(x + 2)$:

	$-\infty$		-2		1		$+\infty$
$x - 1$		$-$		$-$		$+$	
$x + 2$		$-$		$+$		$+$	
$(x - 1)(x + 2)$		$+$		$-$		$+$	

Assume now that $x \in (-2, 1)$, then we have the equation $(x-1)(x+2) = -3$ or equivalently $x^2 + x + 1 = 0$, which does not have real solutions.

Assume that $x \in (-\infty, -2) \cup (1, +\infty)$, then we have that $(x-1)(x+2) = 3$ or equivalently $x^2 + x - 5 = 0$. Solving the second degree equation $x^2 + x - 5 = 0$ we obtain $x = \frac{1 \pm \sqrt{21}}{2}$, but only $\frac{1 + \sqrt{21}}{2}$ belongs to the set $(-\infty, -2) \cup (1, +\infty)$. $\qquad \square$

> **Exercise 1.36.** *Write the following formulas with equivalent expressions that do not involve absolute values:*
>
> (I) $a + b + |a - b|$,
>
> (II) $a + b - |a - b|$,
>
> (III) $a + b + 2c + |a - b| + |a + b - 2c + |a - b||$,
>
> (IV) $a + b + 2c - |a - b| |a + b - 2c - |a - b||$.

Solution to Exercise 1.36.

(I).

$$a + b + |a - b| = \begin{cases} a + b + (a - b) & \text{if } a - b \geq 0, \\ a + b - (a - b) & \text{if } a - b < 0, \end{cases}$$

that is,

$$a + b + |a - b| = \begin{cases} 2a & \text{if } a \geq b, \\ 2b & \text{if } a < b. \end{cases}$$

(II).

$$a + b - |a - b| = \begin{cases} a + b - (a - b) & \text{if } a - b \geq 0, \\ a + b + (a - b) & \text{if } a - b < 0, \end{cases}$$

that is,

$$a + b + |a - b| = \begin{cases} 2b & \text{if } a \geq b, \\ 2a & \text{if } a < b. \end{cases}$$

(III).

$$a + b + 2c + |a - b| + |a + b - 2c + |a - b||$$

$$= \begin{cases} a + b + 2c + a - b + |a + b - 2c + a - b| & \text{if } a - b \geq 0, \\ a + b + 2c - (a - b) + |a + b - 2c - (a - b)| & \text{if } a - b < 0, \end{cases}$$

that is,

$$a + b + 2c + |a - b| + |a + b - 2c + |a - b|| = \begin{cases} 2a + 2c + |2a - 2c| & \text{if } a \geq b, \\ 2b + 2c + |2b - 2c| & \text{if } a < b. \end{cases}$$

Finally,

$$a + b + 2c + |a - b| + |a + b - 2c + |a - b||$$

$$= \begin{cases} 2(a + c + a - c) & \text{if } a \geq b \text{ and } a - c \geq 0, \\ 2(a + c - (a - c)) & \text{if } a \geq b \text{ and } a - c < 0, \\ 2(b + c + b - c) & \text{if } a < b \text{ and } b - c \geq 0, \\ 2(b + c - (b - c)) & \text{if } a < b \text{ and } b - c < 0, \end{cases}$$

or equivalently

$$a + b + 2c + |a - b| + |a + b - 2c + |a - b|| = \begin{cases} 4a & \text{if } a \geq b \text{ and } a \geq c, \\ 4c & \text{if } c > a \geq b, \\ 4b & \text{if } a < b \text{ and } b \geq c, \\ 4c & \text{if } a < b < c. \end{cases}$$

(IV).

$$a + b + 2c - |a - b| |a + b - 2c - |a - b||$$

$$= \begin{cases} a + b + 2c - (a - b)|a + b - 2c - (a - b)| & \text{if } a - b \geq 0, \\ a + b + 2c + (a - b)|a + b - 2c + a - b| & \text{if } a - b < 0, \end{cases}$$

that is,

$$a + b + 2c - |a - b| |a + b - 2c - |a - b|| = \begin{cases} a + b + 2c - (a - b)|2b - 2c| & \text{if } a \geq b, \\ a + b + 2c + (a - b)|2a - 2c| & \text{if } a < b. \end{cases}$$

Finally,

$$a + b + 2c - |a - b| |a + b - 2c - |a - b||$$

$$= \begin{cases} a + b + 2c - 2(a - b)(b - c) & \text{if } a \geq b \text{ and } b - c \geq 0, \\ a + b + 2c + 2(a - b)(b - c) & \text{if } a \geq b \text{ and } b - c < 0, \\ a + b + 2c + 2(a - b)(a - c) & \text{if } a < b \text{ and } a - c \geq 0, \\ a + b + 2c - 2(a - b)(a - c) & \text{if } a < b \text{ and } a - c < 0, \end{cases}$$

or equivalently

$$a + b + 2c - |a - b| |a + b - 2c - |a - b||$$

$$= \begin{cases} a + b + 2c - 2(a - b)(b - c) & \text{if } a \geq b \geq c, \\ a + b + 2c + 2(a - b)(b - c) & \text{if } a \geq b \text{ and } b < c, \\ a + b + 2c + 2(a - b)(a - c) & \text{if } c \leq a < b, \\ a + b + 2c - 2(a - b)(a - c) & \text{if } a < b \text{ and } a < c. \end{cases}$$

□

Exercise 1.37. *Solve the following inequalities:*

(I) $|2x + 3| - 1 < |x|,$

(II) $|x(x - 4)| < |x - 4| - |x|,$

(III) $|2 - |x|| \geq 2 + |x|,$

(IV) $|x - 1| + |x + 1| < 1,$

(V) $|x - 5| < |x + 1|,$

(VI) $|3x - 5| < 3,$

(VII) $|x^2 - 1| < 1,$

(VIII) $|x^2 - x + 1| > 1,$

(IX) $1 < |x - \frac{1}{2}| < 2,$

(X) $x - |x| > 2,$

(XI) $|x^2 - x| + x > 1,$

(XII) $|x + |x - 1|| < 2,$

(XIII) $\dfrac{1}{1 + |x - 1|} < |x - 2|,$

(XIV) $-1 \leq \dfrac{|x^3 - 1|}{x - 1} \leq 2.$

Solution to Exercise 1.37.

(I). Let us see first the case $2x + 3 \geq 0$, that is, $x \geq -\frac{3}{2}$. Hence, $2x + 3 - 1 < |x|$ if, and only if, $2x + 2 < |x|$. Now we have two options $x \geq 0$ or $x < 0$:

- Let $x \geq 0$ (in particular, $x \geq -\frac{3}{2}$), then

$$2x + 2 < x \iff x < -2.$$

But $-\frac{3}{2} > -2$, so the inequality does not have a solution in this case.

- Let $x < 0$ (and $x \geq -\frac{3}{2}$), then

$$2x + 2 < -x \iff 3x < -2 \iff x < -\frac{2}{3}.$$

Thus, $x \in \left[-\frac{3}{2}, -\frac{2}{3}\right)$.

Let us analyze the case $2x + 3 < 0$, that is, when $x < -\frac{3}{2}$. The inequality can now be written as $-(2x + 3) - 1 < |x|$ or equivalently $-2x - 4 < |x|$. Observe that now we do not have to study two cases since $x < -\frac{3}{2} < 0$. Therefore, the inequality now has the form

$$-2x - 4 < -x \iff -4 < x,$$

that is, $x \in \left(-4, -\frac{3}{2}\right)$.

Finally, we see that $x \in \left[-\frac{3}{2}, -\frac{2}{3}\right) \cup \left(-4, -\frac{3}{2}\right)$. From which we conclude that $x \in \left(-4, -\frac{2}{3}\right)$.

(II). Observe that if $x = 4$, then we obtain $0 < -4$ which is absurd. If $x = 0$, then we have that $0 < 4$.

- Assume that $x > 4$; then the inequality can be written as:

$$x(x - 4) < x - 4 - x \iff x^2 - 4x + 4 < 0 \iff (x - 2)^2 < 0.$$

But $(x - 2)^2$ cannot be negative, which means that we have reached a contradiction.

- Assume now that $x < 0$; then the inequality can be written as:

$$x(x - 4) < -x + 4 + x \iff x^2 - 4x - 4 < 0$$

By the second degree equation formula, we have that the roots of $x^2 - 4x - 4 = 0$ are $x = 2 \pm 2\sqrt{2}$. Knowing the roots we can study the sign of $x^2 - 4x - 4$:

	$-\infty$		$2 - 2\sqrt{2}$		$2 + 2\sqrt{2}$		$+\infty$
$x - 2 + 2\sqrt{2}$		$-$		$+$		$+$	
$x - 2 - 2\sqrt{2}$		$-$		$-$		$+$	
$x^2 - 4x - 4$		$+$		$-$		$+$	

Hence, since $x < 0 < 2 + 2\sqrt{2}$, we have that $x \in (2 - 2\sqrt{2}, 0)$.

- Assume that $0 < x < 4$, then the inequality can written as:

$$-x(x - 4) < -x + 4 - x \iff x^2 - 6x + 4 > 0.$$

By the second degree equation formula, we have that the roots of $x^2 - 6x + 4 = 0$ are $x = 3 \pm \sqrt{5}$. As an immediate consequence, we can study the sign of $x^2 - 6x + 4$:

	$-\infty$		$3 - \sqrt{5}$		$3 + \sqrt{5}$		$+\infty$
$x - 3 + \sqrt{5}$		$-$		$+$		$+$	
$x - 3 - \sqrt{5}$		$-$		$-$		$+$	
$x^2 - 6x + 4$		$+$		$-$		$+$	

Since $0 < 3 - \sqrt{5} < 4 < 3 + \sqrt{5}$, it yields $x \in [0, 3 - \sqrt{5})$.

Finally, we get that $x \in (2 - 2\sqrt{2}, 0)$ or $x \in [0, 3 - \sqrt{5})$, or equivalently $x \in (2 - 2\sqrt{2}, 3 - \sqrt{5})$.

(III). First, let us analyze the case $x \geq 0$. Then we have the inequality $|2 - x| \geq 2 + x$. We have two options: $2 - x \leq 0$ or $2 - x > 0$, that is, $x \geq 2$ or $x < 2$:

- Assume that $x \geq 2$; then

$$x - 2 \geq 2 + x \iff 0 \geq 4,$$

which is absurd.

- Assume that $0 \leq x < 2$; then

$$2 - x \geq 2 + x \iff 0 \geq 2x \iff 0 \geq x.$$

Hence, we obtain that $0 \leq x \leq 0$, that is, $x = 0$.

Now let us study the case when $x < 0$ which yields $|2 + x| \geq 2 - x$. Note that we have two options: $2 + x \geq 0$ or $2 + x < 0$, that is, $x \geq -2$ or $x < -2$.

- Assume that $x < -2$; then

$$-2 - x \geq 2 - x \iff -2 \geq 2,$$

which is absurd.

- Assume that $-2 \leq x < 0$; then

$$2 + x \geq 2 - x \iff 2x \geq 0.$$

Thus, $0 \leq x < 0$ which is a contradiction.

Then the inequality $|2 - |x|| \geq 2 + |x|$ holds true only when $x = 0$.

(IV). Let us first study the case when $x - 1 \leq 0$, that is, when $x \leq 1$. In this case, it yields the inequality $-x + 1 + |x + 1| < 1$. Notice that we have two cases: $x + 1 \leq 0$ or $x + 1 > 0$, that is, $x \leq -1$ or $x > -1$.

- Assume that $x \leq -1$. Then

$$-x + 1 - x - 1 < 1 \iff -2x < 1 \iff x > -\frac{1}{2}.$$

Hence, we have that $-\frac{1}{2} < x \leq -1$, which is absurd.

- Assume that $-1 < x \leq 1$; then

$$-x + 1 + x + 1 < 1 \iff 2 < 1,$$

which is impossible.

Now let us analyze the case when $x - 1 > 0$, that is, $x > 1$. In this latter case, the inequality can be written as $x - 1 + |x + 1| < 1$. Note that we have two cases: $x + 1 \leq 0$ or $x + 1 > 0$, that is, $x \leq -1$ or $x > -1$. Thus, we only have the option when $x > 1$, in which case we have the following chain of equivalences:

$$x - 1 + x + 1 < 1 \iff 2x < 1 \iff x < \frac{1}{2}.$$

That is, $1 < x < \frac{1}{2}$, which is absurd.

Therefore the inequality $|x - 1| + |x + 1| < 1$ has no solutions.

(v). Let us first study the case $x - 5 \leq 0$, that is, $x \leq 5$. Then the inequality can be written as $-x + 5 < |x + 1|$. Now we have two options $x + 1 \leq 0$ or $x + 1 > 0$, that is, $x \leq -1$ or $x > -1$.

- Assume that $x \leq -1$; then

$$-x + 5 < -x - 1 \iff 5 < -1,$$

which is absurd.

- Assume that $-1 < x \leq 5$; then

$$-x + 5 < x + 1 \iff 4 < 2x \iff 2 < x.$$

Thus, $x \in (2, 5]$.

Now let us analyze the case when $x - 5 > 0$, that is, $x > 5$. Then the inequality becomes $x - 5 < |x + 1|$. Once again, we have two options $x + 1 \leq 0$ or $x + 1 > 0$, that is, $x \leq -1$ or $x > -1$. However, since $x > 5$ and ($x \leq -1$ or $x > -1$), we only have one possibility: $x > 5$. As a consequence

$$x - 5 < x + 1 \iff -5 < 1.$$

Hence, $x \in (5, +\infty)$.

Finally, we obtain that $x \in (2, 5]$ or $x \in (5, +\infty)$. Therefore, $x \in (2, +\infty)$.

(VI). We have

$$|3x - 5| < 3 \iff -3 < 3x - 5 < 3 \iff 2 < 3x < 8 \iff \frac{2}{3} < x < \frac{8}{3}.$$

Thus, $x \in \left(\frac{2}{3}, \frac{8}{3}\right)$.

(VII). Let us first study the case $x^2 - 1 \leq 0$, that is, $-1 \leq x \leq 1$. Then, we obtain

$$1 - x^2 < 1 \iff 0 < x^2,$$

which is always true when $x \neq 0$, so $x \in [-1, 1] \setminus \{0\}$.

Now let us analyze the case when $x^2 - 1 > 0$, that is, $x < -1$ or $x > 1$. In this case, we obtain

$$x^2 - 1 < 1 \iff x^2 < 2.$$

Hence, $|x| < \sqrt{2}$; and knowing that $x < -1$ or $x > 1$, it yields $x \in \left(-\sqrt{2}, -1\right) \cup \left(1, \sqrt{2}\right)$.

Finally, we get that $x \in [-1, 1] \setminus \{0\}$ or $x \in \left(-\sqrt{2}, -1\right) \cup \left(1, \sqrt{2}\right)$. Thus, $x \in \left(-\sqrt{2}, 0\right) \cup \left(0, \sqrt{2}\right)$.

(VIII). By the second degree equation formula applied to $x^2 - x + 1 = 0$, we obtain the roots $x = \frac{1 \pm \sqrt{-3}}{2}$ which are not real numbers. Moreover, taking any real value, we have that $x^2 - x + 1 > 0$. Therefore, $x^2 - x + 1 > 0$ for any $x \in \mathbb{R}$. Hence,

$$x^2 - x + 1 > 1 \iff x^2 - x > 0 \iff x(x-1) > 0.$$

Let us study the sign of $x(x-1)$:

	$-\infty$		0		1		$+\infty$
x		$-$		$+$		$+$	
$x-1$		$-$		$-$		$+$	
$x(x-1)$		$+$		$-$		$+$	

So $x \in \mathbb{R} \setminus [0,1]$.

(IX). Let us study first the case $x - \frac{1}{2} \le 0$.

• On the one hand,

$$1 < \frac{1}{2} - x \iff x < -\frac{1}{2}.$$

• On the other hand,

$$\frac{1}{2} - x < 2 \iff -\frac{3}{2} < x.$$

Thus, $x \in \left(-\frac{3}{2}, -\frac{1}{2}\right)$.

Now let us analyze the case when $x - \frac{1}{2} > 0$.

• On the one hand,

$$1 < x - \frac{1}{2} \iff \frac{3}{2} < x.$$

• On the other hand,

$$x - \frac{1}{2} < 2 \iff x < \frac{5}{2}.$$

Hence, $x \in \left(\frac{3}{2}, \frac{5}{2}\right)$.

Finally, we obtain that $x \in \left(-\frac{3}{2}, -\frac{1}{2}\right)$ or $x \in \left(\frac{3}{2}, \frac{5}{2}\right)$. Therefore, $x \in \left(-\frac{3}{2}, -\frac{1}{2}\right) \cup \left(\frac{3}{2}, \frac{5}{2}\right)$.

(X). If $x > 0$ we would have that $x - x = 0 > 2$, which means that there would not be any solutions. Let us study the case when $x \le 0$:

$$x + x > 2 \iff x > 1.$$

Then the inequality $x - |x| > 2$ has no solutions.

(XI). First, observe that the roots of $x^2 - x = x(x-1)$ are $x = 0$ and $x = 1$; and we proceed to analyze the sign of $x^2 - x$:

	$-\infty$		0		1		$+\infty$
x		$-$		$+$		$+$	
$x-1$		$-$		$-$		$+$	
x^2-x		$+$		$-$		$+$	

Let us study first the case when $x^2 - x \le 0$, that is, when $0 \le x \le 1$. Hence, we obtain that

$$-x^2 + x + x > 1 \iff 0 > x^2 - 2x + 1 = (x-1)^2,$$

which is absurd.

Now let us study the case when $x^2 - x > 0$, that is, when $x < 0$ or $x > 1$. Then we have that

$$x^2 - x + x > 1 \iff x^2 - 1 > 0 \iff (x-1)(x+1) > 0.$$

So it is enough to study when $(x-1)(x+1) > 0$:

	$-\infty$		-1		1		$+\infty$
$x+1$		$-$		$+$		$+$	
$x-1$		$-$		$-$		$+$	
$(x-1)(x+1)$		$+$		$-$		$+$	

Therefore, $x \in (-\infty, -1) \cup (1, \infty)$.

(XII). Let us study first the case when $x - 1 \le 0$, that is, when $x \le 1$. In this case, the inequality satisfies the following:

$$|x - x + 1| < 2 \iff 1 < 2,$$

and the latter inequality is always true. So $x \in (-\infty, 1]$.

Now let us analyze the case when $x - 1 > 0$, that is, when $x > 1$. In this latter case, the inequality satisfies the following:

$$|x + x - 1| < 2 \iff |2x - 1| < 2.$$

Since $x > 1$, we have, in particular, that $2x - 1 > 0$. Hence,

$$2x - 1 < 2 \iff x < \frac{3}{2}.$$

Thus, $x \in \left(1, \frac{3}{2}\right)$.

Finally, we see that $x \in (-\infty, 1]$ or $x \in \left(1, \frac{3}{2}\right)$. So $x \in \left(-\infty, \frac{3}{2}\right)$.

(XIII). Let us first study the case when $x - 1 \le 0$, that is, when $x \le 1$. Then the inequality satisfies the following:

$$\frac{1}{1 - x + 1} < |x - 2| \iff \frac{1}{2 - x} < |x - 2|.$$

Since $x \leq 1$ we have that $x - 2 \leq -1 < 0$. As a consequence now the inequality satisfies:

$$\frac{1}{2 - x} < 2 - x \iff 1 < (2 - x)^2 \iff 1 < 4 + x^2 - 4x$$

$$\iff 0 < x^2 - 4x + 3 \iff 0 < (x - 1)(x - 3),$$

where in the last inequality we have used the second degree equation formula to obtain the roots of $x^2 - 4x + 3$. Observe that it is enough to study when $(x - 1)(x - 3) > 0$:

	$-\infty$		1		3		$+\infty$
$x - 1$		$-$		$+$		$+$	
$x - 3$		$-$		$-$		$+$	
$(x - 1)(x - 3)$		$+$		$-$		$+$	

Hence, $(x - 1)(x - 3) > 0$ when $x \in (-\infty, 1) \cup (3, \infty)$. But we are also assuming that $x \leq 1$. So $x \in (-\infty, 1)$.

Now let us analyze the case when $x - 1 > 0$, that is, when $x > 1$. Then the inequality satisfies:

$$\frac{1}{1 + x - 1} < |x - 2| \iff \frac{1}{x} < |x - 2|.$$

We have two options $x - 2 \leq 0$ or $x - 2 > 0$, that is, $1 < x \leq 2$ or $x > 2$:

- Assume that $1 < x \leq 2$; then

$$\frac{1}{x} < -x + 2 \iff 1 < -x^2 + 2x \iff x^2 - 2x + 1 < 0 \iff (x - 1)^2 < 0,$$

  which is absurd.

- Assume that $x > 2$; then

$$\frac{1}{x} < x - 2 \iff 1 < x^2 - 2x \iff 0 < x^2 - 2x - 1$$

$$\iff 0 < \left[x - \left(1 + \sqrt{2}\right)\right]\left[x - \left(1 - \sqrt{2}\right)\right].$$

  Thus, $x \in \left(1 + \sqrt{2}, +\infty\right)$.

Finally, we obtain that $x \in (-\infty, 1)$ or $x \in \left(1 + \sqrt{2}, +\infty\right)$, or equivalently $x \in (-\infty, 1) \cup \left(1 + \sqrt{2}, +\infty\right)$.

(XIV). Note that $x \neq 1$. Let us study first the case $x^3 - 1 \leq 0$. To do so, first observe that by Ruffini's Rule (or the Cyclotomic Equation of Exercise 1.6) we can factorize $x^3 - 1$ as $x^3 - 1 = (x - 1)(x^2 + x + 1)$, where $x^2 + x + 1$ has only complex roots (it can be checked using the second degree equation formula) and $x^2 + x + 1 > 0$ for every $x \in \mathbb{R}$. Hence, we are assuming that $x < 1$, and moreover we have to study the inequalities:

$$-1 \le \frac{-(x-1)(x^2+x+1)}{x-1} \le 2 \iff -1 \le -x^2 - x - 1 \le 2$$

$$\iff -3 \le x(x+1) \le 0.$$

- On the one hand, let us see when $x(x+1) \le 0$:

	$-\infty$	-1	0	$+\infty$
$x+1$	$-$		$+$	$+$
x	$-$		$-$	$+$
$x(x+1)$	$+$		$-$	$+$

Hence, we have that $x \in [-1, 0]$.

- On the other hand, let us analyze when $x^2 + x + 3 \ge 0$. Using the second degree equation formula it can be checked that the roots of $x^2 + x + 3$ are complex, but also $x^2 + x + 3 > 0$ for any $x \in \mathbb{R}$.

Since both inequalities must be satisfied at the same time, we have that $x \in [-1, 0]$.

Now let us study the case when $x^3 - 1 > 0$, that is, when $x > 1$. In this latter case, the inequalities satisfy the following:

$$-1 \le \frac{(x-1)(x^2+x+1)}{x-1} \le 2 \iff -1 \le x^2 + x + 1 \le 2.$$

- On the one hand (and as we have seen), $x^2 + x + 2 > 0$ for every $x \in \mathbb{R}$.

- On the other hand, by the second degree equation formula we have that the roots of $x^2 + x - 1$ are $x = \frac{-1 \pm \sqrt{5}}{2}$. We proceed to analyze the sign of $x^2 + x - 1$:

	$-\infty$	$\frac{-1-\sqrt{5}}{2}$	$\frac{-1+\sqrt{5}}{2}$	$+\infty$
$x - \frac{-1-\sqrt{5}}{2}$	$-$		$+$	$+$
$x - \frac{-1+\sqrt{5}}{2}$	$-$		$-$	$+$
$x^2 + x - 1$	$+$		$-$	$+$

Now, we are assuming that $x > 1$ and $1 > \frac{-1+\sqrt{5}}{2}$; hence, there is no $x > 1$ such that $\frac{x^3-1}{x-1} \le 2$.

Since both inequalities must be satisfied at the same time, we have that there is no real number that satisfies both of them.

Finally, we obtain that $x \in [-1, 0]$.

$\square$

> **Exercise 1.38.** *Study for which real numbers it is fulfilled:*
>
> (I) $\frac{|x|+1}{x} < 1$ and $\frac{-2|x|+1}{x} < 1$.   (II) $\left|2x - |2x - 1|\right| = -5x$.

Solution to Exercise 1.38.

(I). First, let us analyze when $\frac{|x|+1}{x} < 1$. Observe that necessarily $x \neq 0$ and we also have two possibilities: $x > 0$ or $x < 0$.

- Assume that $x > 0$, then

$$\frac{x+1}{x} < 1 \iff x + 1 < x \iff 1 < 0,$$

which is absurd.

- Assume that $x < 0$, then

$$\frac{-x+1}{x} < 1 \iff -x + 1 > x \iff 1 > 2x \iff \frac{1}{2} > x.$$

Since $x < 0$ and $x < \frac{1}{2}$, we obtain that $x \in (-\infty, 0)$.

So the inequality $\frac{|x|+1}{x} < 1$ is satisfied when $x \in (-\infty, 0)$.

Finally, we will study when $\frac{-2|x|+1}{x} < 1$. Note that once again x cannot be 0 and we have two options: $x > 0$ or $x < 0$.

- Assume that $x > 0$; then

$$\frac{-2x+1}{x} < 1 \iff -2x + 1 < x \iff 1 < 3x \iff \frac{1}{3} < x.$$

Hence, $x \in \left(\frac{1}{3}, +\infty\right)$.

- Assume that $x < 0$; then

$$\frac{2x+1}{x} < 1 \iff 2x + 1 > x \iff x > -1.$$

Thus, $x \in (-1, 0)$.

Since both inequalities must be satisfied at the same time: $\frac{|x|+1}{x} < 1$ and $\frac{-2|x|+1}{x} < 1$, we have that

$$x \in (-\infty, 0) \cap [(-1, 0) \cup (1/3, +\infty)] = (-1, 0).$$

(II). Notice that we have two possibilities: $2x - 1 \leq 0$ or $2x - 1 > 0$.

- Assume that $2x - 1 \leq 0$, that is, $x \leq \frac{1}{2}$. Then,

$$|2x - (-2x + 1)| = -5x \iff |4x - 1| = -5x.$$

Now we have two options: $4x - 1 \leq 0$ or $4x - 1 > 0$, that is, when $x \leq \frac{1}{4}$ or $\frac{1}{4} < x \leq \frac{1}{2}$.

- Assume that $x \leq \frac{1}{4}$; then

$$-4x + 1 = -5x \iff x = -1.$$

Knowing that $-1 < \frac{1}{4}$, we have that $x \in \{-1\}$.
- Assume that $\frac{1}{4} < x \leq \frac{1}{2}$; then

$$4x - 1 = -5x \iff 9x = 1 \iff x = \frac{1}{9}.$$

Hence, $x = \frac{1}{9}$ but $x = \frac{1}{9} < \frac{1}{4} < x$, which is a contradiction.

- Assume finally that $2x - 1 > 0$, that is, when $x > \frac{1}{2}$. Then,

$$|2x - 2x + 1| = -5x \iff 1 = -5x \iff x = -\frac{1}{5}.$$

However, we have reached a contradiction since otherwise we would have that $\frac{1}{2} < x = -\frac{1}{5}$.

Therefore, the inequality $\big|2x - |2x - 1|\big| = -5x$ is satisfied only when $x = -1$.

$\square$

Exercise 1.39. *If $0 < x < y$ are two real numbers, prove that*

$$\frac{2xy}{x + y} < \sqrt{xy} < \frac{x + y}{2} < \sqrt{x^2 + y^2}.$$

Solution to Exercise 1.39.

Let us prove that $\frac{2xy}{x+y} < \sqrt{xy}$. First, we have that $\frac{2xy}{x+y} = \frac{2}{\frac{1}{x} + \frac{1}{y}}$. Now observe that we have the following chain of equivalences:

$$\frac{2}{\frac{1}{x} + \frac{1}{y}} < \sqrt{xy} \overset{x,y>0}{\iff} \frac{4}{\left(\frac{1}{x} + \frac{1}{y}\right)^2} < xy \iff 4 < xy\left(\frac{1}{x^2} + \frac{1}{y^2} + 2\frac{1}{xy}\right)$$

$$\iff 4 < \frac{y}{x} + \frac{x}{y} + 2 \iff \frac{y}{x} + \frac{x}{y} \geq 2$$

$$\iff \frac{y^2 + x^2}{xy} > 2 \iff y^2 + x^2 - 2xy > 0 \iff (y - x)^2 > 0.$$

Since we always have that $(y-x)^2 > 0$ for $x \neq y$, we obtain that the inequality $\frac{2}{\frac{1}{x}+\frac{1}{y}} < \sqrt{xy}$ is also true for $0 < x < y$.

Let us now show that $\sqrt{xy} < \frac{x+y}{2}$. Observe that we have the following chain of equivalences:

$$\frac{x+y}{2} > \sqrt{xy} \overset{x,y \geq 0}{\Longleftrightarrow} \frac{(x+y)^2}{4} > xy \Longleftrightarrow \frac{x^2+y^2+2xy}{4} > xy$$

$$\Longleftrightarrow x^2+y^2+2xy > 4xy \Longleftrightarrow x^2+y^2-2xy > 0$$

$$\Longleftrightarrow (x-y)^2 > 0.$$

Thus, since the latter inequality is always true for any distinct x, y, we have that $\frac{x+y}{2} > \sqrt{xy}$.

Finally, let us prove that $\frac{x+y}{2} < \sqrt{x^2+y^2}$. We have the following chain of equivalences:

$$\frac{x+y}{2} < \sqrt{x^2+y^2} \overset{x,y \geq 0}{\Longleftrightarrow} \left(\frac{x+y}{2}\right)^2 < x^2+y^2$$

$$\Longleftrightarrow x^2+2xy+y^2 < 4(x^2+y^2)$$

$$\Longleftrightarrow 3x^2+3y^2-2xy > 0$$

$$\Longleftrightarrow (x-y)^2 + 2(x^2+y^2) > 0.$$

So, as the latter inequality is always satisfied, we have that $\frac{x+y}{2} < \sqrt{x^2+y^2}$. □

Exercise 1.40 (Arithmetic-Geometric Mean Inequality). *Prove by induction that: if a_1, a_2, ..., a_n are positive real numbers such that $a_1 a_2 \cdots a_n = 1$, then*

$$a_1 + a_2 + \cdots + a_n \geq n.$$

Deduce that if x_1, x_2, ..., x_n are any non-negative real numbers, then

$$\frac{x_1 + x_2 + \cdots + x_n}{n} \geq \sqrt[n]{x_1 x_2 \cdots x_n},$$

that is, the arithmetic mean is always greater or equal than the geometric mean.

Solution to Exercise 1.40.

Let us show by induction on $n \in \mathbb{N}$ that: if $a_1, \ldots, a_n > 0$ and $a_1 \cdot a_2 \cdots a_n = 1$, then $a_1 + a_2 + \cdots + a_n \geq n$.

- Base case ($n = 1$) is trivial: by hypothesis we have $a_1 = 1$.

- Induction step. Assume that if $x_1, \ldots, x_n$ are positive real numbers such that $x_1 \cdot x_2 \cdots x_n = 1$, then we have that $x_1 + x_2 + \cdots + x_n \geq n$ (Induction Hypothesis). Let $a_1, \ldots, a_n, a_{n+1} > 0$ be such that $a_1 \cdot a_2 \cdots a_n \cdot a_{n+1} = 1$, and let us prove that $a_1 + a_2 + \cdots + a_n + a_{n+1} \geq n+1$.

Assume, without loss of generality, that $a_{n+1} = \max\{a_i : i \in \{1, \ldots, n+1\}\}$ and $a_n = \min\{a_i : i \in \{1, \ldots, n+1\}\}$. Note that $a_{n+1} \geq 1 \geq a_n$ since $a_1 \cdot a_2 \cdots a_n \cdot a_{n+1} = 1$. As $1 = a_1 \cdot a_2 \cdots a_n \cdot a_{n+1} = a_1 \cdot a_2 \cdots (a_n \cdot a_{n+1})$, we have by the Induction Hypothesis that $a_1 + a_2 + \cdots + (a_n \cdot a_{n+1}) \geq n$. On the other hand, since $0 \leq (a_{n+1} - 1)(1 - a_n) = a_n + a_{n+1} - a_n \cdot a_{n+1} - 1$, it yields $a_n \cdot a_{n+1} + 1 \leq a_n + a_{n+1}$. Thus,

$$a_1 + a_2 + \cdots + a_n + a_{n+1} \geq a_1 + a_2 + \cdots + a_n \cdot a_{n+1} + 1 \geq n+1.$$

For the second part of the exercise, let $x_1, \ldots, x_n > 0$. Then,

$$1 = \frac{x_1 \cdots x_n}{x_1 \cdots x_n} = \frac{x_1}{\sqrt[n]{x_1 \cdots x_n}} \cdots \frac{x_n}{\sqrt[n]{x_1 \cdots x_n}}.$$

Hence, applying the first part we obtain

$$n \leq \frac{x_1}{\sqrt[n]{x_1 \cdots x_n}} + \cdots + \frac{x_n}{\sqrt[n]{x_1 \cdots x_n}} = \frac{1}{\sqrt[n]{x_1 \cdots x_n}} (x_1 + \cdots + x_n),$$

that is, $\frac{x_1 + \cdots + x_n}{n} \geq \sqrt[n]{x_1 \cdots x_n}$, as required.

$\square$

Exercise 1.41. *The harmonic mean of positive numbers x_1, x_2, $\ldots$, x_n is the inverse of the arithmetic mean of their inverses, that is,*

$$\frac{n}{1/x_1 + 1/x_2 + \cdots + 1/x_n}.$$

Show that the harmonic mean is less or equal than the geometric mean.

Solution to Exercise 1.41.

Let $x_1, \ldots, x_n > 0$. Notice that

$$1 = \frac{(1/x_1) \cdots (1/x_n)}{\frac{1}{x_1 \cdots x_n}} = \frac{(1/x_1) \cdots (1/x_n)}{\frac{1}{\sqrt[n]{x_1 \cdots x_n}} \cdots \frac{1}{\sqrt[n]{x_1 \cdots x_n}}}.$$

So by the first part of Exercise 1.40 we have

$$n \leq \frac{1/x_1}{\frac{1}{\sqrt[n]{x_1 \cdots x_n}}} + \cdots + \frac{1/x_n}{\frac{1}{\sqrt[n]{x_1 \cdots x_n}}} = \sqrt[n]{x_1 \cdots x_n} \left(\frac{1}{x_1} + \cdots + \frac{1}{x_n} \right).$$

Thus, $\frac{n}{1/x_1 + \cdots + 1/x_n} \leq \sqrt[n]{x_1 \cdots x_n}$.

$\square$

Exercise 1.42. *Let $x \in \mathbb{R}$. Prove that if $|x| \leq \varepsilon$ for every $\varepsilon > 0$, then $x = 0$. Which real numbers x satisfy that $x \leq \varepsilon$ for any $\varepsilon > 0$?*

Solution to Exercise 1.42.

We will prove the first part by contradiction. Assume that $x \in \mathbb{R}$ fulfills $|x| \leq \varepsilon$ for any $\varepsilon > 0$ and $x \neq 0$. Since $x \neq 0$, we have that $|x| > 0$. Now, as $(0, |x|) \neq \varnothing$, there exists $\varepsilon_0 \in (0, |x|)$. Then, we have that $\varepsilon_0 > 0$ and $|x| > \varepsilon_0$ which is absurd.

For the second part notice that for any $x \leq 0$ we have $x \leq 0 < \varepsilon$ for every $\varepsilon > 0$. Assume that there exists $x_0 > 0$ such that $x_0 \leq \varepsilon$ for every $\varepsilon > 0$. As $x_0 > 0$ we have that $|x_0| = x_0 \leq \varepsilon$ for every $\varepsilon > 0$, but the latter is only satisfied if $x_0 = 0$ which is not the case. Hence, $x \leq \varepsilon$ for every $\varepsilon > 0$ if, and only if, $x \leq 0$.

$\square$

Exercise 1.43. *Let p be a nonzero rational number and x an irrational number. Prove that $p + x$ and px are irrational numbers.*

Solution to Exercise 1.43.

Let us show first that $p + x \in \mathbb{R} \setminus \mathbb{Q}$. Assume, by contradiction, that $p + x \in \mathbb{Q}$, then there exist $r, s \in \mathbb{Z}$ with $s \neq 0$ such that $p + x = \frac{r}{s}$. Thus, $x = \frac{r}{s} - p \in \mathbb{Q}$ which is impossible.

Let us now prove that $px \in \mathbb{R} \setminus \mathbb{Q}$. Assume, once again by contradiction, that $px \in \mathbb{Q}$, then there exist $m, n \in \mathbb{Z}$ with $n \neq 0$ such that $px = \frac{m}{n}$. Since $p \neq 0$, we have $x = \frac{m}{n} \cdot \frac{1}{p} \in \mathbb{Q}$ which is absurd.

$\square$

Exercise 1.44. *Let $x, y \in \mathbb{Q}^+$ be such that $\sqrt{x} + \sqrt{y} \in \mathbb{Q}$. Prove that $\sqrt{x}, \sqrt{y}$ are rational numbers.*

Solution to Exercise 1.44.

Let us prove it by contradiction. Assume that $x, y \in \mathbb{Q}^+$ are such that $\sqrt{x} + \sqrt{y} \in \mathbb{Q}$, but $\sqrt{x}$ or $\sqrt{y}$ are irrational numbers. Without loss of generality, assume that $\sqrt{x}$ is an irrational number. Since $\sqrt{x} + \sqrt{y} \in \mathbb{Q}$, then $(\sqrt{x} + \sqrt{y})^2 = x + y + 2\sqrt{xy} \in \mathbb{Q}$. As x, y and 2 are rational numbers, we obtain that $\sqrt{xy} \in \mathbb{Q}$. Observe that $\sqrt{x} + \sqrt{y} \neq 0$, since otherwise both $\sqrt{x}$ and $\sqrt{y}$ would be 0. Now, by Exercise 1.43, as $\sqrt{x} + \sqrt{y} \in \mathbb{Q}$ and $\sqrt{x}$ is irrational, we have that $\sqrt{x}(\sqrt{x} + \sqrt{y}) = x + \sqrt{xy}$ is irrational. But, since x and $\sqrt{xy}$ are rational numbers, we obtain that $x + \sqrt{xy} \in \mathbb{Q}$, a contradiction. $\quad\square$

Exercise 1.45. *Show that $\sqrt{p}$ is irrational if p is prime.*

Solution to Exercise 1.45.

Assume, by contradiction, that $\sqrt{p} \in \mathbb{Q}$. Since $\sqrt{p} > 0$, there exist $a, b \in \mathbb{N}$ such that $\sqrt{p} = \frac{a}{b}$. Thus, $p = \frac{a^2}{b^2}$ if, and only if, $a^2 = pb^2$. Now, given any $n \in \mathbb{N}$ we have that n^2 has an even number of prime divisors. Hence, a^2 has an even number of prime divisors while pb^2 has an odd number of prime divisors as p is prime. But the latter statement contradicts the fact that $a^2 = pb^2$. $\quad\square$

Exercise 1.46. *Prove that $\log_m n$ is irrational if $m, n \in \mathbb{N} \setminus \{1\}$ are such that m and n do not have the same prime divisors.*

Solution to Exercise 1.46.

Assume that $m, n \in \mathbb{N} \setminus \{1\}$ satisfy that m and n do not have the same prime divisors, that is, there exists a prime number $p \in \mathbb{N}$ such that p divides m but not n. Since $m, n > 1$ we have that $\log_m n > 0$. Assume, by contradiction, that $\log_m n \in \mathbb{Q}$, then there exist $a, b \in \mathbb{N}$ such that $\log_m n = \frac{a}{b}$. Thus, $n = m^{a/b}$, which implies that $n^b = m^a$. Now, since p divides m it yields that p divides m^a, and as p does not divide n we obtain that p does not divide n^b. But the latter contradicts the fact that $n^b = m^a$. $\quad\square$

Exercise 1.47 (Gauss Theorem). (I) *Let $r = \frac{p}{q}$ with $mcd(p, q) = 1$ be a rational solution of the polynomial equation with integer coefficients*

$$a_n x^n + a_{n-1} x^{n-1} + \cdots + a_1 x + a_0 = 0.$$

Show that p divides a_0 and q divides a_n. Deduce that in the particular case when $a_n = 1$, we have that r is an integer divisor of a_0.

(II) *Use part (I) to prove that $\sqrt{3} - \sqrt{2}$ and $1 - \sqrt[3]{2} + \sqrt{5}$ are irrational algebraic numbers.*

Solution to Exercise 1.47.

(I). Since $\frac{p}{q}$ is a solution of $a_n x^n + a_{n-1} x^{n-1} + \cdots + a_1 x + a_0 = 0$, we obtain that

$$a_n \left(\frac{p}{q}\right)^n + a_{n-1} \left(\frac{p}{q}\right)^{n-1} + \cdots + a_1 \frac{p}{q} + a_0 = 0.$$

Multiplying both sides of the equality by q^n we have that

$$a_n p^n + a_{n-1} q p^{n-1} + \cdots + a_1 p q^{n-1} + a_0 q^n = 0.$$

Hence, on the one hand,

$$a_0 q^n = p \left(-a_n p^{n-1} - a_{n-1} q p^{n-2} - \cdots - a_1 q^{n-1}\right),$$

where $-a_n p^{n-1} - a_{n-1} q p^{n-2} - \cdots - a_1 q^{n-1} \in \mathbb{Z}$ and $\mathrm{mcd}(p, q^n) = 1$; hence, p divides a_0. On the other hand,

$$a_n p^n = q \left(-a_{n-1} p^{n-1} - \cdots - a_1 p q^{n-2} - a_0 q^{n-1}\right),$$

where $-a_{n-1} p^{n-1} - \cdots - a_1 p q^{n-2} - a_0 q^{n-1} \in \mathbb{Z}$ and $\mathrm{mcd}(q, p^n) = 1$; hence, q divides a_n.

If $a_n = 1$, then q divides 1. So $q = 1$, that is, $r = p \in \mathbb{Z}$.

(II). First, let us show that $\sqrt{2} - \sqrt{3}$ is an irrational algebraic number. To do so, let us see on the one hand that it is the solution of a polynomial equation with integer coefficients. Consider the polynomial that has roots $\sqrt{2} - \sqrt{3}$, $\sqrt{2} + \sqrt{3}$, $-\sqrt{2} - \sqrt{3}$ and $-\sqrt{2} + \sqrt{3}$:

$$\left[x - \left(\sqrt{2} - \sqrt{3}\right)\right] \left[x - \left(\sqrt{2} + \sqrt{3}\right)\right] \left[x - \left(-\sqrt{2} - \sqrt{3}\right)\right] \left[x - \left(-\sqrt{2} + \sqrt{3}\right)\right]$$

$$= \left[x^2 - \left(\sqrt{2} + \sqrt{3}\right) x - \left(\sqrt{2} - \sqrt{3}\right) x - 1\right]$$

$$\cdot \left[x^2 - \left(-\sqrt{2} + \sqrt{3}\right) x - \left(-\sqrt{2} - \sqrt{3}\right) x - 1\right]$$

$$= \left(x^2 - 2\sqrt{2}x - 1\right) \left(x^2 + 2\sqrt{2}x - 1\right)$$

$$= x^4 + 2\sqrt{2}x^3 - x^2 - 2\sqrt{2}x^3 - 8x^2 + 2\sqrt{2}x - x^2 - 2\sqrt{2}x + 1$$

$$= x^4 - 10x^2 + 1.$$

Then $\sqrt{2} - \sqrt{3}$ is a solution of $x^4 - 10x^2 + 1$ which is a polynomial with integer coefficients. Moreover, assume that $\sqrt{2} - \sqrt{3}$ is rational. Then, by Gauss Theorem, since $a_4 = 1$ we have that $\sqrt{2} - \sqrt{3} \in \mathbb{Z}$, but $\sqrt{2} - \sqrt{3} \approx -0,317\ldots$

Now let us prove that $1 - \sqrt[3]{2} + \sqrt{5}$ is an irrational algebraic number. To do so, observe that $1 - \sqrt[3]{2} + \sqrt{5}$ is a solution of

$$x - \left(1 - \sqrt[3]{2} + \sqrt{5}\right) = 0 \iff x - 1 = -\sqrt[3]{2} + \sqrt{5}$$

$$\iff (x-1)^3 = -2 - \sqrt{5}$$

$$\iff x^3 - 3x^2 + 3x + 1 = -\sqrt{5}$$

$$\iff (x^3 - 3x^2 + 3x + 1)^2 = 5$$

$$\iff x^6 - 6x^5 + 15x^4 - 16x^3 + 3x^2 + 6x - 4 = 0,$$

where $x^6 - 6x^5 + 15x^4 - 16x^3 + 3x^2 + 6x - 4$ is a polynomial with integer coefficients. Finally, assume that $1 - \sqrt[3]{2} + \sqrt{5}$ is rational. Then, by Gauss Theorem, as $a_6 = 1$ we have that $1 - \sqrt[3]{2} + \sqrt{5} = p \in \mathbb{Z}$, or equivalently $\sqrt{5} = (1 - p)^3 - 2 \in \mathbb{Z}$, which is a contradiction by Exercise 1.45. $\qquad \square$

Exercise 1.48. *Calculate the supremum and infimum, if they exist, of the following sets, indicating if they are maximum or minimum, respectively:*

(I) $\left\{ \dfrac{1}{n} : n \in \mathbb{N} \right\} \cup \{0\}$,

(II) $\left\{ \dfrac{2n+1}{n} : n \in \mathbb{N} \right\}$,

(III) $\left\{ n \pm \dfrac{1}{n} : n \in \mathbb{N} \right\}$,

(IV) $\left\{ x \in \mathbb{Q} : |x| < \sqrt{2} \right\} \cup \left\{ x \in \mathbb{Q} : \dfrac{1}{x-5} > 7 \right\}$,

(V) $\left\{ \dfrac{1}{n} : n \in \mathbb{N} \right\}$,

(VI) $\left\{ -\dfrac{1}{n} + [1 + (-1)^n]n : n \in \mathbb{N} \right\}$,

(VII) $\left\{ \dfrac{1}{n} + (-1)^n : n \in \mathbb{N} \right\}$,

(VIII) $\{ 2, 2.2, 2.22, 2.222, \ldots \}$,

(IX) $\displaystyle\bigcup_{n=1}^{\infty} \{ x \in \mathbb{R} : n^2 x^2 - n(3n-1)x + (2n^2 - 3n - 2) = 0 \}$,

(X) $\left\{ (-1)^n \dfrac{n^2+1}{n+1} : n \in \mathbb{N} \right\}$,

(XI) $\{x \in \mathbb{R} : x^2 < 9\}$,

(XII) $\{x \in \mathbb{R} : x^2 \le 7\}$,

(XIII) $\{x \in \mathbb{R} : x^2 + x - 1 < 0\}$,

(XIV) $\{x \in \mathbb{R} : x < 0, \ x^2 + x - 1 < 0\}$,

(XV) $\{x \in \mathbb{R} : x^2 + x + 1 \ge 0\}$,

(XVI) $\{x \in \mathbb{R} : |2x + 1| < 5\}$,

(XVII) $\left\{x \in \mathbb{R} : (x^2 + 1)^{-1} > \dfrac{1}{2}\right\}$,

(XVIII) $\{x \in \mathbb{Q} : x^2 \le 7\}$,

(XIX) $\displaystyle\bigcup_{n=1}^{\infty} \left(-\tfrac{1}{n}, \tfrac{1}{n}\right)$,

(XX) $\displaystyle\bigcap_{n=1}^{\infty} \left(-\tfrac{1}{n}, \tfrac{1}{n}\right)$,

(XXI) $\displaystyle\bigcap_{n=1}^{\infty} \left[\tfrac{1}{2n}, \tfrac{1}{2n-1}\right]$.

Solution to Exercise 1.48.

(I). For every $n \in \mathbb{N}$, we have that $\frac{1}{n+1} < \frac{1}{n}$. So the elements of the set $\left\{\frac{1}{n} : n \in \mathbb{N}\right\}$ decrease strictly as n increases. Hence, the supremum is 1 and it is a maximum as it is attained when $n = 1$. The infimum is the limit of the terms $\frac{1}{n}$ which is 0 and, moreover, it is a minimum, since it belongs to the set $\left\{\frac{1}{n} : n \in \mathbb{N}\right\} \cup \{0\}$.

(II). For every $n \in \mathbb{N}$, it is clear that $\frac{2n+1}{n} = 2 + \frac{1}{n}$. Thus, the elements of the set $\left\{\frac{2n+1}{n} : n \in \mathbb{N}\right\}$ decrease strictly as n increases. Therefore, the supremum is 3 and it is a maximum since it is attained when $n = 1$. The infimum is the limit of the terms $\frac{2n+1}{n}$ which is 2, but it is not a minimum as it does not belong to the set $\left\{\frac{2n+1}{n} : n \in \mathbb{N}\right\}$.

(III). For every $n \in \mathbb{N}$, we have that $n - 1 \le n - \frac{1}{n} < n + \frac{1}{n} \le n + 1$. Hence, the set $\left\{n \pm \frac{1}{n} : n \in \mathbb{N}\right\}$ does not have supremum nor infimum.

(IV). Observe that in the inequality $\frac{1}{x-5} > 7$ we must assume that $x > 5$ since otherwise for $x = 5$ we would be dividing by 0 and if $x < 5$ we would have that $\frac{1}{x-5} < 0$. Hence,

$$\left\{ x \in \mathbb{Q} : |x| < \sqrt{2} \right\} \cup \left\{ x \in \mathbb{Q} : \frac{1}{x-5} > 7 \right\}$$

$$= \left\{ x \in \mathbb{Q} : -\sqrt{2} < x < \sqrt{2} \right\} \cup \left\{ x \in \mathbb{Q} : 0 < x - 5 < \frac{1}{7} \right\}$$

$$= \left\{ x \in \mathbb{Q} : -\sqrt{2} < x < \sqrt{2} \right\} \cup \left\{ x \in \mathbb{Q} : 5 < x < \frac{36}{7} \right\}$$

$$= \left[(-\sqrt{2}, \sqrt{2}) \cup (5, 36/7) \right] \cap \mathbb{Q}$$

Thus, the infimum is $-\sqrt{2}$ and it is not a minimum; and the supremum is $\frac{36}{7}$ but it is not a maximum.

(V). Applying the idea of part (I), we have that 1 is the supremum and maximum; and 0 is the infimum but it is not minimum.

(VI). In the set $\left\{ -\frac{1}{n} + [1 + (-1)^n]n : n \in \mathbb{N} \right\}$ we have the following. If $n \in \mathbb{N}$ is even, then we obtain the terms $2n - \frac{1}{n}$, which are positive and the set that they form is not bounded above. If $n \in \mathbb{N}$ is odd, then we obtain the terms $-\frac{1}{n}$, which are negative and increase strictly as n increases along the odd natural numbers. Hence, the set $\left\{ -\frac{1}{n} + [1 + (-1)^n]n : n \in \mathbb{N} \right\}$ does not have supremum (and in particular, there is no maximum), but it has an infimum: -1, which is also a minimum since it is attained when $n = 1$.

(VII). In the set $\left\{ \frac{1}{n} + (-1)^n : n \in \mathbb{N} \right\}$ we have the following. If $n \in \mathbb{N}$ is even, then we obtain the terms $\frac{1}{n} + 1$, which are positive, decrease strictly as n increases along the even natural numbers, and the limit is 1. If $n \in \mathbb{N}$ is odd, then we have the terms $\frac{1}{n} - 1$, which are negative, decrease strictly as n increases along the odd natural numbers, and the limit is -1. Thus, the set $\left\{ \frac{1}{n} + (-1)^n : n \in \mathbb{N} \right\}$ has a supremum: $\frac{3}{2}$ which is a maximum since it is attained when $n = 2$, and it has an infimum: -1 but it is not a minimum.

(VIII). In the set $\{ 2, 2.2, 2.22, 2.222, \dots \}$ the infimum (and minimum) is 2 while the supremum is $2.\overline{2} = 20/9$ but it is not a maximum.

(IX). We begin by solving, for every $n \in \mathbb{N}$, the second degree equation $n^2x^2 - n(3n-1)x + (2n^2 - 3n - 2) = 0$:

$$
\begin{aligned}
x &= \frac{n(3n-1) \pm \sqrt{n^2(3n-1)^2 - 4n^2(2n^2 - 3n - 2)}}{2n^2} \\
&= \frac{n(3n-1) \pm n\sqrt{9n^2 - 6n + 1 - 8n^2 + 12n + 8}}{2n^2} \\
&= \frac{3n-1 \pm \sqrt{n^2 + 6n + 9}}{2n} = \frac{3n-1 \pm \sqrt{(n+3)^2}}{2n} = \frac{3n-1 \pm (n+3)}{2n} \\
&= \begin{cases} \frac{4n+2}{2n} = 2 + \frac{1}{n}, \\ \frac{2n-4}{2n} = 1 - \frac{2}{n}. \end{cases}
\end{aligned}
$$

Thus,

$$
\bigcup_{n=1}^{\infty} \{x \in \mathbb{R} : n^2x^2 - n(3n-1)x + (2n^2 - 3n - 2) = 0\}
$$

$$
= \left\{ 2 + \frac{1}{n} : n \in \mathbb{N} \right\} \cup \left\{ 1 - \frac{2}{n} : n \in \mathbb{N} \right\}.
$$

Then, we obtain the terms $2 + \frac{1}{n}$, which already appeared in part (II), and $1 - \frac{2}{n}$ which are terms that increase strictly as n increases, with limit 1 and when $n = 1$ we have -1. Hence, the set $\bigcup_{n=1}^{\infty} \{x \in \mathbb{R} : n^2x^2 - n(3n-1)x + (2n^2 - 3n - 2) = 0\}$ has a supremum: 3, and it is also a maximum; and the set has an infimum: -1, which is also a minimum since it is attained at $1 - \frac{2}{n}$ when $n = 1$.

(X). In the set $\left\{ (-1)^n \frac{n^2+1}{n+1} : n \in \mathbb{N} \right\}$ we have, when $n \in \mathbb{N}$ is even, the terms $\frac{n^2+1}{n+1}$ which show that the set is not bounded above. Also, when $n \in \mathbb{N}$ is odd, we have the terms $-\frac{n^2+1}{n+1}$ which imply that the set is not bounded below. Thus, $\left\{ (-1)^n \frac{n^2+1}{n+1} : n \in \mathbb{N} \right\}$ is not bounded above nor below.

(XI). $\{x \in \mathbb{R} : x^2 < 9\} = \{x \in \mathbb{R} : -3 < x < 3\} = (-3, 3)$. Thus, the infimum is -3 and the supremum is 3, but there is no minimum nor maximum.

(XII). $\{x \in \mathbb{R} : x^2 \le 7\} = \{x \in \mathbb{R} : -\sqrt{7} \le x \le \sqrt{7}\} = [-\sqrt{7}, \sqrt{7}]$. Hence, the infimum is $-\sqrt{7}$, and in particular, it is a minimum, and the supremum is $\sqrt{7}$ which is also a maximum.

(XIII). First, let us study when $x^2 + x - 1 < 0$. To do so, we solve the second degree equation $x^2 + x - 1 = 0$ which gives us the roots $x = \frac{-1 \pm \sqrt{5}}{2}$. Now let us analyze the sign of $x^2 + x - 1$:

	$-\infty$		$\frac{-1-\sqrt{5}}{2}$		$\frac{-1+\sqrt{5}}{2}$		$+\infty$
$x - \frac{-1-\sqrt{5}}{2}$		$-$		$+$		$+$	
$x - \frac{-1+\sqrt{5}}{2}$		$-$		$-$		$+$	
$x^2 + x - 1$		$+$		$-$		$+$	

That is,

$$\{x \in \mathbb{R} : x^2 + x - 1 < 0\} = \left\{ x \in \mathbb{R} : \frac{-1-\sqrt{5}}{2} < x < \frac{-1+\sqrt{5}}{2} \right\}$$

$$= \left(\frac{-1-\sqrt{5}}{2}, \frac{-1+\sqrt{5}}{2} \right).$$

Hence, the infimum is $\frac{-1-\sqrt{5}}{2}$, the supremum is $\frac{-1+\sqrt{5}}{2}$ and there is no minimum nor maximum.

(XIV). By part (XIII) we have

$$\{x \in \mathbb{R} \mid x < 0, \; x^2 + x - 1 < 0\} = \left\{ x \in \mathbb{R} \mid \frac{-1-\sqrt{5}}{2} < x < 0 \right\}$$

$$= \left(\frac{-1-\sqrt{5}}{2}, 0 \right).$$

Therefore, the infimum is $\frac{-1-\sqrt{5}}{2}$, the supremum is 0 and there is no minimum nor maximum.

(XV). First, observe that $x^2 + x + 1 > 0$ for every $x \in \mathbb{R}$. Indeed, if we solve the second degree equation $x^2 + x + 1 = 0$ we obtain complex roots which are not real numbers and we also see that $x^2 + x + 1$ that takes positive values. Hence, $\{x \in \mathbb{R} : x^2 + x + 1 \geq 0\} = \mathbb{R}$, that is, the set is not bounded above nor below.

(XVI). Let us see first the case $2x + 1 \geq 0$, that is, $x \geq -\frac{1}{2}$. Then we have

$$2x + 1 < 5 \iff x < 2,$$

So $-\frac{1}{2} \leq x < 2$.

Now let us analyze the case when $2x + 1 < 0$, that is, $x < -\frac{1}{2}$. In this case we get

$$-2x - 1 < 5 \iff x > -3.$$

So $-3 < x < -\frac{1}{2}$.

Hence, $\{x \in \mathbb{R} : |2x + 1| < 5\} = (-3, 2)$, that is, the infimum is -3, the supremum is 2 and there is no maximum nor minimum.

(XVII). Since $x^2 + 1 > 0$ for every $x \in \mathbb{R}$, we have that

$$(x^2 + 1)^{-1} > \frac{1}{2} \iff x^2 + 1 < 2 \iff x^2 < 1 \iff -1 < x < 1.$$

Thus, $\left\{ x \in \mathbb{R} : (x^2 + 1)^{-1} > \dfrac{1}{2} \right\} = \{ x \in \mathbb{R} : -1 < x < 1 \} = (-1, 1)$. Therefore, the infimum is -1, the supremum is 1 and there is no minimum nor
maximum.

(XVIII). $\{ x \in \mathbb{Q} : x^2 \leq 7 \} = \{ x \in \mathbb{Q} : -\sqrt{7} \leq x \leq \sqrt{7} \}$. Thus, the
infimum is $-\sqrt{7}$ and the supremum is $\sqrt{7}$. Moreover, there is no minimum
nor maximum since the set that we are considering contains only rational
numbers and neither $-\sqrt{7}$ nor $\sqrt{7}$ are rational numbers.

(XIX). $\displaystyle\bigcup_{n=1}^{\infty} \left(-\frac{1}{n}, \frac{1}{n} \right) = (-1, 1)$. Thus, the infimum is -1, the supremum
is 1 and there is no minimum nor maximum.

(XX). $\displaystyle\bigcap_{n=1}^{\infty} \left(-\frac{1}{n}, \frac{1}{n} \right) = \{0\}$. Hence, 0 is the infimum, the supremum, the
minimum and the maximum.

(XXI). $\displaystyle\bigcap_{n=1}^{\infty} \left[\frac{1}{2n}, \frac{1}{2n-1} \right] = \left[\frac{1}{2}, 1 \right] \cap \left[\frac{1}{4}, \frac{1}{3} \right] \cap \displaystyle\bigcap_{n=3}^{\infty} \left[\frac{1}{2n}, \frac{1}{2n-1} \right] = \varnothing$. In this
case, the infimum and the supremum do not exist, so there is no minimum
nor maximum.

$\square$

Exercise 1.49. *Let A be a nonempty set of real numbers, $s = \sup A$
and $\varepsilon > 0$. Can we guarantee that there is an $a \in A$ such that
$s - \varepsilon < a < s$? If so, prove it. If not, find a counterexample and modify
the previous inequalities so that it is true.*

Solution to Exercise 1.49.

Take $A = \{0\}$. Then $s = 0$ and for any $\varepsilon > 0$ we do not have that
$-\varepsilon < 0 < 0$.

If we change the inequalities $s - \varepsilon < a < s$ as $s - \varepsilon < a \leq s$, then
the existence of a is guaranteed. Indeed, let A be a nonempty subset of real
numbers, $s = \sup A$ and $\varepsilon > 0$. If $s \in A$, we are done. If $s \notin A$, let us prove
that exists $a \in A$ such that $s - \varepsilon < a \leq s$ by contradiction. Assume that for
every $a \in A$, we have that $a \leq s - \varepsilon$ or $a > s$. If there exists $a \in A$ such that
$a > s$, then s is not the supremum of A, and we have reached a contradiction.
So, assume that $a \leq s - \varepsilon$ for every $a \in A$; then $\sup A \leq s - \varepsilon < s = \sup A$,
which is also absurd.

$\square$

Exercise 1.50. *If $A \subset \mathbb{R}$, $-A$ denotes the set of all opposite elements of A. Show that $\inf A + \sup(-A) = 0$ when A is bounded.*

Solution to Exercise 1.50.

Let $a = \inf A$ and $b = \sup(-A)$. On the one hand, $a \leq x$ for every $x \in A$, so $-a \geq -x$ for every $x \in A$, that is, $-a \geq b$. On the other hand, $b \geq -x$ for every $x \in A$, so $-b \leq x$ for every $x \in A$, that is, $-b \leq a$ or equivalently $b \geq -a$. Thus, $b = -a$ which shows that $\inf A + \sup(-A) = 0$, as needed. □

Exercise 1.51. *Let A and B be two nonempty bounded sets of real numbers.*

(I) *Prove that if $A \subset B$, then*

$$\sup A \leq \sup B, \qquad \inf A \geq \inf B.$$

(II) *Prove that if $x \leq y$ for all $x \in A$, $y \in B$, then*

$$\sup A \leq y \quad \text{for every } y \in B,$$
$$x \leq \inf B \quad \text{for every } x \in A,$$

and so $\sup A \leq \inf B$. Note: in this part it is not necessary to say explicitly that A and B are bounded since the statement "$x \leq y$ for all $x \in A$, $y \in B$" implies that A is bounded above and B is bounded below (that is, there exists $\sup A$ and $\inf B$).

(III) *Prove that if $\sup A < \inf B$, then $a < b$ for all $a \in A$, $b \in B$. Justify if the converse is true.*

Solution to Exercise 1.51.

(I). Let us show first that if $A \subset B$, then $\sup(A) \geq \sup(B)$. Let $a = \sup A$ and $b \in B$, that is, for every $x \in A$ and $y \in B$, we have that $x \leq a$ and $y \leq b$. Since $A \subset B$, every $y \in A$ fulfills $b \geq y$, which implies that b is an upper bound of A. But a is the lowest upper bound, then we deduce that $\sup A \leq \sup B$.

Let us prove now that if $A \subset B$, then $\inf A \geq \inf B$. Analogously to the previous case, let $a = \inf A$ and $b = \inf B$, that is, for every $x \in A$ and $y \in B$, we obtain that $x \geq a$ and $y \geq b$. As $A \subset B$, for every $y \in A$, we have $y \geq b$, which shows that b is lower bound of A. Therefore, since a is the largest lower bound we deduce that $\inf A \geq \inf B$.

(II). Let us see first that if $x \leq y$, for every $x \in A$ and $y \in B$, then $\sup A \leq y$ and $x \leq \inf B$.

Let $a = \sup A$. If $a' \geq x$, for every $x \in A$, then $a \leq a'$, since a is the lowest upper bound of A. By hypothesis, any $y \in B$ is an upper bound of A, so $a = \sup A \leq y$ for every $y \in B$.

Analogously, let $b = \inf B$. If $b' \geq y$, for every $y \in B$, then $b' \leq b$ as b is the largest lower bound of B. By hypothesis, any $x \in A$ is a lower bound of B, so $b = \inf B \geq x$ for all $x \in A$.

Therefore, for every $x \in A$ and $y \in B$, we have that $\sup A \leq x \leq y \leq \inf B$.

(III). Let us prove that if $\sup A < \inf B$, then $x < y$ for every $x \in A$ and $y \in B$. If $\sup A < \inf B$, then $x \leq \sup A < \inf B \leq y$ for every $x \in A$ and $y \in B$.

The converse is not true in general. For instance, if we take $A = (0, 1)$ and $B = (1, 2)$ we have that $x < y$ for all $x \in A$ and $y \in B$, but $\sup A = 1 = \inf B$. $\qquad \square$

Exercise 1.52.   (I) *Let A and B be two nonempty bounded sets of real numbers. Define the set*

$$A + B = \{x + y : x \in A, \, y \in B\}.$$

Show that

$$\sup(A + B) = \sup A + \sup B, \qquad \inf(A + B) = \inf A + \inf B.$$

(II) *Let $A = \{x_1, x_2, \ldots, x_n\} \subset \mathbb{R}$, $B = \{y_1, y_2, \ldots, y_n\} \subset \mathbb{R}$, and consider the set*

$$C = \{x_1 + y_1, x_2 + y_2, \ldots, x_n + y_n\}.$$

Prove that

$$\sup C \leq \sup A + \sup B, \qquad \inf C \geq \inf A + \inf B.$$

Give an example that shows that the inequalities may be strict.

(III) *Let A and B be two nonempty bounded sets of real numbers. Define the set*

$$A - B = \{x - y : x \in A, \, y \in B\}.$$

Show that

$$\sup(A - B) = \sup A - \inf B, \qquad \inf(A - B) = \inf A - \sup B.$$

(IV) *Deduce from part* (III) *that*

$$\sup\{|x - y| : x, y \in A\} = \sup A - \inf A.$$

Solution to Exercise 1.52.

(I). First, let us show that $\sup(A + B) = \sup A + \sup B$. Let $a = \sup A$, $b = \sup B$ and $c = \sup(A+B)$, then $a \geq x$, $b \geq y$ and $c \geq x+y$ for every $x \in A$ and $y \in B$. Hence, $a + b \geq x + y$ for every $x + y \in A + B$, that is, $a + b \geq c$. On the other hand, $c \geq x + y$ for every $x \in A$ and every $y \in B$. Now fixing $y \in B$, we have that $c - y \geq x$ for all $x \in A$. Therefore, since a is the least upper bound of A we have that $c - y \geq a$ or, equivalently, $c - a \geq y$. Taking into consideration that y was an arbitrary element of B, we have shown that $c - a$ is an upper bound for B, and therefore $c - a \geq b$, from which $c \geq a + b$. Thus, $c = a + b$.

Let us now show that $\inf(A + B) = \inf A + \inf B$. Let $a = \inf A$, $b = \inf B$ and $c = \inf(A + B)$, then $a \leq x$, $b \leq y$ and $c \leq x + y$ for every $x \in A$ and $y \in B$. Since $a \leq x$ and $b \leq y$ for all $x \in A$ and $y \in B$, we have that $a + b \leq x + y$ for every $x + y \in A + B$, that is, $a + b \leq c$. On the other hand, $c \leq x + y$ for every $x \in A$ and every $y \in B$. Now fixing $y \in B$, we have that $c - y \leq x$ for all $x \in A$. Therefore, since a is the greatest lower bound of A we have that $c - y \leq a$ or, equivalently, $c - a \leq y$. Taking into consideration that y was an arbitrary element of B, we have shown that $c - a$ is an lower bound for B, and therefore $c - a \leq b$, from which $c \leq a + b$. Thus, $c = a + b$.

(II). Let us prove first that $\sup C \leq \sup A + \sup B$. Observe that there exist $i, p \in \{1, \ldots, n\}$ such that $x_i = \sup A$ and $y_p = \sup B$, then $x_i \geq x_j$ and $y_p \geq y_j$ for every $j \in \{1, \ldots, n\}$. Moreover, there exists $q \in \{1, \ldots, n\}$ such that $x_q + y_q = \sup C$. So $x_i + y_p \geq x_j + y_j$ for every $j \in \{1, \ldots, n\}$, that is, $\sup A + \sup B = x_i + y_p \geq x_q + y_q = \sup C$.

Now let us show that $\inf C \geq \inf A + \inf B$. Notice that there exist $i, p \in \{1, \ldots, n\}$ such that $x_i = \inf A$ and $y_p = \inf B$, then $x_i \leq x_j$ and $y_p \leq y_j$ for every $j \in \{1, \ldots, n\}$. Moreover, there exists $q \in \{1, \ldots, n\}$ such that $x_q + y_q = \inf C$. So $\inf A + \inf B = x_i + y_p \leq x_j + y_j$ for every $j \in \{1, \ldots, n\}$.

Let us see that there exist finite sets A and B with the same cardinality, that is, $\inf A + \inf B = x_i + y_p \leq x_q + y_q = \inf C$ such that $\sup C < \sup A + \sup B$ and $\inf C > \inf A + \inf B$. Take $A = \{1, 2\}$ and $B = \{-1, -3\}$, then $C = \{0, -1\}$. Hence, $\sup C = 0 < 1 = 2 - 1 = \sup A + \sup B$ and $\inf C = -1 > -2 = 1 - 3 = \inf A + \inf B$.

(III). First, let us show that $\sup(A - B) = \sup A - \inf B$. By part (I) and Exercise 1.50, $\sup(A-B) = \sup(A+(-B)) = \sup A + \sup(-B) = \sup A - \inf B$.

Now let us see that $\inf(A - B) = \inf A - \sup B$. By part (I) and Exercise 1.50, $\inf(A - B) = \inf(A + (-B)) = \inf A + \inf(-B) = \inf A - \sup B$.

(IV). Observe that for every $x, y \in A$ we have that

$$|x - y| = \begin{cases} x - y & \text{if } x \geq y, \\ y - x & \text{if } x \leq y. \end{cases}$$

Thus, by part (III),

$$\sup A - \inf A = \sup(A - A) = \sup\{\, x - y : x, y \in A \,\}$$
$$= \sup\{\, y - x : x, y \in A \,\} = \sup\{\, |x - y| : x, y \in A \,\}.$$

□

Exercise 1.53. *Given an ordered field* $(\mathbb{K}, +, \cdot, \leq)$*, prove that the following statement is equivalent to the Supremum Axiom:*

Given two non-empty sets $X, Y \subset \mathbb{K}$ *such that* $x \leq y$ *for any* $x \in X$*,* $y \in Y$*, there exists* $s \in \mathbb{K}$ *such that* $x \leq s$ *for every* $x \in X$ *and* $s \leq y$ *for every* $y \in Y$*.*

Solution to Exercise 1.53.

Assume first that $(\mathbb{K}, +, \cdot, \leq)$ satisfies the Supremum Axiom, that is, given any $A \subset \mathbb{K}$ non-empty and bounded above, there exists $\sup A \in \mathbb{K}$. Let $X, Y \subset \mathbb{K}$ non-empty such that $x \leq y$ for any $x \in X$ and $y \in Y$, then X is bounded since any $y \in Y$ is an upper bound of X. Thus, there exists $s = \sup X \in \mathbb{K}$, that is, $x \leq s$ for every $x \in X$. Assume by contradiction that there exists $y_0 \in Y$ such that $y_0 < s$; then $s = \sup X \leq y_0 < s$, which is absurd. So $s \leq y$ for every $y \in Y$.

Assume now that given any non-empty $X, Y \subset \mathbb{K}$ such that $x \leq y$ for any $x \in X$ and $y \in Y$, we have that there exists $s \in \mathbb{K}$ such that $x \leq s$ and $s \leq y$ for every $x \in X$ and $y \in Y$. Take $A \subset \mathbb{K}$ non-empty and bounded above. Now let B be the set of all upper bounds of A, then $a \leq b$ for every $a \in A$ and $b \in B$. Hence, there exists $c \in \mathbb{K}$ such that $a \leq c$ and $c \leq b$ for every $a \in A$ and $b \in B$. Then $c = \sup A$. Indeed, c is an upper bound as $a \leq c$ for all $a \in A$, and c is the smallest of all upper bounds since $c \leq b$ for every $b \in B$. Hence, $(\mathbb{K}, +, \cdot, \leq)$ satisfies the Supremum Axiom. □

Chapter 2

The Field of Complex Numbers

> **Exercise 2.1.** *Write the following numbers in the form $a + ib$, with $a, b \in \mathbb{R}$:*
>
> (I) $(1+i)^3$.
>
> (II) $\dfrac{2+3i}{3-4i}$.
>
> (III) $i^5 + i^{16}$.
>
> (IV) $\frac{1}{2}(1+i)/(1+i^{-8})$.

Solution to Exercise 2.1.

(I). $(1+i)^3 = 1^3 + i^3 + 3 \cdot 1^2 \cdot i + 3 \cdot 1 \cdot i^2 = 1 - i + 3i - 3 = -2 + 2i$.

(II). $\dfrac{2+3i}{3-4i} = \dfrac{(2+3i)(3+4i)}{(3-4i)(3+4i)} = \dfrac{6+8i+9i+12i^2}{3^2+4^2} = \dfrac{-7}{25} + i\dfrac{17}{25}$.

(III). $i^5 + i^{16} = i^1 + i^0 = i + 1 = 1 + i$.

(IV). $\dfrac{1}{2}(1+i)/(1+i^{-8}) = \dfrac{1}{2} \cdot \dfrac{1+i}{1+i^0} = \dfrac{1}{4} + i\dfrac{1}{4}$.

$\square$

> **Exercise 2.2.** *Determine, in each case, all real values x and y satisfying the given relation:*
>
> (I) $x + iy = |x - iy|$.
>
> (II) $x + iy = (x - iy)^2$.
>
> (III) $\displaystyle\sum_{k=0}^{100} i^k = x + iy$.

Solution to Exercise 2.2.

(I). We have

$$x + iy = |x - iy| \iff x + iy = \sqrt{x^2 + y^2} \iff [x = \sqrt{x^2 + y^2} \text{ and } y = 0]$$
$$\iff [x = |x| \text{ and } y = 0] \iff [x \geq 0, \text{ and } y = 0].$$

DOI: 10.1201/9781003400745-2

Therefore, the solution is the set of all pairs $(x, 0)$ with x a nonnegative real number.

(II). We have

$$x + iy = (x - iy)^2 \iff x + iy = x^2 + (-iy)^2 - i2xy$$
$$\iff x + iy = x^2 - y^2 - i2xy$$
$$\iff [x = x^2 - y^2 \text{ and } y = -2xy]$$
$$\iff [x = x^2 - y^2, \text{ and } y = 0 \text{ or } x = -1/2].$$

If $y = 0$, then $x = x^2$, and so $x = 0$ or 1. If $x = -1/2$, then $-\frac{1}{4} = \frac{1}{4} - y^2$, and so $y^2 = \frac{3}{4}$, in which case $y = \pm\frac{3}{\sqrt{2}}$. Therefore, the requested pairs (x, y) are $(0, 0)$, $(1, 0)$, $(-1/2, 3/\sqrt{2})$, and $(-1/2, -3/\sqrt{2})$.

(III). By using the formula of the sum of the terms of a geometric progression (see Exercise 2.7), we obtain:

$$x + iy = \sum_{k=0}^{100} i^k = i^0 \cdot \frac{1 - i^{101}}{1 - i} = \frac{1 - i^{4 \cdot 25 + 1}}{1 - i} = \frac{1 - i^1}{1 - i} = 1.$$

Therefore, $(x, y) = (1, 0)$.

□

Exercise 2.3. *Provide a geometrical description of the set of complex numbers z satisfying each of the following conditions:*

(I) $|z| = 1$.   (III) $|z| \leq 1$.   (V) $z - \bar{z} = i$.

(II) $|z| < 1$.   (IV) $z + \bar{z} = 1$.   (VI) $\left| \dfrac{z - 3}{z + 3} \right| = 2$.

(VII) $|z - a| + |z - b| = r$, where $a, b \in \mathbb{C}$ and $r > 0$ are fixed, and $a \neq b$.

(VIII) $\bar{z} + z = |z|^2$.

(IX) $|z| < 1 - \operatorname{Re} z$.

(X) $\operatorname{Im} \dfrac{z - a}{z - b} = 0$, where $a, b \in \mathbb{C}$ are fixed and different.

(XI) $\operatorname{Re} \dfrac{z - a}{z - b} = 0$, where $a, b \in \mathbb{C}$ are fixed and different.

Solution to Exercise 2.3.

(I). It is the circle with center at the origin and radius 1.

(II). It is the open disk with center at the origin and radius 1.

(III). It is the closed disk with center at the origin and radius 1.

(IV). Let $z = x + iy$. Observe that $z + \bar{z} = 1 \iff 2x = 1 \iff x = \frac{1}{2}$, which is the vertical line with constant abscissa $1/2$.

(V). Let $z = x + iy$. Observe that

$$z - \bar{z} = i \iff x + iy - (x - iy) = i \iff 2yi = i \iff y = \frac{1}{2},$$

which is the horizontal line having constant height $1/2$.

(VI). Clearly, the point $z = -3$ is outside the set. Under this assumption, and letting $z = x + iy$, we have

$$\left| \frac{z-3}{z+3} \right| = 2 \iff |z-3|^2 = 4|z+3|^2 \iff (x-3)^2 + y^2 = 4((x+3)^2 + y^2)$$

$$\iff x^2 + 9 - 6x + y^2 = 4(x^2 + 9 + 6x + y^2)$$

$$\iff x^2 + 30x + 3y^2 + 27 = 0 \iff x^2 + y^2 + 10x + 9 = 0$$

$$\iff (x+5)^2 + y^2 = 16,$$

that is the circle with center at $z_0 = -5$ and radius 4.

(VII). The set is the ellipse whose foci are a, b and whose largest axis has length r.

(VIII). Let $z = x + iy$. Then

$$\bar{z} + z = |z|^2 \iff x - iy + x + iy = x^2 + y^2 \iff 2x = x^2 + y^2$$

$$\iff x^2 - 2x + 1 + y^2 = 1 \iff (x-1)^2 + (y-0)^2 = 1^2.$$

This set is the circle with center at $z_0 = 1$ and radius 1.

(IX). Let $z = x + iy$. Then

$$|z| < 1 - \operatorname{Re} z \iff \sqrt{x^2 + y^2} < 1 - x$$

$$\iff [x^2 + y^2 < (1-x)^2 \text{ and } 1 - x > 0]$$

$$\iff [x^2 + y^2 < 1 + x^2 - 2x \text{ and } x < 1]$$

$$\iff [y^2 < 1 - 2x \text{ and } x < 1].$$

Since the fact $1 - 2x < y^2$ implies $1 - 2x > 0$, and this forces x to satisfy $x < 1/2$, the condition $x < 1$ is superfluous. There the requested set is the plane region lying inside the parabola $y^2 = 1 - 2x$, whose axis is the real axis and whose vertex is $z_0 = 1/2$, this point being the right end of the parabola.

(X). Let $z = x + iy$, $a = \alpha + i\beta$, $b = \gamma + i\delta$. Then $\dfrac{z-a}{z-b} = \dfrac{(z-a)(\overline{z}-\overline{b})}{|z-b|^2}$.
Since $|z-b|^2$ is real and positive, we derive that the condition $\operatorname{Im}\frac{z-a}{z-b} = 0$ is equivalent to $\operatorname{Im}(z-a)(\overline{z}-\overline{b}) = 0$. We have that

$$\operatorname{Im}(z-a)((\overline{z}-\overline{b})) = \operatorname{Im}(|z|^2 - a\overline{z} - \overline{b}z + a\overline{b}) = \operatorname{Im}(-a\overline{z} - \overline{b}z + a\overline{b}),$$

because $|z|^2$ is real. A simple calculation yields

$$-a\overline{z} - \overline{b}z + a\overline{b} = -\alpha x - \beta y - \gamma x - \delta y + \alpha\gamma + \beta\delta + i(\alpha y - \beta x - \gamma y - \delta x - \alpha\delta + \beta\gamma).$$

Then $\operatorname{Im}\frac{z-a}{z-b} = 0$ if, and only if, $(\delta - \beta)x + (\alpha - \gamma)y - \alpha\delta + \beta\gamma = 0$, which is the equation of a straight line.

(XI). Under the same notation of the preceding part, we have that the given condition is equivalent to $\operatorname{Re}((z-a)(\overline{z}-\overline{b})) = 0$. Now, we calculate:

$$\operatorname{Re}((z-a)(\overline{z}-\overline{b})) = \operatorname{Re}(|z|^2 - \overline{a}z - \overline{b}z + a\overline{b}) = |z|^2 + \operatorname{Re}(-\overline{a}z - \overline{b}z + a\overline{b}),$$

because $|z|^2$ is real. Taking advantage of the previous calculation, we obtain:

$$\operatorname{Re}((z-a)(\overline{z}-\overline{b})) = 0 \iff x^2 + y^2 - \alpha x - \beta y - \gamma x - \delta y + \alpha\gamma + \beta\delta = 0.$$

By completing squares, the last equality is equivalent to

$$\left(x - \frac{\alpha+\gamma}{2}\right)^2 + \left(y - \frac{\beta+\delta}{2}\right)^2 = \frac{\alpha^2 + \beta^2 + \gamma^2 + \delta^2 - 2\alpha\gamma - 2\beta\delta}{4} =: \Delta.$$

This is the equation of the circle having center at the point $z_0 := \frac{\alpha+\gamma}{2} + i\frac{\beta+\delta}{2}$ and radius $\Delta^{1/2}$. Note that $\Delta > 0$ because of the Cauchy-Schwarz inequality (see Exercise 1.33). Indeed, $\alpha^2 + \gamma^2 \geq 2\alpha\gamma$ and $\beta^2 + \delta^2 \geq 2\beta\delta$, where the equality holds only if $\alpha = \gamma$ and $\beta = \delta$, respectively. These two equalities cannot take place simultaneously, because $a \neq b$. $\qquad\square$

Exercise 2.4. *If z and w are complex numbers, prove the following:*

(I) $|z-w|^2 \leq (1+|z|^2)(1+|w|^2)$.

(II) *If $z \neq 0$, then $|z+w| = |z| + |w|$ if, and only if, w/z is real and nonnegative.*

Solution to Exercise 2.4.

Recall that $|\alpha|^2 = \alpha \cdot \overline{\alpha}$ and $|\mathrm{Re}\alpha|, |\mathrm{Im}\alpha| \le |\alpha|$ for all $\alpha \in \mathbb{C}$.

(I). We have that

$$|z-w|^2 = (z-w)(\overline{z}-\overline{w}) = z\overline{z} - z\overline{w} - w\overline{z} + w\overline{w} = |z|^2 - (z\overline{w} + w\overline{z}) + |w|^2$$
$$= |z|^2 - 2\mathrm{Re}(z\overline{w}) + |w|^2 \le |z|^2 + |w|^2 + 2|z\overline{w}| = |z|^2 + |w|^2 + 2|zw|.$$

Observe that $(|zw|-1)^2 \ge 0$, so $|zw|^2 + 1 - 2|zw| \ge 0$. This implies $2|zw| \le 1 + |zw|^2$. Therefore,

$$|z-w|^2 \le |z|^2 + |w|^2 + 1 + |zw|^2 = 1 + |z|^2 + |w|^2 + 1 + |z|^2|w|^2$$
$$= (1+|z|^2)(1+|w|^2), \text{ qed.}$$

(II). It is evident that $|z+w| = |z|+|w|$ if, and only if, $|z+w|^2 = (|z|+|w|)^2$. Now, $|z+w|^2 = (z+w)(\overline{z}+\overline{w}) = z\overline{z} + w\overline{w} + z\overline{w} + \overline{z}w = |z|^2 + |w|^2 + 2\mathrm{Re}(w\overline{z})$. Hence, $|z+w| = |z| + |w|$ if, and only if, $\mathrm{Re}(w\overline{z}) = |z| \cdot |w| = |w \cdot \overline{z}|$. But $|\alpha| = \mathrm{Re}\alpha$ if, and only if, $\alpha \ge 0$. Consequently:

$$|z+w| = |z|+|w| \iff w\overline{z} \ge 0 \iff \frac{w\overline{z}}{|z|^2} \ge 0 \iff \frac{w}{z} \ge 0.$$

□

Exercise 2.5. *Let z be a complex number with modulus 1. Compute*
$$|1+z|^2 + |1-z|^2.$$

Solution to Exercise 2.5.

Since $1 = |z|^2 = z\overline{z}$, we get

$$|1+z|^2 + |1-z|^2 = (1+z)(1+\overline{z}) + (1-z)(1-\overline{z})$$
$$= 1 + z\overline{z} + z + \overline{z} + 1 + z\overline{z} - z - \overline{z}$$
$$= 1 + z\overline{z} + 1 + z\overline{z} = 1+1+1+1 = 4.$$

□

Exercise 2.6 (Parallelogram identity). *If z and w are complex numbers, prove that*
$$|z+w|^2 + |z-w|^2 = 2|z|^2 + 2|w|^2.$$
Interpret it geometrically as a property of parallelograms.

Solution to Exercise 2.6.

Taking into account that $|\alpha^2| = \alpha \cdot \overline{\alpha}$ for all $\alpha \in \mathbb{C}$, we calculate:

$$\begin{aligned} |z+w|^2 + |z-w|^2 &= (z+w)(\overline{z}+\overline{w}) + (z-w)(\overline{z}-\overline{w}) \\ &= z\overline{z} + z\overline{w} + w\overline{z} + w\overline{w} + z\overline{z} - z\overline{w} - w\overline{z} + w\overline{w} \\ &= z\overline{z} + z\overline{z} + w\overline{w} + w\overline{w} = 2|z|^2 + 2|w|^2. \end{aligned}$$

The identity says that the sum of the squares of the lengths of the diagonals of a parallelogram equals twice the sum of the squares of the lengths of its sides.

□

Exercise 2.7. *If z and w are complex numbers, prove that*

$$|z - w| = |1 - \overline{z}w|$$

if, and only if, $|z| = 1$ or $|w| = 1$. What numbers z and w satisfy the inequality $|z - w| < |1 - \overline{z}w|$?

Solution to Exercise 2.7.

Recall that $|\alpha^2| = \alpha \cdot \overline{\alpha}$ for all $\alpha \in \mathbb{C}$. Observe the following chain of equivalences:

$$\begin{aligned} |z - w| = |1 - \overline{z}w| &\Longleftrightarrow |z-w|^2 = |1 - \overline{z}w|^2 \\ &\Longleftrightarrow (z-w)(\overline{z}-\overline{w}) = (1-\overline{z}w)(1-\overline{w}z) \\ &\Longleftrightarrow z\overline{z} - w\overline{z} - z\overline{w} + w\overline{w} = 1 + z\overline{z}w\overline{w} - \overline{z}w - z\overline{w} \\ &\Longleftrightarrow |z|^2 + |w|^2 = 1 + |z|^2|w|^2 \\ &\Longleftrightarrow |z|^2|w|^2 + |z|^2 + |w|^2 - 1 = 0 \\ &\Longleftrightarrow (|z|^2 - 1)(|w|^2 - 1) = 0 \\ &\Longleftrightarrow [|z|^2 = 1 \text{ or } |w|^2 = 1] \\ &\Longleftrightarrow [|z| = 1 \text{ or } |w| = 1]. \end{aligned}$$

Analogously, we obtain

$$\begin{aligned} |z - w| < |1 - \overline{z}w| &\Longleftrightarrow (|z|^2 - 1)(|w|^2 - 1) > 0 \\ &\Longleftrightarrow [|z|^2 - 1 > 0 < |w|^2 - 1 \text{ or } |z|^2 - 1 < 0 > |w|^2 - 1] \\ &\Longleftrightarrow [|z|^2, |w|^2 > 1 \text{ or } |z|^2, |w|^2 < 1]. \end{aligned}$$

Then, the inequality is satisfied if, and only if, either both $|z|$ and $|w|$ are less than 1 or both $|z|$ and $|w|$ are greater than 1.

□

Exercise 2.8. *If $a \neq 0$ and c are constant real numbers, and b is a complex number satisfying $ac < |b|^2$, prove that the equation*

$$az\bar{z} + b\bar{z} + \bar{b}z + c = 0$$

represents a circle in the Gauss plane.

Solution to Exercise 2.8.

First, we have the following equivalences:

$$az\bar{z} + b\bar{z} + \bar{b}z + c = 0 \iff z\bar{z} - (-b/a)\bar{z} + \overline{(-b/a)}z + \frac{c}{a} = 0$$

$$\iff z\bar{z} - (-b/a)\bar{z} + \overline{(-b/a)}z + |-b/a|^2 = \frac{|b|^2 - ac}{a^2}.$$

Now, notice that the equation of the circle centered at $\alpha \in \mathbb{C}$ with radius $r > 0$ is $|z - \alpha|^2 = r^2$, which is the same as $(z - \alpha)(\bar{z} - \bar{\alpha}) = r^2$, that is, $z\bar{z} - \alpha\bar{z} - \bar{\alpha}z + |\alpha|^2 = r^2$.

Then the equation given in the exercise represents the circle with center $\alpha := -\dfrac{b}{a}$ and radius $r := \left(\dfrac{|b|^2 - ac}{a^2} \right)^{1/2}$. Note that, thanks to the assumption, we have that r is a (positive) real number.

$\square$

Exercise 2.9. *Let us define the following relation in $\mathbb{C}$. We say that $z \leq w$ if one of the following three cases holds:*

(a) $z = 0$.

(b) $0 < |z| < |w|$.

(c) $0 < |z| = |w|$ and $\operatorname{Arg} z \leq \operatorname{Arg} w$.

Here Arg denotes the principal value of the argument, that is, the one belonging to $(-\pi, \pi]$. Which order axioms does this relation satisfy? Is it a total order? Is it compatible with the sum? Is it compatible with the product?

Solution to Exercise 2.9.

According to the definition, we have $0 \leq z$ for all $z \in \mathbb{C}$; in particular, $0 \leq 0$. If $z \neq 0$, then $z \leq z$ due to case (c). Then the relation is reflexive. If $z \leq w$ and $w \leq z$, then either $z = 0 = w$ or case (iii) holds for the pairs (z, w)

and (w, z), so $|z| = |w|$ and $\text{Arg} z = \text{Arg} w$ in this case, which implies $z = w$. Therefore, the relation is symmetric. If $z \leq w$ and $w \leq t$, then by using $0 \leq \alpha$ for all $\alpha \in \mathbb{C}$ and the fact that (when regarding moduli and arguments) the relation $\leq$ is an order on $\mathbb{R}$, we get that the given relation is transitive. Hence, it is an order relation.

That it is a total order comes from the facts that $0 \leq \alpha$ for all $\alpha \in \mathbb{C}$ and that, given $z, w \in \mathbb{C} \setminus \{0\}$ with $z \neq w$, then exactly one of the cases (a) or (c) holds.

The relation is not compatible with the sum because, for instance, $1 \geq 1$ and $-1 \geq 0$ but $1 + (-1) = 0 < 1 = 1 + 0$. Also, the relation is not compatible with the product: indeed, note that $-1 \geq 1$ and $-1 \geq i$ but $(-1) \cdot (-1) = 1 < i = 1 \cdot i$.

$\square$

Exercise 2.10. *Let us define the following relation in* $\mathbb{C}$. *If* $z = z_1 + i z_2$ *and* $w = w_1 + i w_2$, *then we say that* $z \leq w$ *if one of the following cases holds:*

(a) $z_1 < w_1$.

(b) $z_1 = w_1$ *and* $z_2 \leq w_2$.

Answer the same questions posed in Exercise 2.9.

Solution to Exercise 2.10.

By case (b) it is evident that $z \leq z$ for all $z \in \mathbb{C}$, so reflexivity holds. Symmetry also holds, because if $z \leq w$ and $w \leq z$ then the case (b) must happen for (z, w) and (w, z). If $z \leq w$ and $w \leq t$ (where $z = z_1 + i z_2$, $w = w_1 + i w_2$ and $t = t_1 + i t_2$), then by distinguishing the cases $z_1 < w_1 < t_1$, $z_1 = w_1 < t_1$, $z_1 < w_1 = t_1$, and $z_1 = w_1 = t_1$, we obtain easily $z \leq t$. Of course, the fact that the usual $\leq$ is an order relation on $\mathbb{R}$ has been used. Hence the given relation is an order on $\mathbb{C}$.

This order is total because, for $z, w \in \mathbb{C}$, we have of the following four possibilities: $z_1 < w_1$; $w_1 < z_1$; $z_1 = w_1$ and $z_2 \leq w_2$; $z_1 = w_1$ and $w_2 \leq z_2$. Then we have $z \leq w$, $w \leq z$, $z \leq w$, $w \leq z$, respectively.

This order is compatible with the sum, because both cases (a) and (b) give relations which are compatible with the sum of both real and imaginary parts. However, our order is not compatible with the product. Indeed, note for instance that $i \geq 0$ and again $i \geq 0$, but $i \cdot i = -1 < 0 = 0 \cdot 0$.

$\square$

Exercise 2.11. *Find the modulus and the argument of the following complex numbers:*

(I) $3 + 4i$. (III) $(1+i)^5$. (V) $\dfrac{1+i}{1-i}$.

(II) $(3 + 4i)^{-1}$. (IV) $|3 + 4i|$.

Solution to Exercise 2.11.
We denote by z_j $(j = 1, 2, 3, 4, 5)$ the respective given numbers. Once an argument of each z_j is found, the remaining ones are obtained by adding all $2\pi n$, with $n \in \mathbb{Z}$, the set of all integers.

(I). $|z_1| = (3^2 + 4^2)^{1/2} = 25^{1/2} = 5$. Since z_1 is in the first quadrant of the plane, we have $\arg z_1 = \arctan \dfrac{4}{3}$.

(II). $|z_2| = \left| \dfrac{1}{z_1} \right| = \dfrac{1}{5}$. Since $z_2 = \dfrac{1}{z_1}$, we have $\arg z_2 = -\arg z_1 = -\arctan \dfrac{4}{3}$.

(III). $|z_3| = |1 + i|^5 = (\sqrt{2})^5 = 4\sqrt{2}$. Since $\arg(z^N) = N \cdot \arg z$, we get $\arg z_3 = 5 \cdot \arg(1+i) = 5 \cdot \arctan \dfrac{1}{1} = \dfrac{5\pi}{4}$.

(IV). Observe that $z_4 = |3 + 4i| = 5$ is a positive real number. Then $|z_4| = z_4 = 5$ and $\arg z_4 = 0$.

(V). $|z_5| = \dfrac{|1 + i|}{|1 - i|} = \dfrac{(1^2 + 1^2)^{1/2}}{(1^2 + 1^2)^{1/2}} = \dfrac{\sqrt{2}}{\sqrt{2}} = 1$. And $\arg z_5 = \arg \dfrac{(1+i)(1+i)}{(1-i)(1+i)} = \arg \dfrac{2i}{2} = \arg i = \dfrac{\pi}{2}$. Another way to obtain $\arg z_5$ is using $\arg \dfrac{z}{w} = \arg z - \arg w$.

$\square$

Exercise 2.12. *Find all values of the following roots:*

(I) $\sqrt[3]{1}$. (III) $\sqrt[4]{-1}$. (V) $\sqrt[3]{1 + i}$.

(II) $\sqrt[3]{i}$. (IV) $\sqrt{1 - i}$. (VI) $\sqrt[6]{1 - \sqrt{3}i}$.

Solution to Exercise 2.12.
If $z = r_\alpha = re^{i\alpha}$ (polar expression: $r = |z|$ and $\alpha = \arg z$) and $n \in \mathbb{N}$, then the nth-roots of z are $z_j = r^{1/n}_{\frac{\alpha + 2\pi j}{n}}$ $(j = 0, 1, \ldots, n - 1)$.

(I). Let $z = 1 = 1_0$. Then its three cubic roots are:

$$z_0 = 1,$$

$$z_1 = 1_{\frac{0+2\pi}{3}} = 1_{\frac{2\pi}{3}} = \cos\frac{2\pi}{3} + i\sin\frac{2\pi}{3} = \frac{-1 + i\sqrt{3}}{2},$$

$$z_2 = 1_{\frac{0+4\pi}{3}} = e^{i\frac{4\pi}{3}} = e^{i(2\pi - \frac{2\pi}{3})} = e^{-i\frac{2\pi}{3}} = \overline{z_1} = \frac{-1 - i\sqrt{3}}{2}.$$

(II). Let $z = i = 1_{\pi/2}$. Then its three cubic roots are:

$$z_0 = 1_{\frac{\pi}{2\cdot 3}} = e^{\frac{i\pi}{6}} = \frac{\sqrt{3} + i}{2},$$

$$z_1 = 1_{\frac{\pi/2 + 2\pi}{3}} = e^{i\frac{5\pi}{6}} = \cos\frac{\pi}{6} + i\sin\frac{\pi}{6} = e^{i\pi} \cdot e^{-i\frac{\pi}{6}}$$

$$= -1 \cdot \left(\cos\frac{-\pi}{6} + i\sin\frac{-\pi}{6}\right) = -\cos\frac{-\pi}{6} + i\sin\frac{\pi}{6} = \frac{-\sqrt{3} + i}{2},$$

$$z_2 = 1_{\frac{\pi/2 + 4\pi}{3}} = e^{i\frac{3\pi}{2}} = -i.$$

(III). Let $z = -1 = 1_\pi$. Then its four 4th-roots are:

$$z_0 = 1_{\pi/4} = e^{i\frac{\pi}{4}} = \cos\frac{\pi}{4} + i\sin\frac{\pi}{4} = \frac{1 + i}{\sqrt{2}},$$

$$z_1 = 1_{\frac{\pi + 2\pi}{4}} = 1_{\frac{3\pi}{4}} = e^{i\frac{3\pi}{4}} = \cos\frac{3\pi}{4} + i\sin\frac{3\pi}{4} = \frac{-1 + i}{\sqrt{2}},$$

$$z_2 = 1_{\frac{\pi + 4\pi}{4}} = 1_{\frac{5\pi}{4}} = e^{i\pi} \cdot e^{i\frac{\pi}{4}} = -1 \cdot \left(\cos\frac{\pi}{4} + i\sin\frac{\pi}{4}\right) = \frac{-1 - i}{\sqrt{2}},$$

$$z_3 = 1_{\frac{\pi + 6\pi}{4}} = e^{i\frac{7\pi}{4}} = e^{i2\pi} \cdot e^{-i\frac{\pi}{4}} = 1 \cdot \left(\cos\frac{-\pi}{4} + i\sin\frac{-\pi}{4}\right)$$

$$= \cos\frac{\pi}{4} - i\sin\frac{\pi}{4} = \frac{1 - i}{\sqrt{2}}.$$

(IV). Let $z = 1 - i = \sqrt{2}_{-\pi/4}$. Then its two square roots are;

$$z_0 = \sqrt{2}_{-\pi/8} = \sqrt{2} \cdot \left(\cos\frac{-\pi}{8} + i\sin\frac{-\pi}{8}\right) = \sqrt{2} \cdot \left(\cos\frac{\pi}{8} - i\sin\frac{\pi}{8}\right)$$

$$= \sqrt{2} \cdot \sqrt{\frac{1 + \cos(\pi/4)}{2}} - i\sqrt{2} \cdot \sqrt{\frac{1 - \cos(\pi/4)}{2}} = \sqrt{1 + \frac{1}{\sqrt{2}}} - i\sqrt{1 - \frac{1}{\sqrt{2}}}$$

$$= \sqrt{\frac{\sqrt{2} + 1}{2}} - i\sqrt{\frac{\sqrt{2} - 1}{2}},$$

$$z_1 = -z_0 = -\sqrt{\frac{\sqrt{2} + 1}{2}} + i\sqrt{\frac{\sqrt{2} - 1}{2}}.$$

(v). Let $z = 1 + i = \sqrt{2}_{\pi/4}$. Then its three cubic roots are:

$$z_0 = \sqrt[3]{2}_{\pi/12} = \cos\frac{\pi}{12} + i\sin\frac{\pi}{12} = \sqrt{\frac{1 + \cos(\pi/6)}{2}} + i\sqrt{\frac{1 - \cos(\pi/6)}{2}}$$

$$= \frac{\sqrt{2 + \sqrt{3}}}{2} + i\frac{\sqrt{2 - \sqrt{3}}}{2},$$

$$z_1 = \sqrt[3]{2}_{\frac{\pi/4 + 2\pi}{3}} = \sqrt[3]{2}_{3\pi/4} = \sqrt[3]{2}e^{i\frac{3\pi}{4}} = \sqrt[3]{2} \cdot \left(\cos\frac{3\pi}{4} + i\sin\frac{3\pi}{4}\right)$$

$$= \sqrt[3]{2} \cdot \left(\frac{-1}{\sqrt{2}} + i\frac{1}{\sqrt{2}}\right) = -1 + i,$$

$$z_2 = \sqrt[3]{2}_{\frac{\pi/4 + 4\pi}{3}} = \sqrt[3]{2}_{17\pi/12} = e^{i\pi} \cdot e^{i\frac{5\pi}{12}} = e^{i\pi} \cdot e^{\frac{i\pi}{2}} \cdot e^{-i\frac{\pi}{12}}$$

$$= -1 \cdot i \cdot \left(\cos\frac{-\pi}{12} + i\sin\frac{-\pi}{12}\right) = -\sin\frac{\pi}{12} - i\cos\frac{\pi}{12}$$

$$= -\frac{\sqrt{2 - \sqrt{3}}}{2} - i\frac{\sqrt{2 + \sqrt{3}}}{2}.$$

(VI). Let $z = 1 - \sqrt{3}i = 2_{-\frac{\pi}{3}}$. Then its six 6th-roots are:

$$z_0 = 2_{-\pi/18} = 2\left(\cos\frac{-\pi}{18} + i\sin\frac{-\pi}{18}\right) = 2\cos\frac{\pi}{18} + i2\sin\frac{\pi}{18},$$

$$z_1 = 2_{\frac{-\pi/3 + 2\pi}{6}} = 2_{5\pi/18} = 2\left(\cos\frac{5\pi}{18} + i\sin\frac{5\pi}{18}\right)$$

$$= 2\left(\cos\left(\frac{\pi}{3} - \frac{5}{18}\right) + i\sin\left(\frac{\pi}{3} - \frac{5\pi}{18}\right)\right)$$

$$= 2\left(\cos\frac{\pi}{3}\cos\frac{5\pi}{18} + \sin\frac{\pi}{3}\sin\frac{5\pi}{18}\right) + 2i\left(\sin\frac{\pi}{3}\cos\frac{5\pi}{18} - \cos\frac{\pi}{3}\sin\frac{5\pi}{18}\right)$$

$$= \cos\frac{\pi}{18} + \sqrt{3}\sin\frac{\pi}{18} + i\left(\sqrt{3}\cos\frac{\pi}{18} - \sin\frac{\pi}{18}\right),$$

$$z_2 = 2_{\frac{-\pi/3 + 4\pi}{6}} = 2_{11\pi/18} = 2\left(\cos\left(\frac{2\pi}{3} - \frac{\pi}{18}\right) + i\sin\left(\frac{2\pi}{3} - \frac{\pi}{18}\right)\right)$$

$$= 2\left(\cos\frac{2\pi}{3}\cos\frac{\pi}{18} + \sin\frac{2\pi}{3}\sin\frac{\pi}{18}\right) + 2i\left(\sin\frac{2\pi}{3}\cos\frac{\pi}{18} - \cos\frac{2\pi}{3}\sin\frac{\pi}{18}\right)$$

$$= -\cos\frac{\pi}{18} + \sqrt{3}\sin\frac{\pi}{18} + i\left(\sqrt{3}\cos\frac{\pi}{18} + \sin\frac{\pi}{18}\right),$$

$$z_3 = -z_0 = -2\cos\frac{\pi}{18} - i2\sin\frac{\pi}{18},$$

$$z_4 = -z_1 = -\cos\frac{\pi}{18} - \sqrt{3}\sin\frac{\pi}{18} + i\left(\sin\frac{\pi}{18} - i\sqrt{3}\cos\frac{\pi}{18}\right),$$

$$z_5 = -z_2 = \cos\frac{\pi}{18} - \sqrt{3}\sin\frac{\pi}{18} - i\left(\sqrt{3}\cos\frac{\pi}{18} + \sin\frac{\pi}{18}\right).$$

We have used the fact that, if N is even and z is an Nth-root of w, then $-z$ is also an Nth-root of w. The exact values of $\cos\frac{\pi}{18}$ and $\sin\frac{\pi}{18}$ can be obtained by making use of the formulas for $\cos(3\theta)$ and $\sin(3\theta)$ given in Exercise 2.24, with the choice $\theta = \frac{\pi}{18}$.

□

Exercise 2.13. *Determine all complex numbers satisfying each one of the following properties:*

(I) $z^2 = 3 - 4i$.

(II) $z^2 + zi + 2 = 0$.

(III) $z^4 - 2z^2 + 4 = 0$.

(IV) $z^3 - |z|^2 = 0$.

(V) $z^n = \bar{z}$.

(VI) $z^3 + 8 = 0$ *and* $|z + 1| < 2$.

Solution to Exercise 2.13.

(I). The complex numbers satisfying $z^2 = 3 - 4i$ are the two square roots of $3 - 4i$. We put $z = a + ib$, with $a, b \in \mathbb{R}$. Then $(a + ib)^2 = 3 - 4i$, so that $a^2 - b^2 + i2ab = 3 - 4i$, which implies $\begin{cases} a^2 - b^2 = 3 \\ 2ab = 4. \end{cases}$ Then $ab = -2$, that entails $b = -2/a$ (note that $a \neq 0$ because otherwise $z^2 = 3 - 4i = b^2 \in \mathbb{R}$, which is absurd). Hence $a^2 - \dfrac{4}{a^2} = 3$. Letting $A = a^2$ we get $A^2 - 3A - 4 = 0$. Solving this equation yields $A = \dfrac{3 \pm \sqrt{9 + 16}}{2} = 4$ or -1. But $A = a^2 > 0$, so $A = 4$. We derive that $a = 2$ (so $b = -1$) or $a = -2$ (so $b = 1$). Then the solutions are $z = 2 - ib, -2 + ib$.

(II). We simply solve this second degree equation:

$$z = \frac{-i \pm \sqrt{i^2 - 8}}{2} = \frac{-i \pm \sqrt{-9}}{2} = \frac{-i \pm 3i}{2} = i, -2i.$$

(III). Letting $w := z^2$ the equation becomes $w^2 - 2w + 4 = 0$, or equivalently, $(w - 1)^2 = -3$, which gives $w - 1 = \pm\sqrt{3}i$, that is, $z^2 = 1 + \sqrt{3}i$ or $z^2 = 1 - \sqrt{3}i$. Since $1 + \sqrt{3}i = 2e^{i\pi/3}$ and $1 - \sqrt{3}i = 2e^{-i\pi/3}$, the requested numbers z must be the square roots of the last two numbers:

$$z_1 = \sqrt{2}e^{i\pi/6} = \sqrt{2}\left(\cos\frac{\pi}{6} + i\sin\frac{\pi}{6}\right) = \frac{\sqrt{6}}{2} + i\frac{\sqrt{2}}{2},$$

$$z_2 = -z_1 = -\frac{\sqrt{6}}{2} - i\frac{\sqrt{2}}{2},$$

$$z_3 = \sqrt{2}e^{-i\pi/6} = \overline{z_1} = \frac{\sqrt{6}}{2} - i\frac{\sqrt{2}}{2},$$

$$z_4 = -z_3 = -\frac{\sqrt{6}}{2} + i\frac{\sqrt{2}}{2}.$$

(IV). Observe that $z^3 = |z|^2$, and so $|z|^2 \cdot |z| = |z|^3 = |z|^2$. Then either $|z| = 0$ or $|z| = 1$. In the first case we get the solution $z = 0$. In the second one, we obtain $z^3 = 1$. Therefore, the other solutions are the cubic roots of the unity, that is, $z = 1$, $z = 1 \cdot e^{i2\pi/3} = \cos\frac{2\pi}{3} + i\sin\frac{2\pi}{3} = -\frac{1}{2} + i\frac{\sqrt{3}}{2}$, and $z = 1 \cdot e^{i4\pi/3} = e^{-i2\pi/3} = -\frac{1}{2} - i\frac{\sqrt{3}}{2}$.

(V). Taking moduli in the identity $z^n = \overline{z}$ we get $|z|^n = |z|$, because $|\alpha^n| = |\alpha|^n$ and $|\overline{\alpha}| = |\alpha|$ for all $\alpha \in \mathbb{C}$, $n \in \mathbb{N}$. It is plain that $z = 0$ is a solution. And if $z \neq 0$, then $|z|^{n-1} = 1$. If $n = 1$, the original equation is $z = \overline{z}$, whose set of solutions is the real line $\mathbb{R}$. If $n \geq 2$, then $|z|^{n-1} = 1$ entails $|z| = 1$, so that $z = e^{it}$ with $t \in \mathbb{R}$. Then $(e^{it})^n = e^{-it}$, or equivalently, $e^{i(n+1)t} = 1$. But for $\alpha \in \mathbb{R}$ we have that $e^{i\alpha} = 1$ if, and only if, $\alpha \in 2\pi\mathbb{Z}$, where $\mathbb{Z}$ is the set of integers. Consequently, $t = \frac{2\pi j}{n+1}$ $(j \in \mathbb{Z})$, which yields as solutions the $n+1$ roots of order $n+1$ of the unity: $z = e^{\frac{2\pi ji}{n+1}}$ $(j = 0, 1, 2, \ldots, n)$.

(VI). The points z satisfying $z^3 + 8 = 0$ are the cubic roots of $-8 = 8_\pi$, that is, the three complex numbers $z_j = 2 \cdot e^{i\frac{\pi+2\pi j}{3}}$ $(j = 0, 1, 2)$. Specifically, $z_0 = 2e^{i\pi/3} = 2(\cos\frac{\pi}{3} + i\sin\frac{\pi}{3}) = 1 + i\sqrt{3}$, $z_1 = 2e^{i\pi} = -2$, and $z_2 = 2e^{i5\pi/3} = 2e^{-i\pi/3} = \overline{z_0} = 1 - i\sqrt{3}$. Among z_0, z_1, z_2 only z_1 lies inside the disk having center at -1 and radius 2 (that is, $|z + 1| < 2$), because $|z_0 + 1| = |2 + i\sqrt{3}| = |2 - i\sqrt{3}| = |z_2 + 1| = \sqrt{7} > 2$. Thus, $z_1 = -2$ is the unique complex number satisfying the given conditions.

$\square$

Exercise 2.14. (I) *Let $z \neq 1$ with $|z| = 1$. Prove that $\dfrac{1+z}{1-z}$ is purely imaginary.*

(II) *Let z be a complex number with modulus 1. Prove that $z + z^{-1}$ is real.*

Solution to Exercise 2.14.

(I). Let us denote $w = \dfrac{1+z}{1-z}$. Then

$$\overline{w} = \frac{1+\overline{z}}{1-\overline{z}} = \frac{z(1+\overline{z})}{z(1-\overline{z})} = \frac{z+z\overline{z}}{z-z\overline{z}} = \frac{z+|z|^2}{z-|z|^2} = \frac{z+1}{z-1} = -\frac{1+z}{1-z} = -w.$$

Hence $\overline{w} = -w$, which tells us that w is purely imaginary.

(II). There is $t \in \mathbb{R}$ with $z = e^{it}$. Taking into account that $\cos(-t) = \cos t$ and $\sin(-t) = -\sin t$, we obtain

$$z + z^{-1} = e^{it} + e^{-it} = \cos t + i\sin t + \cos(-t) + i\sin(-t) = 2\cos t \in \mathbb{R}.$$

$\square$

Exercise 2.15. *Let z, w, t be three complex numbers such that $|z| = |w| = |t| = 1$. Prove that they are the vertices of an equilateral triangle inscribed in the circle of radius 1 centered at the origin if and only if $z + w + t = 0$.*

Solution to Exercise 2.15.

Assume that z, w, t are three unimodular points. All that should be proved is that $z + w + t = 0$ if, and only if, $d(z,w) = d(z,t) = d(w,t)$, where $d(a,b) = |a - b|$, the distance from a to b. Recall that $|a|^2 = a\overline{a}$ for all $a \in \mathbb{C}$.

If $z + w + t = 0$, then

$$|z - w|^2 = |2z + t|^2 = (2z + t)(2\overline{z} + \overline{t}) = 4|z|^2 + |t|^2 + 2(z\overline{t} + \overline{z}t) = 5 + 2(z\overline{t} + \overline{z}t)$$

and

$$|t - w|^2 = |2t + z|^2 = (2t + z)(2\overline{t} + \overline{z}) = 4|t|^2 + |z|^2 + 2(z\overline{t} + \overline{z}t) = 5 + 2(z\overline{t} + \overline{z}t).$$

Then $d(z,w) = |z - w| = |t - w| = d(w,t)$ and, analogously,

$$|w - t|^2 = |2w + z|^2 = 5 + 2(z\overline{w} + w\overline{z}) = |2z + w|^2 = |z - t|^2,$$

so $d(w,t) = d(z,t)$.

Conversely, if the three distances are the same, then by the configuration of the three cubic roots of an unimodular complex number, we have that z, w, t are the cubic roots of $z_0 := e^{i3\alpha}$, where $\alpha \in \mathbb{R}$ is such that $z = e^{i\alpha}$. Hence z, w, t are the roots of the polynomial $P(Z) = Z^3 - z_0$, and so $P(Z) = (Z - z)(Z - w)(Z - t)$. The coefficient of Z^2 in this polynomial is 0, but this coefficient is $z + w + t$. Consequently, $z + w + t = 0$, as required.

$\square$

> **Exercise 2.16.** *Prove that the n-th roots of the unity that are different from 1 satisfy the equation:*
>
> $$1 + z + z^2 + \cdots + z^{n-1} = 0.$$

Solution to Exercise 2.16.

The n roots of the unity $z_1 = 1, z_2, \ldots, z_n$ are the solutions of the equation $z^n - 1 = 0$, so that we have the factorization

$$z^n - 1 = \prod_{j=1}^{n}(z - z_j) = (z - 1) \cdot \prod_{j=2}^{n}(z - z_j).$$

Now, we also have $z^n - 1 = (z - 1)(1 + z + z^2 + \cdots + z^{n-1})$. Consequently,

$$1 + z + z^2 + \cdots + z^{n-1} = \prod_{j=2}^{n}(z - z_j),$$

and so $z_2, \ldots, z_n$ are solutions of the equation $1 + z + z^2 + \cdots + z^{n-1} = 0$. $\square$

> **Exercise 2.17.** *Find the principal value of the logarithm of the following complex numbers:*
>
> (I) 1. (II) i. (III) $-i$. (IV) $1 + i$.

Solution to Exercise 2.17.

We denote by $\mathrm{Log}\,z$ the principal value of the logarithm of a complex number $z \neq 0$. Then $\mathrm{Log}\,z = \ln|z| + i \cdot \mathrm{Arg}\,z$, where $\mathrm{Arg}\,z$ denotes the principal value of $\arg z$, that is, the unique value of $\arg z$ belonging to the interval $(-\pi, \pi]$. We have

(I). $\mathrm{Log}\,1 = \ln 1 + i \cdot \mathrm{Arg}\,1 = 0 + i \cdot 0 = 0$.

(II). $\mathrm{Log}\,i = \ln|i| + i \cdot \mathrm{Arg}\,i = \ln 1 + i \cdot \frac{\pi}{2} = i \cdot \frac{\pi}{2}$.

(III). $\mathrm{Log}(-i) = \ln|-i| + i \cdot \mathrm{Arg}(-i) = \ln 1 + i \cdot \frac{-\pi}{2} = -i \cdot \frac{\pi}{2}$.

(IV). $\mathrm{Log}(1 + i) = \ln|1 + i| + i \cdot \mathrm{Arg}(1 + i) = \ln 2 + i \cdot \frac{\pi}{4}$.

$\square$

> **Exercise 2.18.** *Compute the following complex powers:*
>
> (I) $(-i)^i$.   (II) $(1+i)^{1+i}$. (III) $(-1)^i$.   (IV) 2^i.

Solution to Exercise 2.18.

If $z, w \in \mathbb{C}$ and $z \neq 0$, then $z^w = e^{w \cdot \mathrm{Log} z}$, where Log denotes the principal value of the logarithm. Taking this into account, we have

(I). $(-i)^i = e^{i \cdot \mathrm{Log} i} = e^{i \cdot (\ln|-i| + i \mathrm{Arg}(-i))} = e^{i \cdot (\ln 1 + i(-\pi/2))} = e^{i \cdot (0 + i(-\pi/2))} = e^{\pi/2}$.

(II).

$$(1+i)^{1+i} = e^{(1+i)\mathrm{Log}(1+i)} = e^{(1+i)(\ln|1+i| + i\mathrm{Arg}(1+i))} = e^{(1+i)(\ln\sqrt{2} + i\pi/4)}$$
$$= e^{\ln\sqrt{2} + i\pi/4 + i\ln\sqrt{2} - \pi/4} = e^{\ln\sqrt{2} - \pi/4} \cdot e^{i(\pi/4 + \ln\sqrt{2})}$$
$$= \sqrt{2} e^{-\pi/4} \left(\cos\left(\frac{\pi}{4} + \frac{\ln 2}{2} \right) + i \sin\left(\frac{\pi}{4} + \frac{\ln 2}{2} \right) \right).$$

(III). $(-1)^i = e^{i \cdot \mathrm{Log}(-1)} = e^{i \cdot (\ln|-1| + i\mathrm{Arg}(-1))} = e^{i \cdot (0 + i\pi)} = e^{-\pi}$.

(IV). $2^i = e^{i \cdot \mathrm{Log} 2} = e^{i \cdot (\ln|2| + i\mathrm{Arg} 2)} = e^{i \cdot (\ln 2 + i \cdot 0)} = e^{i\ln 2} = \cos(\ln 2) + i \sin(\ln 2)$.

$\square$

> **Exercise 2.19.**   (I) *Prove that $|z^i| < e^\pi$ for all complex number $z \neq$*
> *0.*
>
> (II) *Prove that there is no constant $M > 0$ satisfying $|\cos z| < M$ for all $z \in \mathbb{C}$.*

Solution to Exercise 2.19.

(I). Let $z \in \mathbb{C} \setminus \{0\}$. We have that

$$|z^i| = |e^{i\mathrm{Log} z}| = |e^{i(\ln|z| + i\mathrm{Arg} z)}| = |e^{i\ln|z|}| \cdot |e^{-\mathrm{Arg} z}| = e^{-\mathrm{Arg} z} < e^\pi$$

because $\mathrm{Arg} z > -\pi$.

(II). If $z = iy$ with $y \in \mathbb{R}$ then we have

$$|\cos z| = |\cos(iy)| = \left| \frac{e^{iiy} + e^{-iiy}}{2} \right| = \frac{e^y + e^{-y}}{2} \geq \frac{e^y}{2}.$$

Since $e^y \to +\infty$ as $y \to +\infty$, we derive that, given $M > 0$, there is $y_0 \in \mathbb{R}$ with $e^{y_0} > 2M$. Hence the number $z_0 := iy_0$ satisfies $|\cos z_0| > M$. This concludes the proof.

$\square$

Exercise 2.20. *If $z \neq 0$ and $w = x + iy$, prove the following:*

$$z^w = e^{x\log|z|-y\mathrm{Arg}z}e^{i(y\log|z|+x\mathrm{Arg}z)},$$

where $\mathrm{Arg}z$ denotes the principal value of the argument of z.

Solution to Exercise 2.20.

By definition, $z^w = e^{w \cdot \mathrm{Log}z}$. Recall that $e^{\alpha+\beta} = e^\alpha \cdot e^\beta$ for all $\alpha, \beta \in \mathbb{C}$. Then

$$z^w = e^{(x+iy)(\log|z|+i\mathrm{Arg}z)} = e^{x\log|z|+ix\mathrm{Arg}z+iy\log|z|-y\mathrm{Arg}z}$$

$$= e^{x\log|z|-y\mathrm{Arg}z+i(x\mathrm{Arg}z+y\log|z|)} = e^{x\log|z|-y\mathrm{Arg}z}e^{i(y\log|z|+x\mathrm{Arg}z)}.$$

$\square$

Exercise 2.21. *Let $z, w \in \mathbb{C}$ with $z \neq 0$. Prove that $\mathrm{Log}\,z^w = w\,\mathrm{Log}\,z + 2\pi i n$, where n is an integer and Log denotes the principal value of the logarithm.*

Solution to Exercise 2.21.

Let $w = x + iy$. Recall that $|e^{is}| = 1$ for all $s \in \mathbb{R}$. By the definition of $\mathrm{Log}\,\alpha$ and α^β ($\alpha, \beta \in \mathbb{C}, \alpha \neq 0$), we get

$$\mathrm{Log}z^w = \ln|z^w| + i\mathrm{Arg}(z^w) = \ln|e^{w\,\mathrm{Log}\,z}| + i\mathrm{Arg}(e^{w\,\mathrm{Log}\,z})$$

$$= \ln\left|e^{(x+iy)(\ln|z|+i\mathrm{Arg}z)}\right| + i \cdot \mathrm{Arg}\left(e^{(x+iy)(\ln|z|+i\mathrm{Arg}z)}\right)$$

$$= \ln\left|e^{x\ln|z|-y\mathrm{Arg}z} \cdot e^{i(x\mathrm{Arg}z+y\ln|z|)}\right| + i\mathrm{Arg}\left(e^{x\ln|z|-y\mathrm{Arg}z} \cdot e^{i(x\mathrm{Arg}z+y\ln|z|)}\right)$$

$$= \ln\left(e^{x\ln|z|-y\mathrm{Arg}z}\right) + i\mathrm{Arg}(e^{i\gamma}) = x\ln|z| - y\mathrm{Arg}z + i(\gamma + 2\pi n)$$

for some integer n, where we have set $\gamma := x\mathrm{Arg}z + y\ln|z|$. Since

$$w\,\mathrm{Log}\,z = (x + iy)(\ln|z| + i\mathrm{Arg}z) = x\ln|z| - y\mathrm{Arg}z + i\gamma,$$

we obtain the desired equality. We have used the facts that γ is an argument of $e^{i\gamma}$ and that the difference between two arguments of a given complex number is an integer multiple of 2π.

$\square$

Exercise 2.22. *Let $z, w \in \mathbb{C}$ with $z \neq 0$, and let $\alpha \in \mathbb{R}$. Prove that*

$$(z^w)^\alpha = z^{w\alpha}e^{2\pi i n\alpha},$$

where n is an integer.

Solution to Exercise 2.22.

Let $w = x + iy$. By using a Exercises 2.20 and 2.21, together with the addition law $e^{a+b} = e^a \cdot e^b$ (where $a, b \in \mathbb{C}$), we get that for a certain integer n we have

$$(z^w)^\alpha = e^{\alpha \operatorname{Log}(z^w)} = e^{\alpha(x\ln|z| - y\operatorname{Arg}z + i(x\operatorname{Arg}z + y\ln|z| + 2\pi n))}$$

$$= e^{(\alpha x)\ln|z| - (\alpha y)\operatorname{Arg}z + i((\alpha x)\operatorname{Arg}z + (\alpha y)\ln|z|) + i2\pi n\alpha}$$

$$= e^{(\alpha x)\ln|z| - (\alpha y)\operatorname{Arg}z + i((\alpha x)\operatorname{Arg}z + (\alpha y)\ln|z|)} \cdot e^{i2\pi n\alpha}$$

$$= z^{w\alpha}e^{2\pi i n\alpha},$$

as required.

$\square$

Exercise 2.23.

(I) *If θ and a are real numbers, with $-\pi < \theta \leq \pi$, prove that*

$$(\cos\theta + i\sin\theta)^a = \cos(a\theta) + i\sin(a\theta).$$

(II) *Prove that, in general, the restriction $-\pi < \theta \leq \pi$ is necessary, by choosing $\theta = -\pi$ and $a = \frac{1}{2}$.*

(III) *If a is an integer, prove that the preceding formula holds even without imposing any restrictions over θ. In this case, the formula is known as De Moivre's formula.*

Solution to Exercise 2.23.

(I). Let $z = \cos\theta + i\sin\theta$, where $-\pi < \theta \leq \pi$. Note that $|z| = 1$. Then:

$$(\cos\theta + i\sin\theta)^a = e^{a\,\mathrm{Log}\,z} = e^{a(\ln|z|+i\mathrm{Arg}z)}$$
$$= e^{0+i\theta} = e^{ia\theta} = \cos(a\theta) + i\sin(a\theta).$$

(II). Letting $\theta = -\pi$ and $a = 1/2$, we have

$$(\cos\theta + i\sin\theta)^a = (-1 + i\cdot 0)^{1/2} = (-1)^{1/2} = e^{\frac{1}{2}\mathrm{Log}(-1)}$$
$$= e^{\frac{1}{2}(\ln|-1|+i\mathrm{Arg}(-1))} = e^{\frac{1}{2}(0+i\pi)} = e^{i\pi/2} = i,$$

while

$$\cos(a\theta) + i\sin(a\theta) = \cos\frac{-\pi}{2} + i\sin\frac{-\pi}{2} = \cos\frac{\pi}{2} - i\sin\frac{\pi}{2} = 0 - i\cdot 1 = -i \neq i.$$

(III). Let $\theta \in \mathbb{R}$. If $a = 0$, then

$$(\cos 0 + i\sin 0)^0 = e^{0\cdot\mathrm{Log}(\cos\theta+i\sin\theta)} = e^0 = 1 = 1 + 0\cdot i = \cos(0\cdot\theta) + i\sin(0\cdot\theta),$$

so the formula holds in this case.

If $a \in \mathbb{N}$, then the formula for the product of complex numbers gives

$$(\cos\theta + i\sin\theta)^a = \underbrace{(\cos\theta + i\sin\theta)\cdots(\cos\theta + i\sin\theta)}_{a\text{ times}}$$
$$= \cos(\underbrace{\theta + \theta + \cdots + \theta}_{a\text{ times}}) + i\sin(\underbrace{\theta + \theta + \cdots + \theta}_{a\text{ times}})$$
$$= \cos(a\theta) + i\sin(a\theta).$$

Finally, if $a \in \mathbb{Z}$ is negative, then $a = -b$ with $b \in \mathbb{N}$. We have

$$(\cos\theta + i\sin\theta)^a = (\cos\theta + i\sin\theta)^{-b} = \frac{1}{(\cos\theta + i\sin\theta)^b}$$
$$= \frac{1}{\cos(b\theta) + i\sin(b\theta)} = \frac{\cos(b\theta) - i\sin(b\theta)}{(\cos(b\theta) + i\sin(b\theta))(\cos(b\theta) - i\sin(b\theta))}$$
$$= \frac{\cos(-a\theta) - i\sin(-a\theta)}{\cos^2(b\theta) + \sin^2(b\theta)} = \cos(a\theta) + i\sin(a\theta).$$

$\square$

Exercise 2.24. *Use De Moivre's formula from Exercise 2.23 to obtain the following trigonometrical identities:*

$$\sin 3\theta = 3\cos^2\theta \sin\theta - \sin^3\theta,$$
$$\cos 3\theta = \cos^3\theta - 3\cos\theta \sin^2\theta.$$

Solution to Exercise 2.24.

For each $\theta \in \mathbb{R}$, we have $e^{i\theta} = \cos\theta + i\sin\theta$. On the one hand, we can apply De Moivre's formula to get $(\cos\theta + i\sin\theta)^3 = e^{i3\theta} = \cos 3\theta + i\sin 3\theta$. But, on the other hand, we have that

$$(\cos\theta + i\sin\theta)^3 = \cos^3\theta + 3\cos^2\theta i\sin\theta + 3\cos\theta i^2\sin^2\theta + i^3\sin^3\theta$$
$$= \cos^3\theta - 3\cos\theta\sin^2\theta + i\left(3\cos^2\theta\sin\theta - \sin^3\theta\right).$$

We have used $i^2 = -1$, $i^3 = -i$. It only remains to match real parts and imaginary parts in the two expressions obtained for $(\cos\theta + i\sin\theta)^3$. □

Exercise 2.25. *Express $\cos 4\theta$ and $\sin 4\theta$ through $\cos\theta$ and $\sin\theta$.*

Solution to Exercise 2.25.

We shall use an approach similar to the one of Exercise 2.24, just by replacing 3θ with 4θ. On the one hand, De Moivre's formula gives

$$(\cos\theta + i\sin\theta)^4 = e^{i4\theta} = \cos 4\theta + i\sin 4\theta.$$

On the other hand, we can apply Newton's binomial formula (recall Exercise 1.11)

$$(a + b)^4 = \sum_{j=0}^{4}\binom{4}{j}a^{4-j}b^j = a^4 + 4a^3b + 6a^2b^2 + 4ab^3 + b^4$$

with $a = \cos 4\theta$, $b = i\sin 4\theta$ to obtain, after matching real and imaginary parts, the following:

$$\sin 4\theta = 4\cos^3\theta\sin\theta - 4\sin^3\theta\cos\theta,$$
$$\cos 4\theta = \cos^4\theta - 6\cos^2\theta\sin^2\theta + \sin^4\theta.$$

We have taken into account that $i^2 = -1$, $i^3 = -i$ and $i^4 = 1$. □

Exercise 2.26. *If n is a natural number with $n \geq 2$, prove the following equalities:*

(I) $\cos \frac{2\pi}{n} + \cos \frac{4\pi}{n} + \cdots + \cos \frac{2(n-1)\pi}{n} = -1$.

(II) $\sin \frac{2\pi}{n} + \sin \frac{4\pi}{n} + \cdots + \sin \frac{2(n-1)\pi}{n} = 0$.

Solution to Exercise 2.26.

Let $z := e^{i2\pi/n}$. Then the geometrical sum $z + z^2 + \cdots + z^{n-1}$ equals $z \cdot \dfrac{1 - z^{n-1}}{1 - z}$. Hence

$$\sum_{j=1}^{n-1} \cos \frac{2\pi j}{n} + i \cdot \sum_{j=1}^{n-1} \sin \frac{2\pi j}{n} = \sum_{j=1}^{n-1} \left(\cos \frac{2\pi j}{n} + i \cdot \sin \frac{2\pi j}{n} \right) = \sum_{j=1}^{n-1} e^{\frac{i2\pi j}{n}}$$

$$= \sum_{j=1}^{n-1} z^j = z \cdot \frac{1 - z^{n-1}}{1 - z} = e^{i2\pi/n} \cdot \frac{1 - e^{i2\pi(n-1)/n}}{1 - e^{i2\pi/n}}$$

$$= \frac{e^{i2\pi/n} - e^{i2\pi}}{1 - e^{i2\pi/n}} = \frac{e^{i2\pi/n} - 1}{1 - e^{i2\pi/n}} = -1 = -1 + 0 \cdot i.$$

Then parts (I) and (II) are obtained by matching real and imaginary parts in the extreme members of the preceding equality. $\square$

Exercise 2.27. *Verify that the identity $\cos^2 z + \sin^2 z = 1$ holds for all $z \in \mathbb{C}$.*

Solution to Exercise 2.27.

We shall use that $e^{z+w} = e^z \cdot e^w$ for all $z, w \in \mathbb{C}$. Recall that, by definition, $\cos z = \dfrac{e^{iz} + e^{-iz}}{2}$ and $\sin z = \dfrac{e^{iz} - e^{-iz}}{2i}$. Therefore,

$$\cos^2 z + \sin^2 z = \left(\frac{e^{iz} + e^{-iz}}{2} \right)^2 + \left(\frac{e^{iz} - e^{-iz}}{2i} \right)^2$$

$$= \frac{e^{2iz} + e^{-2iz} + 2e^{iz}e^{-iz}}{4} + \frac{e^{2iz} + e^{-2iz} - 2e^{iz}e^{-iz}}{-4}$$

$$= \frac{1}{4} \cdot (0 + 0 + 2 \cdot e^0 + 2 \cdot e^0) = \frac{4}{4} = 1.$$

$\square$

Exercise 2.28. *If z and w are complex numbers, prove the following:*

$$\cos(z+w) = \cos z \cos w - \sin z \sin w,$$
$$\sin(z+w) = \cos z \sin w + \sin z \cos w.$$

Solution to Exercise 2.28.

It is a simple calculation. Just take into account the definition of cos and sin, and the fact $i^2 = -1$. We have

$$\cos z \cos w - \sin z \sin w = \frac{e^{iz}+e^{-iz}}{2}\cdot\frac{e^{iw}+e^{-iw}}{2} - \frac{e^{iz}-e^{-iz}}{2i}\cdot\frac{e^{iw}-e^{-iw}}{2i}$$

$$= \frac{1}{4}\cdot\left(e^{i(z+w)}+e^{-i(z+w)}+e^{i(z-w)}+e^{i(w-z)}\right.$$

$$\left.+e^{i(z+w)}+e^{-i(z+w)}-e^{i(z-w)}-e^{i(w-z)}\right)$$

$$= \frac{1}{4}\cdot\left(2e^{i(z+w)}+2e^{-i(z+w)}\right)$$

$$= \frac{1}{2}\cdot\left(e^{i(z+w)}+e^{-i(z+w)}\right)$$

$$= \cos(z+w),$$

and

$$\cos z \sin w + \sin z \cos w = \frac{e^{iz}+e^{-iz}}{2}\cdot\frac{e^{iw}-e^{-iw}}{2i} + \frac{e^{iz}-e^{-iz}}{2i}\cdot\frac{e^{iw}+e^{-iw}}{2}$$

$$= \frac{1}{4i}\cdot\left(e^{i(z+w)}-e^{-i(z+w)}+e^{i(w-z)}-e^{-i(z+w)}\right.$$

$$\left.+e^{i(z+w)}+e^{i(z-w)}-e^{i(w-z)}-e^{-i(z+w)}\right)$$

$$= \frac{1}{4i}\cdot\left(2e^{i(z+w)}-2e^{-i(z+w)}\right)$$

$$= \frac{1}{2i}\cdot\left(e^{i(z+w)}-e^{-i(z+w)}\right)$$

$$= \sin(z+w).$$

$\square$

Exercise 2.29. *Define $\tan z = \sin z / \cos z$. Prove that, if $z = x + iy$, then*

$$\tan z = \frac{\sin 2x + i \sinh 2y}{\cos 2x + \cosh 2y}.$$

Solution to Exercise 2.29.

Let $z = x + iy$. Taking into account Exercise 2.28 and the relations $\cos(iy) = \cosh y$ and $\sin(iy) = i \sinh y$, we obtain

$$\cos(x + iy) = \cos x \cosh y - i \sin x \sinh y$$

and

$$\sin(x + iy) = \sin x \cosh y + i \cos x \sinh y.$$

Putting

$$A := \sin(x + iy) \cdot \overline{\cos(x + iy)}$$
$$= (\sin x \cosh y + i \cos x \sinh y) \cdot (\cos x \cosh y + i \sin x \sinh y)$$

and taking into account that $|w|^2 = w \cdot \bar{w}$ for all $w \in \mathbb{C}$, we obtain

$$\tan z = \frac{\sin x \cosh y + i \cos x \sinh y}{\cos x \cosh y - i \sin x \sinh y}$$

$$= \frac{A}{\cos^2 x \cosh^2 y + \sin^2 x \sinh^2 y}$$

$$= \frac{A}{\cos^2 x \cdot \frac{e^{2y} + e^{-2y} + 2}{4} + \sin^2 x \cdot \frac{e^{2y} + e^{-2y} - 2}{4}}$$

$$= \frac{A}{\frac{1}{2} \cdot ((\cos^2 x - \sin^2 x) + (\cos^2 x + \sin^2 x)) \cdot \frac{e^{2y} + e^{-2y}}{2}}$$

$$= \frac{2A}{\cos 2x + \cosh 2y}.$$

We have used the equalities $\cos^2 x + \sin^2 x = 1$ and $\cos^2 x - \sin^2 x = \cos(2x)$.

Therefore, our goal is to prove that $2A = \sin 2x + i \sinh 2y$. Indeed, we have

$$2A = 2 \cdot [\sin x \cos x \cosh^2 y - \sin x \cos x \sinh^2 y$$
$$+ i(\sin^2 x \cosh y \sinh y + \cos^2 \sinh y \cosh y)]$$
$$= 2 \sin x \cos x \cdot (\cosh^2 y - \sinh^2 y) + i \cdot 2 \sinh y \cosh y \cdot (\sin^2 x + \cos^2 x)$$
$$= \sin 2x + i \sinh 2y,$$

where we have used $\cos^2 x + \sin^2 x = 1 = \cosh^2 y - \sinh^2 y$ and $\sin 2x = 2 \sin x \cos x$.

The requested identity has been proved. $\qquad\qquad\qquad\qquad\qquad\qquad\qquad\square$

Exercise 2.30. *Prove that if $z \in \mathbb{C}$ is such that $z + \dfrac{1}{z} = 1$ then $|z| = 1$.*

Solution to Exercise 2.30.

Let $z = re^{it}$ with $r > 0$ and $t \in (-\pi, \pi]$. Then

$$1 = z + \frac{1}{z} = re^{it} + r^{-1}e^{-it} = \left(r + \frac{1}{r}\right)\cos t + i \cdot \left(r - \frac{1}{r}\right)\sin t.$$

Therefore, $\left(r + \frac{1}{r}\right)\cos t = 1$ and $\left(r - \frac{1}{r}\right)\sin t = 0$. If $t = 0$ or π, then $r + \frac{1}{r} = 1$ or -1, which is not possible because $r + \frac{1}{r} > 1$ for all $r > 0$. Hence $t \notin \{0, \pi\}$, so $\sin t \neq 0$, in which case $r - \frac{1}{r} = 0$. Consequently, $r^2 = 1$, and we infer $|z| = r = 1$, as required.

$\square$

Exercise 2.31. *Prove that the distinct points $0, z, w$ of the complex plane $\mathbb{C}$ are the vertices of an equilateral triangle if and only if $z^2 + w^2 = zw$. Infer that $z_1, z_2, z_3 \in \mathbb{C}$ are the vertices of an equilateral triangle if, and only if,*

$$z_1^2 + z_2^2 + z_3^2 = z_1 z_2 + z_1 z_3 + z_2 z_3.$$

Solution to Exercise 2.31.

Since the property of being equilateral is preserved under translation of triangles, the second part can be easily deduced by applying the first one to the points $0 = z_1 - z_1$, $z = z_2 - z_1$, $w = z_3 - z_1$.

Concerning the first part, let $0, z, w$ be three pairwise different points of the complex plane $\mathbb{C}$. By using an homothety followed by a rotation with center at 0, we get that these points are the vertices of an equilateral triangle if and only if $0, 1, s$ are, where $s = \frac{w}{z}$. From the formula and configuration of the nth-roots of a complex number, the last property holds if and only if $s = e^{i\pi/3}$ or $e^{-i\pi/3}$, because the chord joining the points 1 and $e^{i\pi/3}$ (or $e^{-i\pi/3}$) has length $= 1 =$ the radius. Now, observe that

$$1 + s^2 = 1 + e^{\pm i\pi 2/3} = 1 + \cos\frac{2\pi}{3} \pm i\sin\frac{2\pi}{3} = 1 - \frac{1}{2} \pm i\frac{\sqrt{3}}{2} = \frac{1 \pm i\sqrt{3}}{2}$$

$$= \cos\left(\pm\frac{\pi}{3}\right) \pm i\sin\frac{\pi}{3} = e^{\pm i\pi/3} = s.$$

Consequently, $1 + \left(\frac{w}{z}\right)^2 = \frac{w}{z}$, which implies $z^2 + w^2 = zw$, as required.

Conversely, if z, w are nonzero points of $\mathbb{C}$ satisfying $z^2 + w^2 = zw$, then, under the notation above, we get $1 + s^2 = s$, that is, $s^2 - s + 1 = 0$. The solutions of this equation are

$$s = \frac{1 \pm \sqrt{1-4}}{2} = \frac{1 \pm i\sqrt{3}}{2} = e^{\pm i\pi/3}.$$

As said before, this implies that $0, z, w$ are the vertices of an equilateral triangle. $\square$

> **Exercise 2.32.** *Prove that if $z_1 + z_2 + z_3 + z_4 = 0$ and $|z_1| = |z_2| = |z_3| = |z_4|$, then the points z_1, z_2, z_3, and z_4 are either the vertices of a rectangle or identical by pairs.*

Solution to Exercise 2.32.

By dividing each point z_j by their common modulus r, we can assume without loss of generality that $|z_j| = 1$ $(j = 1, 2, 3, 4)$, so that they are points on the unit circle. Let us consider the polynomial

$$P(z) = (z - z_1)(z - z_2)(z - z_3)(z - z_4).$$

By expanding this polynomial, we get

$$P(z) = z^4 + Az^3 + Bz^2 + Cz + D,$$

where $A = -(z_1 + z_2 + z_3 + z_4) = 0$, $B = z_1z_2 + z_1z_3 + z_1z_4 + z_2z_3 + z_2z_4 + z_3z_4$, $C = -(z_1z_2z_3 + z_1z_2z_4 + z_1z_3z_4 + z_2z_3z_4)$ and $D = z_1z_2z_3z_4$.

Since $|z_j| = 1$, we have $1/z_j = \overline{z_j}$ for all $j = 1, 2, 3, 4$. Then

$$C = -z_1z_2z_3z_4 \cdot \left(\frac{1}{z_1} + \frac{1}{z_2} + \frac{1}{z_3} + \frac{1}{z_4} \right) = -z_1z_2z_3z_4 \cdot (\overline{z_1} + \overline{z_2} + \overline{z_3} + \overline{z_4})$$

$$= -z_1z_2z_3z_4 \cdot \overline{(z_1 + z_2 + z_3 + z_4)} = -z_1z_2z_3z_4 \cdot 0 = 0.$$

To summarize, we obtain $P(z) = z^4 + Cz^2 + D$, a polynomial with no odd powers of z. Then $P(z) = P(-z)$ for all $z \in \mathbb{C}$. So, since the zeroes of P are z_1, z_2, z_3, and z_4, we have that

(1) $z_1 = -z_2$ and $z_3 = -z_4$, or

(2) $z_1 = z_2$ and $z_3 = z_4$

(and all the other possibilities, relabeling the z_j's).

Hence, in (2), they are pairwise identical, and in (1) we have that they form a rectangle, since they define a parallelogram with vertices on the unit circle whose diagonal measures are the same: $|z_4 - z_3| = 2 = |z_1 - z_2|$ (draw a picture to convince oneself of this). $\square$

Chapter 3

Sequences of Real Numbers. Convergence

Exercise 3.1. *Use the definition of limit to verify that the limit is the one indicated for the below sequences:*

(I) $\displaystyle\lim_{n\to\infty}\left(2+\frac{1}{n+1}\right)=2,$

(II) $\displaystyle\lim_{n\to\infty}\frac{1}{n^2+1}=0,$

(III) $\displaystyle\lim_{n\to\infty}\frac{2n}{n+1}=2,$

(IV) $\displaystyle\lim_{n\to\infty}\frac{3n+1}{2n+5}=\frac{3}{2},$

(V) $\displaystyle\lim_{n\to\infty}\frac{\alpha+n}{\beta+n}=1,$ with $\alpha,\beta\in\mathbb{R},$

(VI) $\displaystyle\lim_{n\to\infty}\frac{n+3}{n^3+4}=0,$

(VII) $\displaystyle\lim_{n\to\infty}\frac{n^2-1}{2n^2+3}=\frac{1}{2},$

(VIII) $\displaystyle\lim_{n\to\infty}\frac{n^2+2n+2}{n^2+n}=1,$

(IX) $\displaystyle\lim_{n\to\infty}\frac{n}{2n+\sqrt{n+1}}=\frac{1}{2},$

(X) $\displaystyle\lim_{n\to\infty}\frac{1}{n}\sin\frac{n\pi}{4}=0,$

(XI) $\displaystyle\lim_{n\to\infty}\frac{\sin n}{\sqrt{n}}=0,$

(XII) $\displaystyle\lim_{n\to\infty}\left(\sqrt{n^2+n}-n\right)=\frac{1}{2},$

(XIII) $\displaystyle\lim_{n\to\infty}\left(\sqrt[8]{n^2+1}-\sqrt[4]{n+1}\right)=0.$

Solution to Exercise 3.1.

(I). Let $\varepsilon>0$. We need to find $n_0\in\mathbb{N}$ such that $\left|2+\frac{1}{n+1}-2\right|<\varepsilon$ for every $n\geq n_0$. By the Archimedean Property, there exists $n_0\in\mathbb{N}$ such that $\frac{1}{n_0}<\varepsilon$. Hence, for every $n\geq n_0$, we have that

$$\left|2+\frac{1}{n+1}-2\right|=\left|\frac{1}{n+1}\right|=\frac{1}{n+1}<\frac{1}{n}\leq\frac{1}{n_0}<\varepsilon,$$

where we have used the fact that $\frac{1}{n+1}>0$ and $\frac{1}{n+1}<\frac{1}{n}$ for every $n\in\mathbb{N}$.

DOI: 10.1201/9781003400745-3

(II). Let $\varepsilon > 0$. By the Archimedean Property, there exists $n_0 \in \mathbb{N}$ such that $\frac{1}{n_0} < \varepsilon$. Hence, for every $n \geq n_0$, we have that

$$\left| \frac{1}{n^2 + 1} \right| = \frac{1}{n^2 + 1} < \frac{1}{n^2} \leq \frac{1}{n} \leq \frac{1}{n_0} < \varepsilon,$$

where we have used the fact that $\frac{1}{n^2+1} > 0$ and $\frac{1}{n^2+1} < \frac{1}{n^2} \leq \frac{1}{n}$ for every $n \in \mathbb{N}$.

(III). Let $\varepsilon > 0$. By the Archimedean Property, there exists $n_0 \in \mathbb{N}$ such that $\frac{1}{n_0} < \varepsilon$. Hence, for every $n \geq n_0$, we have that

$$\left| \frac{2n}{n+1} - 2 \right| = \left| \frac{1}{n+1} \right| = \frac{1}{n+1} < \frac{1}{n} \leq \frac{1}{n_0} < \varepsilon,$$

where we have used the fact that $\frac{1}{n+1} > 0$ and $\frac{1}{n+1} < \frac{1}{n}$ for every $n \in \mathbb{N}$.

(IV). Let $\varepsilon > 0$. By the Archimedean Property, there exists $n_0 \in \mathbb{N}$ such that $\frac{1}{n_0} < \frac{\varepsilon}{13}$. Hence, for every $n \geq n_0$, we have that

$$\left| \frac{3n+1}{2n+5} - \frac{3}{2} \right| = \left| \frac{6n+2 - (6n+15)}{4n+10} \right| = \left| \frac{-13}{4n+10} \right| = \frac{13}{4n+10}$$
$$< \frac{13}{n} \leq \frac{13}{n_0} < \varepsilon,$$

where we have used the fact that $\frac{-13}{4n+10} < 0$ and $\frac{13}{4n+10} < \frac{13}{4n} < \frac{13}{n}$ for every $n \in \mathbb{N}$.

*The reader may be wondering why we know that n_0 with $\frac{1}{n_0} < \frac{\varepsilon}{13}$ solves the Exercise beforehand. The answer is that we do not know. In a separate sheet of paper, we first make the calculations

$$\left| \frac{3n+1}{2n+5} - \frac{3}{2} \right| = \left| \frac{6n+2 - (6n+15)}{4n+10} \right| = \left| \frac{-13}{4n+10} \right| = \frac{13}{4n+10} < \frac{13}{n},$$

which gives us a clue on what the value of n_0 needs to satisfy by the Archimedean Property.

(V). Let $\varepsilon > 0$. We begin by doing the following calculations on $\left| \frac{\alpha+n}{\beta+n} - 1 \right|$:

$$\left| \frac{\alpha+n}{\beta+n} - 1 \right| = \left| \frac{\alpha-\beta}{\beta+n} \right| = \frac{|\alpha - \beta|}{|\beta + n|}.$$

Observe that, if $\alpha = \beta$, then $\left| \frac{\alpha+n}{\beta+n} - 1 \right| = 0 < \varepsilon$ for every $n \in \mathbb{N}$ (so we can take any $n_0 \in \mathbb{N}$). Thus, assume that $\alpha \neq \beta$. Since β is a fixed number, by the Archimedean Property, there exists $n_1 \in \mathbb{N}$ such that $\beta + n_1 > 0$. Thus, for every $n \geq n_1$, we have that $\beta + n \geq \beta + n_1 > 0$. That is,

$$\left| \frac{\alpha+n}{\beta+n} - 1 \right| = \frac{|\alpha - \beta|}{\beta + n},$$

for every $n \geq n_1$. Once again, by the Archimedean Property, there exists $n_2 \in \mathbb{N}$ with $n_2 \geq n_1$ such that $n_2 > \frac{|\alpha - \beta|}{\varepsilon} - \beta$. Notice that

$$n \geq n_2 > \frac{|\alpha - \beta|}{\varepsilon} - \beta \iff \beta + n \geq \beta + n_2 > \frac{|\alpha - \beta|}{\varepsilon} > 0$$

$$\iff \frac{1}{\beta + n} \leq \frac{1}{\beta + n_2} < \frac{\varepsilon}{|\alpha - \beta|}$$

$$\iff \frac{|\alpha - \beta|}{\beta + n} \leq \frac{|\alpha - \beta|}{\beta + n_2} < \varepsilon. \qquad (3.1)$$

Therefore, taking $n_0 = n_2$ solves the Exercise. Indeed, for every $n \geq n_0$ we have that $n \geq n_1$ and $\frac{|\alpha - \beta|}{\beta + n} < \varepsilon$ by (3.1), which implies that

$$\left| \frac{\alpha + n}{\beta + n} - 1 \right| = \frac{|\alpha - \beta|}{|\beta + n|} = \frac{|\alpha - \beta|}{\beta + n} < \varepsilon.$$

(VI). Let $\varepsilon > 0$. We begin by doing the following calculations on $\left| \frac{n+3}{n^3+4} \right|$:

$$\left| \frac{n+3}{n^3+4} \right| = \frac{n+3}{n^3+4} < \frac{n+3}{n^3},$$

where we have used the fact that $\frac{n+3}{n^3+4} > 0$ and $\frac{n+3}{n^3+4} < \frac{n+3}{n^3}$ for every $n \in \mathbb{N}$. By taking $n \in \mathbb{N}$ so that $n \geq 3$, note that $2n = n + n \geq n + 3$. Hence, for every $n \geq 3$,

$$\left| \frac{n+3}{n^3+4} \right| < \frac{n+3}{n^3} \leq \frac{2n}{n^3} = \frac{2}{n^2}.$$

Now, by the Archimedean Property, there exists $n_1 \in \mathbb{N}$ such that $\frac{1}{n_1} < \frac{\varepsilon}{2}$. Hence, taking $n_0 = \max\{3, n_1\}$ solves the Exercise. Indeed, for every $n \geq n_0$, we have that $n \geq 3$ and $n \geq n_1$. Thus,

$$\left| \frac{n+3}{n^3+4} \right| < \frac{n+3}{n^3} \leq \frac{2}{n^2} \leq \frac{2}{n} \leq \frac{2}{n_0} < \varepsilon.$$

(VII). Let $\varepsilon > 0$. By the Archimedean Property, there exists $n_0 \in \mathbb{N}$ such that $\frac{1}{n_0} < \frac{\varepsilon}{5}$. Hence, for every $n \geq n_0$, we have that

$$\left| \frac{n^2+1}{2n^2+3} - \frac{1}{2} \right| = \left| \frac{2n^2 - 2 - (2n^2+3)}{4n^2+6} \right| = \left| \frac{-5}{4n^2+6} \right| = \frac{5}{4n^2+6}$$

$$\leq \frac{5}{n^2} \leq \frac{5}{n} < \varepsilon,$$

where we have used the fact that $\frac{-5}{4n^2+6} < 0$ and $\frac{5}{4n^2+6} \leq \frac{5}{n^2} \leq \frac{5}{n}$ for every $n \in \mathbb{N}$.

(VIII). Let $\varepsilon > 0$. We need to find $n_0 \in \mathbb{N}$ such that $\left|\frac{n^2+2n+2}{n^2+n} - 1\right| < \varepsilon$ for every $n \geq n_0$. We begin by doing the following calculations on $\left|\frac{n^2+2n+2}{n^2+n} - 1\right|$:

$$\left|\frac{n^2+2n+2}{n^2+n} - 1\right| = \left|\frac{n^2+2n+2-(n^2+n)}{n^2+n}\right| = \left|\frac{n+2}{n^2+n}\right| = \frac{n+2}{n^2+n} \leq \frac{n+2}{n^2}.$$

By taking $n \in \mathbb{N}$ so that $n \geq 2$, note that $2n = n + n \geq n + 2$. Hence, for every $n \geq 2$,

$$\left|\frac{n^2+2n+2}{n^2+n} - 1\right| \leq \frac{n+2}{n^2} \leq \frac{2n}{n^2} = \frac{2}{n}.$$

By the Archimedean Property, there exists $n_1 \in \mathbb{N}$ such that $\frac{1}{n_1} < \frac{\varepsilon}{2}$. Taking $n_0 = \max\{2, n_1\}$ solves the Exercise. Indeed, for every $n \geq n_0$, we have that $n \geq 2$ and $n \geq n_1$, which implies that

$$\left|\frac{n^2+2n+2}{n^2+n} - 1\right| \leq \frac{n+2}{n^2} \leq \frac{2n}{n^2} = \frac{2}{n} < \varepsilon.$$

(IX). Let $\varepsilon > 0$. We begin by doing the following calculations on $\left|\frac{n}{2n+\sqrt{n+1}} - \frac{1}{2}\right|$:

$$\left|\frac{n}{2n+\sqrt{n+1}} - \frac{1}{2}\right| = \left|\frac{2n-(2n-\sqrt{n+1})}{4n+2\sqrt{n+1}}\right| = \left|\frac{\sqrt{n+1}}{4n+2\sqrt{n+1}}\right|$$

$$= \frac{\sqrt{n+1}}{4n+2\sqrt{n+1}} \leq \frac{\sqrt{n+1}}{n},$$

where we have used the fact that $\frac{\sqrt{n+1}}{4n+2\sqrt{n+1}} > 0$ and $\frac{\sqrt{n+1}}{4n+2\sqrt{n+1}} \leq \frac{\sqrt{n+1}}{n}$. By taking $n \in \mathbb{N}$ so that $n \geq 1$, note that $2n = n + n \geq n + 1$. Therefore, if we take the square root we have $\sqrt{2n} \geq \sqrt{n+1}$ for every $n \geq 1$. Hence, for every $n \geq 1$,

$$\left|\frac{n}{2n+\sqrt{n+1}} - \frac{1}{2}\right| \leq \frac{\sqrt{n+1}}{n} \leq \frac{\sqrt{2n}}{n} = \frac{\sqrt{2}}{\sqrt{n}}.$$

By the Archimedean Property, there exists $n_1 \in \mathbb{N}$ such that $\frac{1}{n} < \frac{\varepsilon^2}{2}$. Taking $n_0 = \max\{1, n_1\}$ solves the Exercise. Indeed, for every $n \geq n_0$, we have that $n \geq 1$ and $n \geq n_1$, which implies that

$$\left|\frac{n}{2n+\sqrt{n+1}} - \frac{1}{2}\right| \leq \frac{\sqrt{n+1}}{n} \leq \frac{\sqrt{2n}}{n} = \frac{\sqrt{2}}{\sqrt{n}} < \varepsilon.$$

(X). Let $\varepsilon > 0$. By the Archimedean Property, there exists $n_0 \in \mathbb{N}$ such that $\frac{1}{n_0} < \varepsilon$. Hence, for every $n \geq n_0$, we have that

$$\left|\frac{1}{n}\sin\left(\frac{n\pi}{4}\right)\right| = \left|\frac{1}{n}\right|\left|\sin\left(\frac{n\pi}{4}\right)\right| \leq \left|\frac{1}{n}\right| = \frac{1}{n} < \varepsilon,$$

where we have used the fact that $-1 \leq \sin x \leq 1$ for every $x \in \mathbb{R}$ and $\frac{1}{n} > 0$ for every $n \in \mathbb{N}$.

(XI). Let $\varepsilon > 0$. By the Archimedean Property, there exists $n_0 \in \mathbb{N}$ such that $\frac{1}{n_0} < \varepsilon^2$. Hence, for every $n \geq n_0$, we have that

$$\left| \frac{\sin n}{\sqrt{n}} \right| \leq \frac{1}{\sqrt{n}} < \varepsilon,$$

where we have used the fact that $-1 \leq \sin x \leq 1$ for every $x \in \mathbb{R}$ and $\frac{1}{\sqrt{n}} > 0$ for every $n \in \mathbb{N}$.

(XII). Let $\varepsilon > 0$. We begin by doing the following calculations on $\left| \sqrt{n^2 + n} - n - \frac{1}{2} \right|$:

$$\left| \sqrt{n^2 + n} - n - \frac{1}{2} \right| = \left| \frac{2\sqrt{n^2 + n} - (2n + 1)}{2} \right|.$$

Recall that $x^2 - y^2 = (x - y)(x + y)$ (in fact, this is the Cyclotomic Equation of degree 2 from Exercise 1.6), which implies that $x - y = \frac{x^2 - y^2}{x + y}$ provided that $x + y \neq 0$. If we take $x = 2\sqrt{n^2 + n}$ and $y = n$ we have that $x + y > 0$ and $2\sqrt{n^2 + n} - (2n + 1) = \frac{4n^2 + 4n - (4n^2 + 4n + 1)}{2\sqrt{n^2 + n} + (2n + 1)} = -\frac{1}{2\sqrt{n^2 + n} + 2n + 1}$. By the Archimedean Property, there exists $n_0 \in \mathbb{N}$ such that $\frac{1}{n_0} < \varepsilon$. Hence, for every $n \geq n_0$, we have that

$$\left| \sqrt{n^2 + n} - n - \frac{1}{2} \right| = \left| -\frac{1}{2\sqrt{n^2 + n} + 2n + 1} \right| = \frac{1}{2\sqrt{n^2 + n} + 2n + 1}$$
$$\leq \frac{1}{\sqrt{n^2}} = \frac{1}{|n|} = \frac{1}{n} < \varepsilon,$$

where we have used the fact that $\sqrt{x^2} = |x|$ for every $x \in \mathbb{R}$ and, $\frac{1}{2\sqrt{n^2 + n} + 2n + 1} < 0$ and $\frac{1}{2\sqrt{n^2 + n} + 2n + 1} \leq \frac{1}{\sqrt{n^2}}$ for every $n \in \mathbb{N}$.

(XIII). Let $\varepsilon > 0$. By the Cyclotomic Equation of degree 8 from Exercise 1.6, we have that

$$x^8 - y^8 = (x - y)(x^7 + x^6 y + x^5 y^2 + x^4 y^3 + x^3 y^4 + x^2 y^5 + x y^6 + y^7)$$

which implies that $x - y = \frac{x^8 - y^8}{x^7 + x^6 y + x^5 y^2 + x^4 y^3 + x^3 y^4 + x^2 y^5 + x y^6 + y^7}$ provided that $x + y \neq 0$. If we take $x = \sqrt[8]{n^2 + 1}$ and $y = \sqrt[4]{n + 1}$ we have that $x + y > 0$ and

$$\sqrt[8]{n^2 + 1} - \sqrt[4]{n + 1} = \frac{n^2 + 1 - (n + 1)^2}{x^7 + x^6 y + x^5 y^2 + x^4 y^3 + x^3 y^4 + x^2 y^5 + x y^6 + y^7}$$
$$= \frac{n^2 + 1 - (n^2 + 2n + 1)}{x^7 + x^6 y + x^5 y^2 + x^4 y^3 + x^3 y^4 + x^2 y^5 + x y^6 + y^7}$$
$$= -\frac{2n}{x^7 + x^6 y + x^5 y^2 + x^4 y^3 + x^3 y^4 + x^2 y^5 + x y^6 + y^7}.$$

By the Archimedean Property, there exists $n_0 \in \mathbb{N}$ such that $\frac{1}{n_0} < \left(\frac{\varepsilon}{2}\right)^{4/3}$. Hence, for every $n \geq n_0$, we have that

$$\left|\sqrt[8]{n^2+1} - \sqrt[4]{n+1}\right| = \left|-\frac{2n}{x^7 + x^6 y + x^5 y^2 + x^4 y^3 + x^3 y^4 + x^2 y^5 + xy^6 + y^7}\right|$$

$$= \frac{2n}{x^7 + x^6 y + x^5 y^2 + x^4 y^3 + x^3 y^4 + x^2 y^5 + xy^6 + y^7}$$

$$\leq \frac{2n}{x^7} = \frac{2n}{(n^2+1)^{7/8}} \leq \frac{2n}{(n^2)^{7/8}} = \frac{2n}{n^{7/4}} = \frac{2}{n^{3/4}} < \varepsilon,$$

where we have used the fact that $-\frac{2n}{x^7 + x^6 y + x^5 y^2 + x^4 y^3 + x^3 y^4 + x^2 y^5 + xy^6 + y^7} < 0$ and

$$\frac{2n}{x^7 + x^6 y + x^5 y^2 + x^4 y^3 + x^3 y^4 + x^2 y^5 + xy^6 + y^7} \leq \frac{2n}{x^7}$$

(with $x = \sqrt[8]{n^2+1}$ and $y = \sqrt[4]{n+1}$) for every $n \in \mathbb{N}$. $\qquad \square$

Exercise 3.2. *Use the definition of limit to prove the following statements. Let (s_n) be a sequence and $s \in \mathbb{R}$.*

(I) $\lim\limits_{n \to \infty} s_n = s$ *if, and only if,* $\lim\limits_{n \to \infty} |s_n - s| = 0$.

(II) *Assume that there are a sequence (t_n) that converges to 0 and $n_0 \in \mathbb{N}$ such that $|s_n - s| \leq t_n$ for every $n \geq n_0$, then $\lim\limits_{n \to \infty} s_n = s$.*

Solution to Exercise 3.2.

(I).

($\Leftarrow$) If $\lim\limits_{n \to \infty} s_n = s$, then, by the definition of limit, for every $\varepsilon > 0$, there is $n_0 \in \mathbb{N}$ such that $|s_n - s| < \varepsilon$ for every $n \geq n_0$. In order to have $\lim\limits_{n \to \infty} |s_n - s| = 0$, it is enough to show that $||s_n - s| - 0| < \varepsilon$ for every $n \geq n_0$. But the latter is satisfied since $||s_n - s| - 0| = ||s_n - s|| = |s_n - s| < \varepsilon$.

($\Rightarrow$) If $\lim\limits_{n \to \infty} |s_n - s| = 0$, then by definition of limit we have that for every $\varepsilon > 0$, there exists $n_0 \in \mathbb{N}$ such that $||s_n - s| - 0| < \varepsilon$ for every $n \geq n_0$. But we have already seen that the latter inequality is equivalent to $|s_n - s| < \varepsilon$. Hence $\lim\limits_{n \to \infty} s_n = s$.

(II). Assume that there are a sequence (t_n) such that $\lim\limits_{n\to\infty} t_n = 0$ and $n_0 \in \mathbb{N}$ such that $|s_n - s| \leq t_n$ for every $n \geq n_0$. By definition of limit, for every $\varepsilon > 0$, there exists $n_1 \in \mathbb{N}$ such that $|t_n - 0| < \varepsilon$ for every $n \geq n_1$. If we want to show that (s_n) converges to s, it is enough to find $n_2 \in \mathbb{N}$ such that $|s_n - s| < \varepsilon$ for every $n \geq n_2$. Take $n_2 = \max\{n_0, n_1\}$. For every $n \geq n_2$ we have that $n \geq n_0$ and $n \geq n_1$. Therefore,

$$|s_n - s| \leq t_n = |t_n| = |t_n - 0| < \varepsilon.$$

$\square$

Exercise 3.3. *Of a sequence (s_n) it is known that it is convergent and it alternates positive and negative terms. What does it converge to? Explain your answer and show an example.*

Solution to Exercise 3.3.

We know that (s_n) is a convergent sequence and it alternates positive and negative terms. We will prove that (s_n) converges to 0. To do so, we will prove it by contradiction. We have two cases since the sequence (s_n) is convergent: when the limit is positive and when it is negative.

- Assume that the limit is positive. Then there exists $n_0 \in \mathbb{N}$ such that $s_n > 0$ for every $n \geq n_0$, which is absurd since the sequence alternates positive and negative terms.

- Assume that the limit is negative. Then there is an $n_0 \in \mathbb{N}$ such that $s_n < 0$ for every $n \geq n_0$, which is also absurd.

Hence the limit is 0.

Example: Let $x_n = \dfrac{(-1)^n}{n}$ for every $n \in \mathbb{N}$. Notice that $x_{2k-1} = -\frac{1}{2k-1} < 0$ and $x_{2k} = \frac{1}{2k} > 0$ for every $k \in \mathbb{N}$. Thus, the sequence (x_n) takes negative values when n is odd, and positive values when n is even. Hence the sequence (x_n) alternates positive and negative terms. Moreover,

$$\lim_{k\to\infty} x_{2k-1} = \lim_{k\to\infty} -\frac{1}{2k-1} = 0, \text{ and } \lim_{k\to\infty} x_{2k} = \lim_{k\to\infty} \frac{1}{2k} = 0.$$

Since the subsequence indexed by odd numbers converges to 0 as well as the subsequence indexed by even numbers, we have that (x_n) converges to 0. $\square$

Exercise 3.4. *Calculate, if they exist, the following limits:*

(I) $\displaystyle\lim_{n\to\infty} \frac{1}{n^2+1}$,

(II) $\displaystyle\lim_{n\to\infty} \frac{2n}{n+1}$,

(III) $\displaystyle\lim_{n\to\infty} \frac{n^2-1}{2n^2+3}$,

(IV) $\displaystyle\lim_{n\to\infty} \frac{1}{\sqrt{n+7}}$,

(V) $\displaystyle\lim_{n\to\infty} \frac{\sqrt{n}}{n+1}$,

(VI) $\displaystyle\lim_{n\to\infty} \left(\sqrt{n+1}-\sqrt{n}\right)$,

(VII) $\displaystyle\lim_{n\to\infty} \left(n-\sqrt{n+a}\sqrt{n+b}\right)$, with $a,b \geq -1$,

(VIII) $\displaystyle\lim_{n\to\infty} \frac{\sqrt{n^2+1}}{\sqrt{n+2}}$,

(IX) $\displaystyle\lim_{n\to\infty} \frac{\sqrt{n}}{n^2+1}$,

(X) $\displaystyle\lim_{n\to\infty} \frac{(-1)^n n}{n^2+1}$,

(XI) $\displaystyle\lim_{n\to\infty} \frac{2^n+(-1)^n}{2^{n+1}+(-1)^{n+1}}$,

(XII) $\displaystyle\lim_{n\to\infty} \frac{(-1)^n \sqrt{n}\sin n^n}{n+1}$,

(XIII) $\displaystyle\lim_{n\to\infty} \frac{a^n-b^n}{a^n+b^n}$ with $a,b > 0$,

(XIV) $\displaystyle\lim_{n\to\infty} \frac{b^n}{2^n}$ with $b > 0$,

(XV) $\displaystyle\lim_{n\to\infty} (a^n+b^n)^{1/n}$ with $0 < a < b$,

(XVI) $\displaystyle\lim_{n\to\infty} \frac{n}{b^n}$ with $b > 0$,

(XVII) $\displaystyle\lim_{n\to\infty} \frac{2^{3n}}{3^{2n}}$,

(XVIII) $\displaystyle\lim_{n\to\infty} \frac{2^{n^2}}{n!}$,

(XIX) $\displaystyle\lim_{n\to\infty} n^2 a^n$ with $a \in \mathbb{R}$.

(XX) $\displaystyle\lim_{n\to\infty} \frac{b^n}{n^2}$ with $b \in \mathbb{R}$,

(XXI) $\displaystyle\lim_{n\to\infty} \frac{b^n}{n!}$ with $b \in \mathbb{R}$,

(XXII) $\displaystyle\lim_{n\to\infty} \frac{n!}{n^k}$ with $k \in \mathbb{N}$.

Solution to Exercise 3.4.

(I).

$$\lim_{n\to\infty} \frac{1}{n^2+1} = 0.$$

Indeed, the sequence $\frac{1}{n^2+1}$ is a division of two polynomials where the numerator has degree 0 and n^2+1 has degree 2, that is, the degree of n^2+1 is greater than the degree 0, which implies that the sequence goes to 0 when n tends to ∞.

(II).

$$\lim_{n\to\infty} \frac{2n}{n+1} = 2.$$

Indeed, the sequence $\frac{2n}{n+1}$ is a division of two polynomials of the same degree (which is 1). Hence, the limit is equal to the leading coefficient of the polynomial in the numerator (2) divided by the leading coefficient in the denominator (1).

(III).

$$\lim_{n\to\infty} \frac{n^2-1}{2n^2+3} = \frac{1}{2}.$$

Indeed, the sequence $\frac{n^2-1}{2n^2+3}$ is a division of two polynomials of the same degree (which is 2). Hence, the limit is equal to the leading coefficient of the polynomial in the numerator (1) divided by the leading coefficient in the denominator (2).

(IV).

$$\lim_{n\to\infty} \frac{1}{\sqrt{n+7}} = 0.$$

Indeed, the sequence $\frac{1}{\sqrt{n+7}}$ is a division of sequences, where the sequence in the numerator is the constant sequence 1 and the sequence in the denominator $\sqrt{n+7}$ goes to ∞ when n tends to ∞. Hence, when n tends to ∞, the sequence $\frac{1}{\sqrt{n+7}}$ goes to 0.

(V). For every $n \in \mathbb{N}$, we have that $0 \le \frac{\sqrt{n}}{n+1} \le \frac{\sqrt{n}}{n} = \frac{1}{\sqrt{n}}$. Observe that the sequence $\frac{1}{\sqrt{n}}$ goes to 0 when n tends to ∞ since $\frac{1}{\sqrt{n}}$ is a division of two sequences in which the sequence in the numerator is the constant sequence 1 and the sequence in the denominator goes to ∞ as n goes to ∞. Hence, by the Sandwich Theorem, we have that

$$\lim_{n\to\infty} \frac{\sqrt{n}}{n+1} = 0.$$

(VI).

$$\lim_{n\to\infty} \left(\sqrt{n+1} - \sqrt{n}\right) = \lim_{n\to\infty} \left(\sqrt{n+1} - \sqrt{n}\right) \cdot \frac{\sqrt{n+1} + \sqrt{n}}{\sqrt{n+1} + \sqrt{n}}$$

$$= \lim_{n\to\infty} \frac{1}{\sqrt{n+1} + \sqrt{n}} = 0.$$

If we have a difference of two sequences with squares roots (in this case $\sqrt{n+1}$ and $\sqrt{n}$), it is standard to multiply and divide it by the sum of the square roots. In this case, we end up with a division of two sequences such that the sequence in the numerator is the constant sequence 1 and the sequence in the denominator diverges to ∞ as n goes to ∞.

(VII).

$$\lim_{n\to\infty} \left(n - \sqrt{n+a}\sqrt{n+b}\right) = \lim_{n\to\infty} \left(n - \sqrt{n+a}\sqrt{n+b}\right) \frac{n + \sqrt{n+a}\sqrt{n+b}}{n + \sqrt{n+a}\sqrt{n+b}}$$

$$= \lim_{n\to\infty} \frac{n^2 - (n+a)(n+b)}{n + \sqrt{n+a}\sqrt{n+b}} = \lim_{n\to\infty} \frac{n^2 - [n^2 + (a+b)n + ab]}{n + \sqrt{n+a}\sqrt{n+b}}$$

$$= \lim_{n\to\infty} \frac{-(a+b)n - ab}{n + \sqrt{n+a}\sqrt{n+b}}$$

$$= -(a+b)\lim_{n\to\infty} \frac{n}{n + \sqrt{n+a}\sqrt{n+b}} - ab\lim_{n\to\infty} \frac{1}{n + \sqrt{n+a}\sqrt{n+b}}$$

$$= -(a+b)\lim_{n\to\infty} \frac{n}{n + \sqrt{n+a}\sqrt{n+b}}$$

$$= -(a+b)\lim_{n\to\infty} \frac{n}{n + \sqrt{n^2 + (a+b)n + ab}}$$

$$= -(a+b)\lim_{n\to\infty} \frac{\frac{n}{\sqrt{n^2+(a+b)n+ab}}}{\frac{n}{\sqrt{n^2+(a+b)n+ab}} + 1}$$

$$= -(a+b)\frac{\lim_{n\to\infty} \frac{n}{\sqrt{n^2+(a+b)n+ab}}}{\lim_{n\to\infty} \frac{n}{\sqrt{n^2+(a+b)n+ab}} + 1} = -(a+b)\frac{1}{1+1} = -\frac{a+b}{2},$$

where we have used that $\lim_{n\to\infty} \frac{1}{n+\sqrt{n+a}\sqrt{n+b}} = 0$ (the sequence $\frac{1}{n+\sqrt{n+a}\sqrt{n+b}}$ is a division of two sequences such that the sequence in the numerator is a constant sequence and the sequence in the denominator diverges to ∞ as n goes to ∞) and

$$\lim_{n\to\infty} \frac{n}{\sqrt{n^2 + (a+b)n + ab}} = \lim_{n\to\infty} \frac{n}{\sqrt{n^2}} = \lim_{n\to\infty} \frac{n}{|n|} = \lim_{n\to\infty} \frac{n}{n} = 1$$

(it is straightforward if $a+b = 0$ and $ab = 0$, but otherwise note that $\frac{n^2}{(a+b)n+ab}$ diverges).

(VIII).

$$\lim_{n\to\infty} \frac{\sqrt{n^2 + 1}}{\sqrt{n+2}} = \lim_{n\to\infty} \sqrt{\frac{n^2 + 1}{n + 1}} = \sqrt{\lim_{n\to\infty} \frac{n^2 + 1}{n + 1}} = \infty.$$

Observe that $\frac{n^2+1}{n+1}$ is a division of polynomials in which the degree of the polynomial in the numerator is greater than the degree of the polynomial in the denominator.

(IX). For every $n \in \mathbb{N}$, we have that $0 \le \frac{\sqrt{n}}{n^2+1} \le \frac{\sqrt{n}}{n^2} = \frac{1}{n^{3/2}}$. Since $\lim_{n\to\infty} \frac{1}{n^{3/2}} = 0$, we have by the Sandwich Theorem that

$$\lim_{n\to\infty} \frac{\sqrt{n}}{n^2 + 1} = 0.$$

(X). Let $x_n = \frac{(-1)^n n}{n^2+1}$ for every $n \in \mathbb{N}$. Notice that

$$\lim_{k \to \infty} x_{2k-1} = \lim_{k \to \infty} -\frac{2k-1}{(2k-1)^2+1} = 0, \text{ and } \lim_{k \to \infty} x_{2k} = \lim_{k \to \infty} \frac{2k}{(2k)^2+1} = 0.$$

Since the subsequence indexed by odd numbers converges to 0 as well as the subsequence indexed by even numbers, we have that (x_n) converges to 0, that is,

$$\lim_{n \to \infty} \frac{(-1)^n n}{n^2+1} = 0.$$

(XI). Let $x_n = \frac{2^n+(-1)^n}{2^{n+1}+(-1)^{n+1}}$ for every $n \in \mathbb{N}$. Notice that

$$\lim_{k \to \infty} x_{2k} = \lim_{k \to \infty} \frac{2^{2k}+1}{2^{2k+1}-1} = \lim_{k \to \infty} \frac{2^{2k}+1}{2 \cdot 2^{2k}-1} = \frac{1}{2} \lim_{k \to \infty} \frac{2^{2k}+1}{2^{2k}-\frac{1}{2}} = \frac{1}{2} \cdot 1 = \frac{1}{2}$$

and

$$\lim_{k \to \infty} x_{2k-1} = \lim_{k \to \infty} \frac{2^{2k-1}-1}{2^{2k}+1} = \lim_{k \to \infty} \frac{2^{2k-1}-1}{2 \cdot 2^{2k-1}+1} = \frac{1}{2} \lim_{k \to \infty} \frac{2^{2k-1}-1}{2^{2k-1}+\frac{1}{2}}$$
$$= \frac{1}{2} \cdot 1 = \frac{1}{2}.$$

Since the subsequence indexed by odd numbers converges to $\frac{1}{2}$ as well as the subsequence indexed by even numbers, we have that (x_n) converges to $\frac{1}{2}$, that is,

$$\lim_{n \to \infty} \frac{2^n+(-1)^n}{2^{n+1}+(-1)^{n+1}} = \frac{1}{2}.$$

(XII). For every $n \in \mathbb{N}$, we have that

$$0 \le \left| \frac{(-1)^n \sqrt{n} \sin n^n}{n+1} \right| = \frac{|(-1)^n| \, |\sqrt{n}| \, |\sin n^n|}{|n+1|} \le \frac{\sqrt{n}}{n+1}.$$

Since $\lim_{n \to \infty} \frac{\sqrt{n}}{n+1} = 0$, we have by the Sandwich Theorem and Exercise 3.2 part (I) that

$$\lim_{n \to \infty} \frac{(-1)^n \sqrt{n} \sin n^n}{n+1} = 0.$$

(XIII). We will distinguish three cases: when $a = b$, when $a > b$ and when $a < b$.

Case 1: If $a = b$, then $a^n = b^n$ for every $n \in \mathbb{N}$. Thus,

$$\lim_{n \to \infty} \frac{a^n - b^n}{a^n + b^n} = 0.$$

Case 2: If $a > b$, we will divide the numerator and denominator of $\frac{a^n - b^n}{a^n + b^n}$ by a^n, that is,

$$\lim_{n \to \infty} \frac{a^n - b^n}{a^n + b^n} = \lim_{n \to \infty} \frac{1 - \frac{b^n}{a^n}}{1 + \frac{b^n}{a^n}} = \frac{1 - \lim\limits_{n \to \infty} \frac{b^n}{a^n}}{1 + \lim\limits_{n \to \infty} \frac{b^n}{a^n}} = 1,$$

where we have used that $\lim_{n \to \infty} \frac{b^n}{a^n} = \lim_{n \to \infty} \left(\frac{b}{a}\right)^n = 0$ (note that $0 < \frac{b}{a} < 1$).

Case 3: If $a < b$, we will divide the numerator and denominator of $\frac{a^n - b^n}{a^n + b^n}$ by b^n, that is,

$$\lim_{n \to \infty} \frac{a^n - b^n}{a^n + b^n} = \lim_{n \to \infty} \frac{\frac{a^n}{b^n} - 1}{\frac{a^n}{b^n} + 1} = \frac{\lim\limits_{n \to \infty} \frac{a^n}{b^n} - 1}{\lim\limits_{n \to \infty} \frac{a^n}{b^n} + 1} = -1,$$

where we have used that $\lim_{n \to \infty} \frac{a^n}{b^n} = \lim_{n \to \infty} \left(\frac{a}{b}\right)^n = 0$ (note that $0 < \frac{a}{b} < 1$).

(XIV). We will distinguish three cases: when $b = 2$, when $b > 2$ and when $0 < b < 2$.

- If $b = 2$, then $b^n = 2^n$ for every $n \in \mathbb{N}$. Thus,

$$\lim_{n \to \infty} \frac{b^n}{2^n} = \lim_{n \to \infty} 1 = 1.$$

- If $b > 2$, then $\frac{b}{2} > 1$. Thus,

$$\lim_{n \to \infty} \frac{b^n}{2^n} = \lim_{n \to \infty} \left(\frac{b}{2}\right)^n = \infty.$$

- If $0 < b < 2$, then $0 < \frac{b}{2} < 1$. Thus,

$$\lim_{n \to \infty} \frac{b^n}{2^n} = \lim_{n \to \infty} \left(\frac{b}{2}\right)^n = 0.$$

(XV). We know that $0 < a < b$ which implies that $0 \le a \le b$. Thus, for every $n \in \mathbb{N}$,

$$0 \le a \le b \implies 0 \le a^n \le b^n \implies b^n \le a^n + b^n \le b^n + b^n = 2b^n$$
$$\implies (b^n)^{1/n} \le (a^n + b^n)^{1/n} \le (2b^n)^{1/n}$$
$$\implies b \le (a^n + b^n)^{1/n} \le 2^{1/n} b.$$

Since $\lim_{n \to \infty} 2^{1/n} = 1$, we have by the Sandwich Theorem that

$$\lim_{n \to \infty} (a^n + b^n)^{1/n} = b.$$

(XVI). We will apply the Ratio Test for the case $b \neq 1$:

$$\lim_{n\to\infty} \frac{\frac{n+1}{b^{n+1}}}{\frac{n}{b^n}} = \lim_{n\to\infty} \frac{(n+1)b^n}{nb^{n+1}} = \frac{1}{b} \lim_{n\to\infty} \frac{(n+1)b^n}{nb^n} = \frac{1}{b} \lim_{n\to\infty} \frac{n+1}{n} = \frac{1}{b}.$$

If $b > 1$, then $\frac{1}{b} < 1$, which implies by the Ratio Test that

$$\lim_{n\to\infty} \frac{n}{b^n} = 0.$$

If $b < 1$, then $\frac{1}{b} > 1$, which implies by the Ratio Test that

$$\lim_{n\to\infty} \frac{n}{b^n} = \infty.$$

Finally, assume that $b = 1$, then

$$\lim_{n\to\infty} \frac{n}{b^n} = \lim_{n\to\infty} n = \infty.$$

(XVII).

$$\lim_{n\to\infty} \frac{2^{3n}}{3^{2n}} = \lim_{n\to\infty} \frac{(2^3)^n}{(3^2)^n} = \lim_{n\to\infty} \frac{8^n}{9^n} = \lim_{n\to\infty} \left(\frac{8}{9}\right)^n = 0.$$

(XVIII). We will apply the Ratio Test:

$$\lim_{n\to\infty} \frac{\frac{2^{(n+1)^2}}{(n+1)!}}{\frac{2^{n^2}}{n!}} = \lim_{n\to\infty} \frac{2^{(n+1)^2} n!}{2^{n^2}(n+1)!} = \lim_{n\to\infty} \frac{2^{n^2+2n+1} n!}{2^{n^2}(n+1) \cdot n!}$$

$$= \lim_{n\to\infty} \frac{2^{n^2} 2^{2n+1}}{2^{n^2}(n+1)} = \lim_{n\to\infty} \frac{2^{2n+1}}{n+1} = 2 \lim_{n\to\infty} \frac{4^n}{n+1} = \infty.$$

Thus,

$$\lim_{n\to\infty} \frac{2^{n^2}}{n!} = \infty.$$

*If the reader is not convinced that $\lim_{n\to\infty} \dfrac{4^n}{n+1} = \infty$, it is enough to apply the Ratio Test:

$$= \lim_{n\to\infty} \frac{\frac{4^{n+1}}{n+2}}{\frac{4^n}{n+1}} = \lim_{n\to\infty} \frac{4^{n+1}(n+1)}{4^n(n+2)}$$

$$= 4 \lim_{n\to\infty} \frac{4^n(n+2)}{4^n(n+1)} = 4 \lim_{n\to\infty} \frac{n+2}{n+1} = 4 > 1.$$

(XIX). We will distinguish three cases depending on the value of a:

Case 1: Assume that $|a| < 1$, that is, when $-1 < a < 1$. We will apply the Ratio Test to $|n^2 a^n|$:

$$\lim_{n \to \infty} \frac{|(n+1)^2 a^{n+1}|}{|n^2 a^n|} = \lim_{n \to \infty} \frac{(n^2 + 2n + 1)|a|^n \cdot a}{n^2 |a|^n}$$

$$= |a| \lim_{n \to \infty} \frac{n^2 + 2n + 1}{n^2} = |a| < 1.$$

Hence, by Exercise 3.2 part (I), we have

$$\lim_{n \to \infty} n^2 a^n = 0.$$

Case 2: Assume that $a \geq 1$. Then, for each $n \in \mathbb{N}$, we have $a^n \geq 1$. Hence, for every $n \in \mathbb{N}$, we have that

$$n^2 a^n \geq n^2.$$

Since n^2 diverges to ∞, we have by the Comparison Theorem that $n^2 a^n$ diverges to ∞.

Case 3: Assume that $a \leq -1$. Then,

$$\lim_{k \to \infty} (2k)^2 a^{2k} = 4 \lim_{k \to \infty} k^2 (a^2)^k = \infty$$

and

$$\lim_{k \to \infty} (2k-1)^2 (a)^{2k-1} = -\frac{1}{|a|} \lim_{k \to \infty} (2k-1)^2 (a^2)^k = -\infty,$$

by Case 2. Since the subsequence indexed by even numbers diverges to ∞ and the subsequence indexed by odd numbers diverges to $-\infty$, we have that the limit does not exist.

In summary: If $|a| < 1$, then $\lim_{n \to \infty} n^2 a^n = 0$. If $a \geq 1$, then $n^2 a^n$ diverges to ∞. If $a \leq -1$, then the limit of $n^2 a^n$ does not exist.

(XX). We will distinguish three cases depending on the value of b:

Case 1: Assume that $|b| \leq 1$. For every $n \in \mathbb{N}$, we have that $|b|^n \leq 1$. Thus, for each $n \in \mathbb{N}$,

$$0 \leq \left| \frac{b^n}{n^2} \right| = \frac{|b|^n}{n^2} \leq \frac{1}{n^2}.$$

Since $\lim_{n \to \infty} \frac{1}{n^2} = 0$, we have that $\lim_{n \to \infty} \left| \frac{b^n}{n^2} \right| = 0$ by the Sandwich Theorem. Thus, by Exercise 3.2 part (I), we have that

$$\lim_{n \to \infty} \frac{b^n}{n^2} = 0.$$

Case 2: Assume that $b > 1$. By the Ratio Test applied to $\frac{b^n}{n^2}$, we have

$$\lim_{n \to \infty} \frac{\frac{b^{n+1}}{(n+1)^2}}{\frac{b^n}{n^2}} = \lim_{n \to \infty} \frac{b \cdot b^n n^2}{b^n (n+1)^2} = b \lim_{n \to \infty} \frac{n^2}{n^2 + 2n + 1} = b > 1.$$

Hence,

$$\lim_{n \to \infty} \frac{b^n}{n^2} = \infty.$$

Case 3: Assume that $b < -1$. Then,

$$\lim_{k \to \infty} \frac{b^{2k}}{(2k)^2} = \frac{1}{4} \lim_{k \to \infty} \frac{(b^2)^k}{k^2} = \infty$$

and

$$\lim_{k \to \infty} \frac{b^{2k-1}}{(2k-1)^2} = \frac{1}{b} \lim_{k \to \infty} \frac{(b^2)^k}{(2k-1)^2} = -\infty,$$

by Case 2. Since the subsequence indexed by even numbers diverges to ∞ and the subsequence indexed by odd numbers diverges to $-\infty$, we have that the limit does not exist.

In summary: If $|b| \le 1$, then $\lim_{n \to \infty} \frac{b^n}{n^2} = 0$. If $b > 1$, then $\frac{b^n}{n^2}$ diverges to ∞. If $b < -1$, then the limit of $\frac{b^n}{n^2}$ does not exist.

(XXI). If $b = 0$, then $\frac{b^n}{n!}$ is the constant sequence 0. So, assume that $b \ne 0$. By the Ratio Test applied to $\left| \frac{b^n}{n!} \right|$ we have

$$\lim_{n \to \infty} \left| \frac{\frac{b^{n+1}}{(n+1)!}}{\frac{b^n}{n!}} \right| = \lim_{n \to \infty} \frac{|b||b|^n n!}{|b|^n (n+1) \cdot n!} = |b| \lim_{n \to \infty} \frac{1}{n+1} = 0.$$

Hence, $\lim_{n \to \infty} \left| \frac{b^n}{n!} \right| = 0$, which implies by Exercise 3.2 part (I) that

$$\lim_{n \to \infty} \frac{b^n}{n!} = 0.$$

(XXII). By the Ratio Test we have

$$\lim_{n \to \infty} \frac{\frac{(n+1)!}{(n+1)^k}}{\frac{n!}{n^k}} = \lim_{n \to \infty} \frac{(n+1) \cdot n! n^k}{n!(n+1)^k} = \lim_{n \to \infty} \frac{n^{k+1} + n^k}{(n+1)^k}.$$

Since $n^{k+1} + n^k$ is a polynomial of degree $k + 1$ and $(n+1)^k$ is a polynomial of degree k, we have that $\lim_{n \to \infty} \frac{n^{k+1} + n^k}{(n+1)^k} = \infty$. Hence,

$$\lim_{n \to \infty} \frac{n!}{n^k} = \infty.$$

$\square$

Exercise 3.5. *Study the limit of the sequence*

$$\left(\frac{4+3}{1\cdot 3}, \frac{9-4}{2\cdot 4}, \frac{16+5}{3\cdot 5}, \frac{25-6}{4\cdot 6}, \dots\right).$$

Solution to Exercise 3.5.

First, we will find a general term for the sequence (x_n):

$$\left(\frac{4+3}{1\cdot 3}, \frac{9-4}{2\cdot 4}, \frac{16+5}{3\cdot 5}, \frac{25-6}{4\cdot 6}, \dots\right),$$

that is, we have that

$$x_1 = \frac{4+3}{1\cdot 3} = \frac{(1+1)^2 + (-1)^{1+1}(2+1)}{1(1+2)},$$

$$x_2 = \frac{9-4}{2\cdot 4} = \frac{(2+1)^2 + (-1)^{2+1}(2+2)}{2(2+2)},$$

$$x_3 = \frac{16+5}{3\cdot 5} = \frac{(3+1) + (-1)^{3+1}(3+2)}{3(3+2)},$$

$$x_4 = \frac{25-6}{4\cdot 6} = \frac{(4+1) + (-1)^{4+1}(4+2)}{4(4+2)}.$$

Hence, for every $n \in \mathbb{N}$, we can assume that, for this exercise,

$$x_n = \frac{(n+1)^2 + (-1)^{n+1}(n+2)}{n(n+2)}.$$

Therefore,

$$\lim_{n\to\infty} x_n = \lim_{n\to\infty} \frac{(n+1)^2 + (-1)^{n+1}(n+2)}{n(n+2)}$$

$$= \lim_{n\to\infty} \left(\frac{(n+1)^2}{n(n+2)} + \frac{(-1)^{n+1}(n+2)}{n(n+2)}\right)$$

$$= \lim_{n\to\infty} \left(\frac{n^2 + 2n + 1}{n^2 + 2n} + \frac{(-1)^{n+1}}{n+2}\right) = 1 + 0 = 1.$$

$\square$

Exercise 3.6. *Let (x_n) be a sequence that diverges (to ∞ or $-\infty$). Prove that*

$$\lim_{n\to\infty}\left(1+\frac{1}{x_n}\right)^{x_n}=e.$$

Hint: Work with the definition of limit. To do so, use that

$$\lim_{n\to\infty}\left(1+\frac{1}{n+1}\right)^{n}=\lim_{n\to\infty}\left(1+\frac{1}{n}\right)^{n+1}=e$$

in the case that (x_n) diverges to ∞ and

$$\lim_{n\to\infty}\left(1-\frac{1}{n+1}\right)^{-n}=\lim_{n\to\infty}\left(1-\frac{1}{n}\right)^{-(n+1)}=e$$

in the case that (x_n) diverges to $-\infty$; and keep in mind that

$$[x_n]\le x_n<[x_n]+1.$$

Solution to Exercise 3.6.

Assume that (x_n) diverges to ∞. Since (x_n) diverges to ∞, there exists $n_0\in\infty$ such that $x_n\ge 1$ for every $n\ge n_0$. Recall that $[x_n]\le x_n<[x_n]+1$ for every $n\in\mathbb{N}$. Now, for every $n\ge n_0$, we have that

$$[x_n]\le x_n<[x_n]+1\implies \frac{1}{[x_n]+1}<\frac{1}{x_n}\le\frac{1}{[x_n]}$$

$$\implies 1+\frac{1}{[x_n]+1}<1+\frac{1}{x_n}\le 1+\frac{1}{[x_n]}$$

$$\implies \left(1+\frac{1}{[x_n]+1}\right)^{[x_n]}<\left(1+\frac{1}{x_n}\right)^{x_n}\le\left(1+\frac{1}{[x_n]}\right)^{[x_n]+1}$$

$$\implies \left(1+\frac{1}{[x_n]+1}\right)^{[x_n]}<\left(1+\frac{1}{x_n}\right)^{x_n}\le\left(1+\frac{1}{[x_n]}\right)^{[x_n]+1}.$$

Since (x_n) diverges to ∞, notice that

$$\lim_{n\to\infty}\left(1+\frac{1}{[x_n]+1}\right)^{[x_n]}=e=\lim_{n\to\infty}\left(1+\frac{1}{[x_n]}\right)^{[x_n]+1}.$$

Hence, by the Sandwich Principle, we have that

$$\lim_{n\to\infty}\left(1+\frac{1}{x_n}\right)^{x_n}=e.$$

If (x_n) diverges to $-\infty$, then simply use a similar approach to the case when (x_n) diverges to ∞. $\qquad\square$

Exercise 3.7. *Find the value of*

$$\lim_{n\to\infty} \frac{1}{n}\left[\left(a+\frac{1}{n}\right)^2 + \left(a+\frac{2}{n}\right)^2 + \cdots + \left(a+\frac{n-1}{n}\right)^2\right]$$

with $a \in \mathbb{R}$.

Solution to Exercise 3.7.

$$\lim_{n\to\infty} \frac{1}{n}\left[\left(a+\frac{1}{n}\right)^2 + \left(a+\frac{2}{n}\right)^2 + \cdots + \left(a+\frac{n-1}{n}\right)^2\right]$$

$$= \lim_{n\to\infty} \frac{1}{n}\cdot\frac{(an+1)^2+(an+2)^2+\cdots+(an+(n-1))^2}{n^2}$$

$$= \lim_{n\to\infty} \frac{1}{n^3}\sum_{k=1}^{n-1}(an+k)^2 = \lim_{n\to\infty} \frac{1}{n^3}\left[\sum_{k=1}^{n-1}(an)^2 + \sum_{k=1}^{n-1}k^2 + \sum_{k=1}^{n-1}2ank\right].$$

By Exercise 1.4 parts (II) and (III) we have that

$$\lim_{n\to\infty} \frac{1}{n}\left[\left(a+\frac{1}{n}\right)^2 + \left(a+\frac{2}{n}\right)^2 + \cdots + \left(a+\frac{n-1}{n}\right)^2\right]$$

$$= \lim_{n\to\infty}\left(\frac{1}{n^3}(n-1)(an)^2 + \frac{1}{n^3}\cdot\frac{(n-1)n(2n-1)}{6} + \frac{1}{n^3}\cdot\frac{2an(n-1)n}{2}\right)$$

$$= \lim_{n\to\infty}\left(\frac{a^2(n-1)}{n} + \frac{2n^2-3n+1}{6n^2} + \frac{a(n-1)}{n}\right)$$

$$= a^2 + \frac{2}{6} + a = a^2 + a + \frac{1}{3}.$$

$\square$

Exercise 3.8. *Find the limit of the sequences with the following general terms:*

(I) $\dfrac{\left(1^2+2^2+\cdots+n^2\right)^2}{(1+2+\cdots+n)^3}$,

(II) $\dfrac{1+2^2+3^2+\cdots+n^2}{1+n+n^2+n^3}$,

(III) $\dfrac{3\sqrt[3]{n}-4\sqrt[5]{n^2}}{\sqrt[3]{n}-3\left(4-\sqrt[5]{n}\right)}$,

(IV) $\dfrac{\sqrt{n}}{\sqrt{n + \sqrt{n + \sqrt{n}}}}$,

(V) $\sqrt{4n^2 - 1} - (2n - 1)$,

(VI) $\sqrt[3]{n^3 + an^2} - \sqrt[3]{n^3 - an^2}$,

(VII) $n\left(\sqrt[3]{1 + \dfrac{a}{n}} - 1\right)$,

(VIII) $\dfrac{\sqrt{n^2 + \sqrt{n}} - \sqrt{n^2 - \sqrt{n}}}{n\left(\sqrt[3]{n^3 + \sqrt{n}} - \sqrt[3]{n^3 - \sqrt{n}}\right)}$,

(IX) $\sqrt{n^2 + n + 1} - \sqrt{3n^2 - 1} - 3n$,

(X) $\sqrt{9n^2 - n} - \sqrt[3]{27n^3 - 5n^2}$,

(XI) $(4n + 3) \log \dfrac{n + 1}{n - 2}$,

(XII) $\left(\dfrac{n^2 + 3n - 2}{n^2 + n}\right)^{\frac{n^3 + 2}{2n^2 + 1}}$,

(XIII) $\left(\dfrac{n + 1}{n - 1}\right)^{\frac{n^2 + 2}{n - 3}}$,

(XIV) $\left(1 + \log \dfrac{3n^2 + 2n + 1}{3n^2 + 5n}\right)^{4n + 1}$,

(XV) $\left(\dfrac{1}{n}\right)^{\frac{1}{\log(3/n)}}$,

(XVI) $(2 + 3n^4)^{\frac{1}{3 + 2\log(n+1)}}$,

(XVII) $\left(\dfrac{\log(n^2 + 1)}{\log(n^2 - 1)}\right)^{n^2 \log n}$,

(XVIII) $\sqrt[3]{(n + a)(n + b)(n + c)} - n$
with $a, b, c \in \mathbb{R}$,

(XIX) $\dfrac{2^{2n}(n!)^2\sqrt{n}}{(2n+1)!},$

(XX) $n\left(\dfrac{2\cdot 4\cdot 6\cdot\cdots\cdot(2n-2)}{1\cdot 3\cdot 5\cdot\cdots\cdot(2n-1)}\right)^2,$

(XXI) $\dfrac{\sqrt[n]{(n+1)(n+2)\cdots(n+n)}}{n},$

(XXII) $\dfrac{1}{n}\sqrt[n]{(3n+1)(3n+2)\cdot\cdots\cdot(4n)},$

(XXIII) $\log n - n,$

(XXIV) $\dfrac{1^p+2^p+\cdots+n^p}{n^p}-\dfrac{n}{p+1}$
    with $p\in\mathbb{N},$

(XXV) $\dfrac{\cos 1+\cos\frac{1}{\sqrt{2}}+\cdots+\cos\frac{1}{\sqrt{n}}-n}{\log(n^3+1)},$

(XXVI) $\dfrac{\sqrt{1\cdot 2\cdot 3}+\sqrt{2\cdot 3\cdot 4}+\cdots+\sqrt{n(n+1)(n+2)}}{n^2\sqrt{n}},$

(XXVII) $\dfrac{\log 1-\log 2+\log 3-\cdots+\log(2n-1)-\log(2n)}{\log n},$

(XXVIII) $\dfrac{\sqrt{2!}\tan\frac{1}{2}+\sqrt[3]{3!}\tan\frac{1}{3}+\cdots+\sqrt[n]{n!}\tan\frac{1}{n}}{\sqrt{n^2+1}},$

(XXIX) $\sqrt[n]{\sqrt[n]{\binom{n}{1}\binom{n}{2}\cdot\cdots\cdot\binom{n}{n}}},$

(XXX) $\dfrac{1}{n^2}+\dfrac{1}{(n+1)^2}+\cdots+\dfrac{1}{(n+n)^2},$

(XXXI) $(2^n+3^n)^{1/n},$

(XXXII) $\sin\dfrac{n^2\pi}{2(n^2+1)}+\sin\dfrac{n^2\pi}{2(n^2+2)}+\cdots+\sin\dfrac{n^2\pi}{2(n^2+n)}.$

Solution to Exercise 3.8.

(I).

$$\lim_{n\to\infty} \frac{(1^2 + 2^2 + \cdots + n^2)^2}{(1 + 2 + \cdots + n)^3} = \lim_{n\to\infty} \frac{\left(\displaystyle\sum_{k=1}^{n} k^2\right)^2}{\left(\displaystyle\sum_{k=1}^{n} k\right)^3}.$$

By Exercise 1.4 parts (II) and (III) we have that

$$\lim_{n\to\infty} \frac{(1^2 + 2^2 + \cdots + n^2)^2}{(1 + 2 + \cdots + n)^3} = \lim_{n\to\infty} \frac{\left(\frac{n(n+1)(2n+1)}{6}\right)^2}{\left(\frac{n(n+1)}{2}\right)^3}$$

$$= \lim_{n\to\infty} \frac{\frac{n^2(n+1)^2(2n+1)^2}{36}}{\frac{n^3(n+1)^3}{8}} = \lim_{n\to\infty} \frac{8n^2(n+1)^2(2n+1)^2}{36n^3(n+1)^3}$$

$$= \lim_{n\to\infty} \frac{2(2n+1)^2}{9n(n+1)} = \lim_{n\to\infty} \frac{8n^2 + 8n + 2}{9n^2 + 9n} = \frac{8}{9}.$$

(II). By Exercise 1.4 part (III),

$$\lim_{n\to\infty} \frac{1 + 2^2 + 3^2 + \cdots + n^2}{1 + n + n^2 + n^3} = \lim_{n\to\infty} \frac{\displaystyle\sum_{k=1}^{n} k^2}{1 + n + n^2 + n^3}$$

$$= \frac{\frac{n(n+1)(2n+1)}{6}}{1 + n + n^2 + n^3}$$

$$= \lim_{n\to\infty} \frac{2n^3 + 3n^2 + n}{6n^3 + 6n^2 + 6n + 6}$$

$$= \frac{2}{6}$$

$$= \frac{1}{3}.$$

(III).

$$\lim_{n\to\infty} \frac{3\sqrt[3]{n} - 4\sqrt[5]{n^2}}{\sqrt[3]{n} - 3(4 - \sqrt[5]{n})} = \lim_{n\to\infty} \frac{3n^{1/3} - 4n^{2/5}}{(n-3)^{1/3}(4 - n^{1/5})}$$

$$= \lim_{n\to\infty} \frac{3n^{1/3} - 4n^{2/5}}{4(n-3)^{1/3} - (n-3)^{1/3}n^{1/5}}$$

$$= \lim_{n\to\infty} \frac{3n^{1/3} - 4n^{2/5}}{4(n-3)^{1/3} - [(n-3)^5 n^3]^{1/15}}$$

$$= \lim_{n\to\infty} \frac{3n^{1/3} - 4n^{2/5}}{4n^{1/3}\left(1 - \frac{3}{n}\right)^{1/3} - \left[n^5\left(1 - \frac{3}{n}\right)^5 n^3\right]^{1/15}}$$

$$= \lim_{n\to\infty} \frac{3n^{1/3} - 4n^{2/5}}{4n^{1/3}\left(1 - \frac{3}{n}\right)^{1/3} - n^{8/15}\left(1 - \frac{3}{n}\right)^{1/3}} = 0,$$

where in the last equality we have used the fact that $\dfrac{8}{15} > \dfrac{2}{5} > \dfrac{1}{3}$.

(IV).

$$\lim_{n\to\infty} \frac{\sqrt{n}}{\sqrt{n + \sqrt{n + \sqrt{n}}}} = \lim_{n\to\infty} \frac{n^{1/2}}{\left[n + (n + n^{1/2})^{1/2}\right]^{1/2}}$$

$$= \lim_{n\to\infty} \frac{n^{1/2}}{\left[n + \left(n\left[1 + \frac{n^{1/2}}{n}\right]\right)^{1/2}\right]^{1/2}}$$

$$= \lim_{n\to\infty} \frac{n^{1/2}}{\left[n + n^{1/2}\left(1 + \frac{n^{1/2}}{n}\right)^{1/2}\right]^{1/2}}$$

$$= \lim_{n\to\infty} \frac{n^{1/2}}{\left[n\left(1 + \frac{n^{1/2}}{n}\left(1 + \frac{n^{1/2}}{n}\right)^{1/2}\right)\right]^{1/2}}$$

$$= \lim_{n\to\infty} \frac{n^{1/2}}{n^{1/2}\left[1 + \frac{n^{1/2}}{n}\left(1 + \frac{n^{1/2}}{n}\right)^{1/2}\right]^{1/2}}$$

$$= \lim_{n\to\infty} \frac{1}{\left[1 + \frac{n^{1/2}}{n}\left(1 + \frac{n^{1/2}}{n}\right)^{1/2}\right]^{1/2}} = 1.$$

(V). To obtain the limit we will use the Cyclotomic Equation (Exercise 1.6) with $n = 2$, $x = \sqrt{4n^2 - 1}$ and $y = 2n - 1$. Notice that $x^2 - y^2 = (x - y)(x + y)$ and, if $x + y \neq 0$, we have $x - y = \frac{x^2 - y^2}{x + y}$. Hence, $\sqrt{4n^2 - 1} - (2n - 1) =$

$\frac{\sqrt{4n^2-1}^2-(2n-1)^2}{\sqrt{4n^2-1}+(2n-1)}$. Therefore,

$$\lim_{n\to\infty}\left[\sqrt{4n^2-1}-(2n-1)\right]=\lim_{n\to\infty}\frac{4n^2-1-(4n^2-4n+1)}{\sqrt{4n^2-1}+2n-1}$$

$$=\lim_{n\to\infty}\frac{4n-2}{\left[n^2\left(4-\frac{1}{n^2}\right)\right]^{1/2}+2n-1}=\lim_{n\to\infty}\frac{4n-2}{n\left(4-\frac{1}{n^2}\right)^{1/2}+2n-1}$$

$$=\lim_{n\to\infty}\frac{n\left(4-\frac{2}{n}\right)}{n\left[\left(4-\frac{1}{n^2}\right)^{1/2}+2-\frac{1}{n}\right]}$$

$$=\lim_{n\to\infty}\frac{4-\frac{2}{n}}{\left(4-\frac{1}{n^2}\right)^{1/2}+2-\frac{1}{n}}=1.$$

(VI). To obtain the limit we will use the Cyclotomic Equation (Exercise 1.6) with $n=3$, $x=\sqrt[3]{n^3+an^2}$ and $y=\sqrt[3]{n^3-an^2}$. Notice that $x^3-y^3=(x-y)(x^2+xy+y^2)$ and, if $x^2+xy+y^2\neq 0$, we have $x-y=\frac{x^3-y^3}{x^2+xy+y^2}$. Hence,

$$\sqrt[3]{n^3+an^2}-\sqrt[3]{n^3-an^2}$$

$$=\frac{\left(\sqrt[3]{n^3+an^2}\right)^3-\left(\sqrt[3]{n^3-an^2}\right)^3}{\left(\sqrt[3]{n^3+an^2}\right)^2+\sqrt[3]{n^3+an^2}\sqrt[3]{n^3-an^2}+\left(\sqrt[3]{n^3-an^2}\right)^2}.$$

Therefore,

$$\lim_{n\to\infty}\left(\sqrt[3]{n^3+an^2}-\sqrt[3]{n^3-an^2}\right)$$

$$=\lim_{n\to\infty}\frac{n^3+an^2-(n^3-an^2)}{\left(\sqrt[3]{n^3+an^2}\right)^2+\sqrt[3]{n^3+an^2}\sqrt[3]{n^3-an^2}+\left(\sqrt[3]{n^3-an^2}\right)^2}$$

$$=\lim_{n\to\infty}\frac{n^3+an^2-n^3+an^2}{(n^3+an^2)^{2/3}+(n^3+an^2)^{1/3}(n^3-an^2)^{1/3}+(n^3-an^2)^{2/3}}$$

$$=\lim_{n\to\infty}\frac{2an^2}{\left[n^3u_n\right]^{2/3}+\left[n^3u_n\right]^{1/3}\left[n^3v_n\right]^{1/3}+\left[n^3v_n\right]^{2/3}},$$

with $u_n=1+\frac{an^2}{n^3}=1+\frac{a}{n}$ and $v_n=1-\frac{an^2}{n^3}=1-\frac{a}{n}$. Hence,

$$\lim_{n\to\infty}\left(\sqrt[3]{n^3+an^2}-\sqrt[3]{n^3-an^2}\right)=\lim_{n\to\infty}\frac{2a}{u_n^{2/3}+u_n^{1/3}v_n^{1/3}+v_n^{2/3}}=\frac{2}{3}a,$$

where in the last equality we have used that $\lim_{n\to\infty}u_n=\lim_{n\to\infty}v_n=1$.

(VII). We will use the equivalence $(1+s_n)^\alpha-1\sim\alpha s_n$ provided that s_n converges to 0. In our case $s_n=\frac{a}{n}$, which converges to 0. Thus,

$$\lim_{n\to\infty}n\left(\sqrt[3]{1+\frac{a}{n}}-1\right)=\lim_{n\to\infty}n\left[\left(1+\frac{a}{n}\right)^{1/3}-1\right]=\lim_{n\to\infty}n\cdot\frac{1}{3}\frac{a}{n}=\frac{a}{3}.$$

*Another way of finding the limit of $n\left(\sqrt[3]{1+\frac{a}{n}}-1\right)$ (which involves more calculations) consists of using the Cyclotomic Equation (Exercise 1.6). First,

$$\lim_{n\to\infty} n\left(\sqrt[3]{1+\frac{a}{n}}-1\right)=\lim_{n\to\infty}\left(\sqrt[3]{n^3\left(1+\frac{a}{n}\right)}-n\right).$$

For $n=3$, $x=\sqrt[3]{n^3\left(1+\frac{a}{n}\right)}$ and $y=n$ in Cyclotomic Equation. Notice that $x^3-y^3=(x-y)(x^2+xy+y^2)$ and, if $x^2+xy+y^2\neq 0$, we have $x-y=\frac{x^3-y^3}{x^2+xy+y^2}$. Hence, $\sqrt[3]{n^3\left(1+\frac{a}{n}\right)}-n=\dfrac{n^3\left(1+\frac{a}{n}\right)-n^3}{\left(\sqrt[3]{n^3\left(1+\frac{a}{n}\right)}\right)^2+n\left(\sqrt[3]{n^3\left(1+\frac{a}{n}\right)}\right)+n^2}$. Therefore,

$$\lim_{n\to\infty} n\left(\sqrt[3]{1+\frac{a}{n}}-1\right)=\lim_{n\to\infty}\frac{n^3\left(1+\frac{a}{n}-1\right)}{\left[n\left(1+\frac{a}{n}\right)^{1/3}\right]^2+n\left[n\left(1+\frac{a}{n}\right)^{1/3}\right]+n^2}$$

$$=\lim_{n\to\infty}\frac{an^2}{n^2\left(1+\frac{a}{n}\right)^{2/3}+n^2\left(1+\frac{a}{n}\right)^{2/3}+n^2}$$

$$=\lim_{n\to\infty}\frac{a}{\left(1+\frac{a}{n}\right)^{2/3}+\left(1+\frac{a}{n}\right)^{2/3}+1}=\frac{a}{3}.$$

(VIII). To obtain the limit we will use the Cyclotomic Equation (Exercise 1.6) in the numerator and in the denominator. For the numerator, we will take $n=2$, $x=\sqrt{n^2+\sqrt{n}}$ and $y=\sqrt{n^2-\sqrt{n}}$. Notice that $x^2-y^2=(x-y)(x+y)$ and, if $x+y\neq$, we have $x-y=\frac{x^2-y^2}{x+y}$. Hence, $\sqrt{n^2+\sqrt{n}}-\sqrt{n^2-\sqrt{n}}=\dfrac{n^2+\sqrt{n}-(n^2-\sqrt{n})}{\sqrt{n^2+\sqrt{n}}+\sqrt{n^2-\sqrt{n}}}$. For the denominator, we will take $n=3$, $x=\sqrt[3]{n^3+\sqrt{n}}$ and $y=\sqrt[3]{n^3-\sqrt{n}}$. Notice that $x^3-y^3=(x-y)(x^2+xy+y^2)$ and, if $x^2+xy+y^2\neq 0$, we have $x-y=\frac{x^3-y^3}{x^2+xy+y^2}$. Hence,

$$\sqrt[3]{n^3+\sqrt{n}}-\sqrt[3]{n^3-\sqrt{n}}$$

$$=\frac{n^3+\sqrt{n}-\left(n^3-\sqrt{n}\right)}{\left(\sqrt[3]{n^3+\sqrt{n}}\right)^2+\sqrt[3]{n^3+\sqrt{n}}\,\sqrt[3]{n^3-\sqrt{n}}+\left(\sqrt[3]{n^3-\sqrt{n}}\right)^2}.$$

Therefore,

$$\lim_{n\to\infty}\frac{\dfrac{2\sqrt{n}}{\left(n^2+\sqrt{n}\right)^{1/2}+\left(n^2-\sqrt{n}\right)^{1/2}}}{\dfrac{2n\sqrt{n}}{\left(n^3+\sqrt{n}\right)^{2/3}+\left(n^3+\sqrt{n}\right)^{1/3}\left(n^3-\sqrt{n}\right)^{1/3}+\left(n^3+\sqrt{n}\right)^{2/3}}}$$

$$=\lim_{n\to\infty}\frac{\left[n^3\left(1+\frac{\sqrt{n}}{n^3}\right)\right]^{2/3}+\left[n^3\left(1+\frac{\sqrt{n}}{n^3}\right)\right]^{1/3}\left[n^3\left(1-\frac{\sqrt{n}}{n^3}\right)\right]^{1/3}}{n\left(\left[n^2\left(1+\frac{\sqrt{n}}{n^2}\right)\right]^{1/2}+\left[n^2\left(1-\frac{\sqrt{n}}{n^2}\right)\right]^{1/2}\right)}$$

$$+ \frac{\left[n^3\left(1 - \frac{\sqrt{n}}{n^3}\right)\right]^{2/3}}{n\left(\left[n^2\left(1 + \frac{\sqrt{n}}{n^2}\right)\right]^{1/2} + \left[n^2\left(1 - \frac{\sqrt{n}}{n^2}\right)\right]^{1/2}\right)}$$

$$= \lim_{n\to\infty} \frac{n^2\left(1 + \frac{\sqrt{n}}{n^3}\right)^{2/3} + n^2\left(1 + \frac{\sqrt{n}}{n^3}\right)^{1/3}\left(1 - \frac{\sqrt{n}}{n^3}\right)^{1/3}}{n\left[n\left(1 + \frac{\sqrt{n}}{n^2}\right)^{1/2} + n\left(1 - \frac{\sqrt{n}}{n^2}\right)^{1/2}\right]}$$

$$+ \frac{n^2\left(1 - \frac{\sqrt{n}}{n^3}\right)^{2/3}}{n\left[n\left(1 + \frac{\sqrt{n}}{n^2}\right)^{1/2} + n\left(1 - \frac{\sqrt{n}}{n^2}\right)^{1/2}\right]}$$

$$= \lim_{n\to\infty} \frac{n^2\left[\left(1 + \frac{\sqrt{n}}{n^3}\right)^{2/3} + \left(1 + \frac{\sqrt{n}}{n^3}\right)^{1/3}\left(1 - \frac{\sqrt{n}}{n^3}\right)^{1/3} + \left(1 - \frac{\sqrt{n}}{n^3}\right)^{2/3}\right]}{n^2\left[\left(1 + \frac{\sqrt{n}}{n^2}\right)^{1/2} + \left(1 - \frac{\sqrt{n}}{n^2}\right)^{1/2}\right]}$$

$$= \lim_{n\to\infty} \frac{\left(1 + \frac{\sqrt{n}}{n^3}\right)^{2/3} + \left(1 + \frac{\sqrt{n}}{n^3}\right)^{1/3}\left(1 - \frac{\sqrt{n}}{n^3}\right)^{1/3} + \left(1 - \frac{\sqrt{n}}{n^3}\right)^{2/3}}{\left(1 + \frac{\sqrt{n}}{n^2}\right)^{1/2} + \left(1 - \frac{\sqrt{n}}{n^2}\right)^{1/2}} = \frac{3}{2}.$$

(IX). To obtain the limit we will use the Cyclotomic Equation (Exercises 1.6) with $n = 2$, $x = \sqrt{n^2 + n + 1}$ and $y = \sqrt{3n^2 - 1}$. Notice that $x^2 - y^2 = (x - y)(x^2 + xy + y^2)$ and, if $x^2 + xy + y^2 \neq 0$, we have $x - y = \frac{x^2 - y^2}{x^2 + xy + y^2}$. Hence, $\sqrt{n^2 + n + 1} - \sqrt{3n^2 - 1} = \frac{\left(\sqrt{n^2 + n + 1}\right)^2 - \left(\sqrt{3n^2 - 1}\right)^2}{\sqrt{n^2 + n + 1} + \sqrt{3n^2 - 1}}$. Therefore,

$$\lim_{n\to\infty} \left[\sqrt{n^2 + n + 1} - \sqrt{3n^2 - 1} - 3n\right]$$

$$= \lim_{n\to\infty} \left(\frac{-2n^2 + 2n + 1}{\left[n^2\left(1 + \frac{1}{n} + \frac{1}{n^2}\right)\right]^{1/2} + \left[n^2\left(3 - \frac{1}{n^2}\right)\right]^{1/2}} - 3n\right)$$

$$= \lim_{n\to\infty} \left(\frac{-2n^2 + 2n + 1}{n\left(1 + \frac{1}{n} + \frac{1}{n^2}\right)^{1/2} + n\left(3 - \frac{1}{n^2}\right)^{1/2}} - 3n\right)$$

$$= \lim_{n\to\infty} \frac{-2n^2 + 2n + 1 - 3n^2\left[\left(1 + \frac{1}{n} + \frac{1}{n^2}\right)^{1/2} + \left(3 - \frac{1}{n^2}\right)^{1/2}\right]}{n\left[\left(1 + \frac{1}{n} + \frac{1}{n^2}\right)^{1/2} + \left(3 - \frac{1}{n^2}\right)^{1/2}\right]}$$

$$= \lim_{n\to\infty} \frac{-2n + 2 + \frac{1}{n} - 3n\left[\left(1 + \frac{1}{n} + \frac{1}{n^2}\right)^{1/2} + \left(3 - \frac{1}{n^2}\right)^{1/2}\right]}{\left(1 + \frac{1}{n} + \frac{1}{n^2}\right)^{1/2} + \left(3 - \frac{1}{n^2}\right)^{1/2}} = -\infty.$$

(X). To obtain the limit we will use the Cyclotomic Equation (Exercises 1.6) with $n = 6$, $x = \sqrt{9n^2 - n}$ and $y = \sqrt[3]{27n^3 - 5n^2}$. Notice that $x^6 - y^6 = (x - y)(x^5 + x^4y + x^3y^2 + x^2y^3 + xy^4 + y^5)$ and, if $xx^5 + x^4y + x^3y^2 + x^2y^3 + xy^4 + y^5 \neq 0$, we have

$$x - y = \frac{x^6 - y^6}{x^5 + x^4y + x^3y^2 + x^2y^3 + xy^4 + y^5}.$$

Therefore,

$$\sqrt{9n^2 - n} - \sqrt[3]{27n^3 - 5n^2} = \frac{\left(\sqrt{9n^2 - n}\right)^6 - \left(\sqrt[3]{27n^3 - 5n^2}\right)^6}{x^5 + x^4y + x^3y^2 + x^2y^3 + xy^4 + y^5}.$$

Hence, by taking $u_n = 9 - \frac{1}{n}$ and $v_n = 27 - \frac{5}{n}$, we have that

$$x^5 = \left(n^2 u_n\right)^{5/2} = n^5 u_n^{5/2},$$
$$x^4y = \left(n^2 u_n\right)^2 \left(n^3 v_n\right)^{1/3} = n^5 u_n^2 v_n^{1/3},$$
$$x^3y^2 = \left(n^2 u_n\right)^{3/2} \left(n^3 v_n\right)^{2/3} = n^5 u_n^{3/2} v_n^{2/3},$$
$$x^2y^3 = \left(n^2 u_n\right) \left(n^3 v_n\right) = n^5 u_n v_n,$$
$$xy^4 = \left(n^2 u_n\right)^{1/2} \left(n^2 v_n\right)^{4/3} = n^5 u_n^{1/2} v_n^{4/3},$$
$$y^5 = \left(n^3 v_n\right)^{5/3} = n^5 v_n^{5/3}.$$

Hence,

$$\lim_{n \to \infty} \left(\sqrt{9n^2 - n} - \sqrt[3]{27n^3 - 5n^2}\right)$$
$$= \lim_{n \to \infty} \frac{(9n^2)^3 - n^3 - 3n(9n^2)^2 + 3 \cdot 9n^2 n^2}{x^5 + x^4y + x^3y^2 + x^2y^3 + xy^4 + y^5}$$
$$\frac{-(27n^3)^2 - (5n^2)^2 + 2 \cdot 27 \cdot 5n^3 n^2}{x^5 + x^4y + x^3y^2 + x^2y^3 + xy^4 + y^5}$$
$$= \lim_{n \to \infty} \frac{2 \cdot 5 \cdot 3^3 - 3^5 + (3^3 - 5^2)\frac{1}{n} - \frac{1}{n^2}}{u_n^{5/2} + u_n^2 v_n^{1/3} + u_n^{3/2} v_n^{2/3} + u_n v_n + u_n^{1/2} v_n^{4/3} + v_n^{5/3}},$$

where $u_n = 9 - \frac{1}{n}$ and $v_n = 27 - \frac{5}{n}$. Since $\lim_{n \to \infty} u_n = 9$ and $\lim_{n \to \infty} v_n = 27$, we have that

$$\lim_{n \to \infty} \left(\sqrt{9n^2 - n} - \sqrt[3]{27n^3 - 5n^2}\right)$$
$$= \frac{2 \cdot 5 \cdot 3^3 - 3^5}{3^5 + 9^2 \cdot 3 + 3^3 \cdot 3^2 + 3^2 \cdot 3^3 + 3 \cdot 3^4 + 3^5} = \frac{1}{54}.$$

(XI).

$$\lim_{n\to\infty} (4n+3) \log \frac{n+1}{n-2} = \lim_{n\to\infty} \log \left(\frac{n+1}{n-2}\right)^{4n+3}$$

$$= \lim_{n\to\infty} \log \left(1 + \frac{n+1}{n-2} - 1\right)^{4n+3} = \lim_{n\to\infty} \log \left(1 + \frac{3}{n-2}\right)^{4n+3}$$

$$= \lim_{n\to\infty} \log \left(1 + \frac{1}{\frac{n-2}{3}}\right)^{4n+3} = \lim_{n\to\infty} \log \left[\left(1 + \frac{1}{\frac{n-2}{3}}\right)^{\frac{n-2}{3}}\right]^{\frac{12n+9}{n-2}}$$

$$= \log \left(\lim_{n\to\infty} \left[\left(1 + \frac{1}{\frac{n-2}{3}}\right)^{\frac{n-2}{3}}\right]^{\frac{12n+9}{n-2}}\right)$$

$$= \log \left(\exp \left[\lim_{n\to\infty} \frac{12n+9}{n-2}\right]\right) = 12,$$

where $\exp(\cdot)$ denotes the exponential function $e^{(\cdot)}$.

(XII).

$$\lim_{n\to\infty} \left(\frac{n^2+3n-2}{n^2+n}\right)^{\frac{n^3+2}{2n^2+1}} = \lim_{n\to\infty} \left(1 + \frac{n^2+3n-2}{n^2+n} - 1\right)^{\frac{n^3+2}{2n^2+1}}$$

$$= \lim_{n\to\infty} \left(1 + \frac{2n-2}{n^2+n} - 1\right)^{\frac{n^3+2}{2n^2+1}} = \lim_{n\to\infty} \left(1 + \frac{1}{\frac{n^2+n}{2n-2}} - 1\right)^{\frac{n^3+2}{2n^2+1}}$$

$$= \lim_{n\to\infty} \left[\left(1 + \frac{1}{\frac{n^2+n}{2n-2}}\right)^{\frac{n^2+n}{2n-2}}\right]^{\frac{2n-2}{n^2+n} \cdot \frac{n^3+2}{2n^2+1}}$$

$$= \exp \left(\lim_{n\to\infty} \frac{2n^4-2n^3+4n-4}{2n^4+2n^3+n^2+n}\right) = e.$$

(XIII).

$$\lim_{n\to\infty} \left(\frac{n+1}{n-1}\right)^{\frac{n^2+2}{n-3}} = \lim_{n\to\infty} \left(1 + \frac{n+1}{n-1} - 1\right)^{\frac{n^2+2}{n-3}} = \lim_{n\to\infty} \left(1 + \frac{2}{n-1}\right)^{\frac{n^2+2}{n-3}}$$

$$= \lim_{n\to\infty} \left(1 + \frac{1}{\frac{n-1}{2}}\right)^{\frac{n^2+2}{n-3}} = \lim_{n\to\infty} \left[\left(1 + \frac{1}{\frac{n-1}{2}}\right)^{\frac{n-1}{2}}\right]^{\frac{2}{n-1} \cdot \frac{n^2+2}{n-3}}$$

$$= \exp \left(\lim_{n\to\infty} \frac{2n^2+4}{n^2-4n+3}\right) = e^2.$$

(XIV).

$$\lim_{n\to\infty} \left(1 + \log \frac{3n^2 + 2n + 1}{3n^2 + 5n}\right)^{4n+1}$$

$$= \lim_{n\to\infty} \left[\left(1 + \frac{1}{\log \frac{3n^2+2n+1}{3n^2+5n}}\right)^{\log \frac{3n^2+2n+1}{3n^2+5n}}\right]^{(4n+1)\log \frac{3n^2+2n+1}{3n^2+5n}}$$

$$= \exp\left(\lim_{n\to\infty} (4n+1)\log \frac{3n^2 + 2n + 1}{3n^2 + 5n}\right)$$

$$= \exp\left(\lim_{n\to\infty} \log\left(\frac{3n^2 + 2n + 1}{3n^2 + 5n}\right)^{4n+1}\right)$$

$$= \exp\left(\log\left[\lim_{n\to\infty}\left(\frac{3n^2 + 2n + 1}{3n^2 + 5n}\right)^{4n+1}\right]\right)$$

$$= \lim_{n\to\infty}\left(\frac{3n^2 + 2n + 1}{3n^2 + 5n}\right)^{4n+1} = \lim_{n\to\infty}\left(1 + \frac{3n^2 + 2n + 1}{3n^2 + 5n} - 1\right)^{4n+1}$$

$$= \lim_{n\to\infty}\left(1 + \frac{-3n + 1}{3n^2 + 5n}\right)^{4n+1}$$

$$= \lim_{n\to\infty}\left[\left(1 + \frac{1}{\frac{3n^2+5n}{-3n+1}}\right)^{\frac{3n^2+5n}{-3n+1}}\right]^{\frac{-3n+1}{3n^2+5n}(4n+1)}$$

$$= \exp\left(\lim_{n\to\infty} \frac{-12n^2 + n + 1}{3n^2 + 5n}\right) = e^{-4}.$$

(XV).

$$\lim_{n\to\infty}\left(\frac{1}{n}\right)^{\frac{1}{\log(3/n)}} = \lim_{n\to\infty} \exp\left(\log\left[\left(\frac{1}{n}\right)^{\frac{1}{\log(3/n)}}\right]\right)$$

$$= \lim_{n\to\infty} \exp\left(\frac{1}{\log\left(\frac{3}{n}\right)}\log\left(\frac{1}{n}\right)\right) = \lim_{n\to\infty} \exp\left(\frac{\log(1) - \log(n)}{\log(3) - \log(n)}\right) = e.$$

(XVI).

$$\lim_{n\to\infty}(2+3n^4)^{\frac{1}{3+2\log(n+1)}} = \lim_{n\to\infty}\exp\left(\log\left[(2+3n^4)^{\frac{1}{3+2\log(n+1)}}\right]\right)$$

$$= \lim_{n\to\infty}\exp\left(\frac{1}{3+2\log(n+1)}\cdot\log(2+3n^4)\right)$$

$$= \lim_{n\to\infty}\exp\left(\frac{\log\left[n^4\left(3+\frac{2}{n^4}\right)\right]}{3+2\log\left[n\left(1+\frac{1}{n}\right)\right]}\right)$$

$$\doteq \lim_{n\to\infty}\exp\left(\frac{\log(n^4)+\log\left(3+\frac{2}{n^4}\right)}{3+2\left[\log(n)+\log\left(1+\frac{1}{n}\right)\right]}\right)$$

$$= \lim_{n\to\infty}\exp\left(\frac{4\log(n)+\log\left(3+\frac{2}{n^4}\right)}{3+2\left[\log(n)+\log\left(1+\frac{1}{n}\right)\right]}\right) = e^2.$$

(XVII).

$$\lim_{n\to\infty}\left(\frac{\log(n^2+1)}{\log(n^2-1)}\right)^{n^2\log n} = \lim_{n\to\infty}\left[1+\frac{\log(n^2+1)}{\log(n^2-1)}-1\right]^{n^2\log n}$$

$$= \lim_{n\to\infty}\left[1+\frac{\log(n^2+1)-\log(n^2-1)}{\log(n^2-1)}\right]^{n^2\log n}$$

$$= \lim_{n\to\infty}\left[1+\frac{\log\left(\frac{n^2+1}{n^2-1}\right)}{\log(n^2-1)}\right]^{n^2\log n} = \lim_{n\to\infty}\left[1+\frac{1}{\frac{\log(n^2-1)}{\log\left(\frac{n^2+1}{n^2-1}\right)}}\right]^{n^2\log n}$$

$$= \lim_{n\to\infty}\left[\left(1+\frac{1}{\frac{\log(n^2-1)}{\log\left(\frac{n^2+1}{n^2-1}\right)}}\right)^{\frac{\log(n^2-1)}{\log\left(\frac{n^2+1}{n^2-1}\right)}}\right]^{\frac{\log\left(\frac{n^2+1}{n^2-1}\right)}{\log(n^2-1)}n^2\log n}$$

$$= \exp\left(\lim_{n\to\infty}\frac{\log\left(\frac{n^2+1}{n^2-1}\right)n^2\log n}{\log(n^2-1)}\right)$$

$$= \exp\left(\lim_{n\to\infty}\frac{\log\left(\frac{n^2-1+2}{n^2-1}\right)n^2\log n}{\log\left[n^2(1-\frac{1}{n^2})\right]}\right)$$

$$= \exp\left(\lim_{n\to\infty}\frac{\log\left(1+\frac{2}{n^2-1}\right)n^2\log n}{\log(n^2)+\log\left(1-\frac{1}{n^2}\right)}\right).$$

Now, we will use the equivalence $\log(1+s_n) \sim s_n$ provided that (s_n) converges to 0. In our case $s_n = \frac{2}{n^2-1}$, which converges to 0.

$$\lim_{n\to\infty} \left(\frac{\log(n^2+1)}{\log(n^2-1)}\right)^{n^2\log n} = \exp\left(\lim_{n\to\infty} \frac{\frac{2}{n^2-1}n^2\log n}{\log(n^2) + \log\left(1-\frac{1}{n^2}\right)}\right)$$

$$= \exp\left(\lim_{n\to\infty} \frac{2n^2\log n}{(n^2-1)\left[2\log(n) + \log\left(1-\frac{1}{n^2}\right)\right]}\right)$$

$$= \exp\left(\lim_{n\to\infty} \frac{2n^2\log n}{n^2\left(1-\frac{1}{n^2}\right)\left[2\log(n) + \log\left(1-\frac{1}{n^2}\right)\right]}\right)$$

$$= \exp\left(\lim_{n\to\infty} \frac{2\log n}{\left(1-\frac{1}{n^2}\right)\left[2\log(n) + \log\left(1-\frac{1}{n^2}\right)\right]}\right) = e.$$

(XVIII).

$$\lim_{n\to\infty} \left(\sqrt[3]{(n+a)(n+b)(n+c)} - n\right)$$

$$= \lim_{n\to\infty} \left(\sqrt[3]{n^3\left(1+\frac{a}{n}\right)\left(1+\frac{b}{n}\right)\left(1+\frac{c}{n}\right)} - n\right)$$

$$= \lim_{n\to\infty} \left(n\sqrt[3]{\left(1+\frac{a}{n}\right)\left(1+\frac{b}{n}\right)\left(1+\frac{c}{n}\right)} - n\right)$$

$$= \lim_{n\to\infty} n\left(\sqrt[3]{\left(1+\frac{a}{n}\right)\left(1+\frac{b}{n}\right)\left(1+\frac{c}{n}\right)} - 1\right)$$

$$= \lim_{n\to\infty} n\left(\sqrt[3]{\left(1+\frac{a+b+c}{n} + \frac{ab+ac+bc}{n^2} + \frac{abc}{n^3}\right)} - 1\right)$$

$$= \lim_{n\to\infty} n\left(\left(1+\frac{a+b+c}{n} + \frac{ab+ac+bc}{n^2} + \frac{abc}{n^3}\right)^{1/3} - 1\right).$$

We will use the equivalence $(1+s_n)^\alpha - 1 \sim \alpha s_n$ provided that s_n converges to 0. In our case $s_n = \frac{a+b+c}{n} + \frac{ab+ac+bc}{n^2} + \frac{abc}{n^3}$ which converges to 0.

$$\lim_{n\to\infty} \left(\sqrt[3]{(n+a)(n+b)(n+c)} - n\right)$$

$$= \lim_{n\to\infty} n\left[\frac{1}{3}\left(\frac{a+b+c}{n} + \frac{ab+ac+bc}{n^2} + \frac{abc}{n^3}\right)\right]$$

$$= \lim_{n\to\infty} \frac{1}{3}\left(a+b+c + \frac{ab+ac+bc}{n} + \frac{abc}{n^2}\right) = \frac{a+b+c}{3}.$$

*Another way to find the limit is to apply the Cyclotomic Equation as in the alternative solution of Exercise 3.8 part (VII).

(XIX). We will use Stirling's Formula: $n! \sim n^n e^{-n} \sqrt{2\pi n}$, as follows:

$$\lim_{n\to\infty} \frac{2^{2n}(n!)^2\sqrt{n}}{(2n+1)!} = \lim_{n\to\infty} \frac{2^{2n}\left(n^n e^{-n}\sqrt{2\pi n}\right)^2\sqrt{n}}{(2n+1)^{2n+1}e^{-2n-1}\sqrt{2\pi(2n+1)}}$$

$$= \lim_{n\to\infty} \frac{2^{2n}n^{2n}e^{-2n}2\pi n\sqrt{n}}{(2n+1)^{2n+1}e^{-2n-1}\sqrt{2\pi(2n+1)}}$$

$$= \lim_{n\to\infty} \frac{(2n)^{2n}e^{-2n}2\pi n^{3/2}}{(2n+1)^{2n}(2n+1)e^{-2n}e^{-1}\sqrt{2\pi}\sqrt{2n+1}}$$

$$= e \lim_{n\to\infty} \left(\frac{2n}{2n+1}\right)^{2n} \frac{\sqrt{2\pi}n^{3/2}\cdot}{(2n+1)^{3/2}}$$

$$= e\sqrt{2\pi} \lim_{n\to\infty} \left(\frac{2n}{2n+1}\right)^{2n} \lim_{n\to\infty} \frac{n^{3/2}}{(2n+1)^{3/2}}$$

$$= e\sqrt{2\pi} \lim_{n\to\infty} \left(1 + \frac{2n}{2n+1} - 1\right)^{2n} \lim_{n\to\infty} \frac{n^{3/2}}{2^{3/2}n^{3/2}\left(1+\frac{1}{2n}\right)^{3/2}}$$

$$= e\sqrt{2\pi}\frac{1}{2\sqrt{2}} \lim_{n\to\infty} \left(1 + \frac{1}{-(2n+1)}\right)^{2n} \lim_{n\to\infty} \frac{1}{\left(1+\frac{1}{2n}\right)^{3/2}}$$

$$= \frac{e\sqrt{\pi}}{2} \lim_{n\to\infty} \left[\left(1 + \frac{1}{-(2n+1)}\right)^{-(2n+1)}\right]^{\frac{1}{-(2n+1)}\cdot 2n}$$

$$= \frac{e\sqrt{\pi}}{2} \exp\left(\lim_{n\to\infty} \frac{2n}{-2n-1}\right) = \frac{e\sqrt{\pi}}{2}\cdot e^{-1} = \frac{\sqrt{\pi}}{2}.$$

(XX). In the general term $n\left(\frac{2\cdot4\cdot6\cdots(2n-2)}{1\cdot3\cdot5\cdots(2n-1)}\right)^2$, we will multiply and divide inside the parentheses by $2\cdot4\cdot6\cdots(2n-2)$ as follows:

$$\lim_{n\to\infty} n\left(\frac{2\cdot4\cdot6\cdots(2n-2)}{1\cdot3\cdot5\cdots(2n-1)}\right)^2$$

$$= \lim_{n\to\infty} n\left(\frac{2\cdot4\cdot6\cdots(2n-2)\cdot2\cdot4\cdot6\cdots(2n-2)}{2\cdot4\cdot6\cdots(2n-2)1\cdot3\cdot5\cdots(2n-1)}\right)^2$$

$$= \lim_{n\to\infty} n\left(\frac{2^{n-1}1\cdot2\cdot3\cdots(n-1)\cdot2^{n-1}1\cdot2\cdot3\cdots(n-1)}{1\cdot2\cdot3\cdot4\cdots5\cdot6\cdots(2n-2)(2n-1)}\right)^2$$

$$= \lim_{n\to\infty} n\left[\frac{\left(2^{n-1}(n-1)!\right)^2}{(2n-1)!}\right]^2 = \lim_{n\to\infty} \frac{n2^{4n-4}\left[(n-1)!\right]^4}{\left[(2n-1)!\right]^2}.$$

By Stirling's Formula: $n! \sim n^n e^{-n} \sqrt{2\pi n}$, we have

$$\lim_{n \to \infty} n \left(\frac{2 \cdot 4 \cdot 6 \cdots (2n-2)}{1 \cdot 3 \cdot 5 \cdots (2n-1)} \right)^2$$

$$= \lim_{n \to \infty} \frac{n 2^{4n-4}(n-1)^{4n-4} e^{4-4n} 4\pi^2 (n-1)^2}{(2n-1)^{4n-2} e^{2-4n} 2\pi(2n-1)}$$

$$= \lim_{n \to \infty} \frac{n 2^{4n} 2^{-4}(n-1)^{4n-2} e^{2-4n} e^2 2\pi}{(2n-1)^{4n-1} e^{2-4n}}$$

$$= \frac{\pi e^2}{8} \lim_{n \to \infty} \frac{n 2^{4n}(n-1)^{4n}(n-1)^{-2}}{(2n-1)^{4n}(2n-1)^{-1}}$$

$$= \frac{\pi e^2}{8} \lim_{n \to \infty} 2^{4n} \left(\frac{n-1}{2n-1} \right)^{4n} \frac{n(2n-1)}{(n-1)^2}$$

$$= \frac{\pi e^2}{8} \lim_{n \to \infty} \left(\frac{2(n-1)}{2n-1} \right)^{4n} \frac{2n^2 - n}{n^2 - 2n + 1}$$

$$= \frac{\pi e^2}{8} \lim_{n \to \infty} \left(\frac{2n-2}{2n-1} \right)^{4n} \lim_{n \to \infty} \frac{2n^2 - n}{n^2 - 2n + 1}$$

$$= \frac{\pi e^2}{8} \cdot 2 \lim_{n \to \infty} \left(1 + \frac{2n-2}{2n-1} - 1 \right)^{4n}$$

$$= \frac{\pi e^2}{4} \lim_{n \to \infty} \left(1 + \frac{1}{-2n+1} \right)^{4n}$$

$$= \frac{\pi e^2}{4} \lim_{n \to \infty} \left[\left(1 + \frac{1}{-2n+1} \right)^{-2n+1} \right]^{\frac{1}{-2n+1} \cdot 4n}$$

$$= \frac{\pi e^2}{4} \exp \left(\lim_{n \to \infty} \frac{4n}{-2n+1} \right) = \frac{\pi e^2}{4} \cdot e^{-2} = \frac{\pi}{4}.$$

(XXI).

$$\lim_{n \to \infty} \frac{\sqrt[n]{(n+1)(n+2)\cdots(n+n)}}{n} = \lim_{n \to \infty} \frac{1}{n} \sqrt[n]{(n+1)(n+2)\cdots(2n)}$$

$$= \lim_{n \to \infty} \frac{1}{n} \sqrt[n]{\frac{1 \cdot 2 \cdots n(n+1)(n+2)\cdots(2n)}{1 \cdot 2 \cdots n}} = \lim_{n \to \infty} \frac{1}{n} \sqrt[n]{\frac{(2n)!}{n!}}$$

$$= \lim_{n \to \infty} \frac{1}{n} \frac{\left(\sqrt[2n]{(2n)!} \right)^2}{\sqrt[n]{n!}}.$$

By weak version of Stirling's Formula: $\sqrt[n]{n!} \sim \frac{n}{e}$, we have

$$\lim_{n \to \infty} \frac{\sqrt[n]{(n+1)(n+2)\cdots(n+n)}}{n} = \lim_{n \to \infty} \frac{1}{n} \cdot \frac{\left(\frac{2n}{e} \right)^2}{\frac{n}{e}} = \lim_{n \to \infty} \frac{\frac{4n^2}{e^2}}{\frac{n^2}{e}}$$

$$= \lim_{n \to \infty} \frac{4n^2 e}{n^2 e^2} = \frac{4}{e}.$$

(XXII).

$$\lim_{n\to\infty} \frac{1}{n} \sqrt[n]{(3n+1)(3n+2)\cdots\cdots(4n)}$$

$$= \lim_{n\to\infty} \frac{1}{n} \sqrt[n]{\frac{1\cdot 2\cdots 3n(3n+1)(3n+2)\cdots\cdots(4n)}{1\cdot 2\cdots 3n}}$$

$$= \lim_{n\to\infty} \frac{1}{n} \sqrt[n]{\frac{(4n)!}{(3n)!}}$$

$$= \lim_{n\to\infty} \frac{1}{n} \frac{\left(\sqrt[4n]{(4n)!}\right)^4}{\left(\sqrt[3n]{(3n)!}\right)^3}.$$

By weak version of Stirling's Formula: $\sqrt[n]{n!} \sim \frac{n}{e}$, we have

$$\lim_{n\to\infty} \frac{1}{n} \sqrt[n]{(3n+1)(3n+2)\cdots\cdots(4n)} = \lim_{n\to\infty} \frac{1}{n} \cdot \frac{\left(\frac{4n}{e}\right)^4}{\left(\frac{3n}{e}\right)^3} = \lim_{n\to\infty} \frac{1}{n} \cdot \frac{\frac{4^4 n^4}{e^4}}{\frac{3^3 n^3}{e^3}}$$

$$= \lim_{n\to\infty} \frac{\frac{4^4 n^4}{e^4}}{\frac{3^3 n^4}{e^3}} = \lim_{n\to\infty} \frac{4^4 n^4 e^3}{3^3 n^4 e^4} = \frac{4^4}{3^3 e} \lim_{n\to\infty} \frac{n^4}{n^4} = \frac{4^4}{3^3 e} = \frac{256}{27e}.$$

(XXIII).

$$\lim_{n\to\infty} (\log n - n) = \lim_{n\to\infty} n\left(\frac{\log n}{n} - 1\right) = -\infty.$$

(XXIV). Observe first that

$$\lim_{n\to\infty} \left(\frac{1^p + 2^p + \cdots + n^p}{n^p} - \frac{n}{p+1}\right)$$

$$= \lim_{n\to\infty} \frac{(p+1)\left(1^p + 2^p + \cdots + n^p\right) - n^{p+1}}{(p+1)n^p}.$$

For every $n \in \mathbb{N}$, let $x_n = (p+1)\left(1^p + 2^p + \cdots + n^p\right) - n^{p+1}$ and $y_n = (p+1)n^p$. Notice that (y_n) is strictly increasing, $y_n \neq 0$ for every $n \in \mathbb{N}$ and $\lim_{n\to\infty} y_n = \infty$. Thus, we can apply Stolz's Criterion to $\frac{x_n}{y_n}$ as follows:

$$\lim_{n\to\infty} \frac{x_{n+1} - x_n}{y_{n+1} - y_n}$$

$$= \lim_{n\to\infty} \frac{(p+1)\left(1^p + \cdots + n^p + (n+1)^p\right) - (n+1)^{p+1}}{(p+1)(n+1)^p - (p+1)n^p}$$

$$\frac{-(p+1)\left(1^p + \cdots + n^p\right) + n^{p+1}}{(p+1)(n+1)^p - (p+1)n^p}$$

$$= \lim_{n\to\infty} \frac{(p+1)(n+1)^p - (n+1)^{p+1} + n^{p+1}}{(p+1)(n+1)^p - (p+1)n^p}$$

$$= \lim_{n\to\infty} \frac{(p+1)(n+1)^p - (n+1)^{p+1} + n^{p+1}}{(p+1)[(n+1)^p - n^p]}.$$

By the Binomial Theorem (Exercise 1.11) applied to $(n+1)^p$ and $(n+1)^{p+1}$, we have

$$(n+1)^p = \sum_{j=0}^{p} \binom{p}{j} 1^{p-j} n^j = n^p + \binom{p}{1} n^{p-1} + \cdots + 1.$$

and

$$(n+1)^{p+1} = \sum_{j=0}^{p+1} \binom{p}{j} 1^{p+1-j} n^j = n^{p+1} + \binom{p+1}{1} n^p + \cdots + 1.$$

Hence,

$$\lim_{n\to\infty} \frac{x_{n+1} - x_n}{y_{n+1} - y_n}$$

$$= \lim_{n\to\infty} \frac{(p+1)\left(1^p + \cdots + n^p + (n+1)^p\right) - (n+1)^{p+1}}{(p+1)(n+1)^p - (p+1)n^p}$$

$$\frac{-(p+1)\left(1^p + \cdots + n^p\right) + n^{p+1}}{(p+1)(n+1)^p - (p+1)n^p}$$

$$= \lim_{n\to\infty} \frac{(p+1)\left[n^p + \binom{p}{1}n^{p-1} + \cdots + 1\right]}{(p+1)\left[n^p + \binom{p}{1}n^{p-1} + \cdots + 1 - n^p\right]}$$

$$\frac{-\left[n^{p+1} + \binom{p+1}{1}n^p + \binom{p+1}{2}n^{p-1} + \cdots + 1\right] + n^{p+1}}{(p+1)\left[n^p + \binom{p}{1}n^{p-1} + \cdots + 1 - n^p\right]}$$

$$= \lim_{n\to\infty} \frac{(p+1)n^p + (p+1)pn^{p-1} + \cdots + (p+1)}{(p+1)pn^{p-1} + \cdots + (p+1)}$$

$$\frac{-(p+1)n^p - \frac{(p+1)p}{2}n^{p-1} - \cdots - 1}{(p+1)pn^{p-1} + \cdots + (p+1)}$$

$$= \lim_{n\to\infty} \frac{\frac{(p+1)p}{2}n^{p-1} + \cdots + p}{(p+1)pn^{p-1} + \cdots + (p+1)}$$

$$= \lim_{n\to\infty} \frac{(p+1)p\left(\frac{1}{2}n^{p-1} + \cdots + \frac{1}{p+1}\right)}{(p+1)p\left(n^{p-1} + \cdots + \frac{1}{p}\right)}$$

$$= \lim_{n\to\infty} \frac{\frac{1}{2}n^{p-1} + \cdots + \frac{1}{p+1}}{n^{p-1} + \cdots + \frac{1}{p}}.$$

Notice that $\frac{1}{2}n^{p-1} + \cdots + \frac{1}{p+1}$ in the numerator and $n^{p-1} + \cdots + \frac{1}{p}$ in the denominator are polynomials of the same degree $(p-1)$. Hence,

$$\lim_{n\to\infty} \frac{x_{n+1} - x_n}{y_{n+1} - y_n} = \lim_{n\to\infty} \frac{\frac{1}{2}n^{p-1} + \cdots + \frac{1}{p+1}}{n^{p-1} + \cdots + \frac{1}{p}} = \frac{1}{2}.$$

Therefore,

$$\lim_{n \to \infty} \left(\frac{1^p + 2^p + \cdots + n^p}{n^p} - \frac{n}{p+1} \right) = \frac{1}{2}.$$

(XXV). Observe first that

$$\lim_{n \to \infty} \frac{\cos 1 + \cos \frac{1}{\sqrt{2}} + \cdots + \cos \frac{1}{\sqrt{n}} - n}{\log(n^3 + 1)} = \lim_{n \to \infty} \frac{\left(\sum_{k=1}^n \cos \frac{1}{\sqrt{k}} \right) - n}{\log(n^3 + 1)}.$$

For every $n \in \mathbb{N}$, let $x_n = \left(\sum_{k=1}^n \cos \frac{1}{\sqrt{k}} \right) - n$ and $y_n = \log(n^3 + 1)$. Notice that (y_n) is strictly increasing, $y_n \neq 0$ for every $n \in \mathbb{N}$ and $\lim_{n \to \infty} \log(n^3 + 1) = \infty$. Thus, we can apply Stolz Theorem to $\frac{x_n}{y_n}$ as follows:

$$\lim_{n \to \infty} \frac{x_{n+1} - x_n}{y_{n+1} - y_n} = \lim_{n \to \infty} \frac{\left(\sum_{k=1}^{n+1} \cos \frac{1}{\sqrt{k}} \right) - (n+1) - \left[\left(\sum_{k=1}^n \cos \frac{1}{\sqrt{k}} \right) - n \right]}{\log[(n+1)^3 + 1] - \log(n^3 + 1)}$$

$$= \lim_{n \to \infty} \frac{\cos \left(\frac{1}{\sqrt{n+1}} \right) - 1}{\log \left(\frac{(n+1)^3 + 1}{n^3 + 1} \right)}.$$

Now we will apply the equivalence $\cos(s_n) - 1 \sim -\frac{1}{2} s_n^2$ provided that (s_n) converges to 0. In our case $s_n = \frac{1}{\sqrt{n+1}}$ converges to 0. Hence,

$$\lim_{n \to \infty} \frac{x_{n+1} - x_n}{y_{n+1} - y_n} = \lim_{n \to \infty} \frac{\cos \left(\frac{1}{\sqrt{n+1}} \right) - 1}{\log \left(\frac{(n+1)^3 + 1}{n^3 + 1} \right)} = \lim_{n \to \infty} \frac{-\frac{1}{2(n+1)}}{\log \left(\frac{n^3 + 3n^2 + 3n + 2}{n^3 + 1} \right)}$$

$$= \lim_{n \to \infty} \frac{-\frac{1}{2(n+1)}}{\log \left(1 + \frac{n^3 + 3n^2 + 3n + 2}{n^3 + 1} - 1 \right)} = \lim_{n \to \infty} \frac{-\frac{1}{2(n+1)}}{\log \left(1 + \frac{3n^2 + 3n + 1}{n^3 + 1} \right)}.$$

Let us apply now the equivalence $\log(1 + r_n) \sim r_n$ provided that (r_n) converges to 0. In this latter case, we have that $r_n = \frac{3n^2 + 3n + 1}{n^3 + 1}$ converges to 0. Thus,

$$\lim_{n \to \infty} \frac{x_{n+1} - x_n}{y_{n+1} - y_n} = \lim_{n \to \infty} \frac{-\frac{1}{2(n+1)}}{\log \left(1 + \frac{3n^2 + 3n + 1}{n^3 + 1} \right)} = \lim_{n \to \infty} \frac{-\frac{1}{2(n+1)}}{\frac{3n^2 + 3n + 1}{n^3 + 1}}$$

$$= \lim_{n \to \infty} \frac{-n^3 - 1}{(2n+1)(3n^2 + 3n + 1)} = \lim_{n \to \infty} \frac{-n^3 - 1}{6n^3 + 7n^2 + 5n + 1} = -\frac{1}{6}.$$

Hence,

$$\lim_{n \to \infty} \frac{\cos 1 + \cos \frac{1}{\sqrt{2}} + \cdots + \cos \frac{1}{\sqrt{n}} - n}{\log(n^3 + 1)} = -\frac{1}{6}.$$

(XXVI). Observe first that

$$\lim_{n \to \infty} \frac{\sqrt{1 \cdot 2 \cdot 3} + \sqrt{2 \cdot 3 \cdot 4} + \cdots + \sqrt{n(n+1)(n+2)}}{n^2 \sqrt{n}}$$

$$= \lim_{n \to \infty} \frac{\sum_{k=1}^{n} \sqrt{k(k+1)(k+2)}}{n^2 \sqrt{n}}.$$

For every $n \in \mathbb{N}$, let $x_n = \sum_{k=1}^{n} \sqrt{k(k+1)(k+2)}$ and $y_n = n^2 \sqrt{n}$. Notice that (y_n) is strictly increasing, $y_n \neq 0$ for every $n \in \mathbb{N}$ and $\lim_{n \to \infty} y_n = \infty$. Hence, we can apply Stolz Theorem to $\frac{x_n}{y_n}$ as follows:

$$\lim_{n \to \infty} \frac{x_{n+1} - x_n}{y_{n+1} - y_n} = \lim_{n \to \infty} \frac{\sum_{k=1}^{n+2} \sqrt{k(k+1)(k+2)} - \sum_{k=1}^{n} \sqrt{k(k+1)(k+2)}}{(n+1)^2 \sqrt{n+1} - n^2 \sqrt{n}}$$

$$= \lim_{n \to \infty} \frac{\sqrt{(n+1)(n+2)(n+3)}}{(n+1)^2 \sqrt{n+1} - n^2 \sqrt{n}}$$

$$= \lim_{n \to \infty} \frac{\sqrt{(n+1)(n+2)(n+3)} \cdot \left[(n+1)^2 \sqrt{n+1} + n^2 \sqrt{n}\right]}{\left[(n+1)^2 \sqrt{n+1} - n^2 \sqrt{n}\right] \cdot \left[(n+1)^2 \sqrt{n+1} + n^2 \sqrt{n}\right]}$$

$$= \lim_{n \to \infty} \frac{(n+1)^3 \sqrt{(n+2)(n+3)} + n^2 \sqrt{n(n+1)(n+2)(n+3)}}{(n+1)^5 - n^5}$$

$$= \lim_{n \to \infty} \frac{(n+1)^3 \sqrt{n^2 \left(1 + \frac{2}{n}\right)\left(1 + \frac{3}{n}\right)} + n^2 \sqrt{n^4 \left(1 + \frac{1}{n}\right)\left(1 + \frac{2}{n}\right)\left(1 + \frac{3}{n}\right)}}{(n+1)^5 - n^5}$$

$$= \lim_{n \to \infty} \frac{n(n+1)^3 \sqrt{\left(1 + \frac{2}{n}\right)\left(1 + \frac{3}{n}\right)} + n^4 \sqrt{\left(1 + \frac{1}{n}\right)\left(1 + \frac{2}{n}\right)\left(1 + \frac{3}{n}\right)}}{(n+1)^5 - n^5}$$

$$= \lim_{n \to \infty} \frac{n(n+1)^3 \sqrt{1 + \frac{5}{n} + \frac{6}{n^2}} + n^4 \sqrt{1 + \frac{6}{n} + \frac{11}{n^2} + \frac{6}{n^3}}}{(n+1)^5 - n^5}.$$

We will apply now the equivalence $(1 + s_n)^{\alpha} \sim \alpha s_n + 1$ provided that (s_n) converges to 0. In our case, we will use it twice, for the sequence with general term $\frac{5}{n} + \frac{6}{n^2}$ and for the sequence with general term $\frac{6}{n} + \frac{11}{n^2} + \frac{6}{n^3}$ (note that both of them converge to 0).

$$\lim_{n \to \infty} \frac{x_{n+1} - x_n}{y_{n+1} - y_n}$$

$$= \lim_{n \to \infty} \frac{n(n+1)^3 \left[\frac{1}{2}\left(\frac{5}{n} + \frac{6}{n^2}\right) + 1\right] + n^4 \left[\frac{1}{2}\left(\frac{6}{n} + \frac{11}{n^2} + \frac{6}{n^3}\right) + 1\right]}{(n+1)^5 - n^5}$$

$$= \lim_{n \to \infty} \frac{\frac{1}{2}(n+1)^3 \left(5 + \frac{6}{n}\right) + n(n+1)^3 + 3n^3 + \frac{11}{2}n^2 + 3n + n^4}{(n+1)^5 - n^5}$$

$$= \lim_{n \to \infty} \frac{n^4 + 3n^3 + 3n^2 + n + \frac{1}{2}(n^3 + 3n^2 + 3n + 1)\left(5 + \frac{6}{n}\right)}{n^5 + 5n^4 + 10n^3 + 10n^2 + 5n + 1 - n^5}$$

$$\frac{+3n^3 + \frac{11}{2}n^2 + 3n + n^4}{n^5 + 5n^4 + 10n^3 + 10n^2 + 5n + 1 - n^5}$$

$$= \lim_{n \to \infty} \frac{2n^4 + 6n^3 + \frac{17}{2}n^2 + 4n + \frac{5}{2}n^3 + \frac{15}{2}n^2 + \frac{15}{2}n + \frac{5}{2}}{5n^4 + 10n^3 + 10n^2 + 5n + 1}$$

$$\frac{+3n^2 + 9n + 9 + \frac{3}{n}}{5n^4 + 10n^3 + 10n^2 + 5n + 1}$$

$$= \lim_{n \to \infty} \frac{2n^4 + \frac{17}{2}n^3 + 19n^2 + \frac{41}{2}n + \frac{23}{2} + \frac{3}{n}}{5n^4 + 10n^3 + 10n^2 + 5n + 1} = \frac{2}{5}.$$

Hence,

$$\lim_{n \to \infty} \frac{\sqrt{1 \cdot 2 \cdot 3} + \sqrt{2 \cdot 3 \cdot 4} + \cdots + \sqrt{n(n+1)(n+2)}}{n^2 \sqrt{n}} = \frac{2}{5}.$$

(XXVII). Observe first that

$$\lim_{n \to \infty} \frac{\log 1 - \log 2 + \cdots + \log(2n - 1) - \log(2n)}{\log n}$$

$$= \lim_{n \to \infty} \frac{\sum_{k=1}^{n} \left(\log(2k - 1) - \log(2k)\right)}{\log(n)}.$$

For every $n \in \mathbb{N}$, let $x_n = \sum_{k=1}^{n} \left(\log(2k - 1) - \log(2k)\right)$ and $y_n = \log n$. Notice that (y_n) is strictly increasing, $y_n \neq 0$ for every $n \in \mathbb{N} \setminus \{1\}$ and

$\lim_{n\to\infty} y_n = \infty$. Hence, we can apply the Stolz Theorem to $\frac{x_n}{y_n}$ as follows:

$$\lim_{n\to\infty} \frac{x_{n+1} - x_n}{y_{n+1} - y_n}$$

$$= \lim_{n\to\infty} \frac{\sum_{k=1}^{n+1} (\log(2k-1) - \log(2k)) - \sum_{k=1}^{n} (\log(2k-1) - \log(2k))}{\log(n+1) - \log(n)}$$

$$= \lim_{n\to\infty} \frac{\log(2n+1) - \log(2n+2)}{\log(n+1) - \log(n)} = \lim_{n\to\infty} \frac{\log\left(\frac{2n+1}{2n+2}\right)}{\log\left(\frac{n+1}{n}\right)}$$

$$= \lim_{n\to\infty} \frac{\log\left(1 + \frac{2n+1}{2n+2} - 1\right)}{\log\left(1 + \frac{n+1}{n} - 1\right)} = \lim_{n\to\infty} \frac{\log\left(1 + \frac{1}{-2n-2}\right)}{\log\left(1 + \frac{1}{n}\right)}.$$

We will apply now the equivalence $\log(1 + s_n) \sim s_n$ provided that (s_n) converges to 0. In our case, we will use it twice, for the sequence with general term $\frac{1}{-2n-2}$ and for the sequence with general term $\frac{1}{n}$ (note that both of them converge to 0).

$$\lim_{n\to\infty} \frac{x_{n+1} - x_n}{y_{n+1} - y_n} = \lim_{n\to\infty} \frac{\log\left(1 + \frac{1}{-2n-2}\right)}{\log\left(1 + \frac{1}{n}\right)} = \lim_{n\to\infty} \frac{\frac{1}{-2n-2}}{\frac{1}{n}}$$

$$= \lim_{n\to\infty} \frac{n}{-2n-2} = -\frac{1}{2}.$$

(XXVIII). Observe first that, for every $n \in \mathbb{N} \setminus \{1\}$,

$$\lim_{n\to\infty} \frac{\sqrt{2!}\tan\frac{1}{2} + \sqrt[3]{3!}\tan\frac{1}{3} + \cdots + \sqrt[n]{n!}\tan\frac{1}{n}}{\sqrt{n^2+1}} = \lim_{n\to\infty} \frac{\sum_{k=2}^{n} \sqrt[k]{k!}\tan\frac{1}{k}}{\sqrt{n^2+1}}.$$

For every $n \in \mathbb{N}\setminus\{1\}$, let $x_n = \sum_{k=2}^{n} \sqrt[k]{k!}\tan\frac{1}{k}$ and $y_n = \sqrt{n^2+1}$. Notice that (y_n) is strictly increasing, $y_n \neq 0$ for every $n \in \mathbb{N}$ and $\lim_{n\to\infty} \sqrt{n^2+1} = \infty$. Hence, we can apply the Stolz Theorem to $\frac{x_n}{y_n}$ as follows:

$$\lim_{n\to\infty} \frac{x_{n+1} - x_n}{y_{n+1} - y_n} = \lim_{n\to\infty} \frac{\sum_{k=2}^{n+1} \sqrt[k]{k!}\tan\frac{1}{k} - \sum_{k=2}^{n} \sqrt[k]{k!}\tan\frac{1}{k}}{\sqrt{(n+1)^2+1} - \sqrt{n^2+1}}$$

$$= \lim_{n\to\infty} \frac{\sqrt[n+1]{(n+1)!}\tan\frac{1}{n+1}}{\sqrt{n^2+2n+2} - \sqrt{n^2+1}}.$$

We will now apply the weak version of Stirling's Formula: $\sqrt[n]{n!} \sim \frac{n}{e}$, as follows:

$$\lim_{n\to\infty} \frac{x_{n+1} - x_n}{y_{n+1} - y_n} = \lim_{n\to\infty} \frac{\frac{n+1}{e}\tan\frac{1}{n+1}}{\sqrt{n^2+2n+2} - \sqrt{n^2+1}}.$$

Recall the equivalence $\tan s_n \sim s_n$ provided that (s_n) converges to 0. In our case, $s_n = \frac{1}{n+1}$ which satisfies that (s_n) converges to 0. Hence,

$$\lim_{n\to\infty} \frac{x_{n+1} - x_n}{y_{n+1} - y_n} = \lim_{n\to\infty} \frac{\frac{n+1}{e}\tan\frac{1}{n+1}}{\sqrt{n^2+2n+2} - \sqrt{n^2+1}}$$

$$= \lim_{n\to\infty} \frac{\frac{n+1}{e}\frac{1}{n+1}}{\sqrt{n^2+2n+2} - \sqrt{n^2+1}}$$

$$= \frac{1}{e}\lim_{n\to\infty} \frac{1}{\sqrt{n^2+2n+2} - \sqrt{n^2+1}}$$

$$= \frac{1}{e}\lim_{n\to\infty} \frac{\sqrt{n^2+2n+2} + \sqrt{n^2+1}}{\left(\sqrt{n^2+2n+2} - \sqrt{n^2+1}\right)\left(\sqrt{n^2+2n+2} + \sqrt{n^2+1}\right)}$$

$$= \frac{1}{e}\lim_{n\to\infty} \frac{\sqrt{n^2+2n+2} + \sqrt{n^2+1}}{n^2+2n+2 - (n^2+1)} = \frac{1}{e}\lim_{n\to\infty} \frac{\sqrt{n^2+2n+2} + \sqrt{n^2+1}}{2n+1}$$

$$= \frac{1}{e}\lim_{n\to\infty} \frac{\sqrt{n^2\left(1+\frac{2}{n}+\frac{2}{n^2}\right)} + \sqrt{n^2\left(1+\frac{1}{n^2}\right)}}{2n+1}$$

$$= \frac{1}{e}\lim_{n\to\infty} \frac{n\sqrt{1+\frac{2}{n}+\frac{2}{n^2}} + n\sqrt{1+\frac{1}{n^2}}}{n\left(2+\frac{1}{n}\right)}$$

$$= \frac{1}{e}\lim_{n\to\infty} \frac{\sqrt{1+\frac{2}{n}+\frac{2}{n^2}} + \sqrt{1+\frac{1}{n^2}}}{2+\frac{1}{n}} = \frac{1}{e}\cdot\frac{2}{2} = \frac{1}{e}.$$

Thus,

$$\lim_{n\to\infty} \frac{\sqrt{2!}\tan\frac{1}{2} + \sqrt[3]{3!}\tan\frac{1}{3} + \cdots + \sqrt[n]{n!}\tan\frac{1}{n}}{\sqrt{n^2+1}} = \frac{1}{e}.$$

(XXIX).

$$\lim_{n\to\infty} \sqrt[n]{\sqrt[n]{\binom{n}{1}\binom{n}{2}\cdots\binom{n}{n}}} = \lim_{n\to\infty} \sqrt[n^2]{\prod_{k=1}^{n}\binom{n}{k}} = \lim_{n\to\infty} \left[\prod_{k=1}^{n}\binom{n}{k}\right]^{1/n^2}$$

$$= \lim_{n\to\infty} \exp\left(\log\left[\left[\prod_{k=1}^{n}\binom{n}{k}\right]^{1/n^2}\right]\right) = \lim_{n\to\infty} \exp\left(\frac{1}{n^2}\log\left[\prod_{k=1}^{n}\binom{n}{k}\right]\right)$$

$$= \exp\left(\lim_{n\to\infty} \frac{\log\left[\prod_{k=1}^{n}\binom{n}{k}\right]}{n^2}\right).$$

For every $n \in \mathbb{N}$, let $x_n = \log\left[\prod_{k=1}^{n}\binom{n}{k}\right]$ and $y_n = n^2$. Notice that (y_n) is strictly increasing, $y_n \neq 0$ for every $n \in \mathbb{N}$ and $\lim_{n\to\infty} y_n = \infty$. Therefore, we can apply Stolz Theorem to $\frac{x_n}{y_n}$ as follows:

$$\lim_{n\to\infty} \frac{x_{n+1} - x_n}{y_{n+1} - y_n} = \lim_{n\to\infty} \frac{\log\left[\prod_{k=1}^{n+1}\binom{n+1}{k}\right] - \log\left[\prod_{k=1}^{n}\binom{n}{k}\right]}{(n+1)^2 - n^2}$$

$$= \lim_{n\to\infty} \frac{1}{n^2 + 2n + 1 - n^2} \log\left[\frac{\prod_{k=1}^{n+1}\binom{n+1}{k}}{\prod_{k=1}^{n}\binom{n}{k}}\right]$$

$$= \lim_{n\to\infty} \frac{1}{2n+1} \log\left[\frac{\binom{n+1}{n+1}\prod_{k=1}^{n}\binom{n+1}{k}}{\prod_{k=1}^{n}\binom{n}{k}}\right]$$

$$= \lim_{n\to\infty} \frac{1}{2n+1} \log\left[\frac{\prod_{k=1}^{n}\binom{n+1}{k}}{\prod_{k=1}^{n}\binom{n}{k}}\right]$$

$$= \lim_{n\to\infty} \frac{1}{2n+1} \log\left[\frac{\frac{(n+1)!}{1!n!} \cdot \frac{(n+1)!}{2!(n-1)!} \cdots \frac{(n+1)!}{(n-1)!2!} \cdot \frac{(n+1)!}{n!1!}}{\frac{n!}{1!(n-1)!} \cdot \frac{n!}{2!(n-2)!} \cdots \frac{n!}{(n-1)!1!} \cdot \frac{n!}{n!0!}}\right]$$

$$= \lim_{n\to\infty} \frac{1}{2n+1} \log\left[\frac{1}{n!} \cdot \underbrace{\frac{(n+1)! \cdot (n+1)! \cdots (n+1)! \cdot (n+1)!}{n! \cdot n! \cdots n! \cdot n!}}_{n \text{ times}}\right]$$

$$= \lim_{n\to\infty} \frac{1}{2n+1} \log\left[\frac{1}{n!} \cdot \underbrace{\frac{(n+1)n! \cdot (n+1)n! \cdots (n+1)n! \cdot (n+1)n!}{n! \cdot n! \cdots n! \cdot n!}}_{n \text{ times}}\right]$$

$$= \lim_{n\to\infty} \frac{1}{2n+1} \log\left[\frac{(n+1)^n}{n!}\right] = \lim_{n\to\infty} \frac{1}{2n+1} \log\left[\frac{(n+1)(n+1)^n}{(n+1)n!}\right]$$

$$= \lim_{n\to\infty} \frac{1}{2n+1} \log\left[\frac{(n+1)^{n+1}}{(n+1)!}\right].$$

We will now apply Stirling's Formula: $n! \sim n^n e^{-n}\sqrt{2\pi n}$, as follows:

$$\lim_{n\to\infty} \frac{x_{n+1} - x_n}{y_{n+1} - y_n} = \lim_{n\to\infty} \frac{1}{2n+1} \log\left[\frac{(n+1)^{n+1}}{(n+1)^{n+1}e^{-(n+1)}\sqrt{2\pi(n+1)}}\right]$$

$$= \lim_{n\to\infty} \frac{1}{2n+1} \log\left(\frac{e^{n+1}}{\sqrt{2\pi(n+1)}}\right) = \lim_{n\to\infty} \frac{1}{2n+1} \log\left(\frac{e^{n+1}}{\sqrt{2\pi(n+1)}}\right)$$

$$= \lim_{n\to\infty} \log\left[\left(\frac{e^{n+1}}{\sqrt{2\pi(n+1)}}\right)^{\frac{1}{2n+1}}\right] = \lim_{n\to\infty} \log\left[\frac{e^{\frac{n+1}{2n+1}}}{(2\pi)^{\frac{1}{4n+2}}(n+1)^{\frac{1}{4n+2}}}\right]$$

$$= \log(e^{1/2}) = \frac{1}{2}.$$

In the latter limit we have used that $\lim\limits_{n\to\infty} (2\pi)^{\frac{1}{4n+2}} = 1$, $\lim\limits_{n\to\infty} (n+1)^{\frac{1}{4n+2}} = 1$ and also $\lim\limits_{n\to\infty} e^{\frac{n+1}{2n+1}} = e^{1/2}$. Hence,

$$\lim_{n\to\infty} \frac{\log\left[\prod_{k=1}^{n} \binom{n}{k}\right]}{n^2} = \frac{1}{2}.$$

Thus,

$$\lim_{n\to\infty} \sqrt[n]{\sqrt[n]{\binom{n}{1}\binom{n}{2}\cdots\binom{n}{n}}} = e^{1/2} = \sqrt{e}.$$

(XXX). We will apply the Sandwich Theorem to $\frac{1}{n^2} + \frac{1}{(n+1)^2} + \ldots + \frac{1}{(n+n)^2}$ as follows:

$$0 \leq \frac{1}{n^2} + \frac{1}{(n+1)^2} + \cdots + \frac{1}{(n+n)^2} \leq \underbrace{\frac{1}{n^2} + \frac{1}{n^2} + \cdots + \frac{1}{n^2}}_{n \text{ times}} = n \cdot \frac{1}{n^2} = \frac{1}{n}.$$

Since $\lim_{n\to\infty} \frac{1}{n} = 0$, we have

$$\lim_{n\to\infty} \left(\frac{1}{n^2} + \frac{1}{(n+1)^2} + \cdots + \frac{1}{(n+n)^2} \right) = 0.$$

(XXXI).

$$\begin{aligned}
\lim_{n\to\infty} (2^n + 3^n)^{1/n} &= \lim_{n\to\infty} \exp\left(\log\left[(2^n + 3^n)^{1/n}\right]\right) \\
&= \lim_{n\to\infty} \exp\left(\frac{1}{n}\log(2^n + 3^n)\right) \\
&= \lim_{n\to\infty} \exp\left(\frac{1}{n}\log\left[3^n\left(1 + \frac{2^n}{3^n}\right)\right]\right) \\
&= \lim_{n\to\infty} \exp\left(\frac{1}{n}\left[\log(3^n) + \log\left(1 + \frac{2^n}{3^n}\right)\right]\right) \\
&= \lim_{n\to\infty} \exp\left(\frac{1}{n}\left[n\log(3) + \log\left(1 + \frac{2^n}{3^n}\right)\right]\right) \\
&= \lim_{n\to\infty} \exp\left(\log(3) + \frac{1}{n}\log\left[1 + \frac{2^n}{3^n}\right]\right) \\
&= \lim_{n\to\infty} \left(e^{\log(3)} e^{\frac{1}{n}\log\left(1 + \frac{2^n}{3^n}\right)}\right) \\
&= 3\exp\left(\lim_{n\to\infty} \frac{\log\left(1 + \frac{2^n}{3^n}\right)}{n}\right) = 3 \cdot e^0 = 3,
\end{aligned}$$

where we have used the last limit that $\lim_{n\to\infty} \frac{\log\left(1 + \frac{2^n}{3^n}\right)}{n} = 0$.

(XXXII). Observe that for every $n \in \mathbb{N}$ we have that

$$1 > \frac{n^2}{n^2 + 1} > \frac{n^2}{n^2 + 2} > \ldots > \frac{n^2}{n^2 + n} > 0.$$

Hence, if we multiply the latter chain of inequalities by $\frac{\pi}{2}$ we have

$$\frac{\pi}{2} > \frac{\pi n^2}{2(n^2 + 1)} > \frac{\pi n^2}{2(n^2 + 2)} > \ldots > \frac{\pi n^2}{2(n^2 + n)} > 0.$$

Since the sine function $\sin(\cdot)$ is an increasing function on the interval $\left[0, \frac{\pi}{2}\right]$, we have that

$$n \cdot \sin \frac{\pi n^2}{2(n^2 + n)} = \underbrace{\sin \frac{\pi n^2}{2(n^2 + n)} + \cdots + \sin \frac{\pi n^2}{2(n^2 + n)}}_{n \text{ times}}$$

$$\leq \sin \frac{\pi n^2}{2(n^2 + 1)} + \cdots + \sin \frac{\pi n^2}{2(n^2 + n)}.$$

Notice that $\lim\limits_{n \to \infty} n \cdot \sin \frac{\pi n^2}{2(n^2 + n)} = \infty$ since $\lim_{n \to \infty} \sin \frac{\pi n^2}{2(n^2+n)} = \sin \frac{\pi}{2} = 1$. Hence, by the Comparison Theorem, we have that

$$\lim_{n \to \infty} \left(\sin \frac{\pi n^2}{2(n^2 + 1)} + \sin \frac{\pi n^2}{2(n^2 + 2)} + \cdots + \sin \frac{\pi n^2}{2(n^2 + n)} \right) = \infty.$$

$\square$

Exercise 3.9. *Find the relation between a, b, and c so that*

$$\lim_{n \to \infty} n^a \frac{(n+1)^b - n^b}{(n+1)^c - n^c}$$

is real and different from zero. In this case, find the limit.

Solution to Exercise 3.9.

Notice that $c \neq 0$ since otherwise $(n+1)^c - n^c = 0$, which implies that the denominator in $\frac{(n+1)^b - n^b}{(n+1)^c - n^c}$ is equal to 0 for every $n \in \mathbb{N}$, that is, the sequence is not well defined. Moreover, if $b = 0$, then $n^a \frac{(n+1)^b - n^b}{(n+1)^c - n^c} = 0$ for every $a \in \mathbb{R}$ and $c \neq 0$. Hence, the sequence converges to 0, but we want the limit to be different from 0. Therefore, we will assume that $b, c \neq 0$.

Hence,

$$\lim_{n\to\infty} n^a \frac{(n+1)^b - n^b}{(n+1)^c - n^c} = \lim_{n\to\infty} n^a \frac{\left[n\left(1+\frac{1}{n}\right)\right]^b - n^b}{\left[n\left(1+\frac{1}{n}\right)\right]^c - n^c}$$

$$= \lim_{n\to\infty} n^a \frac{n^b\left(1+\frac{1}{n}\right)^b - n^b}{n^c\left(1+\frac{1}{n}\right)^c - n^c} = \lim_{n\to\infty} n^a \frac{n^b\left[\left(1+\frac{1}{n}\right)^b - 1\right]}{n^c\left[\left(1+\frac{1}{n}\right)^c - 1\right]}$$

$$= \lim_{n\to\infty} n^{a+b-c} \cdot \frac{\left(1+\frac{1}{n}\right)^b - 1}{\left(1+\frac{1}{n}\right)^c - 1}.$$

Let us apply now the equivalence $(1+s_n)^\alpha - 1 \sim \alpha s_n$ provided that (s_n) converges to 0. In our case we will use it twice, for $\left(1+\frac{1}{n}\right)^b - 1$ and $\left(1+\frac{1}{n}\right)^c - 1$, where $s_n = \frac{1}{n}$ converges to 0.

$$\lim_{n\to\infty} n^a \frac{(n+1)^b - n^b}{(n+1)^c - n^c} = \lim_{n\to\infty} n^{a+b-c} \cdot \frac{\left(1+\frac{1}{n}\right)^b - 1}{\left(1+\frac{1}{n}\right)^c - 1} = \lim_{n\to\infty} n^{a+b-c} \cdot \frac{b\frac{1}{n}}{c\frac{1}{n}}$$

$$= \lim_{n\to\infty} n^{a+b-c} \cdot \frac{b}{c} = \frac{b}{c} \lim_{n\to\infty} n^{a+b-c}.$$

Since $c \neq 0$, we have that the limit exists in $\overline{\mathbb{R}}$. But we also want the limit to be a real number different from 0. To do so, we will distinguish three cases depending on the sign of $a+b-c$ in order to calculate $\lim_{n\to\infty} n^{a+b-c}$:

Case 1: If $a+b-c < 0$, then $\lim_{n\to\infty} n^{a+b-c} = 0$.

Case 2: If $a+b-c > 0$, then $\lim_{n\to\infty} n^{a+b-c} = \infty$.

Case 3: If $a+b-c = 0$, then $\lim_{n\to\infty} n^{a+b-c} = \lim_{n\to\infty} 1 = 1$.

Hence, the limit is a real number when $a+b-c = 0$ and, in this case, the limit is different from 0 since $b \neq 0$. In summary, the limit exists and it is a real number different from 0 when $b,c \neq 0$ and $a+b-c = 0$, in which case

$$\lim_{n\to\infty} n^a \frac{(n+1)^b - n^b}{(n+1)^c - n^c} = \frac{b}{c}.$$

$\square$

Exercise 3.10. *Discuss, depending on the value of $a \in \mathbb{R}$, the existence and value of* $\lim_{n\to\infty} \frac{a^n + n}{a^{n-1} + 2n}$.

Solution to Exercise 3.10.

We will distinguish two cases depending on the value of a:

Case 1: If $|a| \leq 1$ (that is, $-1 \leq a \leq 1$), then

$$\lim_{n \to \infty} \frac{a^n + n}{a^{n-1} + 2n} = \lim_{n \to \infty} \frac{n\left(1 + \frac{a^n}{n}\right)}{n\left(2 + \frac{a^{n-1}}{n}\right)} = \lim_{n \to \infty} \frac{1 + \frac{a^n}{n}}{2 + \frac{a^{n-1}}{n}} = \frac{1}{2},$$

where in the last equality we have used that $\lim_{n \to \infty} \frac{a^n}{n} = \lim_{n \to \infty} \frac{a^{n-1}}{n} = 0$ since $|a| \leq 1$.

Case 2: If $|a| > 1$ (that is, $a > 1$ or $a < -1$), then

$$\lim_{n \to \infty} \frac{a^n + n}{a^{n-1} + 2n} = \lim_{n \to \infty} \frac{na\left(\frac{1}{a} + \frac{a^{n-1}}{n}\right)}{n\left(2 + \frac{a^n}{n}\right)} = a \lim_{n \to \infty} \frac{\frac{1}{a} + \frac{a^{n-1}}{n}}{2 + \frac{a^n}{n}}.$$

Dividing the numerator and denominator of $\frac{\frac{1}{a} + \frac{a^{n-1}}{n}}{2 + \frac{a^n}{n}}$ by $\frac{a^{n-1}}{n}$ we have that

$$\lim_{n \to \infty} \frac{a^n + n}{a^{n-1} + 2n} = a \lim_{n \to \infty} \frac{\frac{1}{a} + \frac{a^{n-1}}{n}}{2 + \frac{a^n}{n}} = a \lim_{n \to \infty} \frac{1 + \frac{n}{a^n}}{1 + \frac{n}{a^{n-1}}} = a,$$

where in the last equality we have used that $\lim_{n \to \infty} \frac{n}{a^n} = \lim_{n \to \infty} \frac{n}{a^{n-1}} = 0$ since $|a| > 1$.

In summary: If $|a| \leq 1$, then $\lim_{n \to \infty} \frac{a^n + n}{a^{n-1} + 2n} = \frac{1}{2}$. If $|a| > 1$, then $\lim_{n \to \infty} \frac{a^n + n}{a^{n-1} + 2n} = a$.

$\square$

Exercise 3.11. *Let $u_n = \frac{1}{1+n} + \frac{1}{2+n} + \frac{1}{3+n} + \cdots + \frac{1}{n+n}$. Prove that $\lim_{n \to \infty} u_n$ exists and it is between $\frac{1}{2}$ and 1.*

Solution to Exercise 3.11.

To prove that the limit exists, we will show that the sequence (u_n) is bounded between $\frac{1}{2}$ and 1 and increasing.

We will prove first that the sequence is bounded between $\frac{1}{2}$ and 1. For every $n \in \mathbb{N}$, we have that $\frac{1}{n+n} \leq \frac{1}{k+n} \leq \frac{1}{1+n}$ for each integer $1 \leq k \leq n$. Hence, for every $n \in \mathbb{N}$, we have

$$\frac{1}{1+n} + \cdots + \frac{1}{n+n} \geq \underbrace{\frac{1}{n+n} + \cdots + \frac{1}{n+n}}_{n \text{ times}} = n \cdot \frac{1}{2n} = \frac{1}{2},$$

and

$$\frac{1}{n+1} + \frac{1}{2+n} + \frac{1}{3+n} + \cdots + \frac{1}{n+n} \leq \underbrace{\frac{1}{1+n} + \cdots + \frac{1}{1+n}}_{n \text{ times}} = n \cdot \frac{1}{1+n} < 1.$$

Let us prove now that (u_n) is increasing. For every $n \in \mathbb{N}$, we have

$u_{n+1} - u_n$

$$= \frac{1}{1 + (n+1)} + \cdots + \frac{1}{(n-1) + (n+1)}$$

$$+ \frac{1}{n + (n+1)} + \frac{1}{(n+1) + (n+1)}$$

$$- \left(\frac{1}{1+n} + \frac{1}{2+n} + \cdots + \frac{1}{n+n} \right)$$

$$= \frac{1}{2+n} + \cdots + \frac{1}{n+n} + \frac{1}{2n+1} + \frac{1}{2n+2}$$

$$- \frac{1}{1+n} - \frac{1}{2+n} - \cdots - \frac{1}{n+n}$$

$$= \frac{1}{2n+1} + \frac{1}{2n+2} - \frac{1}{1+n} = \frac{2n+2+2n+1-4n-2}{2(2n+1)(n+1)^2}$$

$$= \frac{1}{2(2n+1)(n+1)^2} > 0.$$

Thus, (u_n) is increasing and, in particular, monotone.

Therefore, since the sequence (u_n) is bounded and monotone, we have by the Monotone Convergence Theorem that (u_n) is convergent. Moreover, the sequence (u_n) is bounded between $\frac{1}{2}$ and 1, so the limit of (u_n) is in the interval $\left[\frac{1}{2}, 1 \right]$.

Moreover, we can go even further here and find the precise value of $\lim_{n \to \infty} u_n$ keeping in mind the definition of the famous Euler-Mascheroni constant:

$$\gamma = \lim_{n \to \infty} (h_n - \log n),$$

where $h_n = \sum_{k=1}^{n} 1/k$. Notice that, if we denote $\gamma_n = h_n - \log n$ for every $n \in \mathbb{N}$ we have that:

$$u_n = \frac{1}{1+n} + \frac{1}{2+n} + \frac{1}{3+n} + \cdots + \frac{1}{n+n}$$

$$= h_{2n} - h_n$$

$$= \gamma_{2n} + \log(2n) - (\gamma_n + \log n)$$

$$= \gamma_{2n} - \gamma_n + \log(2n) - \log n$$

$$= \gamma_{2n} - \gamma_n - \log 2.$$

Next, since (γ_n) is convergent to γ, we have that

$$\lim_{n \to \infty} (\gamma_{2n} - \gamma_n) = 0$$

and, thus,

$$\lim_{n \to \infty} \left(\frac{1}{1+n} + \frac{1}{2+n} + \frac{1}{3+n} + \cdots + \frac{1}{n+n} \right) = \log 2.$$

$\square$

Exercise 3.12. *Study the monotonicity of the sequences*

(I) $\dfrac{1}{\sqrt{n}} + \dfrac{1}{\sqrt{n+1}} + \cdots + \dfrac{1}{\sqrt{n+n}}$,

(II) $\dfrac{n!}{\alpha(\alpha+1)(\alpha+2)\cdots(\alpha+n-1)}$ *with* $\alpha > 0$.

Is it possible to prove the convergence for any of them?

Solution to Exercise 3.12.

(I). For every $n \in \mathbb{N}$, let $x_n = \frac{1}{\sqrt{n}} + \frac{1}{\sqrt{n+1}} + \cdots + \frac{1}{\sqrt{n+n}}$. For each $n \in \mathbb{N}$, we have that

$$x_{n+1} - x_n$$

$$= \frac{1}{\sqrt{n+1}} + \frac{1}{\sqrt{(n+1)+1}} + \cdots + \frac{1}{\sqrt{(n+1)+(n-1)}}$$

$$+ \frac{1}{\sqrt{(n+1)+n}} + \frac{1}{\sqrt{(n+1)+(n+1)}}$$

$$- \left(\frac{1}{\sqrt{n}} + \frac{1}{\sqrt{n+1}} + \cdots + \frac{1}{\sqrt{n+n}} \right)$$

$$= \frac{1}{\sqrt{n+1}} + \frac{1}{\sqrt{n+2}} + \cdots + \frac{1}{\sqrt{n+n}} + \frac{1}{\sqrt{2n+1}} + \frac{1}{\sqrt{2n+2}}$$

$$- \frac{1}{\sqrt{n}} - \frac{1}{\sqrt{n+1}} - \cdots - \frac{1}{\sqrt{n+n}}$$

$$= \frac{1}{\sqrt{2n+1}} + \frac{1}{\sqrt{2n+2}} - \frac{1}{\sqrt{n}}$$

$$= \frac{1}{\sqrt{2n+1}} + \frac{1}{\sqrt{2n+2}} - \frac{1}{2\sqrt{n}} - \frac{1}{2\sqrt{n}}.$$

Notice that

$$\frac{1}{\sqrt{2n+1}} - \frac{1}{2\sqrt{n}} > \frac{1}{\sqrt{2n+2}} - \frac{1}{2\sqrt{n}}.$$

Now, since for every $n \in \mathbb{N}$,

$$4n > 2n+1 \implies \sqrt{4n} > \sqrt{2n+1} \implies \frac{1}{2\sqrt{n}} < \frac{1}{\sqrt{2n+1}},$$

we have that $\frac{1}{\sqrt{2n+2}} - \frac{1}{2\sqrt{n}} > 0$. In particular, $\frac{1}{\sqrt{2n+1}} - \frac{1}{2\sqrt{n}} > 0$ and $\frac{1}{\sqrt{2n+2}} - \frac{1}{2\sqrt{n}} > 0$. Therefore,

$$x_{n+1} - x_n = \frac{1}{\sqrt{2n+1}} - \frac{1}{2\sqrt{n}} + \frac{1}{\sqrt{2n+2}} - \frac{1}{2\sqrt{n}} > 0.$$

Hence, (x_n) is increasing.

Let us study now if the sequence (x_n) is convergent. For every $n \in \mathbb{N}$, we have $\frac{1}{\sqrt{n+k}} \geq \frac{1}{\sqrt{n+n}}$ for every integer $0 \leq k \leq n$. Hence, for every $n \in \mathbb{N}$,

$$\frac{1}{\sqrt{n}} + \frac{1}{\sqrt{n+1}} + \cdots + \frac{1}{\sqrt{n+n}} \geq \underbrace{\frac{1}{\sqrt{n+n}} + \cdots + \frac{1}{\sqrt{n+n}}}_{n+1 \text{ times}} = \frac{n+1}{\sqrt{2n}}.$$

Since $\lim_{n\to\infty} \frac{n+1}{\sqrt{2n}} = \infty$, we have by the Comparison Theorem that

$$\lim_{n\to\infty} \left(\frac{1}{\sqrt{n}} + \frac{1}{\sqrt{n+1}} + \cdots + \frac{1}{\sqrt{n+n}} \right) = \infty.$$

(II). For every $n \in \mathbb{N}$, let $y_n = \frac{n!}{\alpha(\alpha+1)(\alpha+2)\cdots(\alpha+n-1)}$ and observe that $y_n \neq 0$. For each $n \in \mathbb{N}$,

$$\begin{aligned}
\frac{y_{n+1}}{y_n} &= \frac{\frac{(n+1)!}{\alpha(\alpha+1)\ldots(\alpha+(n+1)-2)(\alpha+(n+1)-1)}}{\frac{n!}{\alpha(\alpha+1)\ldots(\alpha+n-1)}} \\
&= \frac{(n+1)!\alpha(\alpha+1)\cdots(\alpha+n-1)}{n!\alpha(\alpha+1)\cdots(\alpha+n-1)(\alpha+n)} = \frac{(n+1)n!}{n!(\alpha+n)} \\
&= \frac{n+1}{\alpha+n}.
\end{aligned}$$

We will distinguish three cases depending on the value of α:

Case 1: Assume that $0 < \alpha < 1$. For every $n \in \mathbb{N}$, we have $n+1 > \alpha+n$, which implies that

$$\frac{y_{n+1}}{y_n} = \frac{n+1}{\alpha+n} > 1,$$

that is, $y_{n+1} > y_n$. Hence, (y_n) is increasing.

Case 2: Assume that $\alpha > 1$. For every $n \in \mathbb{N}$, we have $n+1 < \alpha+n$, which implies that

$$\frac{y_{n+1}}{y_n} = \frac{n+1}{\alpha+n} < 1,$$

that is, $y_{n+1} < y_n$. Hence, (y_n) is decreasing.

Case 3: If $\alpha = 1$, then

$$\frac{y_{n+1}}{y_n} = \frac{n+1}{1+n} = 1,$$

that is, $y_{n+1} = y_n$. Hence, (y_n) is a constant sequence. In fact,

$$y_n = \frac{n!}{1(1+1)(1+2)\cdots(1+n-1)} = \frac{n!}{1 \cdot 2 \cdot 3 \cdots n} = \frac{n!}{n!} = 1,$$

that is, (y_n) is the constant sequence 1.

Let us study now the converge of (y_n). Obviously we will distinguish once again three cases depending on the value of α:

Case 1: Assume that $0 < \alpha < 1$. It can be proved (not elementary) that the sequence (y_n) is not upper bounded. Since it is increasing, we get that it diverges.

Case 2: Assume that $\alpha > 1$. Recall that (y_n) is decreasing and observe that the sequence is bounded from below by 0. Therefore, we have by the Monotone Converge Theorem that the limit of (y_n) exists.

Case 3: If $\alpha = 1$, then the sequence (y_n) is the constant sequence 1 so the limit is 1.

$\square$

Exercise 3.13. *Find, if it exists, the limit of the sequence given by* $s_{n+1} = \frac{n}{2n+1} s_n$.

Solution to Exercise 3.13.

Observe that if $s_{n_0} = 0$ for some $n_0 \in \mathbb{N}$, then $s_n = 0$ for every $n \in \mathbb{N}$, that is, the sequence (s_n) is the 0 sequence so it converges to 0. Thus, assume that $s_n \neq 0$ for every $n \in \mathbb{N}$. We will apply the Ratio Test:

$$\lim_{n \to \infty} \frac{s_{n+1}}{s_n} = \lim_{n \to \infty} \frac{\frac{n}{2n+1} s_n}{s_n} = \lim_{n \to \infty} \frac{n}{2n+1} = \frac{1}{2} < 1,$$

that is, (s_n) converges to 0.

$\square$

Exercise 3.14. *Study the convergence of the following sequences:*

(I) $x_1 = 1$ *and* $x_{n+1} = \frac{x_n}{3} + 4$, $n \in \mathbb{N}$.

(II) $x_1 > 1$ *and* $x_{n+1} = 2 - \frac{1}{x_n}$, $n \in \mathbb{N}$.

(III) $x_1 = \sqrt{2}$ *and* $x_{n+1} = \sqrt{2\sqrt{x_n}}$, $n \in \mathbb{N}$.

(IV) $x_1 > 0$ *and* $x_{n+1} = \sqrt{a + x_n}$, $n \in \mathbb{N}$, *where* $a > 0$.

(V) $x_1 > 1$ *and* $x_{n+1} = \sqrt{1 + x_n^2}$, $n \in \mathbb{N}$.

(VI) $x_1 > 0$ *and* $x_{n+1} = \frac{1}{2}\left(x_n + \frac{a}{x_n}\right)$, $n \in \mathbb{N}$, *where* $a > 0$.

(VII) $x_1 = m$ *and* $x_{n+1} = x_n - \frac{x_n^2 - m}{2x_n}$, $n \in \mathbb{N}$, *where* $m \in \mathbb{N}$.

(VIII) $x_1 < x_2$ *and* $x_n = \frac{1}{2}(x_{n-2} + x_{n-1})$, $n \geq 3$.

(IX) $x_1 < x_2$ *and* $x_n = \frac{1}{3}x_{n-1} + \frac{2}{3}x_{n-2}$, $n \geq 3$.

Solution to Exercise 3.14.

(I). Let us find a general expression for x_n:

- For $n = 2$, $x_2 = \frac{x_1}{3} + 4 = \frac{1}{3} + 4 = \frac{1+12}{3}$.

- For $n = 3$, $x_3 = \frac{x_2}{3} + 4 = \frac{\frac{1+12}{3}}{3} + 4 = \frac{1+12+12\cdot3}{3^2}$.

- For $n = 4$, $x_4 = \frac{x_3}{3} + 4 = \frac{\frac{1+12+12\cdot3}{3^2}}{3} + 4 = \frac{1+12+12\cdot3+12\cdot3^2}{3^3}$.

We claim that $x_n = \frac{1+12\sum_{j=0}^{n-2} 3^j}{3^{n-1}}$ for every $n \in \mathbb{N} \setminus \{1\}$. Indeed, we will prove it by induction on $n \in \mathbb{N} \setminus \{1\}$.

- Base case ($n = 2$): We have already proven the claim for $n = 2$.

- Induction step. Assume that $x_n = \frac{1+12\sum_{j=0}^{n-2} 3^j}{3^{n-1}}$ (Induction Hypothesis) for some $n \in \mathbb{N} \setminus \{1\}$ and let us prove that $x_{n+1} = \frac{1+12\sum_{j=0}^{n-1} 3^j}{3^n}$:

$$x_{n+1} = \frac{x_n}{3} + 4 = \frac{\frac{1+12\sum_{j=0}^{n-2} 3^j}{3^{n-1}}}{3} + 4 = \frac{1 + 12\sum_{j=0}^{n-2} 3^j + 12\cdot3^{n-1}}{3^n}$$
$$= \frac{1 + 12\sum_{j=0}^{n-1} 3^j}{3^n}.$$

By Exercise 1.7 with $x = 3$ we have for every $n \in \mathbb{N} \setminus \{1\}$ that

$$x_n = \frac{1 + 12\cdot\frac{3^{n-1}-1}{2}}{3^{n-1}} = \frac{1 + 6\left(3^{n-1} - 1\right)}{3^{n-1}} = -\frac{5}{3^{n-1}} + 6.$$

Hence, by taking the limit, we have

$$\lim_{n\to\infty} x_n = \lim_{n\to\infty}\left(-\frac{5}{3^{n-1}} + 6\right) = 6.$$

(II). First, we will study if (x_n) is bounded. Observe that $x_1 > 1$, so $1 < 2 - \frac{1}{x_1} = x_2 < 2$. Once again note that $1 < 2 - \frac{1}{x_2} = x_3 < 2$. We claim that $1 < x_n < 2$ for every $n \in \mathbb{N} \setminus \{1\}$. Indeed, we will prove the claim by induction on $n \in \mathbb{N} \setminus \{1\}$.

- Base case ($n = 2$). We have already proven that $1 < x_2 < 2$.

- Induction step. Assume that $1 < x_n < 2$ (Induction Hypothesis) for some $n \in \mathbb{N} \setminus \{1\}$, we will prove that $1 < x_{n+1} < 2$. Since $x_n > 1$ by the Induction Hypothesis, we have that

$$1 > \frac{1}{x_n} > 0 \implies -1 < -\frac{1}{x_n} < 0 \implies 1 < 2 - \frac{1}{x_n} = x_{n+1} < 2.$$

Hence, (x_n) is bounded between 1 and 2.

Second, we will study the monotonicity of (x_n) as follows:

$$x_{n+1} - x_n = 2 - \frac{1}{x_n} - x_n = \frac{2x_n - 1 - x_n^2}{x_n} = -\frac{(x_n - 1)^2}{x_n} < 0,$$

where the last inequality is satisfied since $x_n > 1$ for every $n \in \mathbb{N}$. Therefore, $x_{n+1} < x_n$, which implies that (x_n) is decreasing.

Hence, by the Monotone Convergence Theorem, we have that (x_n) is convergent. Moreover, we can calculate the limit of (x_n). Indeed, let $l = \lim_{n \to \infty} x_n$. Since $\lim_{n \to \infty} x_{n+1} = \lim_{n \to \infty} x_n$ and $\lim_{n \to \infty} x_{n+1} = \lim_{n \to \infty} \left(2 - \frac{1}{x_n} \right)$, we have that $l = 2 - \frac{1}{l}$ which yields $l = 1$.

(III). Let us find a general expression for x_n:

- For $n = 2$, $x_2 = \sqrt{2\sqrt{x_1}} = \sqrt{2\sqrt{2}} = 2^{\frac{1}{2} + \frac{1}{2^2}}$.

- For $n = 3$, $x_3 = \sqrt{2\sqrt{x_2}} = \sqrt{2 \cdot 2^{\frac{1}{2} + \frac{1}{2^2}}} = 2^{\frac{1}{2} + \frac{1}{2^2} + \frac{1}{2^3}}$.

It can be easily seen, by induction, that for every $n \in \mathbb{N}$ we have

$$x_n = 2^{\sum_{k=1}^{n} 1/2^k}.$$

To find the value of $\lim_{n \to \infty} x_n$, let us note that the sequence $s_n = \sum_{k=1}^{n} 1/2^k$ is a geometric sequence, and we are already familiar with it, its value is given by

$$s_n = \sum_{k=1}^{n} 1/2^k = 1 - \frac{1}{2^n} \xrightarrow{n \to \infty} 1,$$

and, therefore,

$$\lim_{n \to \infty} x_n = 2^{\sum_{k=1}^{n} 1/2^k} = 2^1 = 2.$$

(IV). Notice that $x_n > 0$ for every $n \in \mathbb{N}$ since $x_1 > 0$ and $a > 0$.

Now we will study the monotonicity of (x_n). For every $n \in \mathbb{N} \setminus \{1\}$,

$$(x_{n+1} + x_n)(x_{n+1} - x_n) = x_{n+1}^2 - x_n^2 = a + x_n - (a + x_{n-1}) = x_n - x_{n-1}.$$

Since $x_{n+1} + x_n > 0$, we have that

$$\text{sign}(x_{n+1} - x_n) = \text{sign}(x_n - x_{n-1}).$$

Repeating this process, we arrive at

$$\text{sign}(x_{n+1} - x_n) = \text{sign}(x_2 - x_1).$$

Hence, (x_n) is monotone.

Let us study now if (x_n) is increasing or decreasing, and if (x_n) is bounded from below or from above. First,

$$x_{n+1} - x_n = \sqrt{a + x_n} - x_n = \frac{\left(\sqrt{a + x_n} - x_n\right)\left(\sqrt{a + x_n} + x_n\right)}{\sqrt{a + x_n} + x_n}$$

$$= \frac{a + x_n - x_n^2}{\sqrt{a + x_n} + x_n} = -\frac{x_n^2 - x_n - a}{\sqrt{a + x_n} + x_n}.$$

By the second degree equation, we have that the solutions of $x_n^2 - x_n - a = 0$ are $\alpha_1 = \frac{1 + \sqrt{1 + 4a}}{2}$ and $\alpha_2 = \frac{1 - \sqrt{1 + 4a}}{2}$. Hence,

$$x_{n+1} - x_n = -\frac{(x_n - \alpha_1)(x_n - \alpha_2)}{\sqrt{a + x_n} + x_n} = \frac{(\alpha_1 - x_n)(x_n - \alpha_2)}{\sqrt{a + x_n} + x_n}.$$

Notice that $\alpha_1 > 0$ and $\alpha_2 < 0$, so $x_n - \alpha_2 > 0$ since $x_n > 0$ for each $n \in \mathbb{N}$. Thus, since $\sqrt{a + x_n} + x_n > 0$ for each $n \in \mathbb{N}$, we have that

$$\text{sign}(x_{n+1} - x_n) = \text{sign}(\alpha_1 - x_n),$$

for every $n \in \mathbb{N}$. Recall that we have proven that (x_n) is monotone, so the sign of $\alpha_1 - x_n$ remains the same (either ≥ 0 or < 0), independently of $n \in \mathbb{N}$. We will now distinguish two cases:

Case 1: If $\alpha_1 \geq x_1$, then $\alpha_1 \geq x_n$ (the sign does not change). Hence, $x_{n+1} \geq x_n$. Therefore, (x_n) is increasing and notice that it is bounded from above by α_1.

Case 2: If $\alpha_1 < x_1$, then $\alpha_1 < x_n$ for every $n \in \mathbb{N}$ (the sign does not change). Hence, $x_{n+1} < x_n$ for every $n \in \mathbb{N}$. Therefore, (x_n) is decreasing and notice that it is bounded from below by α_1.

In any case, by the Monotone Convergence Theorem, we have that (x_n) converges. Moreover, we can calculate the limit of (x_n) as follows. Since (x_n) converges, let $l = \lim_{n \to \infty} x_n$. As $\lim_{n \to \infty} x_{n+1} = \lim_{n \to \infty} x_n$ and $\lim_{n \to \infty} x_{n+1} = \lim_{n \to \infty} \sqrt{a + x_n}$, we have that

$$l = \sqrt{a + l} \implies l^2 = a + l \implies l^2 - l - a = 0.$$

By the second degree equation, the solutions of $l^2 - l - a = 0$ are $l = \alpha_1$ and $l = \alpha_2$. Since $x_n > 0$ for every $n \in \mathbb{N}$ and $\alpha_2 < 0$, we can dismiss the case of $l = \alpha_2$, which yields $l = \alpha_1$.

(v). We begin by studying the monotonicity of the sequence. Observe that $x_1 > 1$ and $x_{n+1} = \sqrt{1 + x_n^2} > x_n$. Hence, the sequence (x_n) is increasing and it is bounded from below by 1. Assume, by means of contradiction, that the limit of (x_n) exists. Let $l = \lim_{n \to \infty} x_n$. Then, since $\lim_{n \to \infty} x_{n+1} = \lim_{n \to \infty} x_n$ and $\lim_{n \to \infty} x_{n+1} = \lim_{n \to \infty} \sqrt{1 + x_n^2}$, we have that

$$l = \sqrt{1 + l^2} \implies l^2 = 1 + l^2 \implies 0 = 1,$$

that is, a contradiction. Recall that (x_n) is increasing and we have proven that the limit does not exists, so (x_n) diverges to ∞.

(VI). We begin by studying the boundedness of (x_n). First, let us prove that $x_n > 0$ for every $n \in \mathbb{N}$ by induction.

- Base case ($n = 1$). By hypothesis $x_1 > 0$.

- Induction step. Assume that $x_n > 0$ (Induction Hypothesis) for some $n \in \mathbb{N}$ and let us prove that $x_{n+1} > 0$:

$$x_{n+1} = \frac{1}{2}\left(x_n + \frac{a}{x_n}\right) > \frac{a}{2x_n} > 0,$$

where in the last inequality we have used that $a > 0$ and $x_n > 0$ by the Induction Hypothesis.

Now, for every $n \in \mathbb{N}$, we have that

$$x_{n+1} = \frac{1}{2}\left(x_n + \frac{a}{x_n}\right) = \frac{x_n^2 + a}{2x_n} = \frac{(x_n - \sqrt{a})^2 + 2\sqrt{a}x_n}{2x_n}$$

$$= \frac{(x_n - \sqrt{a})^2}{2x_n} + \frac{2\sqrt{a}x_n}{2x_n} = \frac{(x_n - \sqrt{a})^2}{2x_n} + \sqrt{a} \geq \sqrt{a}$$

since $x_n > 0$ for each $n \in \mathbb{N}$. Hence, $x_n \geq \sqrt{a}$ for every $n \in \mathbb{N} \setminus \{1\}$.

Now let us study the monotonicity of (x_n) as follows. For every $n \in \mathbb{N} \setminus \{1\}$, we have

$$x_{n+1} - x_n = \frac{1}{2}\left(x_n + \frac{a}{x_n}\right) - x_n = \frac{1}{2x_n} + \frac{a}{2x_n} - x_n = -\frac{1}{2x_n} + \frac{a}{2x_n}$$

$$= \frac{1}{2}\left(\frac{a}{x_n} - x_n\right) = \frac{1}{2x_n}\left(\frac{a - x_n^2}{x_n}\right) = \frac{1}{2}\left(\sqrt{a} - x_n\right)\left(\sqrt{a} + x_n\right) \leq 0,$$

where in the last inequality we have used that $x_n \geq \sqrt{a} > 0$ for each $n \in \mathbb{N} \setminus \{1\}$, that is, $\sqrt{a} - x_n \leq 0$ and $\sqrt{a} + x_n > 0$. Thus, $x_{n+1} \leq x_n$, that is, (x_n) is decreasing.

Therefore, by the Monotone Convergence Theorem, we have that (x_n) converges. Moreover, we can calculate the limit of (x_n). Indeed, let $l = \lim_{n \to \infty} x_n$. Notice that $l \neq 0$ as $x_n \geq \sqrt{a} > 0$ for every $n \in \mathbb{N}$. Since $\lim_{n \to \infty} x_{n+1} = \lim_{n \to \infty} x_n$ and $\lim_{n \to \infty} x_{n+1} = \lim_{n \to \infty} \frac{1}{2}\left(x_n + \frac{a}{x_n}\right)$, we have that

$$l = \frac{1}{2}\left(l + \frac{a}{l}\right) \implies 2l = \frac{l^2 + a}{l} \implies 2l^2 = l^2 + a \implies l^2 = a.$$

Hence, the limit l is either $\sqrt{a}$ or $-\sqrt{a}$ by the uniqueness of the limit. Since $x_n > 0$ for every $n \in \mathbb{N}$ and $a > 0$, we have $l = \sqrt{a}$.

(VII). We begin by studying the boundedness of (x_n). First, let us prove that $x_n > 0$ for every $n \in \mathbb{N}$ by induction.

- Base case ($n = 1$). By hypothesis $x_1 = m \in \mathbb{N}$, so $x_1 > 0$.

- Induction step. Assume that $x_n > 0$ (Induction Hypothesis) for some $n \in \mathbb{N}$ and let us prove that $x_{n+1} > 0$:

$$x_{n+1} = x_n - \frac{x_n^2 - m}{2x_n} = \frac{2x_n^2 - x_n^2 + m}{2x_n} = \frac{x_n^2 + m}{2x_n} > 0,$$

where in the last inequality we have used that $x_n^2 + m > 0$ since $m > 0$ and the Induction Hypothesis.

Now, for every $n \in \mathbb{N}$, we have that

$$x_{n+1} = x_n - \frac{x_n^2 - m}{2x_n} = \frac{x_n^2 + m}{2x_n} = \frac{(x_n - \sqrt{m})^2 + 2\sqrt{m}x_n}{2x_n}$$

$$= \frac{(x_n - \sqrt{m})^2}{2x_n} + \frac{2\sqrt{m}x_n}{2x_n} = \frac{(x_n - \sqrt{m})^2}{2x_n} + \sqrt{m} \geq \sqrt{m},$$

where in the last inequality we have used that $x_n > 0$ for every $n \in \mathbb{N}$. Hence, notice that $x_n \geq \sqrt{m}$ for every $n \in \mathbb{N}$.

We will study now the monotonicity of (x_n) as follows:

$$x_{n+1} - x_n = x_n - \frac{x_n^2 - m}{2x_n} - x_n = \frac{m - x_n^2}{2x_n}$$

$$= \frac{1}{2x_n} \left(\sqrt{m} - x_n\right) \left(\sqrt{m} + x_n\right) \leq 0,$$

where in the last inequality we have used that $x_n > 0$ and $x_n \geq \sqrt{m} > 0$ for every $n \in \mathbb{N}$, that is, $x_n > 0$, $\sqrt{m} - x_n \leq 0$ and $\sqrt{m} + x_n > 0$ for every $n \in \mathbb{N}$. Thus, $x_{n+1} \leq x_n$ for every $n \in \mathbb{N}$, that is, (x_n) is decreasing.

Therefore, by the Monotone Convergence Theorem, we have that (x_n) converges. Moreover, we can calculate the limit of (x_n). Indeed, let $l = \lim_{n\to\infty} x_n$. Notice that $l \neq 0$ as $x_n \geq \sqrt{m} > 0$ for every $n \in \mathbb{N}$. Since $\lim_{n\to\infty} x_{n+1} = \lim_{n\to\infty} x_n$ and $\lim_{n\to\infty} x_{n+1} = \lim_{n\to\infty} \left(x_n - \frac{x_n^2 - m}{2x_n}\right)$, we have that

$$l = l - \frac{l^2 - m}{2l} \implies \frac{l^2 - m}{2l} = 0 \implies l^2 - m = 0 \implies l^2 = m.$$

Hence, the limit l is either $\sqrt{m}$ or $-\sqrt{m}$ by the uniqueness of the limit. Since $x_n > 0$ for every $n \in \mathbb{N}$ and $m > 0$, we have $l = \sqrt{m}$.

(VIII). Instead of using the equations $x_n = \frac{1}{2}(x_{n-2} + x_{n-1})$, $n \geq 3$, we will be using the equivalent equations $x_{n+2} = \frac{1}{2}(x_n + x_{n+1})$, $n \geq 1$. We will use now the characteristic equation $r^{n+2} = \frac{1}{2}(r^n + r^{n+1})$. Multiplying the latter equation by $\frac{1}{r^n}$ yields:

$$r^2 = \frac{1}{2}(1 + r) \implies 2r^2 - r - 1 = 0.$$

By the second degree equation formula, the solutions of $2r^2 - r - 1 = 0$ are $r = 1$ and $r = -\frac{1}{2}$. Therefore, for every $n \in \mathbb{N}$, we have $x_n = a \cdot 1^n + b\left(-\frac{1}{2}\right)^n = a + b\left(-\frac{1}{2}\right)^n$. Hence, for $n = 1$ and $n = 2$, we have the system of equations

$$\begin{cases} x_1 = a - \frac{1}{2}b \\ x_2 = a + \frac{1}{4}b \end{cases} \implies \begin{cases} -x_1 = -a + \frac{b}{2} \\ x_2 = a + \frac{b}{4} \end{cases} \implies x_2 - x_1 = \frac{3b}{4}.$$

Hence $b = \frac{4(x_2 - x_1)}{3}$, which implies also that

$$a = x_1 + \frac{b}{2} = x_1 + \frac{2(x_2 - x_1)}{3} = \frac{2x_2 + x_1}{3}.$$

Thus, for every $n \in \mathbb{N}$, we have

$$x_n = \frac{x_1 + 2x_2}{3} + \frac{4(x_2 - x_1)}{3}\left(-\frac{1}{2}\right)^n.$$

Therefore, by taking the limit, we arrive at

$$\lim_{n \to \infty} x_n = \lim_{n \to \infty} \left[\frac{x_1 + 2x_2}{3} + \frac{4(x_2 - x_1)}{3}\left(-\frac{1}{2}\right)^n\right] = \frac{x_1 + 2x_2}{3}.$$

(IX). Instead of using the equations $x_n = \frac{1}{3}x_{n-1} + \frac{2}{3}x_{n-2}$, $n \geq 3$, we will be using the equivalent equations $x_{n+2} = \frac{1}{3}x_{n+1} + \frac{2}{3}x_n$, $n \geq 1$. We will use now the characteristic equation $r^{n+2} = \frac{1}{3}r^{n+1} + \frac{2}{3}r$. Multiplying the latter equation by $\frac{1}{r^n}$ yields:

$$r^2 = \frac{1}{3}r + \frac{2}{3} \implies 3r^2 - r - 2 = 0.$$

By the second degree equation formula, the solutions of $3r^2 - r - 2 = 0$ are $r = 1$ and $r = -\frac{2}{3}$. Therefore, for every $n \in \mathbb{N}$, we have $x_n = a \cdot 1^n + b\left(-\frac{2}{3}\right)^n = a + b\left(-\frac{2}{3}\right)^n$. Hence, for $n = 1$ and $n = 2$, we have the system of equations

$$\begin{cases} x_1 = a - \frac{2}{3}b \\ x_2 = a + \frac{4}{9}b \end{cases} \iff \begin{cases} -x_1 = -a + \frac{2b}{3} \\ x_2 = a + \frac{4b}{9} \end{cases} \implies x_2 - x_1 = \frac{10b}{9}.$$

Hence $b = \frac{9(x_2 - x_1)}{10}$, which implies also that

$$a = x_1 + \frac{2b}{3} = x_1 + \frac{3(x_2 - x_1)}{5} = \frac{2x_1 + 3x_2}{5}.$$

Thus, for every $n \in \mathbb{N}$, we have

$$x_n = \frac{2x_1 + 3x_2}{5} + \frac{9(x_2 - x_1)}{10}\left(-\frac{2}{3}\right)^n.$$

Therefore, by taking the limit, we arrive at

$$\lim_{n \to \infty} x_n = \lim_{n \to \infty} \left[\frac{2x_1 + 3x_2}{5} + \frac{9(x_2 - x_1)}{10}\left(-\frac{2}{3}\right)^n\right] = \frac{2x_1 + 3x_2}{5}.$$

$\square$

Exercise 3.15. *Let (x_n) be the sequence defined by*

$$\sqrt{n + \sqrt{(n-1) + \cdots + \sqrt{2 + \sqrt{1}}}}.$$

(I) *Prove that for every $n \in \mathbb{N}$ we have*

$$\sqrt{n} \leq x_n < 2\sqrt{n}.$$

(II) *Prove that the sequence $\frac{x_n}{\sqrt{n}}$ converges to 1.*

Solution to Exercise 3.15.

(I). For this first part, let us first notice that it is straightforward to check that $x_n \geq \sqrt{n}$, since $x_n = \sqrt{n + A}$, with $A > 0$, and since $\sqrt{n + A} \geq \sqrt{n}$ for every $A > 0$, we are done. Now we need to check that $x_n < 2\sqrt{n}$ for every positive integer n. We proceed by induction on n. The base case is $n = 1$, which is clearly true, since $x_1 = 1 < 2\sqrt{1} = 2$. Thus, let us assume that, for certain n, we have that $x_n < 2\sqrt{n}$ (Induction Hypothesis). Let us show the property for $n + 1$. Using the Induction Hypothesis, we have

$$x_{n+1} = \sqrt{(n+1) + x_n} < \sqrt{(n+1) + 2\sqrt{n}}.$$

If we are able to show that $\sqrt{(n+1) + 2\sqrt{n}} \leq 2\sqrt{n+1}$ we will be done. Indeed,

$$\sqrt{(n+1) + 2\sqrt{n}} \leq 2\sqrt{n+1} \Leftrightarrow n + 1 + 2\sqrt{n} \leq 4n + 4 \Leftrightarrow \frac{n+3}{2\sqrt{n}} \geq 1,$$

which is true for every $n \in \mathbb{N}$ (use again induction), and we are finished with this part.

(II). To see that $\lim\limits_{n \to \infty} \dfrac{x_n}{\sqrt{n}} = 1$, notice first that

$$\sqrt{n} \leq x_n < 2\sqrt{n} \Rightarrow 1 \leq \frac{x_n}{\sqrt{n}} < 2,$$

which makes the sequence $\left(\dfrac{x_n}{\sqrt{n}}\right)$ bounded. It can also be seen that $\left(\dfrac{x_n}{\sqrt{n}}\right)$ is decreasing for $n > 2$ and, thus, our sequence is convergent. Next, let

$$\ell = \lim_{n \to} \frac{x_n}{\sqrt{n}} = \lim_{n \to} \frac{x_{n-1}}{\sqrt{n-1}}.$$

Therefore,

$$\ell = \lim_{n \to} \frac{x_n}{\sqrt{n}} = \lim_{n \to} \frac{\sqrt{n + x_{n-1}}}{\sqrt{n}} = \sqrt{1 + \frac{x_{n-1}}{n}}$$

$$= \lim_{n \to} \sqrt{1 + \frac{x_{n-1}}{\sqrt{n-1}} \cdot \frac{\sqrt{n-1}}{\sqrt{n}} \cdot \frac{1}{\sqrt{n}}}$$

$$= \sqrt{1 + \lim_{n \to} \left(\frac{x_{n-1}}{\sqrt{n-1}} \right) \cdot \lim_{n \to} \left(\frac{\sqrt{n-1}}{\sqrt{n}} \right) \cdot \lim_{n \to} \left(\frac{1}{\sqrt{n}} \right)}$$

$$= \sqrt{1 + \ell \cdot 1 \cdot 0} = 1.$$

$\square$

Exercise 3.16. *Let (a_n) and (b_n) be two sequences of real numbers defined by their first term $a_1 = 2$, $b_1 = 2$ and by the relations $a_{n+1} = \frac{a_n + 3b_n}{4}$ and $b_{n+1} = \frac{3a_n + b_n}{4}$. Discuss the veracity of the following propositions:*

(I) *The sequence $u_n = a_n + b_n$ is constant.*

(II) *The sequence $v_n = a_n - b_n$ is a geometric sequence.*

(III) *For every $n \in \mathbb{N}$, it is satisfied that $a_n = 3 - \frac{1}{2^n}$ and $b_n = 3 + \frac{1}{2^n}$.*

Solution to Exercise 3.16.

(I). For every $n \in \mathbb{N}$, we have

$$u_{n+1} - u_n = a_{n+1} + b_{n+1} - (a_n + b_n) = \frac{a_n + 3b_n}{4} + \frac{3a_n + b_n}{4} - (a_n + b_n)$$

$$= \frac{4a_n + 4b_n}{4} - (a_n + b_n) = a_n + b_n - (a_n + b_n) = 0.$$

Therefore, $u_{n+1} = u_n$ for every $n \in \mathbb{N}$, that is, (u_n) is constant. In fact, since $a_1 = 2$ and $b_1 = 2$, we have that $u_1 = 2$, which implies that (u_n) is the constant sequence 2.

(II). For every $n \in \mathbb{N}$, we have

$$\frac{v_{n+1}}{v_n} = \frac{a_{n+1} - b_{n+1}}{a_n - b_n} = \frac{\frac{a_n + 3b_n}{4} - \frac{3a_n + b_n}{4}}{a_n - b_n} = \frac{\frac{-2a_n + 2b_n}{4}}{a_n - b_n} = \frac{1}{2} \frac{b_n - a_n}{a_n - b_n} = -\frac{1}{2}.$$

Therefore, $v_{n+1} = -\frac{1}{2} v_n$, that is, (v_n) is a geometric sequence with common ratio $-\frac{1}{2}$.

(III). We will prove that the proposition does not hold by finding a counterexample. For $n = 1$ we would have that $a_1 = 3 - \frac{1}{2} = \frac{5}{2} \neq 2$ and $b_1 = 3 + \frac{1}{2} = \frac{7}{2} \neq 2$.

$\square$

Exercise 3.17. *Given two numbers a_1 and b_1, $0 < a_1 < b_1$, we define inductively the sequences (a_n) and (b_n) by*

$$a_{n+1} = \sqrt{a_n b_n}, \quad b_{n+1} = \frac{a_n + b_n}{2}, \quad n \in \mathbb{N}.$$

Prove that both converge, and they do so to the same limit.
Hint: Observe first that

$$\sqrt{ab} \leq \frac{a+b}{2}, \quad \text{if } a, b > 0.$$

Solution to Exercise 3.17.

Observe that both sequences (a_n) and (b_n) take positive numbers. Hence, since $\sqrt{ab} \leq \frac{a+b}{2}$ provided that $a, b > 0$ by the Arithmetic-Geometric Mean Inequality (see Exercise 1.40) and $a_1 < b_1$, we have that $a_n \leq b_n$ for every $n \in \mathbb{N}$. Therefore, since $a_n > 0$ for every $n \in \mathbb{N}$, we have that

$$a_{n+1} = \sqrt{a_n b_n} \geq \sqrt{a_n^2} = |a_n| = a_n$$

and

$$b_{n+1} = \frac{a_n + b_n}{2} \leq \frac{2b_n}{2} = b_n.$$

Thus (a_n) is increasing while (b_n) is decreasing. Moreover, the latter and $b_n > 0$ imply that for every $n \in \mathbb{N}$ we have

$$a_{n+1} = \sqrt{a_n b_n} \leq \sqrt{b_n^2} = |b_n| = b_n \leq b_1$$

and

$$b_{n+1} = \frac{a_n + b_n}{2} \geq \frac{2a_n}{2} = a_n \geq a_1.$$

Hence, we have that (a_n) is an increasing sequence that is bounded from above by b_1 and (b_n) is a decreasing sequence that is bounded from below by a_1. So, by the Monotone Convergence Theorem, we have proven that both (a_n) and (b_n) converge.

In order to prove that they have the same limit, let a and b be the limits of (a_n) and (b_n), respectively. Since $\lim_{n\to\infty} b_{n+1} = \lim_{n\to\infty} b_n$ and $\lim_{n\to\infty} b_{n+1} = \lim_{n\to\infty} \frac{a_n + b_n}{2}$, we have that

$$b = \frac{a+b}{2} \implies 2b = a + b \implies a = b.$$

$\square$

Exercise 3.18. *We say that a sequence (s_n) is Cesàro summable with Cesàro sum $l \in \mathbb{R}$ if the sequence of arithmetic means (σ_n) where*

$$\sigma_n = \frac{s_1 + s_2 + \cdots + s_n}{n}$$

converges to l.

(I) *Prove that if a sequence (s_n) converges to l in the classical sense, then it is also Cesàro summable with Cesàro sum l.*

(II) *Show that the reciprocal is not true; that is, find an example of a sequence that is Cesàro summable but is not convergent.*

Solution to Exercise 3.18.

(I). Assume that (s_n) converges to l in the classical sense. Since $y_n = n$ is nonzero, increasing and diverges to ∞, we can apply the Stolz Theorem to $\sigma_n = \frac{\sum_{i=1}^{n} s_i}{n}$ as follows:

$$\lim_{n \to \infty} \frac{\sum_{i=1}^{n+1} s_i - \sum_{i=1}^{n} s_i}{(n+1) - n} = \lim_{n \to \infty} \frac{s_{n+1}}{1} = \lim_{n \to \infty} s_{n+1} = l.$$

Hence,

$$\lim_{n \to \infty} \sigma_n = l.$$

(II). Take $s_n = (-1)^n$ for every $n \in \mathbb{N}$. Clearly, (s_n) is not convergent; in fact, it is alternating. However, notice that for every $n \in \mathbb{N}$ we have $\sum_{i=1}^{2n} (-1)^i = 0$ and $\sum_{i=1}^{2n-1} (-1)^i = -1$. Hence,

$$\lim_{n \to \infty} s_{2n} = \lim_{n \to \infty} \frac{\sum_{i=1}^{2n} (-1)^i}{2n} = 0$$

and

$$\lim_{n \to \infty} s_{2n-1} = \lim_{n \to \infty} \frac{\sum_{i=1}^{2n-1} (-1)^i}{2n-1} = \lim_{n \to \infty} -\frac{1}{2n-1} = 0.$$

Since the subsequences of even and odd terms both converge to 0, we have that (σ_n) converges to 0. $\square$

Exercise 3.19. *Let (x_n) be a sequence. Prove that if $\lim_{n \to \infty} x_{2n}$, $\lim_{n \to \infty} x_{2n-1}$ and $\lim_{n \to \infty} x_{3n}$ exist, then $\lim_{n \to \infty} x_n$ exists and coincides with the previous ones.*

Solution to Exercise 3.19.

Since the limits of (x_{2n-1}), (x_{2n}) and (x_{3n}) exist, we can define $l_1 := \lim_{n\to\infty} x_{2n-1}$, $l_2 := \lim_{n\to\infty} x_{2n}$ and $l_3 := \lim_{n\to\infty} x_{3n}$.

On the one hand, notice that the sequence (x_{6n}) is a subsequence of (x_{2n}) since the indexes $6n$ are multiples of 2, so the sequence (x_{6n}) converges to l_2 (recall that if a sequence converges, then every subsequence converges to the same limit). Moreover, the sequence (x_{6n}) is a subsequence of (x_{3n}) since the indexes $6n$ are multiples of 3, which implies that (x_{6n}) converges to l_3. By the uniqueness of the limit, we have that $l_2 = l_3$.

On the other hand, observe that $(x_{3(2n-1)})$ is a subsequence (x_{2n-1}) since the indexes $3(2n-1)$ are odd numbers, thus $(x_{3(2n-1)})$ converges to l_1. Furthermore, we have that $(x_{3(2n-1)})$ is also a subsequence (x_{3n}) since the indexes $3(2n-1)$ are multiples of 3; hence, $(x_{3(2n-1)})$ converges to l_3. By the uniqueness of the limit, we have that $l_1 = l_3$.

Therefore, we have proven so far that $l_1 = l_3 = l_2$, in particular, $l_1 = l_2$. Let us denote by l the real number $l_1 = l_2$. Observe that l is the limit of the subsequences of odd and even terms of (x_n). Finally, we know that if the subsequences of odd and even terms converge to the same limit, then the sequence converges to the same limit; so (x_n) converges to l and $l_3 = l_2 = l_1 = l$.

$\square$

Exercise 3.20. *For which values of $a \in \mathbb{R}$ is (a^n) a subsequence of $\left(\frac{1}{n}\right)$? And of $\left(\frac{1}{2^n}\right)$? And of $\left(\frac{1}{2n}\right)$? And of $\left(\frac{1}{2n-1}\right)$?*

Solution to Exercise 3.20.

Recall that a subsequence (x_{n_s}) of a sequence (x_n) is a sequence formed by the elements of (x_n) such that the sequence of natural numbers (n_s) is strictly increasing.

First, observe that the sequence (a^n) is of the form: a, a^2, a^3, ..., a^n, ... for any $a \in \mathbb{R}$.

- For the case of $\left(\frac{1}{n}\right)$ we have the sequence 1, $\frac{1}{2}$, $\frac{1}{3}$, $\frac{1}{4}$, ..., $\frac{1}{n}$, ... By definition of subsequence, we need the first element of (a^n) (that is a) to be equal to some element of the sequence $\left(\frac{1}{n}\right)$. Hence, we need $a = \frac{1}{k}$ for some $k \in \mathbb{N}$. Observe that if $a \neq \frac{1}{k}$ for any $k \in \mathbb{N}$, then (a^n) is not a subsequence of $\left(\frac{1}{n}\right)$. Now, for every $n \in \mathbb{N}$, we have $a^n = \frac{1}{k^n}$, where $k^n \in \mathbb{N}$ for any $k \in \mathbb{N}$ and $n \in \mathbb{N}$. Hence, (a^n) is equal to the sequence $\left(\frac{1}{k^n}\right)_{n=1}^{\infty}$, where $\left(\frac{1}{k^n}\right)_{n=1}^{\infty}$ is a subsequence of $\left(\frac{1}{n}\right)$ when $k > 1$ (since $(k^n)_{n=1}^{\infty}$ is strictly increasing only if $k > 1$). Therefore, (a^n) is a subsequence of $\left(\frac{1}{n}\right)$ if, and only if, $a = \frac{1}{k}$ for some $k \in \mathbb{N} \setminus \{1\}$.

- For the case of $\left(\frac{1}{2^n}\right)$ we have the sequence $\frac{1}{2}$, $\frac{1}{4}$, $\frac{1}{8}$, $\frac{1}{16}$ ..., $\frac{1}{2^n}$,.... As in the first case, we need $a = \frac{1}{2^k}$ for some $k \in \mathbb{N}$. If $a \neq \frac{1}{2^k}$ for any $k \in \mathbb{N}$, then (a^n) is not a subsequence of $\left(\frac{1}{2^n}\right)$. Now, $a^n = \frac{1}{(2^k)^n} = \frac{1}{2^{kn}}$ where $kn \in \mathbb{N}$ for any $k \in \mathbb{N}$ and $n \in \mathbb{N}$. Hence, (a^n) is equal to the sequence $\left(\frac{1}{2^{kn}}\right)_{n=1}^{\infty}$, where $\left(\frac{1}{2^{kn}}\right)_{n=1}^{\infty}$ is a subsequence of $\left(\frac{1}{2^n}\right)$ for any $k \in \mathbb{N}$ (as $(kn)_{n=1}^{\infty}$ is strictly increasing for any $k \in \mathbb{N}$). Therefore, (a^n) is a subsequence of $\left(\frac{1}{2^n}\right)$ if, and only if, $a = \frac{1}{2^k}$ for some $k \in \mathbb{N}$.

- For the case of $\left(\frac{1}{2n}\right)$ we have the sequence $\frac{1}{2}$, $\frac{1}{4}$, $\frac{1}{6}$, $\frac{1}{8}$ ..., $\frac{1}{2n}$,.... As in the previous cases, we need $a = \frac{1}{2k}$ for some $k \in \mathbb{N}$. If $a \neq \frac{1}{2k}$ for any $k \in \mathbb{N}$, then (a^n) is not a subsequence of $\left(\frac{1}{2n}\right)$. Note that $a^n = \frac{1}{(2k)^n} = \frac{1}{2^n k^n} = \frac{1}{2(2^{n-1}k^n)}$, where $2(2^{n-1}k^n)$ is a multiple of 2 for any $k \in \mathbb{N}$ and $n \in \mathbb{N}$. Hence, (a^n) is equal to the sequence $\left(\frac{1}{2(2^{n-1}k^n)}\right)_{n=1}^{\infty}$, where $\left(\frac{1}{2(2^{n-1}k^n)}\right)_{n=1}^{\infty}$ is a subsequence of $\left(\frac{1}{2n}\right)$ for any $k \in \mathbb{N}$ (since $(2^{n-1}k^n)_{n=1}^{\infty}$ is strictly increasing for any $k \in \mathbb{N}$). Therefore, (a^n) is a subsequence of $\left(\frac{1}{2n}\right)$ if, and only if, $a = \frac{1}{2k}$ for some $k \in \mathbb{N}$.

- For the case of $\left(\frac{1}{2n-1}\right)$ we have the sequence 1, $\frac{1}{3}$, $\frac{1}{5}$, $\frac{1}{7}$ ..., $\frac{1}{2n-1}$,.... As in the previous cases, we need $a = \frac{1}{2k-1}$ for some $k \in \mathbb{N}$. If $a \neq \frac{1}{2k-1}$ for any $k \in \mathbb{N}$, then (a^n) is not a subsequence of $\left(\frac{1}{2n-1}\right)$. Note that $a^n = \frac{1}{(2k-1)^n}$ where $(2k-1)^n$ is an odd number for any $k \in \mathbb{N}$ and $n \in \mathbb{N}$. Hence, (a^n) is equal to the sequence $\left(\frac{1}{(2k-1)^n}\right)_{n=1}^{\infty}$ which is a subsequence of $\left(\frac{1}{2n-1}\right)$ when $k > 1$. Indeed, if $k = 1$, then (a^n) is the constant sequence 1, that is, (a^n) is not a subsequence of $\left(\frac{1}{2n-1}\right)$. If $k > 1$, then observe that the sequence of natural numbers $(r_n)_{n=1}^{\infty}$ such that $(2r_n - 1)_{n=1}^{\infty} = ((2k-1)^n)_{n=1}^{\infty}$ is strictly increasing since $((2k-1)^n)_{n=1}^{\infty}$ is strictly increasing. Therefore, (a^n) is a subsequence of $\left(\frac{1}{2n-1}\right)$ if, and only if, $a = \frac{1}{2k-1}$ for some $k \in \mathbb{N} \setminus \{1\}$.

□

Exercise 3.21. *Construct a sequence (s_n) with the following property, or show that there does not exist such sequence:*

> *"For every natural number m, (s_n) has a subsequence that converges to m."*

Solution to Exercise 3.21.

Since $\mathbb{Q}$ is countably infinite, there exists a bijection s from $\mathbb{N}$ to $\mathbb{Q}$. For every $n \in \mathbb{N}$, let us define $s_n = s(n)$. Let us show that (s_n) satisfies the property.

Fix $m \in \mathbb{N}$ and take $n_0 \in \mathbb{N}$ such that $s_{n_0} = m$. Note that n_0 is unique since s is a bijection. Since $\mathbb{Q}$ is dense in $\mathbb{R}$, there exists $n_1 \in \mathbb{N}$ such that $0 < |s_{n_1} - m| < 1$. Suppose that we have found $n_1 < \cdots < n_k$ for some $k \in \mathbb{N}$ such that $0 < |s_{n_r} - m| < \frac{1}{r}$ for every integer $1 \le r \le k$. By means of contradiction, assume that $|s_n - m| > \frac{1}{k+1}$ for every integer $n > n_k$. By construction it is clear that $\{1, \ldots, n_k\} \backslash \{n_0\} \ne \emptyset$, so we can take by the Archimedean Property $\delta > 0$ such that $\delta < \min\{|s_n - m| : n \in \{1, \ldots, n_k\} \backslash \{n_0\}\}$. Let $\varepsilon := \min\left\{\delta, \frac{1}{k+1}\right\}$. There exists $n_\varepsilon \in \mathbb{N}$ such that $0 < |s_{n_\varepsilon} - m| < \varepsilon$ since $\mathbb{Q}$ is dense in $\mathbb{Q}$. Moreover, it is clear that $s_{n_\varepsilon} \ne m$, so $n_\varepsilon \ne n_0$. We have two cases for n_ε:

Case 1: If $1 \le n_\varepsilon \le n_k$, then $|s_{n_\varepsilon} - m| > \delta \ge \varepsilon$.

Case 2: If $n > n_k$, then $|s_{n_\varepsilon} - m| > \frac{1}{k+1} \ge \varepsilon$.

In any case we have that $\varepsilon \le |s_{n_\varepsilon} - m| < \varepsilon$, a contradiction. Therefore, there is $n_{k+1} \in \mathbb{N}$ with $n_{k+1} > n_k$ such that $|s_{n_{k+1}} - m| > \frac{1}{k+1}$. By repeating the argument, we have a strictly increasing sequence of natural numbers $(n_k)_{k=1}^\infty$ such that the sequence (s_{n_k}) satisfies that $0 < |s_{n_k} - m| < \frac{1}{k}$ for every $k \in \mathbb{N}$. Hence, it is clear that (s_{n_k}) is a subsequence of s_n and, by the Sandwich Theorem, we have that $\lim_{k \to \infty} |s_{n_k} - m| = 0$, which implies that (s_{n_k}) converges to m by Exercise 3.2 part (I). Finally, observe that an easier example is the following: 1,1,2,1,2,3,1,2,3,4,1,2,3,4,5,1,2,3,4,5,6...

□

Exercise 3.22. *Calculate the superior and inferior limits of the sequences with general term:*

(I) $a + \dfrac{(-1)^n}{n}$ *with* $a \in \mathbb{R}$,

(II) $(-1)^n + \dfrac{1}{n}$,

(III) $\dfrac{(-1)^n}{n} + 1 + (-1)^n$,

(IV) $\left(1 + \dfrac{(-1)^n}{n}\right)^n$,

(V) $\dfrac{(-1)^n n}{2n + 1}$,

(VI) $\dfrac{2n + (-1)^n(n + 2)}{3n + 3}$,

(VII) $(-1)^n \left(3 + \dfrac{2n+1}{3n+2}\right)$,

(VIII) $a - n^{(-1)^n}$ *with* $a \in \mathbb{R}$,

(IX) $\dfrac{(-1)^n(n+1)}{2n+1}$,

(X) $((-1)^n + 1) n^2$,

(XI) $\left(1 + \dfrac{1}{n}\right) \cos \dfrac{n\pi}{2}$,

(XII) $\left(1 - \dfrac{1}{n^2}\right) \sin \dfrac{n\pi}{2}$,

(XIII) $\sin \dfrac{n\pi}{3}$,

(XIV) $\dfrac{n+1}{n+2}\left(\sin \dfrac{n\pi}{4} + \cos \dfrac{n\pi}{4}\right)$,

(XV) $s_n = \begin{cases} 2 & \text{if } n \text{ is a multiple of } 4, \\ 0 & \text{if } n \text{ is even and not a multiple of } 4, \\ 1 & \text{if } n \text{ is odd.} \end{cases}$

Solution to Exercise 3.22.

(I). Since $\lim_{n\to\infty}\left(a + \frac{(-1)^n}{n}\right) = a$, we have that $\liminf_{n\to\infty}\left(a + \frac{(-1)^n}{n}\right) = \limsup_{n\to\infty}\left(a + \frac{(-1)^n}{n}\right) = a$.

(II). Since $\lim_{n\to\infty} \frac{1}{n} = 0$, $\liminf_{n\to\infty}(-1)^n = -1$ and $\limsup_{n\to\infty}(-1)^n = 1$, we have

$$\liminf_{n\to\infty}\left((-1)^n + \frac{1}{n}\right) = \liminf_{n\to\infty}(-1)^n + \lim_{n\to\infty}\frac{1}{n} = -1$$

and

$$\limsup_{n\to\infty}\left((-1)^n + \frac{1}{n}\right) = \limsup_{n\to\infty}(-1)^n + \lim_{n\to\infty}\frac{1}{n} = 1.$$

(III). Since $\lim_{n\to\infty}\left(\frac{(-1)^n}{n} + 1\right) = 1$, $\liminf_{n\to\infty}(-1)^n = -1$ and $\limsup_{n\to\infty}(-1)^n = 1$, we have

$$\liminf_{n\to\infty}\left(\frac{(-1)^n}{n} + 1 + (-1)^n\right) = \liminf_{n\to\infty}(-1)^n + \lim_{n\to\infty}\left(\frac{(-1)^n}{n} + 1\right) = -1 + 1 = 0$$

and

$$\limsup_{n\to\infty} \left(\frac{(-1)^n}{n} + 1 + (-1)^n \right) = \limsup_{n\to\infty}(-1)^n + \lim_{n\to\infty}\left(\frac{(-1)^n}{n} + 1 \right) = 1+1 = 2.$$

(IV). Observe that the subsequences of even and odd terms satisfy, respectively, that

$$\lim_{n\to\infty}\left(1 + \frac{(-1)^{2n}}{2n} \right)^{2n} = \lim_{n\to\infty}\left(1 + \frac{1}{2n} \right)^{2n} = e.$$

and

$$\lim_{n\to\infty}\left(1 + \frac{(-1)^{2n-1}}{2n-1} \right)^{2n-1} = \lim_{n\to\infty}\left(1 + \frac{1}{-(2n-1)} \right)^{2n-1}$$

$$= \lim_{n\to\infty} \frac{1}{\left(1 + \frac{1}{-(2n-1)} \right)^{-(2n-1)}} = e^{-1}.$$

Hence, the subsequential limits of $\left(1 + \frac{(-1)^n}{n} \right)^n$ are e and e^{-1}. Therefore,

$$\liminf_{n\to\infty}\left(1 + \frac{(-1)^n}{n} \right)^n = e^{-1}$$

and

$$\limsup_{n\to\infty}\left(1 + \frac{(-1)^n}{n} \right)^n = e.$$

(V). Since $\lim_{n\to\infty}\frac{n}{2n+1} = \frac{1}{2}$, $\liminf_{n\to\infty}(-1)^n = -1$ and $\limsup_{n\to\infty}(-1)^n = 1$, we have

$$\liminf_{n\to\infty}\frac{(-1)^n n}{2n+1} = \liminf_{n\to\infty}(-1)^n \cdot \lim_{n\to\infty}\frac{n}{2n+1} = -1 \cdot \frac{1}{2} = -\frac{1}{2}$$

and

$$\limsup_{n\to\infty}\frac{(-1)^n n}{2n+1} = \limsup_{n\to\infty}(-1)^n \cdot \lim_{n\to\infty}\frac{n}{2n+1} = 1 \cdot \frac{1}{2} = \frac{1}{2}.$$

(VI). First, since $\lim_{n\to\infty}\frac{n+2}{3n+3} = \frac{1}{3}$, $\liminf_{n\to\infty}(-1)^n = -1$ and $\limsup_{n\to\infty}(-1)^n = 1$, we have

$$\liminf_{n\to\infty}\frac{(-1)^n(n+2)}{3n+3} = \liminf_{n\to\infty}(-1)^n \cdot \lim_{n\to\infty}\frac{n+2}{3n+3} = -1 \cdot \frac{1}{3} = -\frac{1}{3}$$

and

$$\limsup_{n\to\infty}\frac{(-1)^n(n+2)}{3n+3} = \limsup_{n\to\infty}(-1)^n \cdot \lim_{n\to\infty}\frac{n+2}{3n+3} = 1 \cdot \frac{1}{3} = \frac{1}{3}.$$

Finally, since $\lim_{n \to \infty} \frac{2n}{3n+3} = \frac{2}{3}$, we have

$$\liminf_{n \to \infty} \frac{2n + (-1)^n(n+2)}{3n+3} = \liminf_{n \to \infty} \frac{(-1)^n(n+2)}{3n+3} + \lim_{n \to \infty} \frac{2n}{3n+3} = -\frac{1}{3} + \frac{2}{3} = \frac{1}{3}$$

and

$$\limsup_{n \to \infty} \frac{2n + (-1)^n(n+2)}{3n+3} = \limsup_{n \to \infty} \frac{(-1)^n(n+2)}{3n+3} + \lim_{n \to \infty} \frac{n+2}{3n+3} = \frac{1}{3} + \frac{2}{3} = 1.$$

(VII). Since $\lim_{n \to \infty} \left(3 + \frac{2n+1}{3n+2}\right) = 3 + \frac{2}{3} = \frac{11}{3}$, $\liminf_{n \to \infty}(-1)^n = -1$ and $\limsup_{n \to \infty}(-1)^n = 1$, we have

$$\liminf_{n \to \infty}(-1)^n \left(3 + \frac{2n+1}{3n+2}\right) = \liminf_{n \to \infty}(-1)^n \cdot \lim_{n \to \infty}\left(3 + \frac{2n+1}{3n+2}\right)$$
$$= -1 \cdot \frac{11}{3} = -\frac{11}{3}$$

and

$$\limsup_{n \to \infty}(-1)^n \left(3 + \frac{2n+1}{3n+2}\right) = \limsup_{n \to \infty}(-1)^n \cdot \lim_{n \to \infty}\left(3 + \frac{2n+1}{3n+2}\right) = 1 \cdot \frac{11}{3} = \frac{11}{3}.$$

(VIII). Observe that the subsequences of even and odd terms satisfy, respectively, that

$$\lim_{n \to \infty}\left(a - (2n)^{(-1)^{2n}}\right) = \lim_{n \to \infty}(a - 2n) = -\infty$$

and

$$\lim_{n \to \infty}\left(a - (2n-1)^{(-1)^{2n-1}}\right) = \lim_{n \to \infty}\left(a - \frac{1}{2n-1}\right) = a.$$

Hence, the subsequential limits of $\left(a - n^{(-1)^n}\right)$ are $-\infty$ and a. Therefore,

$$\liminf_{n \to \infty}\left(a - n^{(-1)^n}\right) = -\infty$$

and

$$\limsup_{n \to \infty}\left(a - n^{(-1)^n}\right) = a.$$

(IX). Since

$$\lim_{n \to \infty}\frac{n+1}{2n+1} = \frac{1}{2}, \quad \liminf_{n \to \infty}(-1)^n = -1, \quad \text{and} \quad \limsup_{n \to \infty}(-1)^n = 1,$$

we have

$$\liminf_{n \to \infty}\frac{(-1)^n(n+1)}{2n+1} = \liminf_{n \to \infty}(-1)^n \cdot \lim_{n \to \infty}\frac{n+1}{2n+1} = -1 \cdot \frac{1}{2} = -\frac{1}{2}$$

and

$$\limsup_{n\to\infty} \frac{(-1)^n (n+1)}{2n+1} = \limsup_{n\to\infty}(-1)^n \cdot \lim_{n\to\infty} \frac{n}{2n+1} = 1 \cdot \frac{1}{2} = \frac{1}{2}.$$

(x). Observe that the subsequences of even and odd terms satisfy, respectively, that

$$\lim_{n\to\infty} \left((-1)^{2n} + 1\right)(2n)^2 = \lim_{n\to\infty}(1+1)(2n)^2 = \infty.$$

and

$$\lim_{n\to\infty} \left((-1)^{2n-1} + 1\right)(2n-1)^2 = \lim_{n\to\infty}(-1+1)(2n)^2 = \lim_{n\to\infty} 0 = 0$$

Hence, the subsequential limits of $((-1)^n + 1)n^2$ are 0 and ∞. Therefore,

$$\liminf_{n\to\infty} \left((-1)^n + 1\right)n^2 = 0$$

and

$$\limsup_{n\to\infty} \left((-1)^n + 1\right)n^2 = \infty.$$

(xi). Since

$$\lim_{n\to\infty} \left(1 + \frac{1}{n}\right) = 1, \ \liminf_{n\to\infty} \cos \frac{n\pi}{2} = -1, \ \text{and} \ \limsup_{n\to\infty} \cos \frac{n\pi}{2} = 1,$$

we have

$$\liminf_{n\to\infty} \left(1 + \frac{1}{n}\right) \cos \frac{n\pi}{2} = \liminf_{n\to\infty} \cos \frac{n\pi}{2} \cdot \lim_{n\to\infty} \left(1 + \frac{1}{n}\right) = -1 \cdot 1 = -1$$

and

$$\limsup_{n\to\infty} \left(1 + \frac{1}{n}\right) \cos \frac{n\pi}{2} = \limsup_{n\to\infty} \cos \frac{n\pi}{2} \cdot \lim_{n\to\infty} \left(1 + \frac{1}{n}\right) = 1 \cdot 1 = 1.$$

(xii). Since

$$\lim_{n\to\infty} \left(1 - \frac{1}{n^2}\right) = 1, \ \liminf_{n\to\infty} \sin \frac{n\pi}{2} = -1, \ \text{and} \ \limsup_{n\to\infty} \sin \frac{n\pi}{2} = 1,$$

we have

$$\liminf_{n\to\infty} \left(1 - \frac{1}{n^2}\right) \sin \frac{n\pi}{2} = \liminf_{n\to\infty} \sin \frac{n\pi}{2} \cdot \lim_{n\to\infty} \left(1 - \frac{1}{n^2}\right) = -1 \cdot 1 = -1$$

and

$$\limsup_{n\to\infty} \left(1 - \frac{1}{n^2}\right) \sin \frac{n\pi}{2} = \limsup_{n\to\infty} \sin \frac{n\pi}{2} \cdot \lim_{n\to\infty} \left(1 - \frac{1}{n^2}\right) = 1 \cdot 1 = 1.$$

(XIII). Observe that $\sin \frac{\pi}{3} = \frac{\sqrt{3}}{2} = \sin \frac{2\pi}{3}$, $\sin \pi = 0 = \sin 2\pi$ and $\sin \frac{4\pi}{3} = -\frac{\sqrt{3}}{2} = \sin \frac{5\pi}{3}$. Hence,

$$\lim_{n \to \infty} \sin \frac{(6n-5)\pi}{3} = \lim_{n \to \infty} \sin \frac{(6n-4)\pi}{3} = \frac{\sqrt{3}}{2},$$

$$\lim_{n \to \infty} \sin \frac{(6n-3)\pi}{3} = \lim_{n \to \infty} \sin \frac{(6n)\pi}{3} = 0$$

and

$$\lim_{n \to \infty} \sin \frac{(6n-2)\pi}{3} = \lim_{n \to \infty} \sin \frac{(6n-1)\pi}{3} = -\frac{\sqrt{3}}{2}.$$

Thus, the subsequential limits of $\sin \frac{n\pi}{3}$ are $-\frac{\sqrt{3}}{2}$, 0 and $\frac{\sqrt{3}}{2}$. Therefore,

$$\liminf_{n \to \infty} \sin \frac{n\pi}{3} = -\frac{\sqrt{3}}{2}$$

and

$$\limsup_{n \to \infty} \sin \frac{n\pi}{3} = \frac{\sqrt{3}}{2}.$$

(XIV). Observe that $\sin \frac{\pi}{4} + \cos \frac{\pi}{4} = \sqrt{2}$, $\sin \frac{2\pi}{4} + \cos \frac{2\pi}{4} = \sin \frac{8\pi}{4} + \cos \frac{8\pi}{4} = 1$, $\sin \frac{3\pi}{4} + \cos \frac{3\pi}{4} = \sin \frac{7\pi}{4} + \cos \frac{7\pi}{4} = 0$, $\sin \frac{4\pi}{4} + \cos \frac{4\pi}{4} = \sin \frac{6\pi}{4} + \cos \frac{6\pi}{4} = -1$ and $\sin \frac{5\pi}{4} + \cos \frac{5\pi}{4} = -\sqrt{2}$. Hence,

$$\lim_{n \to \infty} \left(\sin \frac{(8n-7)\pi}{4} + \cos \frac{(8n-7)\pi}{4} \right) = \sqrt{2},$$

$$\lim_{n \to \infty} \left(\sin \frac{(8n-6)\pi}{4} + \cos \frac{(8n-6)\pi}{4} \right)$$
$$= \lim_{n \to \infty} \left(\sin \frac{(8n)\pi}{4} + \cos \frac{(8n)\pi}{4} \right) = 1,$$

$$\lim_{n \to \infty} \left(\sin \frac{(8n-5)\pi}{4} + \cos \frac{(8n-5)\pi}{4} \right)$$
$$= \lim_{n \to \infty} \left(\sin \frac{(8n-1)\pi}{4} + \cos \frac{(8n-1)\pi}{4} \right) = 0,$$

$$\lim_{n \to \infty} \left(\sin \frac{(8n-4)\pi}{4} + \cos \frac{(8n-4)\pi}{4} \right)$$
$$= \lim_{n \to \infty} \left(\sin \frac{(8n-2)\pi}{4} + \cos \frac{(8n-2)\pi}{4} \right) = -1$$

and

$$\lim_{n\to\infty} \left(\sin \frac{(8n-3)\pi}{4} + \cos \frac{(8n-3)\pi}{4} \right) = -\sqrt{2}.$$

Thus, the subsequential limits of $\sin \frac{n\pi}{4} + \cos \frac{n\pi}{4}$ are $-\sqrt{2}, -1, 0, 1$ and $\sqrt{2}$, which implies that

$$\liminf_{n\to\infty} \left(\sin \frac{n\pi}{4} + \cos \frac{n\pi}{4} \right) = -\sqrt{2}$$

and

$$\limsup_{n\to\infty} \left(\sin \frac{n\pi}{4} + \cos \frac{n\pi}{4} \right) = \sqrt{2}.$$

Since $\lim_{n\to\infty} \frac{n+1}{n+2} = 1$, we have that

$$\liminf_{n\to\infty} \frac{n+1}{n+2} \left(\sin \frac{n\pi}{4} + \cos \frac{n\pi}{4} \right) = \liminf_{n\to\infty} \left(\sin \frac{n\pi}{4} + \cos \frac{n\pi}{4} \right) \cdot \lim_{n\to\infty} \frac{n+1}{n+2}$$

$$= -\sqrt{2} \cdot 1 = -\sqrt{2}$$

and

$$\limsup_{n\to\infty} \frac{n+1}{n+2} \left(\sin \frac{n\pi}{4} + \cos \frac{n\pi}{4} \right) = \limsup_{n\to\infty} \left(\sin \frac{n\pi}{4} + \cos \frac{n\pi}{4} \right) \cdot \lim_{n\to\infty} \frac{n+1}{n+2}$$

$$= \sqrt{2} \cdot 1 = \sqrt{2}.$$

(xv). Observe that the sequential limits of (s_n) are 0, 1 and 2 since the sets $\{n \in \mathbb{N} : n \text{ is multiple of } 4\}$, $\{n \in \mathbb{N} : n \text{ is even and not a multiple of } 4\}$ and $\{n \in \mathbb{N} : n \text{ is odd } \}$ are infinite sets. Hence,

$$\liminf_{n\to\infty} s_n = 0$$

and

$$\limsup_{n\to\infty} s_n = 2.$$

□

Chapter 4

Continuous Functions

> **Exercise 4.1.** *Find the following limits, if they exist. Use the definition of limit to justify your answer.*
>
> (I) $\lim\limits_{x \to 1}(x^2 + 2x + 1)$.
>
> (II) $\lim\limits_{x \to 2}\dfrac{x^3 - 8}{x - 2}$.
>
> (III) $\lim\limits_{x \to 0}\dfrac{1}{x^2 - 1}$.
>
> (IV) $\lim\limits_{x \to 4}\sqrt{x}$.
>
> (V) $\lim\limits_{x \to 1}\dfrac{x^3 - 1}{(x - 1)(x - 2)}$.
>
> (VI) $\lim\limits_{x \to 0}\dfrac{x + |x|}{x}$.
>
> (VII) $\lim\limits_{x \to 0}\left[x\cos\left(\dfrac{1}{x}\right) + \sin\left(\dfrac{1}{x}\right) + \sin\left(\dfrac{1}{|x|}\right)\right]$.
>
> (VIII) $\lim\limits_{x \to 1}\dfrac{|x - 1|}{x^2 + x - 2}$.

Solution to Exercise 4.1.

(I). Let us justify that

$$\lim_{x \to 1}(x^2 + 2x + 1) = 4$$

using the definition of limit.

Let $f(x) = x^2 + 2x + 1$ for $x \in \mathbb{R}$. Then,

$$|f(x) - 4| = |x^2 + 2x + 1 - 4| = |x^2 + 2x - 3| = |(x - 1)(x + 3)| = |x - 1||x + 3|,$$

where we have used the second degree equation formula to decompose the polynomial $x^2 + 2x - 3$ into $(x - 1)(x + 3)$. In order to find a bound for $|x - 1|$, we restrict x by assuming that

$$|x - 1| < 1 \iff -1 < x - 1 < 1 \iff 0 < x < 2.$$

Thus,

$$3 < x + 3 < 5,$$

which implies that

$$|f(x) - 4| < 5|x - 1|.$$

Now for an arbitrary $\varepsilon > 0$, choose

$$\delta = \min\left\{1, \frac{\varepsilon}{5}\right\}.$$

DOI: 10.1201/9781003400745-4

Then, if $0 < |x - 1| < \delta$, we have that

$$|f(x) - 4| < 5|x - 1| < \varepsilon.$$

(II). Let us justify that

$$\lim_{x \to 2} \frac{x^3 - 8}{x - 2} = \lim_{x \to 2} \frac{(x - 2)(x^2 + 2x + 4)}{x - 2} = \lim_{x \to 2}(x^2 + 2x + 4) = 12$$

using the definition of limit.

Let $f(x) = \frac{x^3 - 8}{x - 2}$ for $x \in \mathbb{R}$. Then,

$$|f(x) - 12| = \left| \frac{x^3 - 8}{x - 2} - 12 \right| = |x^2 + 2x + 4 - 12|$$
$$= |x^2 + 2x - 8| = |(x - 2)(x + 4)| = |x - 2||x + 4|.$$

We now restrict x by assuming that

$$|x - 2| < 1 \iff -1 < x - 2 < 1 \iff 1 < x < 3.$$

Hence,

$$5 < x + 4 < 7,$$

which shows that

$$|f(x) - 12| < 7|x - 2|.$$

Now for any arbitrary $\varepsilon > 0$, choose

$$\delta = \min\left\{1, \frac{\varepsilon}{7}\right\}.$$

Then, if $0 < |x - 2| < \delta$, we have that

$$|f(x) - 12| < 7|x - 2| < \varepsilon.$$

(III). Let us justify that

$$\lim_{x \to 0} \frac{1}{x^2 - 1} = -1$$

using the definition of limit.

Let $f(x) = \frac{1}{x^2 - 1}$ for $x \in \mathbb{R}$. Then,

$$|f(x) - 1| = \left| \frac{1}{x^2 - 1} + 1 \right| = \left| \frac{x^2}{x^2 - 1} \right|$$
$$= \frac{x^2}{|x^2 - 1|} = \frac{x^2}{|(x + 1)(x - 1)|}$$
$$= \frac{x^2}{|x + 1||x - 1|}.$$

We now restrict x by assuming that

$$|x| < \frac{1}{2} \iff -\frac{1}{2} < x < \frac{1}{2},$$

which implies that

$$0 \leq x^2 < \frac{1}{4}, \quad \frac{1}{2} < x+1 < \frac{3}{2} \text{ and } -\frac{3}{2} < x-1 < -\frac{1}{2}.$$

In particular, for $\frac{1}{2} < x+1 < \frac{3}{2}$ we have the following chain of implications

$$\frac{1}{2} < x+1 < \frac{3}{2} \implies \frac{1}{2} < |x+1| < \frac{3}{2} \implies \frac{2}{3} < \frac{1}{|x+1|} < 2$$

and for $-\frac{3}{2} < x-1 < -\frac{1}{2}$ we have

$$-\frac{3}{2} < x-1 < -\frac{1}{2} \implies \frac{1}{2} < |x-1| < \frac{3}{2} \implies \frac{2}{3} < \frac{1}{|x-1|} < 2.$$

Hence,

$$|f(x) - 1| < 4x^2.$$

Now for an arbitrary $\varepsilon > 0$, choose

$$\delta = \min\left\{\frac{1}{2}, \frac{\sqrt{\varepsilon}}{2}\right\}.$$

Then, if $0 < |x| < \delta$, we have

$$|f(x) - 1| < 4x^2 < \varepsilon.$$

(IV). Let us justify that

$$\lim_{x \to 4} \sqrt{x} = 2$$

using the definition of limit.

Let $f(x) = \sqrt{x}$ for $x \in \mathbb{R}$. Then,

$$|f(x) - 2| = |\sqrt{x} - 2| = |x - 4| \left| \frac{\sqrt{x} - 2}{x - 4} \right|$$

$$= |x - 4| \left| \frac{\sqrt{x} - 2}{(\sqrt{x} - 2)(\sqrt{x} + 2)} \right| = |x - 4| \left| \frac{1}{\sqrt{x} + 2} \right|$$

$$= |x - 4| \frac{1}{\sqrt{x} + 2}.$$

We now restrict x by assuming that

$$|x - 4| < 1 \iff -1 < x - 4 < 1 \iff 3 < x < 5 \iff \sqrt{3} < \sqrt{x} < \sqrt{5}$$

$$\iff \sqrt{3} + 2 < \sqrt{x} + 2 < \sqrt{5} + 2$$

$$\iff \frac{1}{\sqrt{5} + 2} < \frac{1}{\sqrt{x} + 2} < \frac{1}{\sqrt{3} + 2}.$$

Hence,

$$|f(x) - 2| < \frac{1}{\sqrt{3}+2}|x - 4|.$$

Now for an arbitrary $\varepsilon > 0$, choose

$$\delta = \min\left\{1, \left(\sqrt{3}+2\right)\varepsilon\right\}.$$

Then, if $0 < |x - 4| < \delta$, we have

$$|f(x) - 2| < \frac{1}{\sqrt{3}+2}|x - 4| < \varepsilon.$$

(v). Let us justify that

$$\lim_{x \to 1} \frac{x^3 - 1}{(x-1)(x-2)} = \lim_{x \to 1} \frac{(x-1)(x^2+x+1)}{(x-1)(x-2)} = \lim_{x \to 1} \frac{x^2+x+1}{x-2} = -3.$$

using the definition of limit.

Let $f(x) = \frac{x^3-1}{(x-1)(x-2)}$ for $x \in \mathbb{R}$. Then,

$$\begin{aligned}
|f(x) + 3| &= \left|\frac{x^3 - 1}{(x-1)(x-2)} + 3\right| = \left|\frac{x^3 - 1 + 3x^2 - 9x + 6}{(x-1)(x-2)}\right| \\
&= \left|\frac{x^3 + 3x^2 - 9x + 5}{(x-1)(x-2)}\right| = \left|\frac{(x-1)^2(x+5)}{(x-1)(x-2)}\right| = \left|\frac{(x-1)(x+5)}{x-2}\right| \\
&= \left|\frac{x+5}{x-2}\right| |x - 1|.
\end{aligned}$$

We now restrict x by assuming that

$$|x - 1| < \frac{1}{2} \iff -\frac{1}{2} < x - 1 < \frac{1}{2} \iff \frac{1}{2} < x < \frac{3}{2}.$$

Hence,

$$\frac{1}{2} < x < \frac{3}{2} \iff \frac{11}{2} < x + 5 < \frac{13}{2} \implies \frac{11}{2} < |x + 5| < \frac{13}{2},$$

and

$$\frac{1}{2} < x < \frac{3}{2} \iff -\frac{3}{2} < x - 2 < -\frac{1}{2}$$

$$\implies \frac{1}{2} < |x - 2| < \frac{3}{2} \iff \frac{2}{3} < \frac{1}{|x - 2|} < 2.$$

Thus,

$$\left|\frac{x+5}{x-2}\right| < 13,$$

which implies that

$$|f(x) + 3| < 13|x - 1|.$$

Now for an arbitrary $\varepsilon > 0$, choose

$$\delta = \min\left\{1, \frac{\varepsilon}{13}\right\}.$$

Then, if $0 < |x - 1| < \delta$, we have

$$|f(x) + 3| < 13|x - 1| < \varepsilon.$$

(VI). Since

$$\lim_{x\to 0^+} \frac{x + |x|}{x} = \lim_{x\to 0^+} \frac{x + x}{x} = \lim_{x\to 0^+} \frac{2x}{x} = \lim_{x\to 0^+} 2 = 2$$

and

$$\lim_{x\to 0^-} \frac{x + |x|}{x} = \lim_{x\to 0^-} \frac{x + (-x)}{x} = \lim_{x\to 0^+} \frac{0}{x} = \lim_{x\to 0^+} 0 = 0,$$

the limit

$$\lim_{x\to 0} \frac{x + |x|}{x}$$

does not exist. Let us justify the latter using the definition of limit.

Let $f(x) = \frac{x+|x|}{x}$ for $x \in \mathbb{R}$. Assume that $\lim_{x\to 0} f(x)$ exists and it is equal to $l \in \mathbb{R}$. Then, for every $0 < \varepsilon < 1$, there exists $\delta > 0$ such that, if $0 < |x| < \delta$, then $|f(x) - l| < \varepsilon$. On the hand, if $0 < x < \delta$, then

$$\varepsilon > |f(x) - l| = \left|\frac{x + |x|}{x} - l\right| = \left|\frac{x + x}{x} - l\right| = |2 - l|.$$

On the other hand, if $-\delta < x < 0$, then

$$\varepsilon > |f(x) - l| = \left|\frac{x + |x|}{x} - l\right| = \left|\frac{x + (-x)}{x} - l\right| = |0 - l| = |l|.$$

Hence,

$$2 - l < \varepsilon \text{ and } l < \varepsilon,$$

which implies that $\varepsilon > 1$, a contradiction.

(VII). Let $f(x) = x\cos\left(\frac{1}{x}\right) + \sin\left(\frac{1}{x}\right) + \sin\left(\frac{1}{|x|}\right)$ for $x \in \mathbb{R}$. Take the sequence $x_n = \frac{1}{n\pi + \frac{\pi}{2}}$ which converges to 0 and notice that

$$f(x_n) = \frac{1}{n\pi + \frac{\pi}{2}}\cos\left(n\pi + \frac{\pi}{2}\right) + \sin\left(n\pi + \frac{\pi}{2}\right) + \sin\left(n\pi + \frac{\pi}{2}\right)$$

$$= \begin{cases} 2 & \text{if } n \in \mathbb{N} \text{ is even,} \\ -2 & \text{if } n \in \mathbb{N} \text{ is odd.} \end{cases}$$

Then, the limit

$$\lim_{x \to 0} \left(x \cos\left(\frac{1}{x}\right) + \sin\left(\frac{1}{x}\right) + \sin\left(\frac{1}{|x|}\right) \right)$$

does not exist. Let us justify the latter using the definition of limit.

Assume that $\lim_{x \to 0} f(x)$ exists and it is equal to $l \in \mathbb{R}$. Then, for every $0 < \varepsilon < 1$, there exists $\delta > 0$ such that, if $0 < |x| < \delta$, then $|f(x) - l| < \varepsilon$. On the hand, by the Archimedean property, there exists $n \in \mathbb{N}$ even such that $0 < x_n = \frac{1}{n\pi + \frac{\pi}{2}} < \delta$. Then,

$$\varepsilon > |f(x_n) - l| = |2 - l|.$$

On the other hand, and again by the Archimedean property, there exists $n \in \mathbb{N}$ odd such that $0 < x_n = \frac{1}{n\pi + \frac{\pi}{2}} < \delta$. Then,

$$\varepsilon > |f(x_n) - l| = |-2 - l|.$$

Hence,

$$2 - l < \varepsilon \text{ and } -2 - l > -\varepsilon$$

which implies that $\varepsilon > 2$, but this is a contradiction.

(VIII). Since

$$\lim_{x \to 1+} \frac{|x - 1|}{x^2 + x - 2} = \lim_{x \to 1+} \frac{x - 1}{(x - 1)(x + 2)} = \lim_{x \to 1+} \frac{1}{x + 2} = \frac{1}{3}$$

and

$$\lim_{x \to 1-} \frac{|x - 1|}{x^2 + x - 2} = \lim_{x \to 1-} \frac{-(x - 1)}{(x - 1)(x + 2)} = \lim_{x \to 1-} \frac{-1}{x + 2} = -\frac{1}{3},$$

we have that the limit

$$\lim_{x \to 1} \frac{|x - 1|}{x^2 + x - 2}$$

does not exist. Let us justify the latter using the definition of limit.

Let $f(x) = \frac{|x-1|}{x^2+x-2}$ for $x \in \mathbb{R}$. Assume that $\lim_{x \to 1} \frac{|x-1|}{x^2+x-2}$ exists and it is equal to $l \in \mathbb{R}$. Then, for every $0 < \varepsilon < \frac{1}{3}$, there exists $\delta > 0$ such that, if $0 < |x - 1| < \delta$, then $|f(x) - l| < \varepsilon$. On the one hand, if $0 < x - 1 < \delta$, then

$$\varepsilon > |f(x) - l| = \left| \frac{|x - 1|}{x^2 + x - 2} - l \right| = \left| \frac{x - 1}{x^2 + x - 2} - l \right| = \left| \frac{1}{3} - l \right|.$$

On the other hand, if $-\delta < x - 1 < 0$, then

$$\varepsilon > |f(x) - l| = \left| \frac{|x - 1|}{x^2 + x - 2} - l \right| = \left| \frac{-(x - 1)}{x^2 + x - 2} - l \right| = \left| -\frac{1}{3} - l \right|.$$

Hence,

$$\frac{1}{3} - l < \varepsilon \text{ and } -\frac{1}{3} - l > -\varepsilon,$$

which implies that $\varepsilon > \frac{1}{3}$, a contradiction.

$\square$

Exercise 4.2. *Find the value of the following limits:*

(I) $\lim\limits_{x\to\infty} \dfrac{x^2+1}{\log x}$.

(II) $\lim\limits_{x\to\infty} \dfrac{x^4+x+2}{e^{x+5}}$.

(III) $\lim\limits_{x\to\infty} \dfrac{x^5 \log x}{\sqrt{x^4+1}}$.

(IV) $\lim\limits_{x\to\infty} \dfrac{(\log x)^3}{e^x-1}$.

(V) $\lim\limits_{x\to\infty} \left(\dfrac{2x}{e^x+4}\right)^x$.

(VI) $\lim\limits_{x\to 0} \dfrac{\pi}{x \cot \frac{\pi x}{2}}$.

(VII) $\lim\limits_{x\to 0} \dfrac{\sqrt{1+x}-\sqrt{1-x}}{x}$.

(VIII) $\lim\limits_{x\to 0} \dfrac{\sqrt{a+x}-\sqrt{a}}{x}$ *with* $a>0$.

(IX) $\lim\limits_{x\to 0} \dfrac{\sqrt{1+x}-1}{\sqrt{1-x}-1}$.

(X) $\lim\limits_{x\to\infty} \sqrt{x+\sqrt{x+\sqrt{x}}}-\sqrt{x}$.

(XI) $\lim\limits_{x\to 1} \dfrac{2x-2}{\sqrt[3]{26+x}-3}$.

(XII) $\lim\limits_{x\to 1} \dfrac{x^p-1}{x^q-1}$ *with* p *any real number and* $q\neq 0$.

(XIII) $\lim\limits_{x\to 0} \dfrac{a^x-b^x}{c^x-d^x}$ *with* $a,b,c,d>0$ *and* $c\neq d$.

(XIV) $\lim\limits_{x\to 0} \dfrac{e^x-e^{\sin x}}{x-\sin x}$.

(XV) $\lim\limits_{x\to 0^+} x^x$.

(XVI) $\lim\limits_{x\to 0^+} x^{x^x}$.

(XVII) $\lim\limits_{x\to\infty} \left(\dfrac{2x^2+3}{2x^2+5}\right)^{8x^2+3}$.

(XVIII) $\lim\limits_{x\to\infty} \left(\dfrac{a^{1/x}+b^{1/x}+c^{1/x}}{3}\right)^x$ *with* $a,b,c>0$.

(XIX) $\lim\limits_{x\to 0} \dfrac{e^x+\sin x-1}{\log(1+x)}$.

(XX) $\lim\limits_{x\to\pi/4} \dfrac{\sec^2 x - 2\tan x}{1+\cos(4x)}$.

(XXI) $\lim\limits_{x\to\pi/4} \tan(2x)\cot\left(\dfrac{\pi}{4}+x\right)$.

(XXII) $\lim\limits_{x\to\pi/6} \dfrac{\sin(x-\pi/6)}{\sqrt{3}-2\cos x}$.

(XXIII) $\lim\limits_{x\to 0} \left(\sin^2\left(\dfrac{\pi}{2-ax}\right)\right)^{\sec^2\left(\frac{\pi}{2-bx}\right)}$ *with* $a\in\mathbb{R}$ *and* $b\neq 0$.

(XXIV) $\lim\limits_{x\to\pi/3} \dfrac{\sin(x-\pi/3)}{1-2\cos x}$.

Solution to Exercise 4.2.

In this exercise it is assumed that the reader is familiar with L'Hôpital's Rule on limits of quotients of differentiable functions. Nevertheless, another approach is indicated in a number of its parts.

(I). Notice that $f(x)=x^2+1$, $g(x)=\log x$ and $\lim\limits_{x\to\infty} \dfrac{x^2+1}{\log x}$ satisfy the hypotheses of L'Hôpital's Rule. Then,

$$\lim_{x\to\infty} \frac{x^2+1}{\log x} = \lim_{x\to\infty} \frac{2x}{1/x} = \lim_{x\to\infty} 2x^2 = \infty.$$

Here it is another way to do it: since $x > \log x > 0$ for all $x > 1$, we have

$$\frac{x^2 + 1}{\log x} = x \cdot \frac{x}{\log x} + \frac{1}{\log x} > x \cdot 1 + 0 = x \to \infty,$$

and so $\dfrac{x^2 + 1}{\log x} \to \infty$ as $x \to \infty$.

(II). Observe that $f(x) = x^4 + x + 2$, $g(x) = e^{x+5}$ and $\lim_{x\to\infty} \frac{x^4+x+2}{e^{x+5}}$ satisfy the hypotheses of L'Hôpital's Rule. Then,

$$\lim_{x\to\infty} \frac{x^4 + x + 2}{e^{x+5}} = \lim_{x\to\infty} \frac{4x^3 + 1}{e^{x+5}} = \lim_{x\to\infty} \frac{12x^2}{e^{x+5}} = \lim_{x\to\infty} \frac{24x}{e^{x+5}} = \lim_{x\to\infty} \frac{24}{e^{x+5}} = 0.$$

Note that we have applied L'Hôpital's Rule several times.

A different approach is sketched: it is known that $\lim_{x\to\infty} a^x \cdot x^b = 0$ for all $b \in \mathbb{R}$ whenever $0 < a < 1$. Then let $a = 1/e$, $b \in \{0, 1, 4\}$, and put

$$\frac{x^4 + x + 2}{e^{x+5}} = \frac{1}{e^5} \cdot \left((1/e)^x \cdot x^4 + (1/e)^x \cdot x + (1/e)^x \cdot 2\right).$$

(III). Notice that $f(x) = x^5 \log x$, $g(x) = \sqrt{x^4 + 1}$ and $\lim_{x\to\infty} \frac{x^5 \log x}{\sqrt{x^4+1}}$ satisfy the hypotheses of L'Hôpital's Rule. Then,

$$\lim_{x\to\infty} \frac{x^5 \log x}{\sqrt{x^4 + 1}} = \lim_{x\to\infty} \frac{5x^4 \log x + x^5 \cdot \frac{1}{x}}{\frac{4x^3}{2\sqrt{x^4+1}}} = \lim_{x\to\infty} \frac{5x^4 \log x + x^4}{\frac{2x^3}{\sqrt{x^4+1}}}$$

$$= \lim_{x\to\infty} \frac{x^4(5\log x + 1)\sqrt{x^4 + 1}}{2x^3}$$

$$= \lim_{x\to\infty} \frac{1}{2}x(5\log x + 1)\sqrt{x^4 + 1} = \infty.$$

(IV). Observe that $f(x) = (\log x)^3$, $g(x) = e^x - 1$ and $\lim_{x\to\infty} \frac{(\log x)^3}{e^x - 1}$ satisfy the hypotheses of L'Hôpital's Rule. Then,

$$\lim_{x\to\infty} \frac{(\log x)^3}{e^x - 1} = \lim_{x\to\infty} \frac{3(\log x)^2 \cdot \frac{1}{x}}{e^x} = \lim_{x\to\infty} \frac{3(\log x)^2}{xe^x} = \lim_{x\to\infty} \frac{6\log x \cdot \frac{1}{x}}{e^x + xe^x}$$

$$= \lim_{x\to\infty} \frac{6\log x}{e^x(x^2 + x)} = \lim_{x\to\infty} \frac{\frac{6}{x}}{e^x(x^2 + x) + e^x(2x + 1)}$$

$$= \lim_{x\to\infty} \frac{6}{e^x(x^3 + 3x^2 + x)} = 0.$$

Note that we have applied L'Hôpital's Rule several times.

We sketch a different way: $\log x < x$ for all $x > 0$, so for $x > 1$ we have

$$0 < \frac{(\log x)^3}{e^x - 1} < \frac{x^3}{e^x - 1} = x^3 \cdot (1/e)^x \cdot \frac{1}{1 - e^{-x}}.$$

Now, as in part (II), it holds that $\lim_{x\to\infty} x^3 \cdot (1/e)^x = 0$. Moreover, $\frac{1}{1-e^{-x}} \to \frac{1}{1-0} = 1$ as $x \to \infty$. It suffices to apply the Sandwich Criterion.

(v). Notice that $f(x) = 2x$, $g(x) = e^x + 4$ and $\lim_{x\to\infty} \frac{2x}{e^x+4}$ satisfy the hypotheses of L'Hôpital's Rule. Then,

$$\lim_{x\to\infty} \frac{2x}{e^x + 4} = \lim_{x\to\infty} \frac{2}{e^x} = 0.$$

Thus,

$$\lim_{x\to\infty} \left(\frac{2x}{e^x + 4} \right)^x = 0.$$

Another way is based on the use of the property $a^x \cdot x^b \to 0$ $(x \to \infty)$ for all reals b whenever $0 < a < 1$, as in part (II); now, take $a = 1/e$ and $b = -1$.

(VI). Let us calculate first the limit

$$\lim_{x\to 0} x \cot \frac{\pi x}{2} = \lim_{x\to 0} \frac{x \cos \frac{\pi x}{2}}{\sin \frac{\pi x}{2}} = \lim_{x\to 0} \frac{\cos \frac{\pi x}{2} - \frac{\pi x}{2} \sin \frac{\pi x}{2}}{\frac{\pi}{2} \cos \frac{\pi x}{2}} = \frac{2}{\pi},$$

where we have used L'Hôpital's Rule for $f(x) = x \cos \frac{\pi x}{2}$ and $g(x) = \sin \frac{\pi x}{2}$ (note that $\lim_{x\to 0} x \cos \frac{\pi x}{2} = 0$ since $\cos \frac{\pi x}{2}$ is bounded between -1 and 1 and $\lim_{x\to 0} x = 0$), and the fact that $\lim_{x\to 0} \frac{\pi x}{2} \sin \frac{\pi x}{2} = 0$ since $\lim_{x\to 0} \frac{\pi x}{2} = 0$ and $\sin \frac{\pi x}{2}$ is bounded between -1 and 1. Hence,

$$\lim_{x\to 0} \frac{\pi}{x \cot \frac{\pi x}{2}} = \pi \cdot \frac{\pi}{2} = \frac{\pi^2}{2}.$$

Another approach is writing $x \cot \frac{\pi x}{2} = \cos \frac{\pi x}{2} \cdot \frac{\pi x/2}{\sin(\pi x/2)} \cdot \frac{2}{\pi}$ and then using $\lim_{t\to 0} \frac{t}{\sin t} = 1$.

(VII). Notice that $f(x) = \sqrt{1+x} - \sqrt{1-x}$, $g(x) = x$ and $\lim_{x\to 0} \frac{\sqrt{1+x}-\sqrt{1-x}}{x}$ satisfy the hypotheses of L'Hôpital's Rule. Then,

$$\lim_{x\to 0} \frac{\sqrt{1+x} - \sqrt{1-x}}{x} = \lim_{x\to 0} \left(\frac{1}{2\sqrt{1+x}} + \frac{1}{2\sqrt{1-x}} \right) = 1.$$

Here it is another way to do it: the required limit is

$$\lim_{x\to 0} \frac{(1+x) - (1-x)}{x(\sqrt{1+x} + \sqrt{1-x})} = \lim_{x\to 0} \frac{2}{\sqrt{1+x} + \sqrt{1-x}} = \frac{2}{\sqrt{1+0} + \sqrt{1-0}} = 1.$$

(VIII). Observe that $f(x) = \sqrt{a+x} - \sqrt{a}$, $g(x) = x$ and $\lim_{x\to 0} \frac{\sqrt{a+x}-\sqrt{a}}{x}$ satisfy the hypotheses of L'Hôpital's Rule for any $a > 0$. Then,

$$\lim_{x\to 0} \frac{\sqrt{a+x} - \sqrt{a}}{x} = \lim_{x\to 0} \frac{1}{2\sqrt{a+x}} = \frac{1}{2\sqrt{a}}.$$

Analogously to part (VII). we also can use the identity

$$\frac{\sqrt{a+x}-\sqrt{a}}{x} = \frac{1}{\sqrt{a+x}+\sqrt{a}}.$$

(IX). Notice that $f(x) = \sqrt{1+x}-1$, $g(x) = \sqrt{1-x}-1$ and $\lim_{x\to 0}\frac{\sqrt{1+x}-1}{\sqrt{1-x}-1}$ satisfy the hypotheses of L'Hôpital's Rule. Then,

$$\lim_{x\to 0}\frac{\sqrt{1+x}-1}{\sqrt{1-x}-1} = \lim_{x\to 0}\frac{\frac{1}{2\sqrt{1+x}}}{-\frac{1}{2\sqrt{1-x}}} = \lim_{x\to 0}-\frac{\sqrt{1-x}}{\sqrt{1+x}} = -1.$$

Analogously to part (VII). we also can use the identity

$$\frac{\sqrt{1+x}-1}{\sqrt{1-x}-1} = \frac{-(\sqrt{1+x}+1)}{\sqrt{1+x}+1}$$

to find the limit.

(X).

$$\lim_{x\to\infty}\sqrt{x+\sqrt{x+\sqrt{x}}}-\sqrt{x}$$

$$= \lim_{x\to\infty}\frac{\left(\sqrt{x+\sqrt{x+\sqrt{x}}}-\sqrt{x}\right)\left(\sqrt{x+\sqrt{x+\sqrt{x}}}+\sqrt{x}\right)}{\sqrt{x+\sqrt{x+\sqrt{x}}}+\sqrt{x}}$$

$$= \lim_{x\to\infty}\frac{x+\sqrt{x+\sqrt{x}}-x}{\sqrt{x+\sqrt{x+\sqrt{x}}}+\sqrt{x}} = \lim_{x\to\infty}\frac{\sqrt{x+\sqrt{x}}}{\sqrt{x+\sqrt{x+\sqrt{x}}}+\sqrt{x}}$$

$$= \lim_{x\to\infty}\frac{\sqrt{x}\sqrt{\frac{1}{\sqrt{x}}+1}}{\sqrt{x}\left(\sqrt{1+\sqrt{\frac{1}{x}+\frac{1}{x^{3/2}}}}+1\right)}$$

$$= \lim_{x\to\infty}\frac{\sqrt{\frac{1}{\sqrt{x}}+1}}{\sqrt{1+\sqrt{\frac{1}{x}+\frac{1}{x^{3/2}}}}+1} = \frac{1}{2}.$$

(XI). Observe that $f(x) = 2x-2$, $g(x) = \sqrt[3]{26+x}-3$ and $\lim_{x\to 1}\frac{2x-2}{\sqrt[3]{26+x}-3}$ satisfy the hypotheses of L'Hôpital's Rule. Then,

$$\lim_{x\to 1}\frac{2x-2}{\sqrt[3]{26+x}-3} = \lim_{x\to 1}\frac{2}{\frac{1}{3(26+x)^{2/3}}} = \lim_{x\to 1}6(26+x)^{2/3} = 54.$$

An alternative way is based on exploiting the special case of the Cyclotomic Identity (Exercise 1.7) $a^3 - b^3 = (a - b)(a^2 + ab + b^2)$ with $a = \sqrt[3]{26 + x}$ and $b = 3$.

(XII). Since $q \neq 0$, notice that $f(x) = x^p - 1$, $g(x) = x^q - 1$ and $\lim_{x \to 1} \frac{x^p - 1}{x^q - 1}$ satisfy the hypotheses of L'Hôpital's Rule. Then,

$$\lim_{x \to 1} \frac{x^p - 1}{x^q - 1} = \lim_{x \to 1} \frac{px^{p-1}}{qx^{q-1}} = \frac{p}{q}.$$

(XIII). Since $a, b, c, d > 0$ and $c \neq d$, observe that $f(x) = a^x - b^x$, $g(x) = c^x - d^x$ and $\lim_{x \to 0} \frac{a^x - b^x}{c^x - d^x}$ satisfy the hypotheses of L'Hôpital's Rule. Then,

$$\lim_{x \to 0} \frac{a^x - b^x}{c^x - d^x} = \lim_{x \to 0} \frac{(\log a)a^x - (\log b)b^x}{(\log c)c^x - (\log d)d^x} = \frac{\log a - \log b}{\log c - \log d}.$$

It is clear that $\frac{\log a - \log b}{\log c - \log d} \in \mathbb{R}$ given that $a, b, c, d > 0$ and $c \neq d$.

We point out another way to do it: observe first that

$$\frac{a^x - b^x}{c^x - d^x} = \frac{(a/b) - 1}{x} \cdot \frac{x}{(c/d)^x - 1} \cdot \left(\frac{b}{d}\right)^x.$$

Since $\left(\frac{b}{d}\right)^x \to 1$ as $x \to 0$, it suffices to show that, for every $l > 0$, it holds $L := \lim_{x \to 0} \frac{l^x - 1}{x} = \log l$. With this aim, note that if $t := \frac{1}{l^x - 1}$ then $x = \log_l(1 + \frac{1}{t})$, and $t \to \infty$ iff $x \to 0$. Recall that $\log_l x = \frac{\log x}{\log l}$ for all $x > 0$. Therefore,

$$L = \lim_{t \to \infty} \frac{1/t}{\log_l(1 + \frac{1}{t})} = \lim_{t \to \infty} \frac{1}{\log_l(1 + \frac{1}{t})^t} = \lim_{t \to \infty} \frac{\log l}{\log(1 + \frac{1}{t})^t} = \frac{\log l}{\log e} = \log l,$$

and we are done.

(XIV). Notice that $f(x) = e^x - e^{\sin x}$, $g(x) = x - \sin x$ and $\lim_{x \to 0} \frac{e^x - e^{\sin x}}{x - \sin x}$ satisfy the hypotheses of L'Hôpital's Rule. Then,

$$\lim_{x \to 0} \frac{e^x - e^{\sin x}}{x - \sin x} = \lim_{x \to 0} \frac{e^x - \cos x e^{\sin x}}{1 - \cos x} = \lim_{x \to 0} \frac{e^x + \sin x e^{\sin x} - \cos^2 x e^{\sin x}}{\sin x}$$

$$= \lim_{x \to 0} \frac{e^x + \cos x e^{\sin x} + \sin x \cos x e^{\sin x} + 2\cos x \sin x e^{\sin x} - \cos^3 x e^{\sin x}}{\cos x}$$

$$= 1.$$

We indicate another approach to find the limit, say L. As in part (XIII), we have that $\lim_{t \to 0} \frac{e^t - 1}{t} = \log e = 1$. Observe that if $t := x - \sin x$ then $t \to 0$ as $x \to 0$. Therefore,

$$L = \lim_{x \to 0} e^{\sin x} \cdot \frac{e^{x - \sin x} - 1}{x - \sin x}$$

$$= \lim_{x \to 0} e^{\sin x} \cdot \lim_{x \to 0} \frac{e^{x - \sin x} - 1}{x - \sin x}$$

$$= e^{\sin 0} \cdot \lim_{t \to 0} \frac{e^t - 1}{t} = e^0 \cdot 1 = 1.$$

(XV). We have

$$\lim_{x \to 0^+} x^x = \lim_{x \to 0^+} e^{\log(x^x)} = \lim_{x \to 0^+} e^{x \log x} = e^{\lim_{x \to 0^+} x \log x}.$$

Let us calculate the limit $\lim_{x \to 0^+} x \log x$. To do so, we will apply the substitution $x = \frac{1}{t}$, which turns the limit $\lim_{x \to 0^+} x \log x$ into

$$\lim_{x \to 0^+} x \log x = \lim_{t \to \infty} \frac{1}{t} \log\left(\frac{1}{t}\right) = \lim_{t \to \infty} -\frac{1}{t} \log t = \lim_{t \to \infty} -\frac{\log t}{t}.$$

Now observe that $f(t) = t$, $g(t) = \log t$ and $\lim_{t \to \infty} -\frac{\log t}{t}$ satisfy the hypotheses of L'Hôpital's Rule. Then,

$$\lim_{t \to \infty} -\frac{\log t}{t} = \lim_{t \to \infty} -\frac{1}{t} = 0.$$

Hence,

$$\lim_{x \to 0^+} x^x = e^{\lim_{x \to 0^+} x \log x} = 1.$$

If we do not want to use L'Hôpital's Rule in the calculation of $\lim_{t \to 0} \frac{\log t}{t}$, just observe that this limit equals $\lim_{x \to \infty} \frac{x}{e^x} = \lim_{x \to \infty} (1/e)^x \cdot x^1$, whose value is 0 because $|1/e| = 1/e < 1$.

(XVI). We have

$$\lim_{x \to 0^+} x^{x^x} = \lim_{x \to 0^+} e^{\log(x^{x^x})} = \lim_{x \to 0^+} e^{x^x \log x} = e^{\lim_{x \to 0^+} x^x \log x}.$$

Hence, by Exercise 4.2 part (XV), we have

$$\lim_{x \to 0^+} x^{x^x} = 0.$$

(XVII). Since $\lim_{x \to \infty} \left(\dfrac{2x^2 + 3}{2x^2 + 5}\right)^{8x^2 + 3}$ is equal to the indeterminate 1^∞, we have

$$\lim_{x \to \infty} \left(\frac{2x^2 + 3}{2x^2 + 5}\right)^{8x^2 + 3}$$

$$= \lim_{x \to \infty} \left(1 + \frac{2x^2 + 3}{2x^2 + 5} - 1\right)^{8x^2 + 3} = \lim_{x \to \infty} \left(1 + \frac{-2}{2x^2 + 5}\right)^{8x^2 + 3}$$

$$= \lim_{x \to \infty} \left(1 + \frac{1}{\frac{2x^2 + 5}{-2}}\right)^{8x^2 + 3} = \lim_{x \to \infty} \left[\left(1 + \frac{1}{\frac{2x^2 + 5}{-2}}\right)^{\frac{2x^2 + 5}{-2}}\right]^{\frac{-2}{2x^2 + 5}(8x^2 + 3)}$$

$$= e^{\lim_{x \to \infty} \frac{-16x^2 - 6}{2x^2 + 5}} = e^{-8}.$$

(XVIII). As $a, b, c > 0$, we have that $\lim_{x \to \infty} \left(\frac{a^{1/x} + b^{1/x} + c^{1/x}}{3} \right)^x$ is equal to the indeterminate 1^∞. So,

$$\lim_{x \to \infty} \left(\frac{a^{1/x} + b^{1/x} + c^{1/x}}{3} \right)^x = \lim_{x \to \infty} \left(1 + \frac{a^{1/x} + b^{1/x} + c^{1/x}}{3} - 1 \right)^x$$

$$= \lim_{x \to \infty} \left(1 + \frac{a^{1/x} + b^{1/x} + c^{1/x} - 3}{3} \right)^x$$

$$= \lim_{x \to \infty} \left(1 + \frac{1}{\frac{-3}{a^{1/x} + b^{1/x} + c^{1/x} - 3}} \right)^x$$

$$= \lim_{x \to \infty} \left[\left(1 + \frac{1}{\frac{-3}{a^{1/x} + b^{1/x} + c^{1/x} - 3}} \right)^{\frac{-3}{a^{1/x} + b^{1/x} + c^{1/x} - 3}} \right]^{\frac{a^{1/x} + b^{1/x} + c^{1/x} - 3}{-3} x}$$

$$= e^{\lim_{x \to \infty} \frac{(a^{1/x} + b^{1/x} + c^{1/x} - 3)x}{-3}}$$

$$= e^{-\frac{1}{3} \lim_{x \to \infty} (a^{1/x} + b^{1/x} + c^{1/x} - 3)x} .$$

Let us calculate now $\lim_{x \to \infty} (a^{1/x} + b^{1/x} + c^{1/x} - 3)x$. Observe that $\lim_{x \to \infty} (a^{1/x} + b^{1/x} + c^{1/x} - 3)x$ is equal to the indeterminate $0 \cdot \infty$. However, we also have that

$$\lim_{x \to \infty} (a^{1/x} + b^{1/x} + c^{1/x} - 3)x = \lim_{t \to 0} \frac{a^t + b^t + c^t - 3}{t},$$

and now we can apply L'Hôpital's Rule:

$$\lim_{t \to 0} \frac{a^t + b^t + c^t - 3}{t}$$

$$= \lim_{t \to 0} [(\log a)a^t + (\log b)b^t + (\log c)c^t]$$

$$= \log a + \log b + \log c = \log(abc).$$

Here it is another way to do it: the limit to be calculated is

$$\lim_{t \to 0} \frac{a^t + b^t + c^t - 3}{t} = \lim_{t \to 0} \left(\frac{a^t - 1}{t} + \frac{b^t - 1}{t} + \frac{c^t - 1}{t} \right) = \log a + \log b + \log c,$$

where we have used that $\lim_{t \to 0} \frac{d^t - 1}{t} = \log d$ for all $d > 0$, according to the calculation done in part (XIII).

(XIX). Notice that $f(x) = e^x + \sin x - 1$, $g(x) = \log(1 + x)$ and $\lim_{x \to 0} \frac{e^x + \sin x - 1}{\log(1+x)}$ satisfy the hypotheses of L'Hôpital's Rule. Then,

$$\lim_{x \to 0} \frac{e^x + \sin x - 1}{\log(1 + x)} = \lim_{x \to 0} \frac{e^x + \cos x}{\frac{1}{1+x}} = 2.$$

(XX). Notice that $f(x) = \sec^2 x - 2\tan x$, $g(x) = 1 + \cos(4x)$ and $\lim_{x\to\pi/4} \frac{\sec^2 x - 2\tan x}{1+\cos(4x)}$ satisfy the hypotheses of L'Hôpital's Rule. Then,

$$
\lim_{x\to\pi/4} \frac{\sec^2 x - 2\tan x}{1 + \cos(4x)} = \lim_{x\to\pi/4} \frac{2\sec^2 x \tan x - 2\sec^2 x}{-4\sin(4x)}
$$

$$
= \lim_{x\to\pi/4} \frac{\sec^2 x(\tan x - 1)}{-2\sin(4x)}
$$

$$
= \lim_{x\to\pi/4} \frac{2\sec^2 x \tan x(\tan x - 1) + \sec^4 x}{-8\cos(4x)} = \frac{1}{2}.
$$

Note that we have applied L'Hôpital's Rule several times.

Here it is another way to do it: as $\sec x = 1/\cos x$, $\tan x = \sin x/\cos x$, $1 + \cos(4x) = 2\cos^2(2x)$ and $2\sin x \cos x = \sin(2x)$, we get that our limit is

$$
\lim_{x\to\pi/4} \frac{1 - \sin(2x)}{\cos^2 x \cdot 2 \cdot \cos^2(2x)} = \lim_{x\to\pi/4} \frac{1 - \sin(2x)}{2\cos^2 x \cdot (1 - \sin^2(2x))}
$$

$$
= \lim_{x\to\pi/4} \frac{1}{2\cos^2 x \cdot (1 + \sin(2x))} = \frac{1}{2 \cdot (1/2) \cdot (1+1)} = \frac{1}{2}.
$$

(XXI). We have

$$
\lim_{x\to\pi/4} \tan(2x) \cot\left(\frac{\pi}{4} + x\right) = \lim_{x\to\pi/4} \frac{\sin(2x)\cos\left(\frac{\pi}{4} + x\right)}{\cos(2x)\sin\left(\frac{\pi}{4} + x\right)}
$$

$$
= \lim_{x\to\pi/4} \frac{\sin(2x)}{\sin\left(\frac{\pi}{4} + x\right)} \cdot \lim_{x\to\pi/4} \frac{\cos\left(\frac{\pi}{4} + x\right)}{\cos(2x)}.
$$

Observe that

$$
\lim_{x\to\pi/4} \frac{\sin(2x)}{\sin\left(\frac{\pi}{4} + x\right)} = 1.
$$

Therefore, it is enough to calculate the limit $\lim_{x\to\pi/4} \frac{\cos\left(\frac{\pi}{4}+x\right)}{\cos(2x)}$. Note that $f(x) = \cos\left(\frac{\pi}{4} + x\right)$, $g(x) = \cos(2x)$ and $\lim_{x\to\pi/4} \frac{\cos\left(\frac{\pi}{4}+x\right)}{\cos(2x)}$ satisfy the hypotheses of L'Hôpital's Rule. Then,

$$
\lim_{x\to\pi/4} \frac{\cos\left(\frac{\pi}{4} + x\right)}{\cos(2x)} = \lim_{x\to\pi/4} \frac{-\sin\left(\frac{\pi}{4} + x\right)}{-2\sin(2x)} = \frac{1}{2}.
$$

Hence,

$$
\lim_{x\to\pi/4} \tan(2x)\cot\left(\frac{\pi}{4} + x\right) = \frac{1}{2}.
$$

(XXII). Notice that $f(x) = \sin(x - \pi/6)$, $g(x) = \sqrt{3} - 2\cos x$ and $\lim_{x\to\pi/6} \frac{\sin(x-\pi/6)}{\sqrt{3}-2\cos x}$ satisfy the hypotheses of L'Hôpital's Rule. Then,

$$
\lim_{x\to\pi/6} \frac{\sin(x - \pi/6)}{\sqrt{3} - 2\cos x} = \lim_{x\to\pi/6} \frac{\cos(x - \pi/6)}{2\sin x} = \frac{1}{2 \cdot (1/2)} = 1.
$$

Another approach is based on the change of variables $t = x - \frac{\pi}{6}$. Note that $t \to 0$ as $x \to \pi/6$. Then $\sin(x - \frac{\pi}{6}) = \sin t = 2\sin\frac{t}{2}\cos\frac{t}{2}$ and $\sqrt{3} - 2\cos x = \sqrt{3} - 2\cos(t + \frac{\pi}{6}) = \sqrt{3} - 2(\cos t \cdot \frac{\sqrt{3}}{2} - \sin t \cdot \frac{1}{2}) = \sqrt{3}(1 - \cos t) - \sin t = 2\sqrt{3}\sin^2\frac{t}{2} - 2\sin\frac{t}{2}\cos\frac{t}{2}$. Finally, divide both numerator and denominator by $\sin\frac{t}{2}$ and take limits as $t \to 0$.

(XXIII). Note that $\lim_{x\to 0}\left(\sin^2\left(\frac{\pi}{2-ax}\right)\right)^{\sec^2\left(\frac{\pi}{2-bx}\right)}$ is equal to the indeterminate 1^∞. Then,

$$\lim_{x\to 0}\left(\sin^2\left(\frac{\pi}{2-ax}\right)\right)^{\sec^2\left(\frac{\pi}{2-bx}\right)} = \lim_{x\to 0}\left(1 + \sin^2\left(\frac{\pi}{2-ax}\right) - 1\right)^{\sec^2\left(\frac{\pi}{2-bx}\right)}$$

$$= \lim_{x\to 0}\left(1 + \frac{1}{\frac{1}{\sin^2\left(\frac{\pi}{2-ax}\right)-1}}\right)^{\sec^2\left(\frac{\pi}{2-bx}\right)}$$

$$= \lim_{x\to 0}\left[\left(1 + \frac{1}{\frac{1}{\sin^2\left(\frac{\pi}{2-ax}\right)-1}}\right)^{\frac{1}{\sin^2\left(\frac{\pi}{2-ax}\right)-1}}\right]^{\left(\sin^2\left(\frac{\pi}{2-ax}\right)-1\right)\sec^2\left(\frac{\pi}{2-bx}\right)}$$

$$= e^{\lim_{x\to 0}\left(\sin^2\left(\frac{\pi}{2-ax}\right)-1\right)\sec^2\left(\frac{\pi}{2-bx}\right)}.$$

Now let us calculate the limit $\lim_{x\to 0}\left(\sin^2\left(\frac{\pi}{2-ax}\right) - 1\right)\sec^2\left(\frac{\pi}{2-bx}\right)$. Note that

$$\lim_{x\to 0}\left(\sin^2\left(\frac{\pi}{2-ax}\right) - 1\right)\sec^2\left(\frac{\pi}{2-bx}\right) = \lim_{x\to 0}\frac{\sin^2\left(\frac{\pi}{2-ax}\right) - 1}{\cos^2\left(\frac{\pi}{2-bx}\right)}.$$

Now we have that $f(x) = \sin^2\left(\frac{\pi}{2-ax}\right) - 1$, $g(x) = \cos^2\left(\frac{\pi}{2-bx}\right)$ and $\lim_{x\to 0}\frac{\sin^2\left(\frac{\pi}{2-ax}\right)-1}{\cos^2\left(\frac{\pi}{2-bx}\right)}$ satisfy the hypotheses of L'Hôpital's Rule. So

$$\lim_{x\to 0}\frac{\sin^2\left(\frac{\pi}{2-ax}\right) - 1}{\cos^2\left(\frac{\pi}{2-bx}\right)} = \lim_{x\to 0}\frac{2\sin\left(\frac{\pi}{2-ax}\right)\cos\left(\frac{\pi}{2-ax}\right)\frac{\pi a}{(2-ax)^2}}{-2\cos\left(\frac{\pi}{2-bx}\right)\sin\left(\frac{\pi}{2-bx}\right)\frac{\pi a}{(2-bx)^2}}$$

$$= \lim_{x\to 0} -\frac{\sin\left(\frac{\pi}{2-ax}\right)\frac{\pi a}{(2-ax)^2}}{\sin\left(\frac{\pi}{2-bx}\right)\frac{\pi b}{(2-bx)^2}} \cdot \lim_{x\to 0}\frac{\cos\left(\frac{\pi}{2-ax}\right)}{\cos\left(\frac{\pi}{2-bx}\right)}$$

$$= -\frac{a}{b} \cdot \lim_{x\to 0}\frac{\cos\left(\frac{\pi}{2-ax}\right)}{\cos\left(\frac{\pi}{2-bx}\right)}.$$

Let us find the value of $\lim\limits_{x\to 0} \dfrac{\cos\left(\frac{\pi}{2-ax}\right)}{\cos\left(\frac{\pi}{2-bx}\right)}$. Notice that $f(x) = \cos\left(\frac{\pi}{2-ax}\right)$,

$g(x) = \cos\left(\frac{\pi}{2-bx}\right)$ and $\lim\limits_{x\to 0} \dfrac{\cos\left(\frac{\pi}{2-ax}\right)}{\cos\left(\frac{\pi}{2-bx}\right)}$ satisfy the hypotheses of L'Hôpital's

Rule. Then,

$$\lim_{x\to 0} \frac{\cos\left(\frac{\pi}{2-ax}\right)}{\cos\left(\frac{\pi}{2-bx}\right)} = \lim_{x\to 0} \frac{\sin\left(\frac{\pi}{2-ax}\right) \frac{\pi a}{(2-ax)^2}}{\sin\left(\frac{\pi}{2-bx}\right) \frac{\pi b}{(2-bx)^2}} = \frac{a}{b}.$$

Hence,

$$\lim_{x\to 0} \left(\sin^2\left(\frac{\pi}{2-ax}\right) - 1\right)\sec^2\left(\frac{\pi}{2-bx}\right) = -\frac{a^2}{b^2}.$$

Therefore,

$$\lim_{x\to 0} \left(\sin^2\left(\frac{\pi}{2-ax}\right)\right)^{\sec^2\left(\frac{\pi}{2-bx}\right)} = e^{-\frac{a^2}{b^2}}.$$

(XXIV). Observe that $f(x) = \cos\left(\frac{\pi}{2-ax}\right)$, $g(x) = \cos\left(\frac{\pi}{2-bx}\right)$ and $\lim_{x\to\pi/3} \frac{\sin(x-\pi/3)}{1-2\cos x}$ satisfy the hypotheses of L'Hôpital's Rule. Then,

$$\lim_{x\to\pi/3} \frac{\sin(x-\pi/3)}{1-2\cos x} = \lim_{x\to\pi/3} \frac{\cos(x-\pi/3)}{2\sin x} = \frac{1}{\sqrt{3}}.$$

□

Exercise 4.3. *Prove that the following limits do not exist:*

  (I) $\lim\limits_{x\to\infty} \cos x.$ (II) $\lim\limits_{x\to 0} \sin\dfrac{1}{x^2}.$

  (III) $\lim\limits_{x\to 0} e^{1/\sin(x)}.$ (IV) $\lim\limits_{x\to\pi/2} \dfrac{\cos x}{\sqrt[3]{(1-\sin x)^2}}.$

Solution to Exercise 4.3.

  (I). Take the sequences $x_n = \frac{\pi}{2} + 2\pi n$ and $y_n = 2\pi n$. Note that both $(x_n)_{n=1}^\infty$ and $(y_n)_{n=1}^\infty$ diverge to ∞, and also

$$\cos(x_n) = 0$$

and

$$\cos(y_n) = 1$$

for all $n \in \mathbb{N}$. Thus, by the sequential limit criterion for limits, the limit $\lim_{x \to \infty} \cos x$ does not exist.

(II). Take the sequences $x_n = \frac{1}{\sqrt{2\pi n}}$ and $y_n = \frac{1}{\sqrt{\frac{\pi}{2} + 2\pi n}}$. It is clear that both $(x_n)_{n=1}^{\infty}$ and $(y_n)_{n=1}^{\infty}$ converge to 0. Moreover,

$$\sin(x_n) = 0$$

and

$$\sin(y_n) = 1$$

for each $n \in \mathbb{N}$. Hence, by the sequential limit criterion for limits, the limit $\lim_{x \to 0} \sin \frac{1}{x^2}$ does not exist.

(III). Let us calculate the one-sided limits of $\lim_{x \to 0^-} e^{1/\sin(x)}$:

$$\lim_{x \to 0^-} e^{1/\sin(x)} = e^{\lim_{x \to 0^-} 1/\sin(x)}$$

and

$$\lim_{x \to 0^+} e^{1/\sin(x)} = e^{\lim_{x \to 0^+} 1/\sin(x)}.$$

Hence, it is enough to find the one-sided limits of $\lim_{x \to 0} 1/\sin(x)$. To do so, let $x_n = \frac{1}{n}$ and $y_n = -\frac{1}{n}$. Notice that $\sin(x_n) > 0$ and $\sin(y_n) < 0$ for all $n \in \mathbb{N}$, since $\sin(x) > 0$ for all $x \in \left(0, \frac{\pi}{2}\right)$ and $\sin(x) < 0$ for all $x \in \left(-\frac{\pi}{2}, 0\right)$. Thus,

$$\lim_{x \to 0^-} 1/\sin(x) = -\infty$$

and

$$\lim_{x \to 0^+} 1/\sin(x) = \infty,$$

which implies that

$$\lim_{x \to 0^-} e^{1/\sin(x)} = 0$$

and

$$\lim_{x \to 0^+} e^{1/\sin(x)} = \infty.$$

Therefore, since the one-sided limits do not coincide, the limit $\lim_{x \to 0} e^{1/\sin(x)}$ does not exist.

(IV). Note that

$$\lim_{x \to \pi/2} \sqrt[3]{(1 - \sin x)^2} = 0.$$

Now, since $\cos x > 0$ for all $x \in \left(0, \frac{\pi}{2}\right)$ and $\cos(x) < 0$ for any $x \in \left(\frac{\pi}{2}, x\right)$. Observe that we have that

$$\lim_{x \to \pi/2^-} \frac{\cos x}{\sqrt[3]{(1 - \sin x)^2}} = \infty$$

and

$$\lim_{x \to \pi/2^+} \frac{\cos x}{\sqrt[3]{(1 - \sin x)^2}} = -\infty.$$

As the one-sided limits do not coincide, the limit $\lim\limits_{x \to \pi/2} \dfrac{\cos x}{\sqrt[3]{(1 - \sin x)^2}}$ does not exist.

□

Exercise 4.4. *Find the following one-sided limits, if they exist:*

(I) $\lim\limits_{x \to 2^\pm} f(x)$ with $f(x) = \begin{cases} x + 1 & \text{if } x \neq 2, \\ 0 & \text{if } x = 2. \end{cases}$

(II) $\lim\limits_{x \to 1^\pm} f(x)$ with $f(x) = \begin{cases} -2x + 3 & \text{if } x \geq 1, \\ 3x - 5 & \text{if } x < 1. \end{cases}$

(III) $\lim\limits_{x \to 0^\pm} \dfrac{\sqrt{1 - \cos(2x)}}{x}$.

(IX) $\lim\limits_{x \to 0^\pm} \dfrac{1}{e^{1/x} + 1}$.

(IV) $\lim\limits_{x \to 0^\pm} \dfrac{\sqrt{1 - \sqrt{1 - x^2}}}{x}$.

(X) $\lim\limits_{x \to 0^\pm} \dfrac{\sin \frac{1}{x}}{e^{1/x} + 1}$.

(V) $\lim\limits_{x \to 0^\pm} \dfrac{|x|}{x^2 + x}$.

(XI) $\lim\limits_{x \to 0^\pm} e^{1/x} \sin \dfrac{1}{x}$.

(VI) $\lim\limits_{x \to 1^\pm} \dfrac{x^2 - 1}{|x - 1|}$.

(XII) $\lim\limits_{x \to 0^\pm} \dfrac{e^{1/x}}{e^{1/x} - 1}$.

(VII) $\lim\limits_{x \to 0^\pm} \dfrac{1}{2 - 2^{1/x}}$.

(XIII) $\lim\limits_{x \to 2^\pm} \dfrac{x^2 - 2x}{x^2 - 4x + 4}$.

(VIII) $\lim\limits_{x \to 1^\pm} (x - 1)e^{x/(x-1)}$.

(XIV) $\lim\limits_{x \to 2^\pm} \dfrac{x^2 + x + 6}{x^2 - 4}$.

Solution to Exercise 4.4.

(I). On the one hand,

$$\lim_{x \to 2^+} f(x) = \lim_{x \to 2} f|_{\mathbb{R} \cap (2, \infty)}(x) = \lim_{x \to 2} f|_{(2, \infty)}(x) = \lim_{x \to 2} (x + 1) = 3.$$

On the other hand,

$$\lim_{x \to 2^-} f(x) = \lim_{x \to 2} f|_{\mathbb{R} \cap (-\infty, 2)}(x) = \lim_{x \to 2} f|_{(-\infty, 2)}(x) = \lim_{x \to 2} (x + 1) = 3.$$

(II). On the one hand,

$$\lim_{x \to 1^+} f(x) = \lim_{x \to 1} f|_{\mathbb{R} \cap (1, \infty)}(x) = \lim_{x \to 1} f|_{(1, \infty)}(x) = \lim_{x \to 1} (-2x + 3) = 1.$$

On the other hand,

$$\lim_{x \to 1^-} f(x) = \lim_{x \to 1} f|_{\mathbb{R} \cap (-\infty,1)}(x) = \lim_{x \to 1} f|_{(-\infty,1)}(x) = \lim_{x \to 1^-} (3x - 5) = -2.$$

(III). On the one hand,

$$\lim_{x \to 0^+} \frac{\sqrt{1 - \cos(2x)}}{x} = \lim_{x \to 0^+} \frac{\sqrt{2\sin^2(x)}}{x} = \lim_{x \to 0^+} \frac{\sqrt{2}|\sin(x)|}{x},$$

where we have used the formula $\sin^2(x) = \dfrac{1 - \cos(2x)}{2}$. Now let $(x_n)_{n=1}^\infty$ be a sequence such that $x_n > 0$ for all $n \in \mathbb{N}$ and $(x_n)_{n=1}^\infty$ converges to 0. Since $\sin(x_n) \sim x_n$, we have

$$\lim_{n \to \infty} \frac{\sqrt{2}|\sin(x_n)|}{x_n} = \lim_{n \to \infty} \frac{\sqrt{2}|x_n|}{x_n} = \lim_{n \to \infty} \frac{\sqrt{2}x_n}{x_n} = \lim_{n \to \infty} \sqrt{2} = \sqrt{2}.$$

Thus, by the sequential limit criterion for limits we have

$$\lim_{x \to 0^+} \frac{\sqrt{1 - \cos(2x)}}{x} = \sqrt{2}.$$

On the other hand,

$$\lim_{x \to 0^-} \frac{\sqrt{1 - \cos(2x)}}{x} = \lim_{x \to 0^-} \frac{\sqrt{2\sin^2(x)}}{x} = \lim_{x \to 0^-} \frac{\sqrt{2}|\sin(x)|}{x},$$

where once again we have used $\sin^2(x) = \dfrac{1 - \cos(2x)}{2}$. Now take $(x_n)_{n=1}^\infty$ be a sequence such that $x_n < 0$ for any $n \in \mathbb{N}$ and $(x_n)_{n=1}^\infty$ converges to 0. Then,

$$\lim_{n \to \infty} \frac{\sqrt{2}|\sin(x_n)|}{x_n} = \lim_{n \to \infty} \frac{\sqrt{2}|x_n|}{x_n} = \lim_{n \to \infty} -\frac{\sqrt{2}x_n}{x_n} = \lim_{n \to \infty} -\sqrt{2} = -\sqrt{2}.$$

Thus, by the sequential limit criterion for limits we obtain

$$\lim_{x \to 0^-} \frac{\sqrt{1 - \cos(2x)}}{x} = -\sqrt{2}.$$

(IV). Since $x^2 - y^2 = (x - y)(x + y)$ for any $x, y \in \mathbb{R}$ (or the Cyclotomic Equation, recall Exercise 1.7) we get for $x = 1$ and $y = \sqrt{1 - x^2}$ that

$$1 - \sqrt{1 - x^2} = \frac{1^2 - (\sqrt{1 - x^2})^2}{1 + \sqrt{1 - x^2}} = \frac{1 - (1 - x^2)}{1 + \sqrt{1 - x^2}} = \frac{x^2}{1 + \sqrt{1 - x^2}}.$$

Hence, on the one hand,

$$\lim_{x \to 0^+} \frac{\sqrt{1 - \sqrt{1 - x^2}}}{x} = \lim_{x \to 0^+} \frac{\sqrt{\frac{x^2}{1+\sqrt{1-x^2}}}}{x} = \lim_{x \to 0^+} \frac{|x| \cdot \frac{1}{\sqrt{1+\sqrt{1-x^2}}}}{x}$$

$$= \lim_{x \to 0^+} \frac{x \cdot \frac{1}{\sqrt{1+\sqrt{1-x^2}}}}{x} = \lim_{x \to 0^+} \frac{1}{\sqrt{1 + \sqrt{1 - x^2}}} = \frac{1}{\sqrt{2}}.$$

On the other hand,

$$\lim_{x \to 0^-} \frac{\sqrt{1 - \sqrt{1 - x^2}}}{x} = \lim_{x \to 0^-} \frac{\sqrt{\frac{x^2}{1+\sqrt{1-x^2}}}}{x} = \lim_{x \to 0^-} \frac{|x| \cdot \frac{1}{\sqrt{1+\sqrt{1-x^2}}}}{x}$$

$$= \lim_{x \to 0^-} -\frac{x \cdot \frac{1}{\sqrt{1+\sqrt{1-x^2}}}}{x} = \lim_{x \to 0^-} -\frac{1}{\sqrt{1 + \sqrt{1 - x^2}}}$$

$$= -\frac{1}{\sqrt{2}}.$$

(v). On the one hand,

$$\lim_{x \to 0^+} \frac{|x|}{x^2 + x} = \lim_{x \to 0^+} \frac{x}{x^2 + x} = \lim_{x \to 0^+} \frac{x}{x(x + 1)} = \lim_{x \to 0^+} \frac{1}{x + 1} = 1.$$

On the other hand,

$$\lim_{x \to 0^-} \frac{|x|}{x^2 + x} = \lim_{x \to 0^-} -\frac{x}{x^2 + x} = \lim_{x \to 0^-} -\frac{x}{x(x + 1)} = \lim_{x \to 0^-} -\frac{1}{x + 1} = -1.$$

(vi). On the one hand,

$$\lim_{x \to 1^+} \frac{x^2 - 1}{|x - 1|} = \lim_{x \to 1^+} \frac{x^2 - 1}{x - 1} = \lim_{x \to 1^+} \frac{(x + 1)(x - 1)}{x - 1} = \lim_{x \to 1^+} (x + 1) = 2.$$

On the other hand,

$$\lim_{x \to 1^-} \frac{x^2 - 1}{|x - 1|} = \lim_{x \to 1^-} -\frac{x^2 - 1}{x - 1} = \lim_{x \to 1^-} -\frac{(x + 1)(x - 1)}{x - 1}$$

$$= \lim_{x \to 1^-} -(x + 1) = -2.$$

(vii). On the one hand, first observe that

$$\lim_{x \to 0^+} \frac{1}{x} = \infty,$$

which implies that

$$\lim_{x \to 0^+} 2^{1/x} = \infty.$$

Hence,

$$\lim_{x \to 0^+} \frac{1}{2 - 2^{1/x}} = 0.$$

On the other hand, note that

$$\lim_{x \to 0^-} \frac{1}{x} = -\infty,$$

which yields

$$\lim_{x \to 0^-} 2^{1/x} = 0.$$

Hence,

$$\lim_{x \to 0^-} \frac{1}{2 - 2^{1/x}} = \frac{1}{2}.$$

(VIII). On the one hand, since $x \to 1^+$ iff $t := x - 1 \to 0^+$ iff $s := 1/t \to \infty$, we get

$$\lim_{x \to 1^+} (x - 1)e^{x/(x-1)} = \lim_{t \to 0^+} te^{1+(1/t)} = e \cdot \lim_{s \to \infty} \frac{e^s}{s} = \infty.$$

On the other hand, observe that

$$\lim_{x \to 1^+} \frac{x}{x - 1} = -\infty,$$

which gives us

$$\lim_{x \to 1^+} e^{x/(x-1)} = 0.$$

Hence,

$$\lim_{x \to 1^+} (x - 1)e^{x/(x-1)} = 0 \cdot 0 = 0.$$

(IX). On the one hand, note that

$$\lim_{x \to 0^+} \frac{1}{x} = \infty,$$

which yields

$$\lim_{x \to 0^+} e^{1/x} = \infty.$$

Hence,

$$\lim_{x \to 0^+} \frac{1}{e^{1/x} + 1} = \infty.$$

On the other hand, observe that

$$\lim_{x \to 0^-} \frac{1}{x} = -\infty,$$

which yields

$$\lim_{x \to 0^-} e^{1/x} = 0.$$

Hence,

$$\lim_{x \to 0^-} \frac{1}{e^{1/x} + 1} = 1.$$

(X). On the one hand, since $\sin \dfrac{1}{x}$ is bounded and by Exercise 4.4 part (IX), we have

$$\lim_{x \to 0^+} \frac{\sin \frac{1}{x}}{e^{1/x} + 1} = \lim_{x \to 0^+} \sin \frac{1}{x} \cdot 0 = \lim_{x \to 0^+} 0 = 0.$$

On the other hand, since $\lim\limits_{x \to 0^-} \sin \dfrac{1}{x}$ does not exist and by Exercise 4.4 part (IX), we obtain that

$$\lim_{x \to 0^-} \frac{\sin \frac{1}{x}}{e^{1/x} + 1} = \lim_{x \to 0^+} \sin \frac{1}{x} \cdot 1 = \lim_{x \to 0^+} \sin \frac{1}{x}$$

does not exist.

(XI). On the one hand, take the sequences $x_n = \frac{1}{n\pi}$ and $y_n = \frac{1}{\frac{\pi}{2} + 2n\pi}$. Then $x_n > 0 < y_n$ for all $n \in \mathbb{N}$, and both sequences tend to 0. Therefore, $\lim_{n \to \infty} e^{1/x_n} \sin \frac{1}{x_n} = \lim_{n \to \infty} 0 = 0$ but $\lim_{n \to \infty} e^{1/y_n} \sin \frac{1}{x_n} = \lim_{n \to \infty} e^{(\pi/2) + 2n\pi} = \infty \neq 0$. Hence $\lim\limits_{x \to 0^+} e^{1/x} \sin \dfrac{1}{x}$ does not exist.

On the other hand, as $\lim\limits_{x \to 0^-} e^{1/x} = 0$ and since $\sin \frac{1}{x}$ is bounded, they yield

$$\lim_{x \to 0^+} e^{1/x} \sin \frac{1}{x} = 0.$$

(XII). On the one hand, since $\lim\limits_{x \to 0^+} e^{1/x} = \infty$, we have

$$\lim_{x \to 0^+} \frac{e^{1/x}}{e^{1/x} - 1} = \lim_{x \to 0^+} \frac{e^{1/x}}{e^{1/x} \left(1 - \frac{1}{e^{1/x}}\right)} = \lim_{x \to 0^+} \frac{1}{1 - \frac{1}{e^{1/x}}} = 1.$$

On the other hand, as $\lim\limits_{x \to 0^-} e^{1/x} = 0$, it yields

$$\lim_{x \to 0^-} \frac{e^{1/x}}{e^{1/x} - 1} = \frac{0}{-1} = 0.$$

(XIII). On the one hand,

$$\lim_{x \to 2^+} \frac{x^2 - 2x}{x^2 - 4x + 4} = \lim_{x \to 2^+} \frac{x(x-2)}{(x-2)^2} = \lim_{x \to 2^+} \frac{x}{x-2} = \infty.$$

On the other hand,

$$\lim_{x \to 2^-} \frac{x^2 - 2x}{x^2 - 4x + 4} = \lim_{x \to 2^-} \frac{x(x-2)}{(x-2)^2} = \lim_{x \to 2^-} \frac{x}{x-2} = -\infty.$$

(XIV). On the one hand, since $\lim_{x \to 2^+} (x^2 + x + 6) = 12$ and $\lim_{x \to 2^+} (x+2) = 4$, we have

$$\lim_{x \to 2^+} \frac{x^2 + x + 6}{x^2 - 4} = \lim_{x \to 2^+} \frac{x^2 + x + 6}{(x-2)(x+2)} = \infty.$$

On the other hand, as $\lim_{x \to 2^-} (x^2 + x + 6) = 12$ and $\lim_{x \to 2^-} (x+2) = 4$, they yield

$$\lim_{x \to 2^-} \frac{x^2 + x + 6}{x^2 - 4} = \lim_{x \to 2^-} \frac{x^2 + x + 6}{(x-2)(x+2)} = -\infty.$$

$\square$

Exercise 4.5. *Given $p \geq 2$ an integer, calculate the limit when x tends to 1 of the function*

$$\frac{(\sqrt{x} - 1)(\sqrt[3]{x} - 1) \cdots (\sqrt[p]{x} - 1)}{(x-1)^{p-1}}.$$

Hint: Express this function as a product of $p - 1$ factors.

Solution to Exercise 4.5.

Note that

$$\lim_{x \to 1} \frac{(\sqrt{x} - 1)(\sqrt[3]{x} - 1) \cdots (\sqrt[p]{x} - 1)}{(x-1)^{p-1}} = \lim_{x \to 1} \frac{\sqrt{x} - 1}{x-1} \cdot \frac{\sqrt[3]{x} - 1}{x-1} \cdots \frac{\sqrt[p]{x} - 1}{x-1}.$$

For each integer $2 \leq n \leq p$, apply the Cyclotomic Equation (Exercise 1.7): For $n = 2$, we have $(\sqrt{x})^2 - 1^2 = (\sqrt{x} - 1)(\sqrt{x} + 1)$, that is,

$$\sqrt{x} - 1 = \frac{x-1}{\sqrt{x} + 1}.$$

For $n = 3$, we have $(\sqrt[3]{x})^3 - 1^3 = (\sqrt[3]{x} - 1)\left(\sqrt[3]{x^2} + \sqrt[3]{x} + 1\right)$, that is,

$$\sqrt[3]{x} - 1 = \frac{x-1}{\sqrt[3]{x^2} + \sqrt[3]{x} + 1}.$$

Repeating this process until $n = p$, we obtain that

$$\left(\sqrt[p]{x}\right)^p - 1^p = \left(\sqrt[p]{x} - 1\right)\left(\sqrt[p]{x^{p-1}} + \sqrt[p]{x^{p-2}} + \cdots + \sqrt[p]{x} + 1\right),$$

that is,

$$\sqrt[p]{x} - 1 = \frac{x-1}{\sqrt[p]{x^{p-1}} + \sqrt[p]{x^{p-2}} + \cdots + \sqrt[p]{x} + 1}.$$

Hence,

$$\lim_{x \to 1} \frac{\left(\sqrt{x} - 1\right)\left(\sqrt[3]{x} - 1\right) \cdot \cdots \cdot \left(\sqrt[p]{x} - 1\right)}{(x-1)^{p-1}}$$

$$= \lim_{x \to 1} \frac{\frac{x-1}{\sqrt{x}+1}}{x-1} \cdot \frac{\frac{x-1}{\sqrt[3]{x^2}+\sqrt[3]{x}+1}}{x-1} \cdot \cdots \cdot \frac{\frac{x-1}{\sqrt[p]{x^{p-1}}+\sqrt[p]{x^{p-2}}+\cdots \sqrt[p]{x}+1}}{x-1}$$

$$= \lim_{x \to 1} \frac{x-1}{(x-1)\left(\sqrt{x}+1\right)} \cdot \frac{x-1}{(x-1)\left(\sqrt[3]{x^2}+\sqrt[3]{x}+1\right)}$$

$$\cdots \cdot \frac{x-1}{(x-1)\left(\sqrt[p]{x^{p-1}}+\sqrt[p]{x^{p-2}}+\cdots+\sqrt[p]{x}+1\right)}$$

$$= \lim_{x \to 1} \frac{1}{\left(\sqrt{x}+1\right)\left(\sqrt[3]{x^2}+\sqrt[3]{x}+1\right) \cdot \cdots \cdot \left(\sqrt[p]{x^{p-1}}+\sqrt[p]{x^{p-2}}+\cdots+\sqrt[p]{x}+1\right)}$$

$$= \frac{1}{2 \cdot 3 \cdots p} = \frac{1}{p!}.$$

□

Exercise 4.6. *Study the continuity of the following functions:*

(I) $f(x) = \begin{cases} 0 & \text{if } \sin x \le 0, \\ 1/e & \text{if } \cos(2x) = 0 \text{ and } \sin x > 0, \\ (2\sin x)^{1/\cos(2x)} & \text{otherwise.} \end{cases}$

(II) $f(x) = \begin{cases} 2(1+e^{-1/x})^{-1} & \text{if } x \ne 0, \\ 2 & \text{if } x = 0. \end{cases}$

(III) $f(x) = \begin{cases} \frac{1}{x}\sin\frac{1}{x} & \text{if } x \ne 0, \\ 0 & \text{if } x = 0. \end{cases}$

(IV) $f(x) = \begin{cases} \frac{x+|x|(1+x)}{x}\sin\frac{1}{x} & \text{if } x \ne 0, \\ 1 & \text{if } x = 0. \end{cases}$

$$(\text{v}) \ \ f(x) = \begin{cases} x^n & \text{if } x \in \mathbb{R} \setminus \mathbb{Q}, \\ 0 & \text{if } x \in \mathbb{Q}, \end{cases} \text{ where } n \in \mathbb{N} \text{ is fixed.}$$

Solution to Exercise 4.6.

(I). For this part, it is a good idea to look at the graph of the function while doing the calculations.

First, observe that

$$\sin x \leq 0 \iff x \in [\pi(2k+1), \pi(2k+2)] \text{ for every } k \in \mathbb{Z}.$$

Indeed, the equation $\sin x = 0$ is satisfied only when $x = \pi k$, and note that on the interval $[0, 2\pi]$ we have that $\sin x < 0$ if $x \in (\pi, 2\pi)$. Since $\sin x$ is a periodic function with period 2π, it yields the sets of points where $\sin x \leq 0$. Now, since $f(x) = 0$ if $\sin x \leq 0$, we have that f is continuous at every $x \in (\pi(2k+1), \pi(2k+2))$ for any $k \in \mathbb{Z}$.

Second, notice that

$$\cos(2x) \leq 0 \iff x \in \left[\frac{\pi}{4} + \frac{\pi}{2}k, \frac{\pi}{4} + \frac{\pi}{2}(k+1)\right] \text{ for every } k \in \mathbb{Z}.$$

Also,

$$\sin x > 0 \iff x \in (2\pi k, \pi(2k+1)) \text{ for every } k \in \mathbb{Z}.$$

To see the latter, simply apply a similar argument as before. Thus,

$$\cos(2x) \leq 0 \text{ and } \sin x > 0 \iff x \in \left(\frac{\pi}{4} + \frac{\pi}{2}4k, \frac{\pi}{4} + \frac{\pi}{2}(4k+1)\right)$$
$$\text{for every } k \in \mathbb{Z}.$$

As $f(x) = \frac{1}{e}$ if $\cos(2x) \leq 0$ and $\sin x > 0$, we have that f is continuous at every $x \in \left(\frac{\pi}{4} + \frac{\pi}{2}4k, \frac{\pi}{4} + \frac{\pi}{2}(4k+1)\right)$ for every $k \in \mathbb{Z}$.

Thirdly, if

$$x \in \mathbb{R} \setminus \left(\{[\pi(2k+1), \pi(2k+2)] : k \in \mathbb{Z}\} \right.$$
$$\left. \cup \left\{ \left[\frac{\pi}{4} + \frac{\pi}{2}4k, \frac{\pi}{4} + \frac{\pi}{2}(4k+1)\right] : k \in \mathbb{Z} \right\} \right),$$

then $f(x) = (2\sin x)^{1/\cos(2x)}$ where $\cos(2x) \neq 0$ so $\dfrac{1}{\cos(2x)} \in \mathbb{R}$ and $\sin x > 0$ in which case $(2\sin x)^{1/\cos(2x)} \in \mathbb{R}$. In particular, we obtain that f is continuous at such points x.

Finally, let us study when

$$x \in \left\{ \pi(2k+1), \pi(2k+2), \frac{\pi}{4} + \frac{\pi}{2}4k, \frac{\pi}{4} + \frac{\pi}{2}(4k+1) : k \in \mathbb{Z} \right\}.$$

Since f is a periodic function with period 2π, it is enough to study the continuity at 0, $\frac{\pi}{4}$, $\frac{3\pi}{4}$ and π.

For $x = 0$:

$$\lim_{x \to 0^+} f(x) = \lim_{x \to 0^+} (2\sin x)^{1/\cos(2x)} = 0,$$

$$\lim_{x \to 0^-} f(x) = \lim_{x \to 0^-} (2\sin x)^{1/\cos(2x)} = 0,$$

and $f(0) = 0$. Hence, f is continuous at 0.

For $x = \frac{\pi}{4}$:

$$\lim_{x \to \pi/4^+} f(x) = \lim_{x \to \pi/4^+} (2\sin x)^{1/\cos(2x)} = 0,$$

$$\lim_{x \to \pi/4^-} f(x) = \lim_{x \to \pi/4^-} (2\sin x)^{1/\cos(2x)} = \infty,$$

and $f\left(\frac{\pi}{4}\right) = \frac{1}{e}$. Thus, f is not continuous at $\frac{\pi}{4}$.

For $x = \frac{3\pi}{4}$:

$$\lim_{x \to 3\pi/4^+} f(x) = \lim_{x \to 3\pi/4^+} (2\sin x)^{1/\cos(2x)} = \infty,$$

$$\lim_{x \to 3\pi/4^-} f(x) = \lim_{x \to 3\pi/4^-} (2\sin x)^{1/\cos(2x)} = 0,$$

and $f\left(\frac{3\pi}{4}\right) = \frac{1}{e}$. Therefore, f is not continuous at $\frac{3\pi}{4}$.

For $x = \pi$:

$$\lim_{x \to \pi^+} f(x) = \lim_{x \to \pi^+} 0 = 0,$$

$$\lim_{x \to \pi^-} f(x) = \lim_{x \to \pi^-} (2\sin x)^{1/\cos(2x)} = 0,$$

and $f(\pi) = 0$. Hence, f is continuous at π.

(II). It is clear that if $x \neq 0$, then $f(x) = 2(1 + e^{-1/x})^{-1}$ is continuous. Let us study the remaining case $x = 0$:

$$\lim_{x \to 0^+} f(x) = \lim_{x \to 0^+} f|_{\mathbb{R} \cap (0,\infty)}(x) = \lim_{x \to 0^+} f|_{(0,\infty)}(x) = \lim_{x \to 0^+} 2(1 + e^{-1/x})^{-1} = 2,$$

$$\lim_{x \to 0^-} f(x) = \lim_{x \to 0^-} f|_{\mathbb{R} \cap (-\infty,0)}(x) = \lim_{x \to 0^-} f|_{(-\infty,0)}(x)$$
$$= \lim_{x \to 0^-} 2(1 + e^{-1/x})^{-1} = 0,$$

and $f(0) = 2$. Thus, f is not continuous at 0.

(III). It is obvious that if $x \neq 0$, then $f(x) = \frac{1}{x}\sin\frac{1}{x}$ is continuous.

Let us analyze the case when $x = 0$. Consider the sequence $x_n = \frac{1}{\frac{\pi}{2}+2\pi n}$ for every $n \in \mathbb{N}$. Notice that $(x_n)_{n=1}^{\infty}$ converges to 0,

$$\lim_{n\to\infty} f(x_n) = \lim_{n\to\infty} \frac{1}{x_n}\sin\frac{1}{x_n} = \lim_{n\to\infty}\left(\frac{\pi}{2}+2\pi n\right)\sin\left(\frac{\pi}{2}+2\pi n\right)$$

$$= \lim_{n\to\infty}\left(\frac{\pi}{2}+2\pi n\right) = \infty$$

and $f(0) = 0$. Thus, by the sequential criterion for continuous functions, f is not continuous at 0.

(IV). It is clear that if $x \neq 0$, then $f(x) = \frac{x+|x|(1+x)}{x}\sin\frac{1}{x}$ is continuous.

It remains to study the case when $x = 0$. Take the sequence $x_n = \frac{1}{\frac{3\pi}{2}+2\pi n}$ for every $n \in \mathbb{N}$. Observe that $(x_n)_{n=1}^{\infty}$ converges to 0, $x_n > 0$ for all $n \in \mathbb{N}$,

$$\lim_{n\to\infty} f(x_n) = \lim_{n\to\infty} \frac{x_n+|x_n|(1+x_n)}{x_n}\sin\frac{1}{x_n} = \lim_{n\to\infty} \frac{x_n+x_n(1+x_n)}{x_n}\sin\frac{1}{x_n}$$

$$= \lim_{n\to\infty}(1+x_n)\sin\frac{1}{x_n} = \lim_{n\to\infty}\left(1+\frac{1}{\frac{3\pi}{2}+2\pi n}\right)\sin\left(\frac{3\pi}{2}+2\pi n\right)$$

$$= \lim_{n\to\infty} -\left(1+\frac{1}{\frac{3\pi}{2}+2\pi n}\right) = -1$$

and $f(0) = 1$. Thus, by the sequential criterion for continuous functions, f is not continuous at 0.

(v). Let us study first the case when $x = 0$. Take $(x_m)_{m=1}^{\infty}$ any sequence of real numbers that converges to 0. Then,

$$f(x_m) = \begin{cases} x_m^n & \text{if } x_m \in \mathbb{R}\setminus\mathbb{Q}, \\ 0 & \text{if } x_m \in \mathbb{Q}. \end{cases}$$

As $(x_m)_{m=1}^{\infty}$ converges to 0, we have that for every $0 < \varepsilon < 1$, there exists $m_0 \in \mathbb{N}$ such that $|x_m| < \varepsilon$ for any integer $m \geq m_0$. Hence, for every integer $m \geq m_0$, it yields $|x_m^n| = |x_m|^n < \varepsilon^n < \varepsilon$, that is, $(x_m^n)_{m=1}^{\infty}$ converges to 0. Hence, it is clear that any subsequence of $(f(x_m))_{m=1}^{\infty}$ converges to 0, which is equivalent to

$$\lim_{m\to\infty} f(x_m) = 0.$$

Since $f(0) = 0$, we have by the sequential criterion for continuous functions that f is continuous at 0.

Let us assume for the rest of this part that $x \neq 0$. We have two cases:

Case 1: If $x \in \mathbb{R}\setminus\mathbb{Q}$, by the density of $\mathbb{Q}$ in $\mathbb{R}$, there exists a sequence of rational numbers $(x_m)_{m=1}^{\infty}$ that converges to x. Then, for every $m \in \mathbb{N}$, we have $f(x_m) = 0$, which implies that $\lim_{m\to\infty} f(x_m) = 0$. However, $f(x) = x^n \neq 0$. Thus, by the sequential criterion for continuous functions, f is not continuous at x.

Case 2: If $x \in \mathbb{Q}$, by the density of $\mathbb{R} \setminus \mathbb{Q}$ in $\mathbb{R}$, there exists a sequence of irrational numbers $(x_m)_{m=1}^{\infty}$ that converges to x. Hence,

$$\lim_{m \to \infty} f(x_m) = \lim_{m \to \infty} x_m^n = x^n \neq 0,$$

where we have used the fact that $g(x) = x^n$ is a continuous function on $\mathbb{R}$. Finally, since $f(x) = 0$, we have by the sequential criterion for continuous functions that f is not continuous at x.

$\square$

Exercise 4.7. *For each one of the following functions, find all the points of discontinuity.*

(I) $\sqrt{1-x^2}$.

(II) $\sin(e^{-x^2})$.

(III) $\log(1 + \sin x)$.

(IV) $e^{-1/(1-2x)}$.

(V) $\sin \dfrac{1}{(x-1)^2}$.

(VI) $\sin\left(\dfrac{1}{\cos x}\right)$.

(VII) $(1 - \sin^2 x)^{-1/2}$.

(VIII) $\cot(1 - e^{-x^2})$.

(IX) $\cos \dfrac{1}{x}$.

Solution to Exercise 4.7.

(I). Let $f(x) = \sqrt{1-x^2}$. In order to study the continuity of f, we must first find the domain of f. Notice that f takes values in $\mathbb{R}$ only if

$$1 - x^2 \geq 0 \iff x^2 \leq 1 \iff |x| \leq 1.$$

Thus, $\text{dom}(f) = [-1, 1]$ and f is clearly continuous on this domain.

(II). Let $f(x) = \sin(e^{-x^2})$ and note that $\text{dom}(f) = \mathbb{R}$ and f is continuous on its domain.

(III). Let $f(x) = \log(1 + \sin x)$ and let us find the domain of f. Observe that f takes real values if, and only if:

$$1 + \sin x > 0 \iff \sin x > -1.$$

Thus, it is enough to study when $\sin x = -1$. Recall that

$$\sin x = -1 \iff x = \frac{3\pi}{2} + 2k\pi \text{ for every } k \in \mathbb{Z}.$$

Thus, $\text{dom}(f) = \mathbb{R} \setminus \left\{ \dfrac{3\pi}{2} + 2k\pi : k \in \mathbb{Z} \right\}$ and f is continuous on its domain.

(IV). Let $f(x) = e^{-1/(1-2x)}$ and let us find the domain of f. Notice that f takes real values if, and only if, $1 - 2x \neq 0$. Thus, it is enough to study when $1 - 2x = 0$, which occurs only when $x = \frac{1}{2}$. Hence, $\text{dom}(f) = \mathbb{R} \setminus \left\{ \dfrac{1}{2} \right\}$ and f is continuous on its domain.

(V). Let $f(x) = \sin \frac{1}{(x-1)^2}$ and let us find its domain. Observe that f takes real values when $(x-1)^2 \neq 0$. Hence, we must solve the equation $(x-1)^2 = 0$, which yields $x = 1$. Thus, $\text{dom}(f) = \mathbb{R} \setminus \{1\}$ and f is continuous on its domain.

(VI). Let $f(x) = \sin \left(\frac{1}{\cos x} \right)$ and let us proceed to find the domain of f. Notice that we only have to study when $\cos x = 0$, which occurs only when $x = \frac{\pi}{2} + k\pi$ for any $k \in \mathbb{Z}$. Hence, $\text{dom}(f) = \mathbb{R} \setminus \left\{ \frac{\pi}{2} + k\pi \colon k \in \mathbb{Z} \right\}$ and f is continuous on its domain.

(VII). Let $f(x) = (1 - \sin^2 x)^{-1/2}$ and let us find the domain of f. Observe that f takes real values only when:

$$1 - \sin^2 x > 0 \iff \sin^2 x < 1 \iff |\sin x| < 1.$$

Thus, it is enough to know when $\sin x \in \{-1, 1\}$, which only occurs when $x = \frac{\pi}{2} + k\pi$ for any $k \in \mathbb{Z}$. Hence, $\text{dom}(f) = \mathbb{R} \setminus \left\{ \frac{\pi}{2} + k\pi \colon k \in \mathbb{Z} \right\}$ and f is continuous on its domain.

(VIII). Let $f(x) = \cot(1 - e^{-x^2}) = \dfrac{\cos(1 - e^{-x^2})}{\sin(1 - e^{-x^2})}$ and let us find the domain of f. To do so, we must analyze when $\sin(1 - e^{-x^2}) = 0$:

$$\sin(1 - e^{-x^2}) = 0 \iff 1 - e^{-x^2} = k\pi \text{ for some } k \in \mathbb{Z}$$
$$\iff e^{-x^2} = 1 - k\pi \text{ for some } k \in \mathbb{Z}.$$

Now, since the exponential is always positive, we need the integer k to satisfy the inequality

$$1 - k\pi > 0 \iff k < \frac{1}{\pi}.$$

Thus, since $0 < \frac{1}{\pi} < 1$, we need $k \leq 0$; in which case:

$$e^{-x^2} = 1 - k\pi \iff e^{x^2} = \frac{1}{1 - k\pi} \iff x^2 = \log\left(\frac{1}{1 - k\pi} \right) = -\log(1 - k\pi).$$

As $x^2 \geq 0$ for any value of x, we need $0 < 1 - k\pi \leq 1$ so that the equality $x^2 = -\log(1 - k\pi)$ is well defined. Let us show when $0 < 1 - k\pi \leq 1$:

$$0 < 1 - k\pi \leq 1 \iff -1 < -k\pi \leq 0 \iff 0 \leq k\pi < 1 \iff 0 \leq k < \frac{1}{\pi} < 1.$$

Thus, $k = 0$, which yields:

$$x^2 = \log 1 = 0 \iff x = 0.$$

Therefore, $\text{dom}(f) = \mathbb{R} \setminus \{0\}$ and f is continuous on its domain.

(IX). Let $f(x) = \cos \frac{1}{x}$. Notice that in this case it is clear that $\text{dom}(f) = \mathbb{R} \setminus \{0\}$ and f is continuous on its domain.

$\square$

Exercise 4.8. *Find the values of $c > 0$ for which the function $f : \mathbb{R} \to \mathbb{R}$, defined by*

$$f(x) = \begin{cases} 3 - \sqrt{c}x & \text{if } x \leq c, \\ \frac{x-c}{\sqrt{x}-\sqrt{c}} & \text{if } x > c, \end{cases}$$

is continuous.

Solution to Exercise 4.8.

On the one hand, if $x < c$, then $f(x) = 3 - \sqrt{c}x$ is continuous for any $c > 0$. On the other hand, if $x > c$, then $\sqrt{x} > \sqrt{c}$, so that $f(x) = \dfrac{x-c}{\sqrt{x}-\sqrt{c}}$ is continuous for any $c > 0$.

Therefore, it is enough to study what happens when $x = c$. Note that

$$f(c) = 3 - \sqrt{c}\,c = \lim_{x \to c^-} (3 - \sqrt{c}x) = \lim_{x \to c^-} f(x)$$

and

$$\lim_{x \to c^+} f(x) = \lim_{x \to c^+} \frac{x-c}{\sqrt{x}-\sqrt{c}} = \lim_{x \to c^+} (\sqrt{x} + \sqrt{c}) = 2\sqrt{c}.$$

Thus, f will be continuous at $x = c$ if

$$3 - \sqrt{c}\,c = 2\sqrt{c} \iff 2\sqrt{c} + \sqrt{c}\,c = 3 \iff \sqrt{c}(2+c) = 3 \implies c(2+c)^2 = 9$$
$$\iff c(c^2 + 4c + 4) = 9 \iff c^3 + 4c^2 + 4c - 9 = 0$$
$$\iff (c-1)(c^2 + 5c + 9) = 0,$$

where in latter chain of equivalences we have used Ruffini's Rule. Now, since $c^2 + 5c + 9 = 0$ does not have real solutions (to see this, simply use the second degree equation formula), we have that $c = 1$. $\qquad\square$

Exercise 4.9. (I) *Find a function defined on $\mathbb{R}$ that is discontinuous on $\left\{\dfrac{1}{n} : n \in \mathbb{N}\right\}$, but continuous everywhere else.*

(II) *Find a function defined on $\mathbb{R}$ that is discontinuous on $\left\{\dfrac{1}{n} : n \in \mathbb{N}\right\} \cup \{0\}$, but continuous everywhere else.*

Solution to Exercise 4.9.

(I). Take

$$f(x) = \begin{cases} \frac{1}{n} & \text{if } x = \frac{1}{n} \text{ for some } n \in \mathbb{N}, \\ 0 & \text{otherwise.} \end{cases}$$

Note that $f(x) = 0$ on $\mathbb{R} \setminus \left\{ \dfrac{1}{n} : n \in \mathbb{N} \right\}$, that is, f is identically 0 on the

intervals $(-\infty, 0)$, $(1, \infty)$, and $\left(\dfrac{1}{n+1}, \dfrac{1}{n} \right)$ for any $n \in \mathbb{N}$. Therefore, f is

clearly continuous on $\mathbb{R} \setminus \left\{ \dfrac{1}{n} : n \in \mathbb{N} \right\}$.

Let us study the continuity at $x = \frac{1}{n}$ where n is some natural number. We
have that

$$\lim_{x \to 1/n^-} f(x) = 0 = \lim_{x \to 1/n^+} f(x)$$

and $f\left(\dfrac{1}{n} \right) = \dfrac{1}{n} > 0$. Hence, f is not continuous at $\frac{1}{n}$ for any $n \in \mathbb{N}$.

It remains to analyze the continuity of f at 0. Let $(x_k)_{k=1}^{\infty}$ be a sequence
of real numbers that converges to 0. Then,

$$f(x_k) = \begin{cases} \dfrac{1}{n} & \text{if } x_k = \dfrac{1}{n} \text{ for some } n \in \mathbb{N}, \\ 0 & \text{otherwise.} \end{cases}$$

Notice that by construction we have

$$\lim_{k \to \infty} f(x_k) = 0,$$

so, by the sequential criterion for continuous functions, f is continuous at 0.

(II). Take

$$f(x) = \begin{cases} n+1 & \text{if } x = \dfrac{1}{n} \text{ for some } n \in \mathbb{N}, \\ 1 & \text{if } x = 0, \\ 0 & \text{otherwise.} \end{cases}$$

It is clear that $f(x) = 0$ on $\mathbb{R} \setminus \left\{ \dfrac{1}{n} : n \in \mathbb{N} \right\}$, or equivalently, f is identically 0

on the intervals $(-\infty, 0)$, $(1, \infty)$, and $\left(\dfrac{1}{n+1}, \dfrac{1}{n} \right)$ for any $n \in \mathbb{N}$. This shows

that f is continuous on $\mathbb{R} \setminus \left\{ \dfrac{1}{n} : n \in \mathbb{N} \right\}$.

We will now analyze the continuity at $x = \frac{1}{n}$ where $n \in \mathbb{N}$. It is clear that

$$\lim_{x \to 1/n^-} f(x) = 0 = \lim_{x \to 1/n^+} f(x)$$

and $f\left(\dfrac{1}{n} \right) = n > 0$. So, f is not continuous at $\frac{1}{n}$ for any $n \in \mathbb{N}$.

Finally, let us study the continuity of f at 0. Take the sequence $(x_n)_{n=1}^{\infty} = \left(\frac{1}{n} \right)_{n=1}^{\infty}$ which converges to 0. Note that $f(x_n) = n$ for any $n \in \mathbb{N}$, so the
sequence $(f(x_n))_{n=1}^{\infty}$ diverges to ∞. Hence, by the sequential criterion for
continuous functions, f is not continuous at 0. $\qquad \square$

Exercise 4.10. *Give an example of a function $f \colon \mathbb{R} \to \mathbb{R}$ that is discontinuous at every point of $\mathbb{R}$, but so that the function $|f|$ is continuous in all of $\mathbb{R}$.*

Solution to Exercise 4.10.

Take

$$f(x) = \begin{cases} 1 & \text{if } x \in \mathbb{Q}, \\ -1 & \text{if } x \in \mathbb{R} \setminus \mathbb{Q}. \end{cases}$$

Observe that $|f|(x) = 1$ for every $x \in \mathbb{R}$, that is, $|f|$ is continuous everywhere. It remains to show that f is discontinuous at every real number. Fix $a \in \mathbb{R}$ arbitrary, Then,

$$\lim_{x \to a, \ x \in \mathbb{Q}} f(x) = \lim_{x \to a} 1 = 1 \neq -1 = \lim_{x \to a, \ x \in \mathbb{R} \setminus \mathbb{Q}} (-1) = \lim_{x \to a, \ x \in \mathbb{Q}} f(x).$$

Hence, f is not continuous at any real number.

$\square$

Exercise 4.11. *Let f and g be continuous functions from $\mathbb{R}$ to $\mathbb{R}$, and assume that $f(r) = g(r)$ for every $r \in \mathbb{Q}$. Is it true that $f = g$?*

Solution to Exercise 4.11.

Let $a \in \mathbb{R}$ be arbitrary. Since $\mathbb{Q}$ is dense in $\mathbb{R}$, there exists a sequence of rational numbers $(r_n)_{n=1}^{\infty}$ such that $\lim_{n \to \infty} r_n = a$. As f and g are continuous functions, we have that $\lim_{n \to \infty} f(r_n) = f\left(\lim_{n \to \infty} r_n\right)$ and $\lim_{n \to \infty} g(r_n) = g\left(\lim_{n \to \infty} r_n\right)$. Hence, as $f(r_n) = g(r_n)$ for any $n \in \mathbb{N}$ by hypothesis, we have

$$f(a) = f\left(\lim_{n \to \infty} r_n\right) = \lim_{n \to \infty} f(r_n) = \lim_{n \to \infty} g(r_n) = g\left(\lim_{n \to \infty} r_n\right) = g(a).$$

Since a is arbitrary, we have proven that $f = g$.

$\square$

Exercise 4.12. *Suppose that f is defined on all $\mathbb{R}$ and satisfies the following two properties:*

(a) $\lim_{x \to 0} f(x) = 1$,

(b) $f(x + y) = f(x)f(y)$ for any $x, y \in \mathbb{R}$.

Prove the following:

(I) f is continuous on all $\mathbb{R}$.

(II) $f(x) > 0$ *for every* $x \in \mathbb{R}$.

(III) $f(rx) = (f(x))^r$ *if* $r \in \mathbb{Q}$. *In particular,* $f(r) = (f(1))^r$ *for every* $r \in \mathbb{Q}$.

(IV) *If* $f(1) = 1$, *then* f *is constant.*

(V) *If* $f(1) > 1$, *then* f *is strictly increasing, and also*

$$\lim_{x \to +\infty} f(x) = \infty \text{ and } \lim_{x \to -\infty} f(x) = 0.$$

(VI) *If* $0 < f(1) < 1$, *then* f *is strictly decreasing, and also*

$$\lim_{x \to +\infty} f(x) = 0 \text{ and } \lim_{x \to -\infty} f(x) = \infty.$$

(Observe that the exponential functions a^x *satisfy these properties. This exercise, together with Exercise 4.11, suggest us a method to define the exponential functions on all* $\mathbb{R}$.)

Solution to Exercise 4.12.

(I). Take $a \in \mathbb{R}$ arbitrary. Then,

$$\lim_{x \to a} f(x) = \lim_{x \to 0} f(x + a) = \lim_{x \to 0} f(x)f(a) = f(a) \lim_{x \to 0} f(x) = f(a),$$

where in the last equality we have used property (a). Thus, we have proven that f is continuous at a.

(II). Take $x \in \mathbb{R}$ arbitrary. Then,

$$f(x) = f\left(\frac{x}{2} + \frac{x}{2}\right) = \left(f\left(\frac{x}{2}\right)\right)^2 \geq 0.$$

Now it is enough to show that $f(x) \neq 0$ for any $x \in \mathbb{R}$. To do so, let us study first the value of f at 0. By property (b) we know that $f(0) = f(0+0) = f(0)f(0) = f(0)^2$. Thus, $f(0)(f(0) - 1) = 0$, which implies that $f(0) \in \{0, 1\}$. Assume by means of contradiction that $f(0) = 0$, then for any $x \in \mathbb{R}$ we have $f(x) = f(x + 0) = f(x)f(0) = 0$, that is, f is identically 0. But the latter contradicts property (a). So $f(0) = 1$.

Assume that there exists $x_0 \in \mathbb{R}$ such that $f(x_0) = 0$. Then,

$$1 = f(0) = f(x_0 - x_0) = f(x_0)f(-x_0) = 0,$$

which is absurd. Hence, $f(x) > 0$ for any $x \in \mathbb{R}$.

(III). We will divide this part into several cases:

Case 1: $f(0x) = (f(0))^1$. As $f(0) = 1$ (see Exercise 4.12 part (II)), we have $f(0x) = f(0) = 1 = (f(x))^0$, where $(f(x))^0$ is well defined since $f(x) > 0$ by Exercise 4.12 part (II).

Case 2: $f(nx) = (f(x))^n$ if $n \in \mathbb{N}$. Indeed,

$$f(nx) = f\left(\sum_{k=1}^{n} x\right) = \prod_{k=1}^{n} f(x) = (f(x))^n.$$

Case 3: $f(-x) = (f(x))^{-1}$. Indeed,

$$1 = f(0) = f(x - x) = f(x + (-x)) = f(x)f(-x),$$

so, as $f(x) \neq 0$ for any $x \in \mathbb{R}$, we have $f(-x) = \frac{1}{f(x)} = (f(x))^{-1}$.

Case 4: $f(mx) = (f(x))^m$ if m is a negative integer.

$$f(mx) = f(-|m|x) = (f(|m|x))^{-1} = ((f(x))^{|m|})^{-1} = (f(x))^{-|m|} = (f(x))^m,$$

where we have applied Case 3 in the equality $f(|m|x) = (f(x))^{|m|}$. Note that we have proven so far that $f(mx) = (f(x))^m$ if $m \in \mathbb{Z}$.

Case 5: $f\left(\dfrac{1}{n}x\right) = (f(x))^{1/n}$ if $n \in \mathbb{N}$:

$$f(x) = f\left(\sum_{k=1}^{n} \frac{1}{n}x\right) = \prod_{k=1}^{n} f\left(\frac{1}{n}x\right) = \left(f\left(\frac{1}{n}x\right)\right)^n,$$

so, since $f(x) > 0$, we can take the n-th root on both sides to obtain $f\left(\dfrac{1}{n}x\right) = (f(x))^{1/n}$.

Case 6: $f(rx) = (f(x))^r$ if $r \in \mathbb{Q}$. To prove this, it is enough to analyze the case when $r = \frac{p}{q}$, where $p \in \mathbb{Z}$ and $q \in \mathbb{N}$:

$$f(rx) = f\left(\frac{p}{q}x\right) = f\left(p\frac{1}{q}x\right) = \left(f\left(\frac{1}{q}x\right)\right)^p$$
$$= \left((f(x))^{1/q}\right)^p = ((f(x)))^{p/q} = (f(x))^r,$$

where we have used the last statement of Case 4 and Case 5.

In particular, we have as an immediate consequence:

$$f(r) = f(r \cdot 1) = (f(1))^r,$$

for any $r \in \mathbb{Q}$.

(IV). Let $x \in \mathbb{R}$. Since $\mathbb{Q}$ is dense in $\mathbb{R}$, there is a sequence $(r_n)_{n=1}^{\infty}$ of rational numbers that converges to x. Hence, by the second statement of

Exercise 4.12 part (III), Exercise 4.12 part (I) and the sequential criterion for continuous functions, we have

$$f(x) = f\left(\lim_{n\to\infty} r_n\right) = \lim_{n\to\infty} f(r_n) = \lim_{n\to\infty} (f(1))^{r_n} = \lim_{n\to\infty} 1^{r_n} = \lim_{n\to\infty} 1 = 1.$$

Since x is arbitrary, it yields that f is the constant function 1.

(v). Let $x < y$ be real numbers. Since $\mathbb{Q}$ is dense in $\mathbb{R}$, notice that there exist strictly increasing sequences of rational numbers $(r_n)_{n=1}^\infty$ and $(s_n)_{n=1}^\infty$ that converge, respectively, to x and y. Moreover, we can assume that there exists $a > 0$ such that $r_n < s_n - a$ for every $n \in \mathbb{N}$. Recall that $f(1) > 1$, and so $f(1)^{-a} < 1$. Then,

$$f(x) = f\left(\lim_{n\to\infty} r_n\right) = \lim_{n\to\infty} f(r_n) = \lim_{n\to\infty} (f(1))^{r_n} \le \lim_{n\to\infty} f(1)^{s_n - a}$$
$$= f(1)^{-a} \cdot \lim_{n\to\infty} f(1)^{s_n}$$
$$< \lim_{n\to\infty} (f(1))^{s_n} = \lim_{n\to\infty} f(s_n) = f\left(\lim_{n\to\infty} s_n\right) = f(y).$$

Thus, f is strictly increasing.

Now take $(r_n)_{n=1}^\infty$ an strictly increasing sequence of rational numbers that diverges to ∞. In order to show that $\lim_{x\to+\infty} f(x) = \infty$, it is enough to show by the continuity of f that $\lim_{n\to\infty} f(r_n) = \infty$:

$$\lim_{n\to\infty} f(r_n) = \lim_{n\to\infty} (f(1))^{r_n} = \infty.$$

Similarly, take $(r_n)_{n=1}^\infty$ an strictly decreasing sequence of rational numbers that diverges to $-\infty$. As above, it is enough to prove that $\lim_{n\to\infty} f(r_n) = 0$ to conclude that $\lim_{x\to-\infty} f(x) = 0$:

$$\lim_{n\to\infty} f(r_n) = \lim_{n\to\infty} (f(1))^{r_n} = 0.$$

(VI). Let $x < y$ be real numbers. As $\mathbb{Q}$ is dense in $\mathbb{R}$, there exist strictly decreasing sequences of rational numbers $(r_n)_{n=1}^\infty$ and $(s_n)_{n=1}^\infty$ that converge, respectively, to x and y. In fact, as before, we can assume that there exists $a > 0$ such that $r_n < s_n - a$ for every $n \in \mathbb{N}$. Recall that $f(1) < 1$, and so $f(1)^{-a} > 1$. Then,

$$f(x) = f\left(\lim_{n\to\infty} r_n\right) = \lim_{n\to\infty} f(r_n) = \lim_{n\to\infty} (f(1))^{r_n} \ge \lim_{n\to\infty} f(1)^{s_n - a}$$
$$= f(1)^{-a} \cdot \lim_{n\to\infty} f(1)^{s_n}$$
$$> \lim_{n\to\infty} (f(1))^{s_n} = \lim_{n\to\infty} f(s_n) = f\left(\lim_{n\to\infty} s_n\right) = f(y).$$

Thus, f is strictly decreasing.

Take $(r_n)_{n=1}^\infty$ a strictly increasing sequence of rational numbers that diverges to ∞. To prove that $\lim\limits_{x\to+\infty} f(x) = 0$, it is sufficient to show that $\lim\limits_{n\to\infty} f(r_n) = 0$:

$$\lim_{n\to\infty} f(r_n) = \lim_{n\to\infty} (f(1))^{r_n} = 0.$$

Analogously, take $(r_n)_{n=1}^\infty$ a strictly increasing sequence of rational numbers that diverges to $-\infty$. To show that $\lim\limits_{x\to-\infty} f(x) = \infty$, it is enough to prove that $\lim\limits_{n\to-\infty} f(r_n) = \infty$:

$$\lim_{n\to\infty} f(r_n) = \lim_{n\to\infty} (f(1))^{r_n} = \infty.$$

$\square$

Exercise 4.13. *For each of the following polynomial functions, find an integer n such that $f(x) = 0$ for some x between n and $n+1$:*

(I) $f(x) = x^3 - x + 3$.   (III) $f(x) = x^5 + x + 1$.

(II) $f(x) = 4x^2 - 4x + 1$.

Solution to Exercise 4.13.

(I). Note that $f(-2) = -3 < 0$ and $f(-1) = 3 > 0$. Then, since f is continuous function on $[-2,-1]$, by Bolzano's Theorem, there exists $x \in (-2,-1)$ such that $f(x) = 0$.

(II). Observe that $f(x) = (2x-1)^2 = 0$ if, and only, $x = \frac{1}{2}$. Thus, for $n = 0$, we have that the desired result.

(III). Notice that $f(-1) = 1 > 0$ and $f(-2) = -33 < 0$. Hence, as f is continuous on $[-2,-1]$, we have by Bolzano's Theorem that there exists $x \in (-2,-1)$ such that $f(x) = 0$.

$\square$

Exercise 4.14. *Prove:*

(I) *The equation $x2^x = 1$ has at least one positive solution not bigger than 1.*

(II) *The equation*

$$x^{179} + \frac{163}{1+x^2+\sin^2 x} = 119$$

has at least one real solution.

(III) *The equation* $\sin x = x - 1$ *has at least one real solution.*

(IV) *The equation* $\frac{\pi}{2} - x - \sin x = 0$ *has a solution on* $\left[0, \frac{\pi}{2}\right]$.

(V) *The equation* $x \sin x - \frac{\pi}{4} = 0$ *has at least two solutions on* $[0, \pi]$.

Solution to Exercise 4.14.

(I). Let $f(x) = x2^x - 1$. Note that $f(0) = -1 < 0$ and $f(1) = 1 > 0$. Since f is a continuous function on $[0, 1]$, by Bolzano's Theorem there exists an $c \in (0, 1)$ such that $f(c) = 0$, that is, c is a positive solution of $x2^x = 1$ not bigger than 1.

(II). Let $f(x) = x^{179} + \dfrac{163}{1 + x^2 + \sin^2 x} - 119$. Notice that $f(0) = 44 > 0$ and $f(1) = -118 + \frac{163}{2 + \sin^2(-1)} \le -120 + \frac{163}{2} = -\frac{77}{2} < 0$. Since $1 + x^2 + \sin^2 x > 0$ for any $x \in \mathbb{R}$, we have that f is continuous on $[0, 1]$. So, by Bolzano's Theorem, there exists $c \in (0, 1)$ such that $f(c) = 0$, that is, c is a real solution of the required equation.

(III). Let $f(x) = \sin x - x + 1$. Notice that $f(3) = \sin(3) - 3 + 1 \le 1 - 2 = -1 < 0$ and $f(0) = 1$. Since f is continuous on $[0, 3]$, there exists $c \in (0, 3)$ such that $f(c) = 0$ by Bolzano's Theorem, that is, c is a real solution of $\sin x - x + 1$.

(IV). Let $f(x) = \frac{\pi}{2} - x - \sin x$. Observe that $f(0) = \frac{\pi}{2} > 0$ and $f\left(\frac{\pi}{2}\right) = -1 < 0$. As f is continuous on $\left[0, \frac{\pi}{2}\right]$, there exists a $c \in \left(0, \frac{\pi}{2}\right)$ such that $f(c) = 0$ by Bolzano's Theorem, that is, c is a solution of the equation $\frac{\pi}{2} - x - \sin x$.

(V). Let $f(x) = x \sin x - \frac{\pi}{4}$. Note that $f(0) = -\frac{\pi}{4} < 0$, $f(\pi) = -\frac{\pi}{4} < 0$ and $f\left(\frac{\pi}{2}\right) = \frac{\pi}{2} - \frac{\pi}{4} = \frac{\pi}{4} > 0$. Since f is continuous on $[0, \pi]$, we have by Bolzano's Theorem applied on $\left[0, \frac{\pi}{2}\right]$ and on $\left[\frac{\pi}{2}, \pi\right]$, that there exist $c_1 \in \left(0, \frac{\pi}{2}\right)$ and $c_2 \in \left(\frac{\pi}{2}, 1\right)$ such that $f(c_1) = 0 = f(c_2)$, that is, $c_1, c_2 \in [0, \pi]$ are two distinct solutions of $x \sin x - \frac{\pi}{4} = 0$.

$\square$

Exercise 4.15. *Let d be a direction on the plane and T a triangle. Prove that there exists a straight line with direction d that divides T into two parts of equal area.*

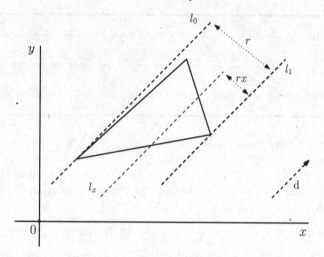

FIGURE 4.1: Representation of Exercise 4.15.

Solution to Exercise 4.15.

Assume that d is not the direction of the vector $(0,1)$. Since T is clearly bounded, we can take l_0 and l_1 straight lines with direction d that are, respectively, "above" and "below" the triangle T. Take $r > 0$ the distance between l_0 and l_1. For any $x \in [0,1]$, let l_x be the straight line with direction d that is "above" the line l_1 and at distance rx from l_1.

Now define $f\colon [0,1] \to \mathbb{R}$ as follows: for any $x \in [0,1]$, $f(x)$ is the area of the triangle "below" the straight line l_x (if T is "above" l_x, then we simply take $f(x) = 0$ for convenience). Then, by construction, f is continuous on $[0,1]$, $f(0) = 0$ and $f(1) = A(T)$, where $A(T)$ denotes the area of T. Therefore, by Bolzano's Intermediate Value Theorem, there exists $c \in (0,1)$ such that $f(c) = \frac{A(T)}{2}$. Finally, by construction, observe that the straight line l_c is as needed.   □

Exercise 4.16. *Let f be a continuous function from $[0,1]$ to $\mathbb{R}$ such that $f(0) = f(1)$. Prove that there exists a point $c \in \left[0, \dfrac{1}{2}\right]$ such that $f(c) = f\left(c + \dfrac{1}{2}\right)$. Hint: Consider $g(x) = f(x) - f\left(x + \dfrac{1}{2}\right)$.*

Solution to Exercise 4.16.

Consider the function $g(x) = f(x) - f\left(x + \dfrac{1}{2}\right)$. By definition, since $f(0) = f(1)$, notice that

$$g(0) = f(0) - f\left(\frac{1}{2}\right) = f(1) - f\left(\frac{1}{2}\right)$$

and

$$g\left(\frac{1}{2}\right) = f\left(\frac{1}{2}\right) - f(1).$$

Thus,

$$g(0)g\left(\frac{1}{2}\right) = \left(f(1) - f\left(\frac{1}{2}\right)\right)\left(f\left(\frac{1}{2}\right) - f(1)\right) = -\left(f(1) - f\left(\frac{1}{2}\right)\right)^2 \leq 0.$$

We have two possible cases:

Case 1: Assume that $g(0) \cdot g\left(\dfrac{1}{2}\right) = 0$, then

$$f(0) - f\left(\frac{1}{2}\right) = f(1) - f\left(\frac{1}{2}\right) = 0,$$

which implies $f(0) = f\left(\dfrac{1}{2}\right)$. Hence, $c = 0$ is as needed.

Case 2: Assume that $g(0)g\left(\dfrac{1}{2}\right) < 0$, then by Bolzano's Intermediate Value Theorem, there exists $c \in \left(0, \dfrac{1}{2}\right)$ such that

$$0 = g(c) = f(c) - f\left(c + \frac{1}{2}\right),$$

that is, $f(c) = f\left(c + \dfrac{1}{2}\right)$. We are done.

$\square$

Exercise 4.17 (Brouwer's Fixed Point Theorem). *Let I be a nonempty, closed and bounded interval. Prove that if $f: I \to I$ is continuous function, then there exists a point $x_0 \in I$ such that $f(x_0) = x_0$.*

Solution to Exercise 4.17.

If $I = [a, a]$ for some $a \in \mathbb{R}$, then by definition of f we have $f(a) = a$. So, assume for the rest of this exercise that $I = [a, b]$ with $a < b$.

If $f(a) = a$ or $f(b) = b$, we are done. So, assume that $f(a) \neq a$ and $f(b) \neq b$. In particular, since the image of f is contained in $I = [a,b]$, we have that $f(a) > a$ and $f(b) < b$. Now consider the function $g(x) = f(x) - x$ defined on I. Clearly, g is a continuous function such that $g(a) = f(a) - a > 0$ and $g(b) = f(b) - b < 0$. Hence, by Bolzano's Theorem, there exists $x_0 \in (a,b) \subset I$ such that $0 = g(x_0) = f(x_0) - x_0$, that is, $x_0 \in I$ and $f(x_0) = x_0$, as needed. $\square$

Exercise 4.18. *Let I be an nonempty interval. Prove that if $f \colon I \to \mathbb{R}$ is a continuous function and $t_1, t_2, \ldots, t_n \in I$ where $n \in \mathbb{N}$, then there exists a point $c \in I$ such that $f(c) = \frac{f(t_1) + f(t_2) + \cdots + f(t_n)}{n}$.*

Solution to Exercise 4.18.

Assume, without loss of generality, that

$$f(t_1) = \min\{f(t_i) \colon 1 \le i \le n\}$$

and

$$f(t_n) = \max\{f(t_i) \colon 1 \le i \le n\}.$$

Then,

$$f(t_1) = \frac{nf(t_1)}{n} \le \frac{f(t_1) + \cdots + f(t_n)}{n}$$

$$\le \frac{nf(t_n)}{n} = f(t_n).$$

Thus, since f is continuous on $[t_1, t_n]$, by Bolzano's Intermediate Value Theorem, there exists $c \in [t_1, t_n] \subset I$ such that

$$f(c) = \frac{f(t_1) + f(t_2) + \cdots + f(t_n)}{n},$$

as needed.

$\square$

Exercise 4.19. *Let f be a bounded function defined on a nonempty, closed and bounded interval I. If $S \subset I$ is nonempty, the number*

$$\omega_f(S) = \sup\{f(x) - f(y) \colon x, y \in S\}$$

is known as the oscillation of f on S. If $x \in I$, the oscillation of f at x is defined as the number

$$\omega_f(x) = \lim_{h \to 0^+} \omega_f((x - h, x + h) \cap I).$$

> *Prove that the latter limit always exists and that $\omega_f(x) = 0$ if, and only if, f is continuous at x.*

Solution to Exercise 4.19.

First of all, notice that by definition of $\omega_f(S)$ we have that $\omega_f(S) \geq 0$ for any nonempty $S \subset I$. Moreover, for any nonempty $S_1, S_2 \subset I$, if $S_1 \subset S_2$, then

$$\omega_f(S_1) = \sup\{f(x) - f(y)\colon x, y \in S_1\}$$
$$\leq \sup\{f(x) - f(y)\colon x, y \in S_2\} = \omega_f(S_2).$$

Now let us define the function $g(h) = \omega_f((x - h, x + h) \cap I)$ and observe that the latter inequality implies that g is a decreasing function since

$$(x - h_1, x + h_1) \cap I \subset (x - h_2, x + h_2) \cap I$$

provided that $h_1 \leq h_2$. Then, since we know that $\omega_f((x - h, x + h) \cap I) \geq 0$ for any $h \geq 0$, we have that the limit $\lim\limits_{h \to 0^+} \omega_f((x - h, x + h) \cap I) = \omega_f(x)$ always exists.

Assume that f is continuous at x. Since I is a closed and bounded interval, we have that f is in fact uniformly continuous by Heine's Theorem. Thus, for any $\varepsilon > 0$, there exists $\delta > 0$ such that $|f(z) - f(w)| < \frac{\varepsilon}{2}$ for any $z, w \in I$ such that $|z - w| < \delta$. Hence, for any $0 < h < \delta$, we have that

$$f(z) - f(w) \leq |f(z) - f(w)| < \frac{\varepsilon}{2},$$

for every $z, w \in (x - h, x + h) \cap I$ (note that $|z - w| < h < \delta$). Therefore,

$$\omega_f((x - h, x + h) \cap I) = \sup\{f(z) - f(w)\colon z, w \in (x - h, x + h) \cap I\} \leq \frac{\varepsilon}{2} < \varepsilon.$$

where $\varepsilon > 0$ is arbitrary. So, we obtain

$$\omega_f(x) = \lim\limits_{h \to 0^+} \omega_f((x - h, x + h) \cap I) = 0.$$

Assume now that $\omega_f(x) = 0$ at some $x \in I$. Fix $\varepsilon > 0$. Since

$$\omega_f(x) = \lim\limits_{h \to 0^+} \omega_f((x - h, x + h) \cap I) = 0,$$

there exists an $h > 0$ such that

$$\omega_f((x - h, x + h) \cap I) = \sup\{f(z) - f(w)\colon z, w \in (x - h, x + h) \cap I\} < \varepsilon.$$

This means that $|f(z) - f(w)| < \varepsilon$ for any $z, w \in (x - h, x + h) \cap I$. In particular, for any $y \in I$ such that $|x - y| < h$ it yields

$$|f(x) - f(y)| < \varepsilon.$$

Since $\varepsilon > 0$ is arbitrary, we have proven that f is continuous at x. $\square$

Exercise 4.20. *Assume that* $f \colon \mathbb{R} \to \mathbb{R}$ *is continuous and*

$$\lim_{x \to -\infty} f(x) = \lim_{x \to \infty} f(x) = 0.$$

Prove that f *is bounded, and it reaches a maximum or a minimum. Give an example to indicate that both a maximum and a minimum do not necessarily have to be attained.*

Solution to Exercise 4.20.

Let us first show that f is bounded. Fix $\varepsilon > 0$. Since $\lim_{x \to -\infty} f(x) = \lim_{x \to \infty} f(x) = 0$, there exists $M > 0$ such that $|f(x)| < \varepsilon$ for every $x \in (-\infty, M) \cup (M, \infty)$. Now, since $[-M, M]$ is a closed and bounded interval, and f is a continuous function on $[-M, M]$, we have that f reaches a maximum and a minimum on $[-M, M]$ by Weierstrass' Theorem; in particular, there exist $A, B \in \mathbb{R}$ such that $A \leq f(x) \leq B$ for every $x \in [-M, M]$. Therefore, for any $x \in \mathbb{R}$, we have

$$|f(x)| \leq \max\{\varepsilon, |A|, |B|\},$$

which shows that f is bounded.

Now let us prove that f reaches a maximum or a minimum.

Fix any $M > 0$. Then, since $[-M, M]$ is a closed, bounded interval and $|f|$ is continuous on $[-M, M]$, there exists

$$A = \max\{|f(x)| \colon x \in [-M, M]\}.$$

If $|f(x)| \leq A$ for any $x \in \mathbb{R}$, then $|f|$ reaches an absolute maximum. If not, since $\lim_{x \to -\infty} |f(x)| = \lim_{x \to \infty} |f(x)| = 0$, take $M' > M$ such that $|f(x)| \leq A$ for every $x \in (-\infty, -M') \cup (M', \infty)$. Now, once again, as $[-M', M']$ is a closed, bounded interval and $|f|$ is continuous on $[-M', M']$, there exists

$$A' = \max\{|f(x)| \colon x \in [-M', M']\}.$$

Notice that $A' > A$. Therefore, by construction, we have that $|f(x)| \leq A'$ for every $x \in \mathbb{R}$, which implies that $|f|$ reaches an absolute maximum in any case. Let $x_0 \in \mathbb{R}$ be a point where $|f|$ reaches an absolute maximum. Then $|f(x)| \leq |f(x_0)|$ for all $x \in \mathbb{R}$. By using $\pm f(x) \leq |f(x)|$ and distinguishing the two cases $f(x_0) \geq 0$, $f(x_0) < 0$, we obtain that f reaches an absolute maximum at x_0 in the first case, while in the second one f reaches an absolute minimum at x_0.

Finally, we will give an example that illustrates that both a maximum and a minimum do not have to be reached. Consider the function

$$f(x) = e^{-x^2}.$$

It is clear that f is continuous on $\mathbb{R}$ and $\lim\limits_{x \to -\infty} f(x) = \lim\limits_{x \to \infty} f(x) = 0$. Moreover, $0 < f(x) \leq 1$ and $f(0) = 1$, that is, f reaches a maximum but not a minimum. If we now take

$$g(x) = -f(x) = -e^{-x^2},$$

then we have an example of a continuous function with $\lim\limits_{x \to -\infty} g(x) = \lim\limits_{x \to \infty} g(x) = 0$ that reaches its minimum but no its maximum. $\qquad \square$

Exercise 4.21. *Check which one from the following list of functions is uniformly continuous:*

(I) $f(x) = \dfrac{1}{x}$ *with* $x \in [1, \infty)$.   (VII) $f(x) = \sin\dfrac{1}{x}$ *with* $x > 0$.

(II) $f(x) = \dfrac{1}{x}$ *with* $x \in (0, 1)$.

(VIII) $f(x) = \sin\dfrac{1}{x}$ *with* $x > 1$.

(III) $f(x) = x^2$ *with* $x \in \mathbb{R}$.

(IV) $f(x) = \sqrt{x}$ *with* $x > 0$.   (IX) $f(x) = \dfrac{1}{1 - x^2}$ *with* $x \in$

(V) $f(x) = \dfrac{1}{x^2}$ *with* $x > 0$. $(-1, 1)$.

(VI) $f(x) = \dfrac{1}{x^2}$ *with* $x > 1$.   (X) $f(x) = x\sin\dfrac{1}{x}$ *with* $x \in (0, 1)$.

Solution to Exercise 4.21.

(I). Let $(x_n)_{n=1}^{\infty}$ and $(y_n)_{n=1}^{\infty}$ be two sequences in $[1, \infty)$ that satisfy $\lim\limits_{n \to \infty}(x_n - y_n) = 0$. Then,

$$0 \leq |f(x_n) - f(y_n)| = \left|\frac{1}{x_n} - \frac{1}{y_n}\right| = \left|\frac{y_n - x_n}{x_n y_n}\right| = \frac{|y_n - x_n|}{x_n y_n} \leq |x_n - y_n|,$$

where we have used that $x_n y_n \geq 1$ since $x_n, y_n \geq 1$ for every $n \in \mathbb{N}$. Thus, by the Sandwich Principle, $\lim\limits_{n \to \infty} |f(x_n) - f(y_n)| = 0$. So, by the sequential criterion for uniformly continuous functions, we have that f is uniformly continuous.

(II). Take $x_n = \dfrac{1}{n+1}$ and $y_n = \dfrac{1}{n}$ for every $n \in \mathbb{N}$. Notice that $(x_n)_{n=1}^{\infty}$ and $(y_n)_{n=1}^{\infty}$ are two sequences in $(0, 1)$ such that

$$\lim_{n \to \infty}(x_n - y_n) = \lim_{n \to \infty}\left(\frac{1}{n+1} - \frac{1}{n}\right) = 0.$$

But

$$\lim_{n\to\infty} (f(x_n) - f(y_n)) = \lim_{n\to\infty} \left(\frac{1}{1/(n+1)} - \frac{1}{1/n} \right)$$
$$= \lim_{n\to\infty} (n+1-n) = \lim_{n\to\infty} 1 = 1.$$

Thus, by the sequential criterion for uniformly continuous functions, f is not uniformly continuous.

(III). Take $x_n = n + \dfrac{1}{n}$ and $y_n = \dfrac{1}{n}$ for any $n \in \mathbb{N}$. Note that $(x_n)_{n=1}^\infty$ and $(y_n)_{n=1}^\infty$ are two sequences in $\mathbb{R}$ such that

$$\lim_{n\to\infty} (x_n - y_n) = \lim_{n\to\infty} \left(n + \frac{1}{n} - n \right) = \lim_{n\to\infty} \frac{1}{n} = 0.$$

But also

$$\lim_{n\to\infty} (f(x_n) - f(y_n)) = \lim_{n\to\infty} \left(\left(n + \frac{1}{n} \right)^2 - n^2 \right) = \lim_{n\to\infty} \left(n^2 + \frac{1}{n^2} + 2 - n^2 \right)$$
$$= \lim_{n\to\infty} \left(\frac{1}{n^2} + 2 \right) = 2.$$

Therefore, by the sequential criterion for uniformly continuous functions, f is uniformly continuous.

(IV). Let $(x_n)_{n=1}^\infty$ and $(y_n)_{n=1}^\infty$ be two sequences in $(0, \infty)$ that satisfy $\lim_{n\to\infty} (x_n - y_n) = 0$. Then, for every $n \in \mathbb{N}$,

$$|f(x_n) - f(y_n)|^2 = |\sqrt{x_n} - \sqrt{y_n}|^2 = |\sqrt{x_n} - \sqrt{y_n}||\sqrt{x_n} - \sqrt{y_n}|$$
$$\leq |\sqrt{x_n} - \sqrt{y_n}|(|\sqrt{x_n}| + |\sqrt{y_n}|) = |\sqrt{x_n} - \sqrt{y_n}|(\sqrt{x_n} + \sqrt{y_n})$$
$$= |\sqrt{x_n} - \sqrt{y_n}||\sqrt{x_n} + \sqrt{y_n}| = |(\sqrt{x_n} - \sqrt{y_n})(\sqrt{x_n} + \sqrt{y_n})|$$
$$= |x_n - y_n|,$$

where we have used that $x_n, y_n > 0$ for every $n \in \mathbb{N}$. Therefore, since the square root is an increasing function, it yields immediately that $0 \leq |f(x_n) - f(y_n)| \leq \sqrt{|x_n - y_n|}$. Hence, by the Sandwich Principle, we have that $\lim_{n\to\infty} (f(x_n) - f(y_n)) = 0$. Thus, by the sequential criterion for uniformly continuous functions, f is uniformly continuous.

(V). Take $x_n = \dfrac{1}{n}$ and $y_n = \dfrac{1}{n+1}$ for any $n \in \mathbb{N}$. Observe that $(x_n)_{n=1}^\infty$ and $(y_n)_{n=1}^\infty$ are two sequences in $(0, \infty)$ such that

$$\lim_{n\to\infty} (x_n - y_n) = \lim_{n\to\infty} \left(\frac{1}{n} - \frac{1}{n+1} \right) = 0.$$

Then,

$$\lim_{n\to\infty} (f(x_n) - f(y_n)) = \lim_{n\to\infty} \left(\frac{1}{(1/n)^2} - \frac{1}{(1/(n+1))^2} \right)$$
$$= \lim_{n\to\infty} (n^2 - n^2 - 2n - 1) = \lim_{n\to\infty} (-2n - 1) = -\infty.$$

So, by the sequential criterion for uniformly continuous functions, f is not uniformly continuous.

(VI). Let $(x_n)_{n=1}^\infty$ and $(y_n)_{n=1}^\infty$ be two sequences in $(1, \infty)$ that satisfy $\lim_{n\to\infty} (x_n - y_n) = 0$. Then, for every $n \in \mathbb{N}$,

$$0 \leq |f(x_n) - f(y_n)| = \left| \frac{1}{x_n^2} - \frac{1}{y_n^2} \right| = \left| \frac{y_n^2 - x_n^2}{x_n^2 y_n^2} \right| = \frac{|y_n^2 - x_n^2|}{x_n^2 y_n^2}$$

$$= \frac{|(y_n + x_n)(y_n - x_n)|}{x_n^2 y_n^2} = \frac{(y_n + x_n)|y_n - x_n|}{x_n^2 y_n^2} \leq 2|x_n - y_n|,$$

where we have used that $x_n, y_n > 1$ for any $n \in \mathbb{N}$. Hence, by the Sandwich Principle, we have that $\lim_{n\to\infty} (f(x_n) - f(y_n)) = 0$. Therefore, by the sequential criterion for uniformly continuous functions, f is uniformly continuous.

(VII). Take $x_n = \dfrac{1}{2\pi n}$ and $y_n = \dfrac{1}{\frac{\pi}{2} + 2\pi n}$ for every $n \in \mathbb{N}$. It is clear that $(x_n)_{n=1}^\infty$ and $(y_n)_{n=1}^\infty$ are two sequences in $(0, \infty)$ such that

$$\lim_{n\to\infty} (x_n - y_n) = \lim_{n\to\infty} \left(\frac{1}{2\pi n} - \frac{1}{\frac{\pi}{2} + 2\pi n} \right) = 0.$$

But

$$\lim_{n\to\infty} (f(x_n) - f(y_n)) = \lim_{n\to\infty} \left(\sin(2\pi n) - \sin\left(\frac{\pi}{2} + 2\pi n \right) \right) = \lim_{n\to\infty} (-1) = -1.$$

So, by the sequential criterion for uniformly continuous functions, f is not uniformly continuous.

(VIII). Let $(x_n)_{n=1}^\infty$ and $(y_n)_{n=1}^\infty$ be two sequences in $(1, \infty)$ that satisfy $\lim_{n\to\infty} (x_n - y_n) = 0$. Then, for every $n \in \mathbb{N}$,

$$0 \leq |f(x_n) - f(y_n)| = \left| \sin \frac{1}{x_n} - \sin \frac{1}{y_n} \right|$$

$$= \left| 2 \cos \left(\frac{1}{2} \left(\frac{1}{x_n} + \frac{1}{y_n} \right) \right) \sin \left(\frac{1}{2} \left(\frac{1}{x_n} - \frac{1}{y_n} \right) \right) \right|$$

$$\leq 2 \left| \sin \left(\frac{y_n - x_n}{2 x_n y_n} \right) \right| \leq 2 \left| \frac{y_n - x_n}{2 x_n y_n} \right| \leq |x_n - y_n|,$$

where we have used that $x_n > 1 < y_n$, $|\sin t| \leq |t|$ for all $t \in \mathbb{R}$, $\sin A - \sin B = 2 \cos \left(\dfrac{A + B}{2} \right) \sin \left(\dfrac{A - B}{2} \right)$, and the fact that the cosine function is bounded between -1 and 1. Now, thanks to the Sandwich Principle, we have that $\lim_{n\to\infty} (f(x_n) - f(y_n)) = 0$. Thus, by the sequential criterion for uniformly continuous functions, f is uniformly continuous.

(IX). Take $x_n = \sqrt{1 - \dfrac{1}{n}}$ and $y_n = \sqrt{1 - \dfrac{1}{n+1}}$ for every $n \in \mathbb{N}$. It is clear that $(x_n)_{n=1}^{\infty}$ and $(y_n)_{n=1}^{\infty}$ are two sequences in $(-1, 1)$ such that

$$\lim_{n \to \infty} (x_n - y_n) = \lim_{n \to \infty} \left(\sqrt{1 - \frac{1}{n}} - \sqrt{1 - \frac{1}{n+1}} \right) = 1 - 1 = 0.$$

But

$$\lim_{n \to \infty} (f(x_n) - f(y_n)) = \lim_{n \to \infty} \left(\frac{1}{1 - \left(\sqrt{1 - \frac{1}{n}}\right)^2} - \frac{1}{1 - \left(\sqrt{1 - \frac{1}{n+1}}\right)^2} \right)$$

$$= \lim_{n \to \infty} (n - (n+1)) = \lim_{n \to \infty} -1 = -1.$$

Hence, by the sequential criterion for uniformly continuous functions, f is not uniformly continuous.

(X). Let $(x_n)_{n=1}^{\infty}$ and $(y_n)_{n=1}^{\infty}$ be two sequences in $(0, 1)$ that satisfy $\lim_{n \to \infty} (x_n - y_n) = 0$. Then, for every $n \in \mathbb{N}$,

$$0 \le |f(x_n) - f(y_n)| = \left| x_n \sin \frac{1}{x_n} - y_n \sin \frac{1}{y_n} \right|$$

$$= \left| x_n \sin \frac{1}{x_n} - y_n \sin \frac{1}{x_n} + y_n \sin \frac{1}{x_n} - y_n \sin \frac{1}{y_n} \right|$$

$$= \left| (x_n - y_n) \sin \frac{1}{x_n} + y_n \left(\sin \frac{1}{x_n} - \sin \frac{1}{y_n} \right) \right|$$

$$\le |x_n - y_n| \left| \sin \frac{1}{x_n} \right| + |y_n| \left| \sin \frac{1}{x_n} - \sin \frac{1}{y_n} \right|$$

$$\le |x_n - y_n| + \left| 2 \cos \left(\frac{1}{2} \left(\frac{1}{x_n} + \frac{1}{y_n} \right) \right) \sin \left(\frac{1}{2} \left(\frac{1}{x_n} - \frac{1}{y_n} \right) \right) \right|$$

$$\le |x_n - y_n| + 2 \left| \sin \left(\frac{y_n - x_n}{2 x_n y_n} \right) \right|,$$

where we have used that $\sin A - \sin B = 2 \cos \left(\dfrac{A+B}{2} \right) \sin \left(\dfrac{A-B}{2} \right)$, the fact that the sine and cosine functions are bounded between -1 and 1, and $y_n \le 1$ for any $n \in \mathbb{N}$. Recall by the proof of Exercise 4.21 part (VIII) that

$$\lim_{n \to \infty} \left| \sin \left(\frac{y_n - x_n}{2 x_n y_n} \right) \right| = 0$$

provided that $\lim_{n \to \infty} (x_n - y_n) = 0$. So, by the Sandwich Principle, we have that $\lim_{n \to \infty} (f(x_n) - f(y_n)) = 0$. Hence, by the sequential criterion for uniformly continuous functions, f is uniformly continuous. $\qquad \square$

Chapter 5

Differentiable Functions

> **Exercise 5.1.** *Study the differentiability of the following functions and find $f'(x)$, $f''(x)$, where possible.*
>
> (I) $f(x) = \sqrt{|x|}$,
>
> (II) $f(x) = \begin{cases} x^2 & \text{If } x \leq 0, \\ -x^2 & \text{If } x > 0, \end{cases}$
>
> (III) $f(x) = |x|^3$.

Solution to Exercise 5.1.

(I). We can write f as follows:

$$f(x) = \begin{cases} \sqrt{x} & \text{if } x \geq 0, \\ \sqrt{-x} & \text{if } x < 0. \end{cases}$$

As $\lim_{x \to 0^+} f(x) = \lim_{x \to 0^-} f(x) = 0$, f is continuous at zero. Then,

$$f'(x) = \begin{cases} \frac{1}{2\sqrt{x}} & \text{if } x > 0, \\ -\frac{1}{2\sqrt{-x}} & \text{if } x < 0. \end{cases}$$

Observe that $\lim_{x \to 0^+} f'(x)$ and $\infty \neq \lim_{x \to 0^-} f'(x) = -\infty$, so f is not differentiable at zero. Another way to see this is considering the quotient $\frac{f(x) - f(0)}{x - 0} = \frac{f(x)}{x}$ and showing that it tends to ∞ when $x \to 0^+$.

(II). $f(x) = \begin{cases} x^2 & \text{if } x \leq 0, \\ -x^2 & \text{if } x > 0. \end{cases}$

As $\lim_{x \to 0^+} f(x) = \lim_{x \to 0^-} f(x) = 0$, f is continuous at zero. Then,

$$f'(x) = \begin{cases} 2x & \text{if } x < 0, \\ -2x & \text{if } x > 0. \end{cases}$$

Since $\lim_{x \to 0^+} f'(x) = \lim_{x \to 0^-} f'(x) = 0$, f is differentiable at zero and $f'(0) = 0$. Moreover,

$$f''(x) = \begin{cases} 2 & \text{if } x < 0, \\ -2 & \text{if } x > 0. \end{cases}$$

DOI: 10.1201/9781003400745-5

Notice that $\lim_{x\to 0+} f''(x) = -2 \neq 2 = \lim_{x\to 0-} f''(x)$, so f' is not differentiable at zero. Another way: $f'(x) = 2|x|$, which is not differentiable at the origin.

(III). We can write f as follows:

$$f(x) = \begin{cases} x^3 & \text{if } x \geq 0, \\ -x^3 & \text{if } x < 0. \end{cases}$$

As $\lim_{x\to 0+} f(x) = \lim_{x\to 0-} f(x) = 0$, f is continuous at zero. Then,

$$f'(x) = \begin{cases} 3x^2 & \text{if } x > 0, \\ -3x^2 & \text{if } x < 0. \end{cases}$$

Since $\lim_{x\to 0+} f'(x) = \lim_{x\to 0-} f'(x) = 0$ f is differentiable at zero and $f'(0) = 0$. As for the second derivative we have

$$f''(x) = \begin{cases} 6x & \text{if } x > 0, \\ -6x & \text{if } x < 0. \end{cases}$$

Since $\lim_{x\to 0+} f''(x) = \lim_{x\to 0-} f''(x) = 0$ then f' is differentiable at zero and $f''(0) = 0$.

$\square$

Exercise 5.2. *Find $f'(x)$, simplifying if possible, in the following cases:*

(I) $f(x) = \dfrac{\sin x + \cos x}{\sin x - \cos x}$,

(II) $f(x) = (x^2 + 1) \arctan x$,

(III) $f(x) = \log(x + \sqrt{x^2 + 1})$,

(IV) $f(x) = \log(x + \sqrt{x^2 - 1})$,

(V) $f(x) = \log \sqrt{\dfrac{1 - \cos x}{1 + \cos x}}$,

(VI) $f(x) = x^{\frac{1}{\log x}}$,

(VII) $f(x) = \arctan \sqrt{\dfrac{\sin a \sin x}{1 - \cos a \cos x}}$

(VIII) $f(x) = x \arcsin x + \sqrt{1 - x^2}$.

Solution to Exercise 5.2.

(I). $f(x) = \dfrac{\sin x + \cos x}{\sin x - \cos x}$:

$$f'(x) = \frac{(\cos x - \sin x)(\sin x - \cos x) - (\cos x + \sin x)^2}{(\sin x - \cos x)^2} = \frac{-2}{1 - \sin 2x}.$$

(II). $f(x) = (x^2 + 1) \arctan x$:

$$f'(x) = 2x \arctan x + (x^2 + 1)\frac{1}{1 + x^2} = 2x \arctan x + 1.$$

(III). $f(x) = \log(x + \sqrt{x^2 + 1})$:

$$f'(x) = \frac{1}{x + \sqrt{x^2 + 1}}\left(1 + \frac{2x}{2}(x^2 + 1)^{-\frac{1}{2}}\right)$$

$$= \frac{1}{x + \sqrt{x^2 + 1}}\left(\frac{\sqrt{x^2 + 1} + x}{\sqrt{x^2 + 1}}\right)$$

$$= \frac{1}{\sqrt{x^2 + 1}}.$$

(IV). $f(x) = \log(x + \sqrt{x^2 - 1})$:

$$f'(x) = \frac{1}{x + \sqrt{x^2 - 1}}\left(1 + \frac{2x}{2}(x^2 - 1)^{-\frac{1}{2}}\right)$$

$$= \frac{1}{x + \sqrt{x^2 - 1}}\left(\frac{\sqrt{x^2 - 1} + x}{\sqrt{x^2 - 1}}\right)$$

$$= \frac{1}{\sqrt{x^2 - 1}}.$$

(V). $f(x) = \log\sqrt{\dfrac{1 - \cos x}{1 + \cos x}} = \frac{1}{2}\log\dfrac{1 - \cos x}{1 + \cos x}$:

$$f'(x) = \frac{1}{2}\frac{1}{\frac{1-\cos x}{1+\cos x}}\frac{\sin x(1 + \cos x) - (1 - \cos x)(-\sin x)}{(1 + \cos x)^2}$$

$$= \frac{1}{2}\cdot\frac{1 + \cos x}{1 - \cos x}\cdot\frac{\sin x + \sin x \cos x + \sin x - \sin x \cos x}{(1 + \cos x)^2}$$

$$= \frac{\sin x}{(1 + \cos x)(1 - \cos x)} = \frac{\sin x}{1 - \cos^2 x} = \frac{1}{\sin x}.$$

(VI). $f(x) = x^{\frac{1}{\log x}}$, $f(x) = e^{\log(f(x))} = e^{\frac{1}{\log x}\log x} = e$, then $f'(x) = 0$.

(VII). $f(x) = \arctan\sqrt{\dfrac{\sin a \sin x}{1 - \cos a \cos x}}$: Let us denote $g(x) = \dfrac{\sin a \sin x}{1 - \cos a \cos x}$. Then:

$$g'(x) = \frac{\sin a \cos x(1 - \cos a \cos x) - \sin a \sin x \cos a \sin x}{(1 - \cos a \cos x)^2}$$

$$= \frac{\sin a \cos x - \sin a \cos a \cos^2 x - \sin a \cos a \sin^2 x}{(1 - \cos a \cos x)^2}$$

$$= \frac{\sin a \cos x - \sin a \cos a}{(1 - \cos a \cos x)^2}$$

$$= \frac{\sin a(\cos x - \cos a)}{(1 - \cos a \cos x)^2},$$

and

$$f'(x) = \frac{1}{1 + \dfrac{\sin a \sin x}{1 - \cos a \cos x}} \cdot \frac{1}{2} \left(\frac{\sin a \sin x}{1 - \cos a \cos x} \right)^{-\frac{1}{2}} g'(x)$$

$$= \frac{1}{2} \sqrt{\frac{1 - \cos a \cos x}{\sin a \sin x}} \cdot \frac{1 - \cos a \cos x}{1 - \cos a \cos x + \sin a \sin x} g'(x)$$

$$= \frac{1}{2} \frac{1}{\sqrt{\sin a \sin x (1 - \cos a \cos x)}} \cdot \frac{(1 - \cos a \cos x)^2}{1 - \cos(x + a)} g'(x)$$

$$= \frac{1}{2} \frac{\sin a (\cos x - \cos a)}{[1 - \cos(x + a)]\sqrt{\sin a \sin x (1 - \cos a \cos x)}}.$$

(VIII). $f(x) = x \arcsin x + \sqrt{1 - x^2}$:

$$f'(x) = \arcsin x + \frac{x}{\sqrt{1 - x^2}} + \frac{1}{2}(1 - x^2)^{-\frac{1}{2}}(-2x) = \arcsin x.$$

□

Exercise 5.3. *Find the number of real solutions of the following equations. For each solution find an interval whose endpoints are consecutive integers:*

(I) $3x^5 + 15x - 8 = 0$,

(II) $2x^3 - 9x^2 + 12x = -1$,

(III) $x^5 - 5x = 1$,

(IV) $e^x = 1 + x$,

(V) $x^5 + 2x + 1 = 0$.

Solution to Exercise 5.3.

(I). $3x^5 + 15x - 8 = 0$. Let us define the function $f(x) = 3x^5 + 15x - 8$. Observe that $f(0) = -8 < 0$ and $f(1) = 10 > 0$. Hence by Bolzano's theorem, since f is continuous in $[0, 1]$ and $f(0) \cdot f(1) < 0$, then there exists $c \in (0, 1)$ such that $f(c) = 0$.

On the other hand f is strictly increasing since $f'(x) = 15x^4 + 15 > 0$ $\forall x \in \mathbb{R}$, which shows that the equation has only one solution.

(II). $2x^3 - 9x^2 + 12x = -1$. Consider the function $f(x) = 2x^3 - 9x^2 + 12x + 1$. Notice that $f(0) = 1 > 0$ and $f(-1) = -32 < 0$, that is $f(0) \cdot f(-1) < 0$. Applying Bolzano's theorem to the continuous function f in $[-1, 0]$, there exists $c \in (-1, 0)$ such that $f(c) = 0$.

To see whether the equation has more roots we study the sign of $f'(x) = 6x^2 - 18x + 12 = 6(x^2 - 3x + 2) = 6(x-1)(x-2)$:

$f'(x)$	$-\infty$	$+$	1	$-$	2	$+$	$+\infty$
$f(x)$		$\nearrow$		$\searrow$		$\nearrow$	

Since f is strictly increasing in $(-\infty, 1)$, the equation $f(x) = 0$ can only have a root in $(-\infty, 1)$. The function f attains its minimum in $(1, \infty)$ at $x = 2$, that is, $f(x) \geq f(2) = 5$ for all $x \in (1, \infty)$, proving that the equation $f(x) = 0$ has only one solution.

(III). $x^5 - 5x = 1$. Define the continuous function $f(x) = x^5 - 5x - 1$. The sign of $f'(x) = 5x^4 - 5 = 5(x^2 - 1)(x^2 + 1)$, shown in the following table:

$f'(x)$	$-\infty$	$+$	-1	$-$	1	$+$	$+\infty$
$f(x)$		$\nearrow$		$\searrow$		$\nearrow$	

tells us that f is strictly increasing in $(-\infty, -1)$ and $(1, \infty)$ and strictly decreasing in $(-1, 1)$. This proves that f has, at most, one root in each of those three intervals, that is, f has at most three roots. In fact f has exactly three roots, which is shown below as a consequence of Bolzano's theorem:

- In $(-2, -1)$ since $f(-2) \cdot f(-1) = (-21) \cdot 3 < 0$.

- In $(-1, 0)$ since $f(-1) \cdot f(0) = 3 \cdot (-1) < 0$.

- In $(1, 2)$ since $f(1) \cdot f(2) = (-5) \cdot 21 < 0$.

(IV). $e^x = 1 + x$. Consider the function $f(x) = e^x - 1 - x$. The sign of $f'(x) = e^x - 1$ is shown below:

$f'(x)$	$-\infty$	$-$	0	$+$	$+\infty$
$f(x)$		$\searrow$		$\nearrow$	

The function f is clearly strictly decreasing in $(-\infty, 0)$ and strictly increasing in $(0, \infty)$ and hence, $f(x) < f(0) = 0$ for all $x \in (-\infty, 0)$ and $f(x) > f(0) = 0$ for all $x \in (0, \infty)$, proving that the equation $f(x) = 0$ has only one root, namely $x = 0$.

(V). $x^5 + 2x + 1 = 0$. Define the continuous function $f(x) = x^5 + 2x + 1$. Since $f'(x) = 5x^4 + 2 > 0$ for all $x \in \mathbb{R}$, then, f is strictly increasing, showing that $f(x) = 0$ has at most one solution. In fact it has exactly one since $f(0) \cdot f(-1) = 1 \cdot (-2) < 0$, so by Bolzano's theorem, the solution of $f(x) = 0$ is in $(-1, 0)$. $\qquad\square$

Exercise 5.4. *Prove the following inequalities:*

(I) $x - \frac{x^3}{3} < \arctan x < x - \frac{x^3}{6}$, $\quad$ (IV) $2x < \sin 2x + \tan x$, $x \in$ $x \in (0, 1]$ $\qquad\qquad\qquad\qquad\qquad (0, \frac{\pi}{2})$,

(II) $e^x > ex$, $x \neq 1$, $\qquad\qquad\qquad$ (V) $\tan x > x + \frac{x^3}{3}$, $x \in (0, \frac{\pi}{2})$,

(III) $\frac{x}{1+x} < \log(1 + x) < x$, $x >$ $\quad$ (VI) $e^x \geq 1 + x + \frac{x^2}{2}$, $x \geq 0$, $-1, x \neq 0$,

$\qquad\qquad\qquad\qquad\qquad\qquad$ (VII) $x - \frac{x^3}{6} < \sin x < x$, $x > 0$.

Solution to Exercise 5.4.

(I). $x - \frac{x^3}{3} \overset{(1)}{<} \arctan x \overset{(2)}{<} x - \frac{x^3}{6}$, $x \in (0, 1]$.

Let us prove (1) first:

If $f(x) = \arctan x - x + \frac{x^3}{3}$, we have to prove that $f(x) > 0$ $\forall x \in (0, 1]$. Observe that,

$$f'(x) = \frac{1}{1+x^2} - 1 + x^2 = \frac{1 - -1 - x^2 + x^2 + x^4}{1+x^2}$$
$$= \frac{x^4}{1+x^2} > 0, \forall x \in \mathbb{R}.$$

Therefore, f is strictly increasing in $(0, 1]$. In particular, $f(x) > f(0) = 0$ $\forall x \in (0, 1]$.

Next we prove (2):

If now $f(x) = \arctan x - x + \frac{x^3}{6}$, we have to show that $f(x) < 0$ $\forall x \in (0, 1]$.

$$f'(x) = \frac{1}{1+x^2} - 1 + \frac{x^2}{2} = \frac{2 - 2(1+x^2) + x^2(1+x^2)}{2(1+x^2)}$$
$$= -\frac{x^4 - x^2}{2(1+x^2)} = \frac{x^2(x-1)(x+1)}{2(x^2+1)} \leq 0, \forall x \in (0, 1].$$

Therefore, f is decreasing in $(0, 1]$. In particular, $f(x) < f(0) = 0$ $\forall x \in (0, 1)$.

(II). $e^x > ex$, $x \neq 1$.

If $f(x) = e^x - ex$, we have to prove that $f(x) > 0$ for all $x \neq 1$.

For $x \in (1, \infty)$ we have that $f'(x) = e^x - e > 0$, so f is strictly increasing in $(1, \infty)$. In particular, $f(x) > f(1) = 0$ for every $x \in (1, \infty)$. Since clearly $f'(x) < 0$ for all $x \in (-\infty, 1)$, f is strictly decreasing so $f(x) > f(1) = 0$ for $x \in (-\infty, 1)$.

(III). $\frac{x}{1+x} \overset{(1)}{<} \log(1+x) \overset{(2)}{<} x$, $x > -1, x \neq 0$.

We prove (1) first:

If $f(x) = \frac{x}{1+x} - \log(1+x)$, we have to show that $f(x) < 0$. The sign of the derivative of f, namely

$$f'(x) = \frac{1+x-x}{(1+x)^2} - \frac{1}{1+x} = \frac{1-(1+x)}{(1+x)^2}$$
$$= -\frac{x}{(1+x)^2}.$$

is given in the following table:

$f'(x)$	-1	$+$	0	$-$	$+\infty$
$f(x)$		$\nearrow$		$\searrow$	

Hence f is strictly decreasing in $(0,\infty)$, and strictly increasing in $(-1,0)$. In particular, $f(x) < f(0) = 0$ for all $x \in (-1,\infty) \setminus \{0\}$.

Next we prove (2):

Now we consider the function $f(x) = \log(1+x) - x$. We need to show that $f(x) < 0$ for $x > -1$ with $x \neq 0$. Since $f'(x) = \frac{1}{1+x} - 1 = \frac{-x}{1+x}$ the sign of f' behaves as follows:

$f'(x)$	-1	$+$	0	$-$	$+\infty$
$f(x)$		$\nearrow$		$\searrow$	

Thus f is strictly decreasing in $(0,\infty)$, and strictly increasing in $(-1,0)$. In particular, $f(x) < f(0) = 0$ for all $x > -1$ with $x \neq 0$.

(IV). $2x < \sin 2x + \tan x$, $x \in \left(0, \frac{\pi}{2}\right)$. If $f(x) = \sin 2x + \tan x - 2x$ we have to prove that $f(x) > 0 \ \forall x \in \left(0, \frac{\pi}{2}\right)$. Then,

$$f'(x) = 2\cos 2x + \frac{1}{\cos^2 x} - 2$$
$$= \frac{2\cos^2 x(\cos 2x - 1) + 1}{\cos^2 x}$$
$$= \frac{2\cos^2 x(-2\operatorname{sen}^2 x) + 1}{\cos^2 x}$$
$$= \frac{1 - (2\sin x \cos x)^2}{\cos^2 x}$$
$$= \frac{1 - \sin^2 2x}{\cos^2 x}.$$

Observe that the equation $f'(x) = 0$ has only one root in the interval $\left(0, \frac{\pi}{2}\right)$ at $x = \frac{\pi}{4}$. Also, $f'(x) > 0 \ \forall x \in \left(0, \frac{\pi}{2}\right) \setminus \{\frac{\pi}{4}\}$, so f is strictly increasing in $\left(0, \frac{\pi}{2}\right)$. Therefore, $f(x) > f(0) = 0 \ \forall x \in \left(0, \frac{\pi}{2}\right)$.

(V). $\tan x > x + \frac{x^3}{3}$. If $f(x) = \tan x - x - \frac{x^3}{3}$, we need to prove that $f(x) > 0 \;\forall x \in \left(0, \frac{\pi}{2}\right)$. Since

$$f'(x) = \frac{1}{\cos^2 x} - 1 - x^2 = \frac{\sin^2 x - x^2 \cos^2 x}{\cos^2 x} = \tan^2 x - x^2$$

the equation $f'(x) = 0$ is equivalent to $\tan x = \pm x$ which, in its turn, has only one root in $\left(0, \frac{\pi}{2}\right)$ at $x = 0$. Now since $\tan x > x > 0 \;\forall x \in \left(0, \frac{\pi}{2}\right)$, we have that $f'(x) > 0 \;\forall x \in \left(0, \frac{\pi}{2}\right)$. Therefore, f is strictly increasing and $f(x) > f(0) = 0$ $\forall x \in \left(0, \frac{\pi}{2}\right)$.

(VI). $e^x \geq 1 + x + \frac{x^2}{2}$, $x \geq 0$. If $f(x) = e^x - 1 - x - \frac{x^2}{2}$ then $f'(x) = e^x - 1 - x$ and $f''(x) = e^x - 1$. It is straightforward that $f''(x) < 0$ in $(-\infty, 0)$ and $f''(x) > 0$ in $(0, \infty)$. Therefore, $f'(x)$ has a minimum at $x = 0$, from which $f'(x) \geq 0$ for every $x \in (0, \infty)$. Since f is increasing in $(0, \infty)$, it follows that $f(x) \geq f(0) = 0$ for all $x \in (0, \infty)$.

(VII). $x - \frac{x^3}{6} \overset{(1)}{<} \sin x \overset{(2)}{<} x$, $x > 0$.
We prove (1) first:

If $f(x) = \sin x - x + \frac{x^2}{2}$, we have to show that $f(x) > 0 \;\forall x > 0$. We have that $f'(x) = \cos x - 1 + x$ and $f''(x) = 1 - \sin x \geq 0 \;\forall x \in \mathbb{R}$. Then $f'(x)$ is increasing in $(0, \infty)$. In particular, $f'(x) > f'(0) = 0$ for all $x \in \mathbb{R}$. Therefore, f is increasing on $(0, \infty)$ and $f(x) > f(0) = 0$ for every $x \in (0, \infty)$.

Next we prove (2):

Now consider the function $f(x) = x - \sin x$. We have to prove that $f(x) > 0$ for each $x > 0$. Since $f'(x) = 1 - \cos x \geq 0$ for all $x \in \mathbb{R}$, it follows that f is increasing in $(0, \infty)$ and $f(x) > f(0) = 0$ for every $x > 0$. $\qquad\square$

Exercise 5.5. *Prove that* $\arctan x - \arctan y < x - y$, *if* $x > y$. *Deduce that the* $\arctan$ *function is Lipschitz in* $\mathbb{R}$.

Solution to Exercise 5.5.

Let us define the function $f(x) = x - \arctan x$ for every $x \in \mathbb{R}$. Then

$$f'(x) = 1 - \frac{1}{1+x^2} = \frac{x^2}{1+x^2}$$

for all $x \in \mathbb{R}$. Since $f'(x) > 0$ for all $x \in \mathbb{R} \setminus \{0\}$, f is strictly increasing in $\mathbb{R}$. Hence, for every $x, y \in \mathbb{R}$ with $x > y$ we have

$$f(x) > f(y) \Leftrightarrow x - \arctan x > y - \arctan y \Leftrightarrow x - y > \arctan x - \arctan y.$$

To prove that $g(x) = \arctan x$ is Lipschitz notice that g is strictly increasing in $\mathbb{R}$. Thus, if $x, y \in \mathbb{R}$ and $x \geq y$ then

$$|\arctan x - \arctan y| = \arctan x - \arctan y \leq x - y = |x - y|,$$

proving that g is Lipschitz in $\mathbb{R}$.

$\square$

Exercise 5.6. *Prove that* $\arcsin x + \arccos x = \frac{\pi}{2}$ *for all* $x \in [-1, 1]$.

Solution to Exercise 5.6.

Let $f(x) = \arcsin x + \arccos x$ with $x \in [-1, 1]$. Then $f'(x) = \frac{1}{\sqrt{1-x^2}} - \frac{1}{\sqrt{1-x^2}} = 0$. Since $f' \equiv 0$ in $[-1, 1]$, then f must be constant. Hence, as $f(0) = 0 + \frac{\pi}{2} = \frac{\pi}{2}$, the equality holds for all $x \in [-1, 1]$.

$\square$

Exercise 5.7. *Prove that* $\arccos \frac{1}{\sqrt{1+x^2}} = \arctan x$ *for all* $x \geq 0$. *What happens if* $x < 0$?

Solution to Exercise 5.7.

If $f(x) = \arccos \frac{1}{\sqrt{1+x^2}} - \arctan x$, then

$$f'(x) = \frac{1}{\sqrt{1 - \left(\frac{1}{\sqrt{1+x^2}}\right)^2}} \cdot \left(\frac{x}{\sqrt{1+x^2}(1+x^2)}\right) - \frac{1}{1+x^2}$$

$$= \frac{x}{(1+x^2) \cdot \sqrt{1+x^2} \cdot \sqrt{1 - \frac{1}{1+x^2}}} - \frac{1}{1+x^2}$$

$$= \frac{x}{(1+x^2) \cdot |x|} - \frac{1}{1+x^2}$$

$$= \begin{cases} 0 & \text{if } x \geq 0, \\ \frac{-2}{1+x^2} & \text{if } x < 0. \end{cases}$$

Therefore, since $f'(x) = 0$ for all $x \geq 0$, f must be constant in $[0, \infty)$, from which $f(x) = f(0) = 0$ for every $x \geq 0$. At the other end of the scale, f is strictly decreasing in $(-\infty, 0]$, so $f(x) < f(0) = 0$ if $x < 0$. In particular, $\arccos \frac{1}{\sqrt{1+x^2}} < \arctan x$ for all $x < 0$.

$\square$

Exercise 5.8. *Let $f, g : [0, 1] \to \mathbb{R}$ be continuous in $[0, 1]$, differentiable in $(0, 1)$, with $f(0) = 0$, $g(0) = 2$ and $|f'(x)| \leq 1$, $|g'(x)| \leq 1$ for all $x \in (0, 1)$. Prove that $f(x) \leq 1 \leq g(x)$ for all $x \in [0, 1]$.*

Solution to Exercise 5.8.

First, notice that $-1 \leq f'(x) \leq 1$ and $-1 \leq g(x) \leq 1$ for all $x \in [0, 1]$. Now fix $x \in (0, 1]$. Applying the mean value theorem to f in the interval $[0, x]$, there exist $\eta \in (0, x)$ such that

$$1 \geq f'(\eta) = \frac{f(x) - f(0)}{x - 0} = \frac{f(x)}{x},$$

from which $f(x) \leq x \leq 1$. On the other hand, by the mean value theorem applied to g in $[0, x]$ there exists $\xi \in (0, x)$ such that

$$-1 \leq g'(\xi) = \frac{g(x) - g(0)}{x - 0} = \frac{g(x) - 2}{x}.$$

Then $g(x) \geq -x + 2 \geq 1$. Since x was an arbitrary element of $[0, 1]$ we have shown that $f(x) \leq 1 \leq g(x)$ for all $x \in [0, 1]$.

$\square$

Exercise 5.9. *Let f be differentiable in (a, b) and continuous in $[a, b]$ with $f(a) = f(b) = 0$. Prove that for all $\lambda \in \mathbb{R}$ there exists $c \in (a, b)$ such that $f'(c) = \lambda f(c)$.*

Solution to Exercise 5.9.

Let $g(x) = f(x) \cdot e^{-\lambda x}$. We have that $g(a) = g(b)$ and g is continuous on $[a, b]$ and differentiable in (a, b). Then, by Rolle's theorem, there exists $c \in (a, b)$ such that $g'(c) = 0$. Therefore,

$$0 = g'(c) = f'(c)e^{-\lambda c} - \lambda f(c)e^{-\lambda c} = [f'(c) - \lambda f(c)]\, e^{-\lambda c}.$$

Since $e^{-\lambda c} \neq 0$ we must have $f'(c) - \lambda f(c) = 0$, that is $f'(c) = \lambda f(c)$.

$\square$

Exercise 5.10. *Let f be continuous in $[a, b]$ and twice differentiable in the open interval (a, b). Suppose the segment connecting the points $A = (a, f(a))$ and $B = (b, f(b))$ meet the graph of the function f at a third point P different from A and B. Prove that $f''(c) = 0$ for some $c \in (a, b)$.*

Solution to Exercise 5.10.
Consider the function $g : [a, b] \to \mathbb{R}$ given by

$$g(x) = \frac{x - b}{a - b} f(a) + \frac{a - x}{a - b} f(b).$$

It is straightforward that $g(a) = f(a)$ and $g(b) = f(b)$. As a matter of fact, the graph of g is the segment joining the points A and B. Next define the function $h(x) = f(x) - g(x)$ for $x \in [a, b]$. As the segment connecting the points A and B intersects the graph of f at a third point different from A and B, then the equation $h(x) = 0$ has at least a solution in (a, b), namely $h(\eta) = 0$. Applying now the mean value theorem to h in the intervals $[a, \eta]$ and $[\eta, b]$, there exist $x_0 \in (a, \eta)$ and $x_1 \in (\eta, b)$ such that:

$$h'(x_0) = \frac{h(\eta) - h(a)}{\eta - a} = \frac{0 - 0}{\eta - a} = 0$$

and

$$h'(x_1) = \frac{h(b) - h(\eta)}{b - \eta} = \frac{0 - 0}{b - \eta} = 0.$$

Since $h'(x)$ is continuous in $[x_0, x_1]$ and differentiable in (x_0, x_1), by the mean value theorem, there exists $c \in (x_0, x_1) \subset (a, b)$ such that

$$h''(c) = \frac{h'(x_1) - h'(x_0)}{x_1 - x_0} = 0.$$

Therefore, $h''(x) = f''(x) - g''(x) = f''(x)$, then $h''(c) = f''(c) = 0$. $\square$

Exercise 5.11. *If f is three times differentiable in $[a, b]$ and*

$$f(a) = f'(a) = f(b) = f'(b) = 0,$$

prove that $f'''(c) = 0$ for some $c \in (a, b)$.

Solution to Exercise 5.11.
Applying the mean value theorem to f in $[a, b]$ there exists $c \in (a, b)$ such that

$$f'(c) = \frac{f(b) - f(a)}{b - a} = 0.$$

Since f' is continuous in $[a, c]$ and $[c, b]$ and differentiable in (a, c) and in (c, b), by the mean value theorem there exists $x_0 \in (a, c)$ and $x_1 \in (c, b)$ such that

$$f''(x_0) = \frac{f'(c) - f'(a)}{c - a} = 0$$

and
$$f''(x_1) = \frac{f'(b) - f'(c)}{b - c} = 0.$$

Finally, we can still apply the mean value theorem to f''' in $[a, b]$ since f'' is continuous in $[x_0, x_1]$ and differentiable in (x_0, x_1), so there exists $c \in (x_0, x_1) \subset (a, b)$ such that

$$f'''(c) = \frac{f''(x_1) - f''(x_0)}{x_1 - x_0} = 0.$$

□

Exercise 5.12. *Let f be a non-negative function and three times differentiable in $(0, 1)$. If f vanishes at least at two points of $(0, 1)$, then $f'''(c) = 0$ for some $c \in (0, 1)$.*

Solution to Exercise 5.12.

Assume that f vanishes at $a, b \in (0, 1)$. Since f is non-negative, f has relative minimums at $x = a$ and $x = b$ with $a < b$. Therefore, $f'(a) = f'(b) = 0$. Applying next the mean value theorem to f in $[a, b]$ there exists $x_0 \in (a, b)$ such that

$$f'(x_0) = \frac{f(b) - f(a)}{b - a} = 0.$$

Applying now the mean value theorem to f' in the intervals $[a, x_0]$ and $[x_0, b]$, there exists $x_1 \in (a, x_0)$ and $x_2 \in (x_0, b)$ such that

$$f''(x_1) = \frac{f(x_0) - f(a)}{x_0 - a} = 0$$

and

$$f''(x_2) = \frac{f(b) - f(x_0)}{b - x_0} = 0.$$

Finally, since f'' satisfies the conditions that allow us to apply the mean value theorem in $[x_1, x_2]$, there exists $c \in (x_1, x_2) \subset (0, 1)$ such that

$$f'''(c) = \frac{f(x_2) - f(x_1)}{x_2 - x_1} = 0.$$

□

Exercise 5.13. *Prove Peano's generalized mean value theorem: Let $f, g, h : \mathbb{R} \to \mathbb{R}$ be continuous in $[a, b]$ and differentiable in (a, b). Then there exists $c \in (a, b)$ such that*

$$\begin{vmatrix} f'(c) & g'(c) & h'(c) \\ f(a) & g(a) & h(a) \\ f(b) & g(b) & h(b) \end{vmatrix} = 0.$$

Prove also that Cauchy's generalized mean value theorem is a particular case of Peano's generalized mean value theorem.

Solution to Exercise 5.13.

Consider the function

$$F(x) = \begin{vmatrix} f(x) & g(x) & h(x) \\ f(a) & g(a) & h(a) \\ f(b) & g(b) & h(b) \end{vmatrix}$$

for $x \in [a, b]$. Then F is clearly continuous in $[a, b]$ and differentiable in (a, b). Also it is straightforward to show that, for every $x \in (a, b)$, we have

$$F'(x) = \begin{vmatrix} f'(x) & g'(x) & h'(x) \\ f(a) & g(a) & h(a) \\ f(b) & g(b) & h(b) \end{vmatrix}.$$

Since $F(a) = 0$ and $F(b) = 0$, applying Rolle's theorem there exists $c \in (a, b)$ such that $F'(c) = 0$. Therefore, the Peano's generalized mean value theorem holds.

Now putting $h(x) \equiv 1$ in the Peano's generalized mean value theorem there exists $c \in (a, b)$ such that

$$\begin{aligned} 0 &= \begin{vmatrix} f'(c) & g'(c) & 0 \\ f(a) & g(a) & 1 \\ f(b) & g(b) & 1 \end{vmatrix} \\ &= f'(c)g(a) + g'(c)f(b) - g'(c)f(a) - f'(c)g(b) \\ &= f'(c)[g(a) - g(b)] + g'(c)[f(b) - f(a)], \end{aligned}$$

proving Cauchy's generalized mean value theorem.

$\square$

Exercise 5.14. *Study the following limits and calculate them if possible:*

(I) $\displaystyle \lim_{x \to \infty} \frac{\log x}{x^\varepsilon}, \ \varepsilon > 0,$

(II) $\displaystyle \lim_{x \to 0^+} x^a \log x, \ a > 0,$

(III) $\displaystyle \lim_{x \to 0} \left(\frac{1}{x^2} - \frac{\cot x}{x} \right),$

(IV) $\displaystyle \lim_{x \to 1} (2 - x)^{\tan\left(\frac{\pi x}{2}\right)},$

(V) $\displaystyle \lim_{x \to 0^+} (\log \cot x)^{\tan x},$

(VI) $\displaystyle \lim_{x \to 1} x^{\frac{1}{1-x}},$

(VII) $\displaystyle \lim_{x \to 0} \left(\frac{1}{\log(x + \sqrt{1 + x^2})} - \frac{1}{\log(1 + x)} \right),$

(VIII) $\displaystyle \lim_{x \to 0} \left(\cot x - \frac{1}{x} \right),$

(IX) $\displaystyle \lim_{x \to 0} \log(1 + \sin^2 x) \cot \log^2(1 + x),$

(X) $\displaystyle \lim_{x \to 0} e^{\frac{1}{1 - \cos x}} \sin x,$

(XI) $\displaystyle \lim_{x \to \infty} \frac{2 + 2x + \sin 2x}{(2x + \sin 2x)e^{\sin x}},$

(XII) $\displaystyle \lim_{x \to 0^+} x^{\frac{1}{\log(e^x - 1)}},$

(XIII) $\displaystyle \lim_{x \to 0} \left(\frac{1}{x} - \frac{1}{e^x - 1} \right),$

(XIV) $\displaystyle \lim_{x \to 0^+} x^{\sin x} \left(\frac{1}{x^2} - \frac{\cosh x}{x \sinh x} \right),$

(XV) $\displaystyle \lim_{x \to 1} \left(\frac{1}{\log x} - \frac{1}{x - 1} \right),$

(XVI) $\displaystyle \lim_{x \to 0} \frac{x(e^x + 1) - 2(e^x - 1)}{x^3},$

(XVII) $\displaystyle \lim_{x \to 0} \frac{\sin 3x^2}{\log \cos(2x^2 - x)},$

(XVIII) $\displaystyle \lim_{x \to 0} \frac{e - (1 + x)^{\frac{1}{x}}}{x}.$

Solution to Exercise 5.14.

(I). If $\varepsilon > 0$, using L'Hôpital's Rule,

$$\lim_{x \to \infty} \frac{\log x}{x^\varepsilon} = \lim_{x \to \infty} \frac{\frac{1}{x}}{\varepsilon x^{\varepsilon - 1}} = \lim_{x \to \infty} \frac{1}{\varepsilon x^\varepsilon} = 0.$$

(II). If $a > 0$, using L'Hôpital's Rule,

$$\lim_{x \to 0^+} x^a \log x = \lim_{x \to 0^+} \frac{\log x}{x^{-a}} = \lim_{x \to 0^+} \frac{\frac{1}{x}}{-a x^{a-1}} = \lim_{x \to 0^+} \frac{1}{-a x^a} = -\infty.$$

(III). Using L'Hôpital's Rule repeatedly, we have

$$\lim_{x \to 0} \left(\frac{1}{x^2} - \frac{\cot x}{x} \right) = \lim_{x \to 0} \left(\frac{1}{x^2} - \frac{\cos x}{x \sin x} \right)$$

$$= \lim_{x \to 0} \frac{\sin x - x \cos x}{x^2 \sin x}$$

$$= \lim_{x \to 0} \frac{\cos x - \cos x + x \sin x}{2x \operatorname{sen} x + x^2 \cos x}$$

$$= \lim_{x \to 0} \frac{\sin x}{2 \sin x + x \cos x}$$

$$= \lim_{x \to 0} \frac{\cos x}{3 \cos x - x \sin x} = \frac{1}{3}.$$

(IV).

$$\lim_{x \to 1} (2 - x)^{\tan\left(\frac{\pi x}{2}\right)} = \lim_{x \to 1} e^{\log\left[(2-x)^{\tan\left(\frac{\pi x}{2}\right)} \right]}$$

$$= e^{\lim_{x \to 1} \tan\left(\frac{\pi x}{2}\right) \log\left[1 + (1 - x) \right]}$$

$$= e^{\lim_{x \to 1} (1 - x) \tan\left(\frac{\pi x}{2}\right)}.$$

Observe that in the last equality we can use the equivalence $\log[1 + (1 - x)] \sim 1 - x$ as $x \to 1$. Let us calculate the limit of the exponent separately. We shall make use of L'Hôpital's Rule repeatedly:

$$\lim_{x \to 1} (1 - x) \tan\left(\frac{\pi x}{2}\right) = \lim_{x \to 1} \frac{\tan\left(\frac{\pi x}{2}\right)}{(1 - x)^{-1}}$$

$$= \lim_{x \to 1} \frac{\frac{\pi}{2} \frac{1}{\cos^2\left(\frac{\pi x}{2}\right)}}{(1 - x)^{-2}} = \lim_{x \to 1} \frac{\pi (1 - x)^2}{2 \cos^2\left(\frac{\pi x}{2}\right)}$$

$$= \lim_{x \to 1} \frac{-2\pi (1 - x)}{2 \cdot 2 \cos\left(\frac{\pi x}{2}\right) \left(- \sin\left(\frac{\pi x}{2}\right) \right) \frac{\pi}{2}}$$

$$= \lim_{x \to 1} \frac{1 - x}{\cos\left(\frac{\pi x}{2}\right) \sin\left(\frac{\pi x}{2}\right)}$$

$$= \lim_{x \to 1} \frac{1}{\frac{\pi}{2} \sin^2\left(\frac{\pi x}{2}\right) - \frac{\pi}{2} \cos^2\left(\frac{\pi x}{2}\right)} = \frac{2}{\pi}.$$

Then, $\lim_{x \to 1} (2 - x)^{\tan\left(\frac{\pi x}{2}\right)} = e^{\frac{2}{\pi}}$.

(v). First, notice that $\tan x \to 0$ as $x \to 0^+$. Moreover, if $x \in (0, \pi/2)$ then $\tan x > 0$. Hence, considering the change of variable $y = \tan x$,

$$\lim_{x \to 0^+} (\log \cot x)^{\tan x} = \lim_{x \to 0} \left[\log \left(\frac{1}{\tan x} \right) \right]^{\tan x} = \lim_{y \to 0^+} \left[\log \left(y^{-1} \right) \right]^y$$

$$= \lim_{y \to 0^+} e^{\log[\log(y^{-1})]^y} = e^{\lim_{y \to 0^+} y \log [\log (y^{-1})]}.$$

Now we calculate the limit of the exponent separately, using L'Hôpital's Rule:

$$\lim_{y \to 0^+} y \log \left[\log \left(y^{-1} \right) \right] = \lim_{y \to 0^+} \frac{\log \left[\log \left(y^{-1} \right) \right]}{y^{-1}}$$

$$= \lim_{y \to 0^+} \frac{\frac{-y^{-2}}{y^{-1}}}{\frac{\log(y^{-1})}{-y^{-2}}} = \lim_{y \to 0^+} \frac{y}{-\log y} = 0.$$

Therefore, $\lim\limits_{x \to 0^+} (\log \cot x)^{\tan x} = e^0 = 1$.

(VI). Assuming that $\lim\limits_{x \to 1} x^{\frac{1}{1-x}} = L$, we have (using L'Hôpital's Rule) that

$$\log L = \lim_{x \to 1} \log x^{\frac{1}{1-x}} = \lim_{x \to 1} \frac{\log x}{1 - x} = \lim_{x \to 1} \frac{1}{-x} = -1.$$

Then, as $\log L = -1$ it follows that $L = e^{-1}$. Hence $\lim\limits_{x \to 1} x^{\frac{1}{1-x}} = e^{-1}$.

(VII). To calculate this limit we will use several times de equivalence $\log(f(x)) \sim f(x) - 1$ as $x \to a$ whenever $\lim\limits_{x \to a} f(x) = 1$. Let us put $L = \lim\limits_{x \to 0} \left(\dfrac{1}{\log(x + \sqrt{1 + x^2})} - \dfrac{1}{\log(1 + x)} \right)$ for short. If $x > -1$ and $x \neq 0$ we have

$$\frac{1}{\log(x + \sqrt{1 + x^2})} - \frac{1}{\log(1 + x)} = \frac{\log(1 + x) - \log(x + \sqrt{1 + x^2})}{\log(x + \sqrt{1 + x^2}) \cdot \log(1 + x)}$$

$$= \frac{\log \frac{1+x}{x + \sqrt{1+x^2}}}{\log(x + \sqrt{1 + x^2}) \cdot \log(1 + x)}.$$

Therefore, since

$$\log \frac{1 + x}{x + \sqrt{1 + x^2}} \sim \frac{1 + x}{x + \sqrt{1 + x^2}} - 1, \quad \text{as } x \to 0,$$

$$\log(x + \sqrt{1 + x^2}) \sim x + \sqrt{1 + x^2} - 1, \quad \text{as } x \to 0 \text{ and,}$$

$$\log(1 + x) \sim x, \quad \text{as } x \to 0,$$

then

$$L = \lim_{x \to 0} \frac{\frac{1+x}{x+\sqrt{1+x^2}} - 1}{x\left(x + \sqrt{1+x^2} - 1\right)} = \lim_{x \to 0} \frac{1 - \sqrt{1+x^2}}{x\left(x + \sqrt{1+x^2}\right)\left(x + \sqrt{1+x^2} - 1\right)}$$

$$= \lim_{x \to 0} \frac{-x^2}{x\left(x + \sqrt{1+x^2}\right)\left(1 + \sqrt{1+x^2}\right)\left(x + \sqrt{1+x^2} - 1\right)}$$

$$= \lim_{x \to 0} \frac{-x\left(x + \sqrt{1+x^2} + 1\right)}{\left(x + \sqrt{1+x^2}\right)\left(1 + \sqrt{1+x^2}\right)\left[(x + \sqrt{1+x^2})^2 - 1\right]}$$

$$= \lim_{x \to 0} \frac{-x\left(x + \sqrt{1+x^2} + 1\right)}{\left(x + \sqrt{1+x^2}\right)\left(1 + \sqrt{1+x^2}\right)\left(2x^2 + 2x\sqrt{1+x^2}\right)}$$

$$= \lim_{x \to 0} \frac{-\left(x + \sqrt{1+x^2} + 1\right)}{\left(x + \sqrt{1+x^2}\right)\left(1 + \sqrt{1+x^2}\right)\left(2x + 2\sqrt{1+x^2}\right)} = -\frac{1}{2}.$$

(VIII). Using L'Hôpital's Rule twice, we have

$$\lim_{x \to 0} \left(\cot x - \frac{1}{x}\right) = \lim_{x \to 0} \frac{x - \tan x}{x \tan x}$$

$$= \lim_{x \to 0} \frac{\tan^2 x}{(1 + \tan^2 x)x + \tan x}$$

$$= \lim_{x \to 0} \frac{2 \tan x (1 + \tan^2 x)}{1 + \tan^2 x + 2 \tan x (1 + \tan^2 x)x + 1 + \tan^2 x}$$

$$= 0.$$

(IX). First, define $g(x) = (\sin^2 x + 1)(1 + \tan^2(\log^2(x+1)))$. Using the fact that $g(0) = 1$ and $g'(0) = 0$ we have, using L'Hôpital's Rule, that

$$\lim_{x \to 0} \log(1 + \sin^2 x) \cot \log^2(1 + x) = \lim_{x \to 0} \frac{\log(1 + \sin^2 x)}{\tan \log^2(1 + x)}$$

$$= \lim_{x \to 0} \frac{2 \sin x \cos x \frac{1}{1 + \sin^2 x}}{(1 + \tan^2(\log^2(x + 1)))2 \log(x + 1)\frac{1}{x+1}}$$

$$= \lim_{x \to 0} \frac{(x + 1) \sin x \cos x}{(\sin^2 x + 1)(1 + \tan^2(\log^2(x + 1))) \log(x + 1)}$$

$$= \lim_{x \to 0} \frac{\sin x \cos x - (x + 1) \sin^2 x + (x + 1) \cos^2 x}{g'(x) \log(x + 1) + \frac{1}{x+1} g(x)} = 1.$$

Alternatively, we can use the equivalences

$$\log(1 + \sin^2 x) \sim \sin^2 x \sim x^2, \quad \text{as } x \to 0 \text{ and}$$

$$\tan \log^2(1 + x) \sim \log^2(1 + x) \sim x^2, \quad \text{as } x \to 0$$

to prove straightforwardly that the limit is 1.

(x). Using L'Hôpital's Rule, we have

$$\lim_{x \to 0+} e^{\frac{1}{1-\cos x}} \sin x = \lim_{x \to 0+} \frac{\sin x}{e^{\frac{1}{\cos x - 1}}}$$

$$= \lim_{x \to 0+} \frac{(\cos x - 1)^2 \cos x}{(\sin x) e^{\frac{1}{\cos x - 1}}} = \infty.$$

Similarly $\lim_{x \to 0-} e^{\frac{1}{1-\cos x}} \sin x = -\infty$.

(xi). Let $g(x) = \frac{2+2x+\sin 2x}{(2x+\sin 2x)}$, $h(x) = e^{-\sin x}$ and $f(x) = g(x)h(x)$. Since

$$\lim_{x \to \infty} g(x) = \lim_{x \to \infty} \frac{2 + 2x + \sin 2x}{(2x + \sin 2x)} = \lim_{x \to \infty} \frac{2 + \frac{2}{x} + \frac{\sin 2x}{x}}{2 + \frac{\sin 2x}{x}} = 1 \neq 0,$$

if $\lim_{x \to \infty} f(x)$ existed, then $\lim_{x \to \infty} h(x)$ would exist as well and

$$\lim_{x \to \infty} h(x) = \frac{\lim_{x \to \infty} f(x)}{\lim_{x \to \infty} g(x)} = \lim_{x \to \infty} f(x).$$

However $\lim_{x \to \infty} h(x)$ does not exist. To see this consider the divergent sequences (x_n) and (y_n) with $x_n = \left(2n + \frac{1}{2}\right)\pi$ and $y_n = \left(2n + \frac{3}{2}\right)\pi$ for every $n \in \mathbb{N}$. Then

$$\lim_n h(x_n) = \frac{1}{e} \neq e = \lim_n h(y_n).$$

(xii). Let us write $L = \lim_{x \to 0+} x^{\frac{1}{\log(e^x - 1)}}$. Then (using L'Hôpital's Rule twice):

$$\log L = \lim_{x \to 0+} \frac{1}{\log(e^x - 1)} \log x$$

$$= \lim_{x \to 0+} \frac{e^x - 1}{xe^x}$$

$$= \lim_{x \to 0+} \frac{e^x}{e^x + xe^x} = 1.$$

Hence $L = e$.

(xiii). By L'Hôpital's Rule (applied twice), we have

$$\lim_{x \to 0} \left(\frac{1}{x} - \frac{1}{e^x - 1}\right) = \lim_{x \to 0} \frac{e^x - 1 - x}{x(e^x - 1)}$$

$$= \lim_{x \to 0} \frac{e^x - 1}{e^x - 1 + xe^x}$$

$$= \lim_{x \to 0} \frac{e^x}{2e^x + xe^x} = \frac{1}{2}.$$

(XIV). We calculate first $\lim\limits_{x\to 0^+} x^{\sin x}$, for which the equivalence $\sin x \sim x$ as $x \to 0$ (and L'Hôpital's Rule) shall be useful:

$$\lim_{x\to 0^+} x^{\sin x} = e^{\lim\limits_{x\to 0^+}\log\left(x^{\sin x}\right)} = e^{\lim\limits_{x\to 0^+}\sin x\log x} = e^{\lim\limits_{x\to 0^+} x\log x}.$$

Since

$$\lim_{x\to 0^+} x\log x = \lim_{x\to 0^+} \frac{\log x}{\frac{1}{x}} = \lim_{x\to 0^+} \frac{\frac{1}{x}}{-\frac{1}{x^2}} = \lim_{x\to 0^+} (-x) = 0,$$

we have $\lim\limits_{x\to 0^+} x^{\sin x} = 1$. Hence, by L'Hôpital's Rule, we have

$$\lim_{x\to 0^+} x^{\sin x}\left(\frac{1}{x^2} - \frac{\cosh x}{x\sinh x}\right) = \lim_{x\to 0^+}\left(\frac{1}{x^2} - \frac{\cosh x}{x\sinh x}\right)$$

$$= \lim_{x\to 0^+}\left(\frac{1}{x^2} - \frac{e^x + e^{-x}}{x(e^x - e^{-x})}\right)$$

$$= \lim_{x\to 0^+}\frac{e^x - e^{-x} - x(e^x + e^{-x})}{x^2(e^x - e^{-x})}$$

$$= \lim_{x\to 0^+}\frac{e^x + e^{-x} - (e^x + e^{-x}) - x(e^x - e^{-x})}{2x(e^x - e^{-x}) + x^2(e^x + e^{-x})}$$

$$= \lim_{x\to 0^+}\frac{-e^x + e^{-x}}{2(e^x - e^{-x}) + x(e^x + e^{-x})}$$

$$= \lim_{x\to 0^+}\frac{-e^x - e^{-x}}{3(e^x + e^{-x}) + x(e^x - e^{-x})} = -\frac{1}{3}.$$

(XV). Using L'Hôpital's Rule:

$$\lim_{x\to 1}\left(\frac{1}{\log x} - \frac{1}{x-1}\right) = \lim_{x\to 1}\frac{x - 1 - \log x}{(x-1)\log x} = \lim_{x\to 1}\frac{1 - \frac{1}{x}}{\frac{1}{x}(x-1) + \log x}$$

$$= \lim_{x\to 1}\frac{x-1}{x - 1 + \log x} = \lim_{x\to 1}\frac{1}{1 + \frac{1}{x}} = \frac{1}{2}.$$

(XVI). Using L'Hôpital's Rule twice we have that:

$$\lim_{x\to 0}\frac{x(e^x + 1) - 2(e^x - 1)}{x^3} = \lim_{x\to 0}\frac{e^x + 1 + xe^x - 2e^x}{3x^2}$$

$$= \lim_{x\to 0}\frac{-e^x + e^x + xe^x}{6x} = \lim_{x\to 0}\frac{e^x}{6} = \frac{1}{6}.$$

(XVII). Again, with the help of L'Hôpital's Rule, we have

$$\lim_{x \to 0} \frac{\sin(3x^2)}{\log \cos(2x^2 - x)} = \lim_{x \to 0} \frac{6x \cos(3x^2)}{\frac{(-\sin(2x^2 - x))(4x - 1)}{\cos(2x^2 - x)}}$$

$$= \lim_{x \to 0} \frac{6x \cos(3x^2) \cos(2x^2 - x)}{-(4x - 1) \sin(2x^2 - x)} = \lim_{x \to 0} \frac{6x}{-(4x - 1) \sin(2x^2 - x)}$$

$$= \lim_{x \to 0} \frac{6}{-(4x - 1)^2 \cos(2x^2 - x) - 4 \sin(2x^2 - x)} = -6.$$

Observe that in the third equality we have used the fact that

$$\lim_{x \to 0} \cos(3x^2) \cos(2x^2 - x) = 1.$$

(XVIII). If $f(x) = (1 + x)^{\frac{1}{x}}$ then $\log(f(x)) = \frac{1}{x} \log(1 + x)$. Differentiating this identity we obtain

$$\frac{1}{f(x)} f'(x) = -\frac{1}{x^2} \log(1 + x) + \frac{1}{x} \left(\frac{1}{1 + x} \right) = \frac{\frac{1}{1+x} x - \log(1 + x)}{x^2},$$

from which

$$f'(x) = f(x) \frac{\frac{1}{1+x} x - \log(1 + x)}{x^2}.$$

Notice that $\lim_{x \to 0} f(x) = e$. Therefore, and using L'Hôpital's Rule repeatedly,

$$\lim_{x \to 0} \frac{e - (1 + x)^{\frac{1}{x}}}{x} = \lim_{x \to 0} [-f'(x)]$$

$$= \lim_{x \to 0} f(x) \frac{-\frac{1}{1+x} x + \log(1 + x)}{x^2}$$

$$= \lim_{x \to 0} f(x) \lim_{x \to 0} \frac{-\frac{1}{(1+x)^2} + \frac{1}{1+x}}{2x}$$

$$= \lim_{x \to 0} f(x) \lim_{x \to 0} \frac{1}{2(1 + x)^2} = \frac{e}{2}.$$

□

Exercise 5.15. *Find the intervals where the following functions are monotone:*

(I) $f(x) = 4x^3 - 21x^2 + 18x + 20$, (IV) $f(x) = \cos x - x$,

(II) $f(x) = \sin x + \cos x$, $x \in [0, 2\pi]$,

(V) $f(x) = 3x^4 - 4x^3 - 12x^2 + 12$,

(III) $f(x) = \dfrac{2x}{1 + x^2}$,

(VI) $f(x) = x^2 e^{-x}$.

Solution to Exercise 5.15.

(I). Observe that $f'(x) = 12x^2 - 42x + 18 = 6(2x - 1)(x - 3)$. Hence f' vanishes at $x = 3$ and $\frac{1}{2}$, which determines the sign of f' as pointed out in the following table.

$f'(x)$	$-\infty$	$+$	$\frac{1}{2}$	$-$	3	$+$	$+\infty$
$f(x)$		↗		↘		↗	

Clearly, f is increasing in $\left(-\infty, \frac{1}{2}\right) \cup (3, \infty)$ and decreasing in $\left(\frac{1}{2}, 3\right)$.

(II). $f(x) = \sin x + \cos x$, $x \in [0, 2\pi]$. The critical points of f satisfy the equation $f'(x) = \cos x - \sin x = 0$, or equivalently $\sin x = \cos x$, from which $x = \frac{\pi}{4} + \pi k$ for $k \in \mathbb{N}$. Since the domain of f is $[0, 2\pi]$, we have just two critical points at $x = \frac{\pi}{4}$ and $x = \frac{5\pi}{4}$. It is straightforward to check that the sign of f' is given by the following table.

$f'(x)$	0	$+$	$\frac{\pi}{4}$	$-$	$\frac{5\pi}{4}$	$+$	2π
$f(x)$		↗		↘		↗	

Therefore, f is increasing in $\left(0, \frac{\pi}{4}\right) \cup \left(\frac{5\pi}{4}, 2\pi\right)$ and decreasing in $\left(\frac{\pi}{4}, \frac{5\pi}{4}\right)$.

(III). Observe that if $f(x) = \frac{2x}{1+x^2}$ then

$$f'(x) = \frac{2(1 + x^2) - 4x^2}{(1 + x^2)^2} = \frac{2 - 2x^2}{(1 + x^2)^2}.$$

It is easy to see that the roots of $f'(x)$ are $x = \pm 1$. Then the sign of f' can be determined and we do it in the following table.

$f'(x)$	$-\infty$	$-$	-1	$+$	1	$-$	$+\infty$
$f(x)$		↘		↗		↘	

Clearly, f is increasing in $(-1, 1)$ and decreasing in $(-\infty, -1) \cup (1, \infty)$.

(IV). If $f(x) = \cos x - x$, then $f'(x) = -\sin x - 1 \leq 0 \ \forall x \in \mathbb{R}$. Clearly f is decreasing for all $\mathbb{R}$.

(V). Here $f(x) = f(x) = 3x^4 - 4x^3 - 12x^2 + 12$. Then $f'(x) = 12x^3 - 12x^2 - 24x = 12x(x^2 - x - 2)$. Thus f' vanishes only at $x = -1, 0, 2$. According to this, the sign of f' is easy to determine as we do in the following table.

$f'(x)$	$-\infty$	$-$	-1	$+$	0	$-$	2	$+$	$+\infty$
$f(x)$		↘		↗		↘		↗	

Then it is straightforward that f is increasing in $(-1, 0) \cup (2, \infty)$ and decreasing in $(-\infty, -1) \cup (0, 2)$.

(VI). Now $f(x) = x^2 e^{-x}$. Then $f'(x) = 2xe^{-x} + x^2 e^{-x}(-1) = e^{-x}x(2-x) = 0$. The only roots of f' are $x = 0$ and $x = 2$. The sign of f' is given in the following table.

$f'(x)$	$-\infty$	$-$	0	$+$	2	$-$	$+\infty$
$f(x)$		$\searrow$		$\nearrow$		$\searrow$	

Therefore, f is increasing in $(0,2)$ and decreasing in $(-\infty,0) \cup (2,\infty)$. $\qquad\square$

Exercise 5.16. *Find the relative extrema of the following functions:*

(I) $f(x) = 3\sqrt[3]{x^2} - x^2$,

(II) $f(x) = (x-1)^{\frac{2}{3}} + (x+1)^{\frac{2}{3}}$,

(III) $f(x) = 2x^3 - 15x^2 - 82x + 8$,

(IV) $f(x) = 2\sin x + \cos 2x$,

(V) $f(x) = \log\left[1 + (x^2 - x - 2)^2\right]$,

(VI) $f(x) = e^x \sin x$, $x \in [-2,2]$.

Solution to Exercise 5.16.

(I). Here $f(x) = 3\sqrt[3]{x^2} - x^2$. The critical points of f are the solutions of the equation $f'(x) = 2x^{-\frac{1}{3}} - 2x = 0$, or equivalently, $x^4 = 1$. Then f' vanishes at $x = \pm 1$. The critical points determine the sign of f' as shown in the following table.

$f'(x)$	$-\infty$	$+$	-1	$-$	0	$+$	1	$-$	$+\infty$
$f(x)$		$\nearrow$		$\searrow$		$\nearrow$		$\searrow$	

We see clearly that, f has relative minimum at $x = 0$ and relative maximums at $x = 1$ and $x = -1$.

(II). Now $f(x) = \sqrt[3]{(x-1)^2} + \sqrt[3]{(x+1)^2}$. Then

$$f'(x) = \frac{2}{3}\left[\frac{1}{(x-1)^{\frac{1}{3}}} + \frac{1}{(x+1)^{\frac{1}{3}}}\right]$$

for all $x \neq \pm 1$. To obtain the critical points of f we solve the equation $f'(x) = 0$ or equivalently $-\frac{1}{x-1} = \frac{1}{x+1}$, from which $2x = 0$. Thus $x = 0$ is the only critical point. The latter shows that the sign of f' is constant in each of the intervals $(-\infty,-1)$, $(-1,0)$, $(0,1)$ and $(1,\infty)$. The sign of f' can be calculated by mere inspection and is shown in the following table.

$f'(x)$	$-\infty$	$-$	-1	$+$	0	$-$	1	$+$	$+\infty$
$f(x)$		$\searrow$		$\nearrow$		$\searrow$		$\nearrow$	

Then f has two relative minimums at $x = -1$ and $x = 1$ whereas it has a relative maximum at $x = 0$.

(III). Here $f(x) = 2x^3 - 15x^2 - 82x + 8$. Then $f'(x) = 6x^2 - 30x - 82 = 0$ has two solutions, namely $x = \dfrac{15 \pm \sqrt{717}}{6}$. The sign of f' can be easily calculated as shown in the following table.

$f'(x)$	$-\infty$	$+$	$\dfrac{15 - \sqrt{717}}{6}$	$-$	$\dfrac{15 + \sqrt{717}}{6}$	$+$	$+\infty$
$f(x)$		↗		↘		↗	

Clearly f has a relative minimum at $x = \dfrac{15 + \sqrt{717}}{6}$ and a relative maximum at $x = \dfrac{15 - \sqrt{717}}{6}$.

(IV). Now $f(x) = 2\sin x + \cos 2x = 2\sin x + 2\cos^2 - 1$. Then

$$f'(x) = 2\cos x - 4\cos x \sin x = 2\cos x(1 - 2\sin x).$$

The critical points of f satisfy $f'(x) = 0$ or, equivalently, $\cos x = 0$ or $\sin x = \frac{1}{2}$. Hence, the roots of f' are the points $x = \frac{\pi}{2} + \pi k$, $x = \frac{\pi}{6} + 2\pi k$ and $x = \frac{2\pi}{3} + 2\pi k$ with $k \in \mathbb{N}$. Since f is 2π-periodic, it suffices to study the sign of f' within the interval $[0, 2\pi]$, where f has four critical points, namely $\frac{\pi}{6}, \frac{\pi}{2}$, $\frac{2\pi}{3}$ and $\frac{3\pi}{2}$. The sign of f' can be easily calculated in $[0, 2\pi]$ as shown in the following table.

$f'(x)$	0	$+$	$\frac{\pi}{6}$	$-$	$\frac{\pi}{2}$	$+$	$\frac{2\pi}{3}$	$-$	$\frac{3\pi}{2}$	$+$	2π
$f(x)$		↗		↘		↗		↘		↗	

Therefore, f has relative minimums at $x = \frac{\pi}{2} + \pi k$ with $k \in \mathbb{N}$ and relative maximums at $x = \frac{\pi}{6} + 2\pi k$ and $x = \frac{2\pi}{3} + 2\pi k$ for $k \in \mathbb{N}$.

(V). Here $f(x) = \log\left[1 + (x^2 - x - 2)^2\right]$. We have that

$$f'(x) = \frac{2(2x - 1)(x^2 - x - 2)}{1 + (x^2 - x - 2)^2} = 0$$

yields $x = \frac{1}{2}$, $x = -1$ and $x = 2$. It can be easily seen that the sign of f' is as pointed out in the following table.

$f'(x)$	$-\infty$	$-$	-1	$+$	$\frac{1}{2}$	$-$	2	$+$	$+\infty$
$f(x)$		↘		↗		↘		↗	

It is straightforward that f has two relative minimums at $x = -1$ and $x = 2$, and a local maximum at $x = \frac{1}{2}$.

(VI). Now $f(x) = e^x \sin x$ with $x \in [-2, 2]$. The zeros of the derivative satisfy the equation $f'(x) = e^x(\sin x + \cos x) = 0$, which is equivalent to

$\sin x = -\cos x$. Then f' vanishes in $[-2,2]$ at the single point $x = -\frac{\pi}{4}$. The sign of f' behaves as follows.

$f'(x)$	-2	$-$	$-\frac{\pi}{4}$	$+$	2
$f(x)$		↘		↗	

Therefore, f has a relative minimum at $x = -\frac{\pi}{4}$.

$\square$

Exercise 5.17. *Find the absolute maximums and minimums, if they exist, in the following cases:*

(I) $f(x) = x^3 - x^2 - 8x + 1$, *in* $[-2,2]$,

(II) $f(x) = x^5 + x + 1$, *in* $[-1,1]$,

(III) $f(x) = \arcsin(1+x)$, *in its domain*,

(IV) $f(x) = \dfrac{1}{2x^4 - x + 1}$, *in* $(0,1]$, $[0,1]$ *and* $\mathbb{R}$,

(V) $f(x) = e^{-x^2}$, *in* $[-1,1]$, $(0,1)$ *and* $\mathbb{R}$,

(VI) $f(x) = x^2 \log x$, *in* $[e^{-1}, e]$ *and* $(0,\infty)$.

Solution to Exercise 5.17.

(I). We have to study the function $f(x) = x^3 - x^2 - 8x + 1$, in $[-2,2]$. As $f'(x) = 3x^2 - 2x - 8 = (3x+4)(x-2)$, the critical points of f are $x = -\frac{4}{3}, 2$. The sign of f' then behaves as shown in the following table.

$f'(x)$	$-\infty$	$+$	$\frac{-4}{3}$	$-$	2	$+$	$+\infty$
$f(x)$		↗		↘		↗	

Therefore, the function f has a local maximum at $x = -\frac{4}{3}$ and a local minimum at $x = 2$. Also, $f(2) = -11$, $f(-2) = 5$, $f\left(-\frac{4}{3}\right) = \frac{203}{27} > f(-2)$. Then, in $[-2,2]$, f attains its absolute minimum at $x = 2$ and its absolute maximum at $x = -\frac{4}{3}$.

(II). Now we consider $f(x) = x^5 + x + 1$, in $[-1,1]$. As $f'(x) = 5x^4 + 1 > 0$ for all $x \in \mathbb{R}$, f is strictly increasing. Then, in $[-1,1]$ f attains its absolute minimum at $x = -1$ and its absolute minimum at $x = 1$.

(III). Next we study $f(x) = \arcsin(1+x)$, in $[-2,0]$. Since $f'(x) = \dfrac{1}{\sqrt{1-(1+x)^2}} = \dfrac{1}{\sqrt{-x(x+2)}} > 0$ for every $x \in (-2,0)$, it turns out that f is strictly increasing in $[-2,0]$. Then f attains its absolute minimum in $[-2,0]$ at $x = -2$ and its absolute maximum at $x = 0$.

(IV). In this case we have to study the function $f(x) = \dfrac{1}{2x^4 - x + 1}$ in three different intervals, namely, $(0, 1]$, $[0, 1]$ and $\mathbb{R}$. The equation $f'(x) = \dfrac{1 - 8x^3}{(2x^4 - x + 1)^2} = 0$ is equivalent to $1 - 8x^3 = 0$ from which $x = \frac{1}{2}$. Then the sign of f' behaves as given in the following table.

$f'(x)$	$-\infty$	$+$	$\frac{1}{2}$	$-$	$+\infty$
$f(x)$		$\nearrow$		$\searrow$	

Clearly f attains its absolute maximum at $x = \frac{1}{2}$ in $\mathbb{R}$, and therefore in $(0, 1]$ and $[0, 1]$ as well. On the other hand, $f(0) = 1 > \frac{1}{2} = f(1)$. Hence f attains its absolute minimum at $x = 1$ in $(0, 1]$ and in $[0, 1]$. Finally, f does not attain its minimum in $\mathbb{R}$. To see this observe that $\lim\limits_{x \to \pm 1} f(x) = 0$ but $f(x) > 0$ for all $x \in \mathbb{R}$.

(V). Now we have to study $f(x) = e^{-x^2}$, in $[-1, 1]$, $(0, 1)$ and $\mathbb{R}$. Since $f'(x) = -2xe^{-x^2}$, f' vanishes only at $x = 0$. The behavior of the sign of f' is shown below.

$f'(x)$	$-\infty$	$+$	0	$-$	$+\infty$
$f(x)$		$\nearrow$		$\searrow$	

Since f is strictly decreasing in $(0, 1)$, f cannot attain neither its absolute maximum nor its absolute minimum in $(0, 1)$. Clearly f attains its absolute maximum at $x = 0$ in $\mathbb{R}$, and therefore in $[-1, 1]$ too. However f does not attain its minimum in $\mathbb{R}$. To see this observe that $\lim\limits_{x \to \pm 1} f(x) = 0$ but $f(x) > 0$ for all $x \in \mathbb{R}$. Finally, f is strictly increasing in $[-1, 0]$ and strictly decreasing in $[0, 1]$. Since, additionally, f is an even function, f attains its absolute minimum in $[-1, 1]$ at $x = -1$, and $x = 1$.

(VI). Finally, we consider the function $f(x) = x^2 \log x$, in $[e^{-1}, e]$ and $(0, \infty)$. The equation $f'(x) = x(2 \log x + 1) = 0$ has only one solution, namely $x = \frac{1}{\sqrt{e}}$. The sign of f' is shown in the table below.

$f'(x)$	0	$-$	e^{-1}	$-$	$\frac{1}{\sqrt{e}}$	$+$	e	$+$
$f(x)$		$\searrow$		$\searrow$		$\nearrow$		$\nearrow$

It is straightforward that f attains its absolute minimum in $(0, \infty)$ and $[e^{-1}, e]$ at $x = \frac{1}{\sqrt{e}}$. On the other hand, $f(e^{-1}) = -\frac{1}{e^2} < e^2 = f(e)$, so f attains its absolute maximum in $[e^{-1}, e]$ at $x = e$. Since $\lim\limits_{x \to \infty} f(x) = \infty$, f does not attain its maximum in $(0, \infty)$.

$\square$

Exercise 5.18. *Let*

$$f(x) = \begin{cases} x^4 \sin^2 \frac{1}{x} & \text{if } x \neq 0, \\ 0 & \text{if } x = 0. \end{cases}$$

(I) *Prove that f has a local minimum at 0.*

(II) *Prove that* $f'(0) = f''(0) = 0$, *and that* $f'''(0)$ *does not exist.*

Solution to Exercise 5.18.

(I). As a matter of fact, f attains its absolute minimum at $x = 0$ since $f(x) \geq 0$ and $f(0) = 0$.

(II). First, we show that f is differentiable at $x = 0$. Notice that $\sin^2 \frac{1}{x}$, for $x \neq 0$, is a bounded function whereas $\lim_{x \to 0} x^3 = 0$. Then

$$f'(0) = \lim_{x \to 0} \frac{f(x) - f(0)}{x - 0} = \lim_{x \to 0} \frac{f(x)}{x} = \lim_{x \to 0} x^3 \sin^2 \frac{1}{x} = 0.$$

Also, if $x \neq 0$ then

$$f'(x) = 4x^3 \sin^2 \frac{1}{x} - 2x^2 \sin \frac{1}{x} \cos \frac{1}{x} = 4x^3 \sin^2 \frac{1}{x} - x^2 \sin \frac{2}{x}.$$

As for the second derivative at $x = 0$ we have

$$f''(0) = \lim_{x \to 0} \frac{f'(x) - f'(0)}{x - 0} = \lim_{x \to 0} \frac{f'(x)}{x}$$

$$= \lim_{x \to 0} \left(4x^2 \sin^2 \frac{1}{x} - x \sin \frac{2}{x} \right) = 0.$$

Notice that in the last equality we have used that the functions $\sin^2 \frac{1}{x}$ and $\sin \frac{2}{x}$, for $x \neq 0$, are both bounded, whereas $\lim_{x \to 0} x^2 = \lim_{x \to 0} x = 0$. If $x \neq 0$ then

$$f''(x) = 12x^2 \sin^2 \frac{1}{x} - 8x \sin \frac{1}{x} \cos \frac{1}{x} - 2x \sin \frac{2}{x} + 2 \cos \frac{2}{x}$$

$$= 12x^2 \sin^2 \frac{1}{x} - 6x \sin \frac{2}{x} + 2 \cos \frac{2}{x}.$$

Finally, we prove that $f'''(0)$ does not exist. If $x \neq 0$ then

$$\frac{f''(x) - f''(0)}{x - 0} = \frac{f''(x)}{x} = 12x \sin^2 \frac{1}{x} - 6 \sin \frac{2}{x} + \frac{2}{x} \cos \frac{2}{x} = 12x \sin^2 \frac{1}{x} + g(x),$$

where $g(x) = -6 \sin \frac{2}{x} + \frac{2}{x} \cos \frac{2}{x}$. Since $\lim_{x \to 0} 12x \sin^2 \frac{1}{x} = 0$, the existence of $f'''(0)$ would imply that $\lim_{x \to 0} g(x) = f'''(0)$. However $\lim_{x \to 0} g(x)$ does not exist.

To see this consider the sequence (x_n) with $x_n = \frac{2}{(n+\frac{1}{2})\pi}$ for all $n \in \mathbb{N}$. Then $\lim_n x_n = 0$ but $g(x_n) = -6(-1)^n$, which does not converge.

$\square$

> **Exercise 5.19.** *Let $f(x) = ax - \frac{x^3}{1+x^2}$. Prove that f is increasing in $\mathbb{R}$ if and only if $a \geq \frac{9}{8}$.*

Solution to Exercise 5.19.

The function f is increasing in $\mathbb{R}$ if and only if $f'(x) \geq 0$ for all $x \in \mathbb{R}$. First, we obtain the derivative of f:

$$f'(x) = a - \frac{3x^2(1+x^2) - x^3(2x)}{(1+x^2)^2} = a - \frac{3x^2 + 3x^4 - 2x^4}{(1+x^2)^2}$$

$$= \frac{a(1+x^2)^2 - 3x^2 - x^4}{(1+x^2)^2} = \frac{(a-1)x^4 + (2a-3)x^2 + a}{(1+x^2)^2}.$$

The denominator of f' is always positive whereas in the numerator we have the polynomial $q(x) = (a-1)x^4 + (2a-3)x^2 + a$. Assume first that $a \geq \frac{9}{8}$ then

$$(a-1)x^4 + (2a-3)x^2 + a \geq (\frac{9}{8} - 1)x^4 + (\frac{9}{8} - 3)x^2 + \frac{9}{8}$$

$$= \frac{x^4 - 6x^2 + 9}{8} = \frac{(x^2-3)^2}{8} \geq 0$$

for all $x \in \mathbb{R}$. Therefore, f is increasing in $\mathbb{R}$ if $a \geq \frac{9}{8}$.

On the other hand, if f is increasing in $\mathbb{R}$ then $f'(x) \geq 0$ for all $x \in \mathbb{R}$. Since the denominator of f' is positive, then $q(x) \geq 0$ for all $x \in \mathbb{R}$. Observe that a necessary condition for the biquadratic equation $q(x) = 0$ to have a real root is that its discriminant

$$\Delta = (2a-3)^2 - 4a(a-1) = 9 - 8a$$

has to be non-negative. This is the reason behind the fact that $a \geq \frac{9}{8}$. However the formal proof of the latter should be worked out carefully.

We are going to show that if $a < \frac{9}{8}$ then q is negative in some interval and therefore we must have $a \geq \frac{9}{8}$. We decompose the proof in 4 different cases:

- If $a < 0$ then $q(0) = a < 0$ then q is negative in some neighborhood of 0.

- If $a = 0$ then $q(x) = -x^4 - 3x^2 = -x^2(x+3) < 0$ for $x \in (-\infty, -3)$.

- If $a = 1$ then $q(x) = -x^2 + 1 < 0$ for $x \in (-1, 1)$.

- Finally, if $0 < a < \frac{9}{8}$ and $a \neq 1$ then every solution λ of $q(x) = 0$ must satisfy either

$$\lambda^2 = \frac{-2a + 3 - \sqrt{(2a-3)^2 - 4a(a-1)}}{2(a-1)} = \frac{-2a - 3 - \sqrt{9 - 8a}}{2(a-1)}$$

or

$$\lambda^2 = \frac{-2a + 3 + \sqrt{(2a-3)^2 + 4a(a-1)}}{2(a-1)} = \frac{-2a + 3 + \sqrt{9 - 8a}}{2(a-1)}.$$

Since

$$\frac{-2a + 3 - \sqrt{9 - 8a}}{2(a-1)} = \frac{(-2a+3)^2 - (9-8a)}{2(a-1)(-2a+3+\sqrt{9-8a})}$$

$$= \frac{2a}{-2a + 3 + \sqrt{9 - 8a}} > 0$$

whenever $a \in (0, 9/8) \setminus \{1\}$, we have that

$$\lambda = \sqrt{\frac{-2a + 3 - \sqrt{9 - 8a}}{2(a-1)}}$$

is a well-defined positive real number that satisfies $q(\lambda) = 0$. Also, $q'(\lambda) = -2\lambda\sqrt{9 - 8a} < 0$, which means that q is strictly decreasing in a neighborhood of λ, namely $(\lambda - \delta, \lambda + \delta)$ with $\delta > 0$. In particular, $q(x) < q(\lambda) = 0$ for $x \in (\lambda, \lambda + \delta)$.

□

Exercise 5.20. *Which number is greater, e^π or π^e? Prove that if $x > e$, then $e^x > x^e$.*

Solution to Exercise 5.20.

Let us define the function $f(x) = \frac{x}{\log x} - e$ with $x \in (1, \infty)$. The function f is continuous in $(1, \infty)$ and $f'(x) = \frac{\log x - 1}{\log^2 x}$. Hence $f'(x) = 0$ only at $x = e$. The sign of f' behaves as pointed out in the table.

$f'(x)$	1	$-$	e	$+$	$+\infty$
$f(x)$		$\searrow$		$\nearrow$	

We deduce that f has a minimum at $x = e$. Therefore, if $x > e$

$$\frac{x}{\log x} - e > f(e) = 0 \Leftrightarrow x > \log x^e \Leftrightarrow e^x > x^e.$$

Since $\pi > e$, we have that $e^\pi > \pi^e$.

□

> **Exercise 5.21.** *Write $x^4 + x^3 - 3x^2 + 4x - 4$ as a sum of powers of $x - 1$. Write $x^4 - 11x^3 + 43x^2 - 60x + 14$ as a sum of powers of $x - 3$.*

Solution to Exercise 5.21.

Let us write first $f(x) = x^4 + x^3 - 3x^2 + 4x - 4$ as a sum of powers of $x - 1$. Observe that f must coincide with the Taylor polynomial $P_{4,1,f}(x)$ because f is a polynomial of degree 4.

$$f(x) = x^4 + x^3 - 3x^2 + 4x - 4 \Rightarrow f(1) = -1,$$
$$f'(x) = 4x^3 + 3x^2 - 6x + 4 \Rightarrow f'(1) = 5,$$
$$f''(x) = 12x^2 + 6x - 6 \Rightarrow f''(1) = 12,$$
$$f'''(x) = 24x + 6 \Rightarrow f'''(1) = 30,$$
$$f^{(IV)}(x) = 24 \Rightarrow f^{(IV)}(1) = 24.$$

Then

$$f(x) = f(1) + \frac{f'(1)}{1!}(x - 1) + \frac{f''(1)}{2!}(x - 1)^2 + \frac{f'''(1)}{3!}(x - 1)^3$$
$$+ \frac{f^{(IV)}(1)}{4!}(x - 1)^4$$
$$= -1 + 5(x - 1) + \frac{12}{2!}(x - 1)^2 + \frac{30}{3!}(x - 1)^3 + \frac{24}{4!}(x - 1)^4$$
$$= -1 + 5(x - 1) + 6(x - 1)^2 + 5(x - 1)^3 + (x - 1)^4.$$

Now, we write $f(x) = x^4 - 11x^3 + 43x^2 - 60x + 14$ as a sum of powers of $x - 3$ having in mind that, again, $f = P_{4,1,f}$.

$$f(x) = x^4 - 11x^3 + 43x^2 - 60x + 14 \Rightarrow f(3) = 5,$$
$$f'(x) = 4x^3 - 33x^2 + 86x - 60 \Rightarrow f'(3) = 9,$$
$$f''(x) = 12x^2 - 66x + 86 \Rightarrow f''(3) = -4,$$
$$f'''(x) = 24x - 66 \Rightarrow f'''(3) = 6,$$
$$f^{(IV)}(x) = 24 \Rightarrow f^{(IV)}(3) = 24.$$

Then

$$f(x) = 5 + 9(x - 3) + \frac{-4}{2!}(x - 3)^2 + \frac{6}{3!}(x - 3)^3 + \frac{24}{4!}(x - 3)^4$$
$$= 5 + 9(x - 3) - 2(x - 3)^2 + (x - 3)^3 + (x - 3)^4.$$

$\square$

Exercise 5.22. *Find the Maclaurin polynomial of order n of the following functions:*

(I) $f(x) = e^{x^2}$,

(IV) $f(x) = \log \frac{1}{\sqrt{1-x}}$,

(II) $f(x) = (1 + e^x)^2$,

(V) $f(x) = \log \frac{1+x}{1-x}$,

(III) $f(x) = (x + x^2)e^x$,

(VI) $f(x) = (1+x)\log(1+x)$.

Solution to Exercise 5.22.

(I). $f(x) = e^{x^2}$. Since $e^x = \sum_{j=0}^{\infty} \frac{x^j}{j!}$ we have that the Maclaurin series of f is given by

$$f(x) = e^{x^2} = \sum_{j=0}^{\infty} \frac{(x^2)^j}{j!} = \sum_{j=0}^{\infty} \frac{x^{2j}}{j!}.$$

Clearly $f(0) = 1$, $f^{(2j-1)}(0) = 0$ and $\frac{f^{(2j)}(0)}{(2j)!} = \frac{1}{j!}$ for all $j \in \mathbb{N}$. Then $P_{1,0,f}(x) = 1$ and

$$P_{n,0,f}(x) = 1 + x^2 + \frac{x^4}{2!} + \cdots + \frac{x^{2k}}{k!}$$

both if n is even with $n = 2k$ and if n is odd with $n = 2k+1$ ($n \in \mathbb{N}$).

(II). $f(x) = (1+e^x)^2$. Using again that $e^x = \sum_{k=0}^{\infty} \frac{x^k}{k!}$ we can easily obtain the Maclaurin series of f as follows:

$$f(x) = (1+e^x)^2 = 1 + 2e^x + e^{2x} = 1 + 2\sum_{k=0}^{\infty} \frac{x^k}{k!} + \sum_{k=0}^{\infty} \frac{(2x)^k}{k!} = 3 + \sum_{k=1}^{\infty} \frac{2 + 2^k}{k!} x^k.$$

Then

$$P_{n,0,f}(x) = 3 + \sum_{k=1}^{n} \frac{2 + 2^k}{k!} x^k.$$

(III). $f(x) = (x + x^2)e^x$. Using $e^x = \sum_{k=0}^{\infty} \frac{x^k}{k!}$ once more we get

$$f(x) = (x + x^2)\sum_{k=0}^{\infty} \frac{x^k}{k!} = \left[\sum_{k=0}^{\infty} \frac{x^{k+1}}{k!}\right] + \left[\sum_{k=0}^{\infty} \frac{x^{k+2}}{k!}\right]$$

$$= \left[x + x^2 + \sum_{k=3}^{\infty} \frac{x^k}{(k-1)!}\right] + \left[x^2 + \sum_{k=3}^{\infty} \frac{x^k}{(k-2)!}\right]$$

$$= x + 2x^2 + \sum_{k=3}^{\infty} \left[\frac{1}{(k-1)!} + \frac{1}{(k-2)!}\right] x^k.$$

Therefore, $P_{1,0,f}(x) = x$, $P_{2,0,f}(x) = x + 2x^2$ and, for $n \geq 3$,

$$P_{n,0,f}(x) = x + 2x^2 + \sum_{k=3}^{n} \left[\frac{1}{(k-1)!} + \frac{1}{(k-2)!} \right] x^k.$$

(IV). $f(x) = \log \frac{1}{\sqrt{1-x}}$. It will be easier to handle the function in the form $f(x) = -\frac{1}{2} \log(1-x)$. Now we calculate the derivatives of f at $x = 0$:

$$f'(x) = \frac{1}{2(1-x)},$$

$$f''(x) = \frac{1}{2(1-x)^2},$$

$$f'''(x) = \frac{1}{(1-x)^3},$$

$$f^{(IV)}(x) = \frac{3}{(1-x)^4}.$$

The pattern for the kth derivative seems to be

$$f^{(k)}(x) = \frac{1}{2} \frac{(k-1)!}{(1-x)^k},$$

which can be proved easily by induction. Hence $f^{(k)}(0) = \frac{(k-1)!}{2}$ for all $k \in \mathbb{N}$ and

$$P_{n,0,f}(x) = f(0) + \frac{f'(0)}{1!} x + \frac{f''(0)}{2!} x^2 + \cdots + \frac{f^{(n)}(0)}{n!} x^n$$

$$= \frac{x}{2} + \frac{x^2}{4} + \frac{x^3}{6} + \frac{x^4}{8} + \cdots + \frac{x^n}{2n}.$$

(v). $f(x) = \log \frac{1+x}{1-x}$. Since $f(x) = \log(1+x) - \log(1-x)$, using the Maclaurin series of $\log(1+x)$, namely $\log(1+x) = \sum_{k=1}^{\infty} \frac{(-1)^{k+1}}{k} x^k$, we obtain

$$f(x) = \left[\sum_{k=1}^{\infty} \frac{(-1)^{k+1}}{k} x^k \right] - \left[\sum_{k=1}^{\infty} \frac{(-1)^{k+1}}{k} (-x)^k \right]$$

$$= \sum_{k=1}^{\infty} \frac{[(-1)^{k+1} + 1]}{k} x^k = \sum_{k=1}^{\infty} \frac{2}{2k-1} x^{2k-1}.$$

Clearly $f(0) = f^{(2j)}(0) = 0$ and $\frac{f^{(2j-1)}(0)}{(2j-1)!} = \frac{2}{2j-1}$ for all $j \in \mathbb{N}$. Then

$$P_{n,0,f}(x) = 2 \left(x + \frac{x^3}{3} + \cdots + \frac{x^{2k-1}}{2k-1} \right)$$

both if n is even with $n = 2k$ and if n is odd with $n = 2k - 1$ ($k \in \mathbb{N}$).

(VI). $f(x) = (1 + x) \log(1 + x)$. Using once again that $\log(1 + x) = \sum_{k=1}^{\infty} \frac{(-1)^{k+1}}{k} x^k$, we obtain the Maclaurin series of f as follows.

$$f(x) = \left[\sum_{k=1}^{\infty} \frac{(-1)^{k+1}}{k} x^k \right] + \left[\sum_{k=1}^{\infty} \frac{(-1)^{k+1}}{k} x^{k+1} \right]$$

$$= \left[x + \sum_{k=2}^{\infty} \frac{(-1)^{k+1}}{k} x^k \right] + \left[\sum_{k=2}^{\infty} \frac{(-1)^k}{k-1} x^k \right]$$

$$= x + \sum_{k=2}^{\infty} \left[\frac{(-1)^{k+1}}{k} + \frac{(-1)^k}{k-1} \right] x^k$$

$$= x + \sum_{k=2}^{\infty} \frac{(-1)^k}{k^2 - k} x^k.$$

Therefore, $f(0) = 0$, $f'(0) = 1$ and $\frac{f^{(k)}}{k!}(0) = \frac{(-1)^k}{k^2-k}$ for all $k \geq 2$, from which $P_{1,0,f}(x) = x$ and

$$P_{n,0,f}(x) = x + \sum_{k=2}^{n} \frac{(-1)^k}{k^2 - k} x^k$$

for every $n \in \mathbb{N}$ with $n \geq 2$.

$\square$

Exercise 5.23. *Find the Taylor polynomial of order n in each of the following cases:*

(I) $f(x) = (2 - x)^{-1}$, *in powers of* $x - 1$,

(II) $f(x) = \sin \frac{3x}{2}$, *in powers of* $x - \pi$,

(III) $f(x) = \log x$, *in powers of* $x - 2$,

(IV) $f(x) = e^x$, *in powers of* $x - 1$,

(V) $f(x) = \sqrt{1 + x}$, *in powers of* $x - 3$,

(VI) $f(x) = \log 2x - \frac{1}{x-1}$, *in powers of* $x - 2$,

Solution to Exercise 5.23.

(I). $f(x) = (2-x)^{-1}$ in powers of $x-1$. It can be easily seen by induction that $f^{k)}(x) = k!(2-x)^{-(k+1)}$ for all $k \in \mathbb{N}$. Then $f(1) = 1$, $f^{k)}(1) = k!$ for every $k \in \mathbb{N}$ and

$$P_{n,1,f}(x) = f(1) + \sum_{k=1}^{n} \frac{f^{(k)}(1)}{k!}(x-1)^k = \sum_{k=0}^{n}(x-1)^k.$$

(II). $f(x) = \sin \frac{3x}{2}$, in powers of $x-\pi$. First, we need to calculate $f^{(k)}(\pi)$ for every $k \in \mathbb{N}$. A formula for $f^{(k)}(x)$ will be useful for that purpose. Let us calculate some derivatives of f searching for a pattern. We will use the formula $\sin(x + \pi/2) = \cos x$:

$$f'(x) = \frac{3}{2}\cos\frac{3x}{2} = \frac{3}{2}\sin\left(\frac{3x}{2} + \frac{\pi}{2}\right),$$

$$f''(x) = \left(\frac{3}{2}\right)^2 \cos\left(\frac{3x}{2} + \frac{\pi}{2}\right) = \left(\frac{3}{2}\right)^2 \sin\left(\frac{3x}{2} + \pi\right),$$

$$f'''(x) = \left(\frac{3}{2}\right)^3 \cos\left(\frac{3x}{2} + \pi\right) = \left(\frac{3}{2}\right)^3 \sin\left(\frac{3x}{2} + \frac{3\pi}{2}\right).$$

Now it is straightforward to prove by induction that $f^{(k)}(x) = \left(\frac{3}{2}\right)^k \sin\left(\frac{3x}{2} + \frac{n\pi}{2}\right)$ for every $k \in \mathbb{N}$, and hence

$$f^{(k)}(\pi) = \left(\frac{3}{2}\right)^k \sin\left(\frac{k+3}{2}\pi\right) = \begin{cases} \left(\frac{3}{2}\right)^{2j}(-1)^{j+1} & \text{if } k \text{ is even and } k = 2j, \\ 0 & \text{if } k \text{ is odd.} \end{cases}$$

Since $f(\pi) = -1$, we finally arrive at $P_{1,\pi,f}(x) = -1$ and

$$P_{n,\pi,f}(x) = -1 + \sum_{j=1}^{k} \frac{(-1)^{j+1}3^{2j}}{2^{2j}(2j)!}(x-\pi)^{2j}$$

both if n is even with $n = 2k$ and if n is odd with $n = 2k+1$ ($k \in \mathbb{N}$).

(III). $f(x) = \log x$, in powers of $x-2$. It is easy to check by induction that $f^{(k)}(x) = \frac{(-1)^{k+1}(k-1)!}{x^n}$ for all $k \in \mathbb{N}$. Hence $f^{(k)}(2) = \frac{(-1)^{k+1}(k-1)!}{2^n}$ for every $k \in \mathbb{N}$ and since $f(2) = \log 2$, we finally obtain

$$P_{n,2,f(x)} = f(2) + \sum_{k=1}^{n} \frac{f^{(k)}(2)}{k!}(x-2)^k = \log 2 + \sum_{k=1}^{n} \frac{(-1)^{k+1}}{k \cdot 2^k}(x-2)^k.$$

(IV). $f(x) = e^x$, in powers of $x-1$. Since $f(1) = e$, $f^{(k)}(x) = e^x$ and $f^{(k)}(1) = e$ for all $k \in \mathbb{N}$, we have that

$$P_{n,1,f(x)} = e + e(x-1) + \frac{e}{2}(x-1)^2 + \frac{e}{3!}(x-1)^3 + \cdots + \frac{e}{n!}(x-1)^n.$$

(V). $f(x) = \sqrt{1+x}$, in powers of $x-3$. First, we find a formula to calculate the derivatives of f:

$$f'(x) = \frac{1}{2}(1+x)^{-\frac{1}{2}},$$

$$f''(x) = -\frac{1}{2^2}(1+x)^{-\frac{3}{2}},$$

$$f'''(x) = \frac{3}{2^3}(1+x)^{-\frac{5}{2}},$$

$$f^{(IV)}(x) = -\frac{3\cdot 5}{2^4}(1+x)^{-\frac{7}{2}}.$$

We deduce from the first 4 derivatives of f calculated above that

$$f^{(k)}(x) = \frac{(-1)^{k+1}\cdot(2k-3)!!}{2^k}(1+x)^{-\frac{2k-1}{2}},$$

where

$$(2k-3)!! = \begin{cases} 1\cdot 3\cdot 5\cdots(2k-3) & \text{if } k \geq 2, \\ 1 & \text{if } k = 1. \end{cases}$$

The formula obtained for $f^{k)}(x)$ can be proved easily by induction. Then

$$f^{(k)}(3) = \frac{(-1)^{k+1}\cdot(2k-3)!!}{2^k}2^{-(2k-1)} = \frac{2(-1)^{k+1}\cdot(2k-3)!!}{8^k}.$$

Finally,

$$P_{n,3,f}(x) = f(3) + \sum_{k=1}^{n}\frac{f^{(k)}(3)}{k!}(x-3)^k = 2 + \sum_{k=1}^{n}\frac{2(-1)^{k+1}\cdot(2k-3)!!}{8^k k!}(x-3)^k.$$

(VI). $f(x) = \log 2x - \frac{1}{x-1}$, in powers of $x-2$. We can prove easily by induction that

$$f^{(k)}(x) = \frac{(-1)^{k+1}(k-1)!}{x^k} + \frac{(-1)^{k+1}k!}{(x-1)^{k+1}}$$

for every $k \in \mathbb{N}$. Then $f^{(k)}(2) = (-1)^{k+1}(k-1)!\frac{1+k2^k}{2^k}$. Since $f(2) = \log 4 - 1$, we have

$$P_{n,2,f}(x) = f(2) + \sum_{k=1}^{n}\frac{f^{(k)}(2)}{k!}(x-2)^k$$

$$= \log 4 - 1 + \sum_{k=1}^{n}(-1)^{k+1}\frac{1+k2^k}{k2^k}(x-2)^k.$$

$\square$

Exercise 5.24. *Prove that $x - \frac{1}{6}x^3 + \frac{1}{120}x^5$ approaches the value of $\sin x$ with an error less than 10^{-4} whenever $|x| \leq \frac{\pi}{4}$.*

Solution to Exercise 5.24.

Observe that if $f(x) = \sin x$ then $P_{6,0,f}(x) = x - \frac{1}{6}x^3 + \frac{1}{120}x^5$. Then the error made when approaching $\sin x$ by $P_{6,0,f}(x)$ is given by Taylor's remainder $R_{6,0,f}(x)$. Since

$$R_{6,0,f}(x) = \frac{f^{(7)}(c)}{7!}x^7 = \frac{-\cos c}{7!}x^7$$

for some c between 0 and x and $|x| \leq \frac{\pi}{4}$, we have that

$$|R_{6,0,f}(x)| = \left| \frac{-\cos c}{7!}x^7 \right| \leq \left| \frac{x^7}{7!} \right| \leq \left| \frac{\pi^7}{7! \cdot 4^7} \right| \approx 3.6576 \cdot 10^{-5} < \frac{1}{10^4}.$$

Then

$$\left| \sin x - \left(x - \frac{1}{6}x^3 + \frac{1}{120}x^5 \right) \right| = |R_{6,0,f}(x)| < 10^{-4}.$$

$\square$

Exercise 5.25. *Estimate, using Taylor's theorem, the error made when approaching $\sin(e^x - 1)$ by $x + \frac{1}{2}x^2$ for $|x| \leq \frac{1}{10}$.*

Solution to Exercise 5.25.

If $f(x) = \sin(e^x - 1)$ then $P_{2,0,f}(x) = x + \frac{1}{2}x^2$. The error we have to estimate is $|f(x) - P_{2,0,f}(x)| = |R_{2,0,f}(x)|$. The third derivative is needed to calculate $R_{2,0,f}(x)$, and is given by

$$f'''(x) = e^x[(1 - e^{2x})\cos(e^x - 1) - 3e^x \sin(e^x - 1)].$$

For any $x \in \mathbb{R}$ there exists c between 0 and x such

$$|R_{2,0,f(x)}| = \left| \frac{f'''(c)}{3!}x^3 \right|$$

$$= \left| \frac{e^c[(1 - e^{2c})\cos(e^c - 1) - 3e^c \sin(e^c - 1)]}{3!}x^3 \right|$$

$$\leq \frac{e^c(1 + 4e^{2c})}{6}|x|^3.$$

Observe that in the last step we have used the triangle inequality together with the fact that $|\sin(e^2 - 1)| \leq 1$ and $|\cos(e^2 - 1)| \leq 1$. Now, if $|x| \leq \frac{1}{10}$ then $|c| \leq \frac{1}{10}$ and we arrive at

$$|R_{2,0,f(x)}| \leq \frac{e^{\frac{1}{10}}(1 + 4e^{\frac{1}{5}})}{6 \cdot 10^3} \approx 4.0917 \cdot 10^{-4} < 5 \cdot 10^{-4}.$$

$\square$

Exercise 5.26. *Prove that* $1 - 9x^2 + 27x^4$ *approaches* $\cos^2 3x$ *with an error less than* $4 \cdot 10^{-5}$, *if* $|x| \leq \frac{1}{10}$.

Solution to Exercise 5.26.

If $f(x) = \cos^2 3x$ then it can be proved that $P_{5,0,f(x)} = 1 - 9x^2 + 27x^4$. Hence we have to estimate $|f(x) - P_{5,0,f}(x)| = |R_{5,0,f}(x)|$. Since $f^{(VI)}(x) = -18 \cdot 6^4 \cos 6x$, for any $x \in \mathbb{R}$ there exists, by Taylor's Theorem, a c between 0 and x such that

$$|R_{5,0,f}(x)| = \left| \frac{f^{(VI)}(c)}{6!} x^6 \right| = \frac{18 \cdot 6^4}{6!} |\cos 3c| |x|^6 \leq \frac{18 \cdot 6^4}{6!} |x|^6 = \frac{2 \cdot 3^4}{5} |x|^6.$$

If, in addition we have $|x| \leq \frac{1}{10}$, then

$$|R_{5,0,f}(x)| \leq \frac{2 \cdot 3^4}{5} \left(\frac{1}{10} \right)^6 = \left(\frac{9}{5} \right)^2 10^{-5} < 2^2 \cdot 10^{-5} = 4 \cdot 10^{-5}.$$

$\square$

Exercise 5.27. *Prove that* $1 + x + \frac{1}{2}x^2$ *approaches* $e^{\sin x}$ *with an error less than* $3 \cdot |x|^3$.

Solution to Exercise 5.27.

If $f(x) = e^{\sin x}$ then it can be seen that $P_{2,0,f}(x) = 1 + x + \frac{1}{2}x^2$. Hence for every $x \in \mathbb{R}$ we have that

$$|f(x) - P_{2,0,f}(x)| = |R_{2,0,f}(x)|.$$

On the other hand, for each $x \in \mathbb{R}$ there exists c between 0 and x such that $R_{2,0,f}(x) = \frac{f'''(c)}{3!} x^3$. We need to obtain the third derivative of f to estimate Taylor's remainder, and that is given by

$$f'''(x) = e^{\sin x} \cos x (\cos^2 x - 3 \sin x - 1).$$

Observe that

$$|f'''(x)| = |e^{\sin x} \cos x (\cos^2 x - 3 \sin x - 1)|$$
$$\leq e^{\sin x} |\cos x| (\cos^2 x + 3 |\sin x| + 1) \leq 5e.$$

Then

$$|R_{2,0,f(x)}| = \left| \frac{f'''(c)}{3!} x^3 \right| \leq \frac{5e}{3!} |x|^3 = \frac{5e}{6} |x|^3 \leq \frac{5}{2} \cdot |x|^3 \leq 3 \cdot |x|^3.$$

$\square$

Exercise 5.28. *Find in each case the largest $p \in \mathbb{N}$ such that $\lim_{x \to a} \frac{f(x)}{(x-a)^p}$ is finite (then, it is said that $f(x)$ is an infinitesimal of order p as $x \to a$):*

(I) $f(x) = \sin x$, $a = 0$,

(II) $f(x) = \log(1 + x)$, $a = 0$,

(III) $f(x) = 1 - x + \log x$, $a = 1$,

(IV) $f(x) = 1 - \cos x$, $a = 0$,

(V) $f(x) = \tan x - \sin x$, $a = 0$,

(VI) $f(x) = \sqrt{x} - 2$, $a = 4$,

(VII) $f(x) = e^x - 1$, $a = 0$,

(VIII) $f(x) = \cos x - e^{-\frac{x^2}{2}}$, $a = 0$,

(IX) $f(x) = \sin x^2 - \log(1 + x^2)$, $a = 0$.

Solution to Exercise 5.28.

We will use Young's formula.

(I). $\lim\limits_{x \to 0} \dfrac{\sin x}{x^p}$. We will use the fact that $\sin x = x - x^3 + o(x^3)$. If $p = 1$,

$$\lim_{x \to 0} \frac{\sin x}{x} = \lim_{x \to 0} \frac{x - x^3 + o(x^3)}{x} = \lim_{x \to 0} \left[1 - x^2 + \frac{o(x^3)}{x} \right] = 1.$$

If $p = 2$,

$$\lim_{x \to 0} \frac{\sin x}{x^2} = \lim_{x \to 0} \frac{x - x^3 + o(x^3)}{x^2}$$
$$= \lim_{x \to 0} \left[\frac{1}{x} - x + \frac{o(x^3)}{x^2} \right] = \lim_{x \to 0} \frac{1}{x}.$$

Since $\lim\limits_{x \to 0} \dfrac{1}{x}$ does not exist (as a real number), $\lim\limits_{x \to 0} \dfrac{\sin x}{x^2}$ does not exist either.

(II). $\lim\limits_{x \to 0} \dfrac{\log(1 + x)}{x^p}$. Now we use $\log(1 + x) = x - \frac{1}{2}x^2 + o(x^2)$. If $p = 1$,

$$\lim_{x \to 0} \frac{\log(1 + x)}{x} = \lim_{x \to 0} \frac{x - \frac{1}{2}x^2 + o(x^2)}{x} = \lim_{x \to 0} \left[1 - \frac{1}{2}x + \frac{o(x^2)}{x} \right] = 1.$$

If $p = 2$,

$$\lim_{x \to 0} \frac{\log(1 + x)}{x^2} = \lim_{x \to 0} \frac{x - \frac{1}{2}x^2 + o(x^2)}{x^2}$$
$$= \lim_{x \to 0} \left[\frac{1}{x} - \frac{1}{2} + \frac{o(x^2)}{x^2} \right] = \lim_{x \to 0} \frac{1}{x} - \frac{1}{2}.$$

Since $\lim\limits_{x \to 0} \dfrac{1}{x}$ does not exist, $\lim\limits_{x \to 0} \dfrac{\log(1 + x)}{x^2}$ does not exist either.

(III). $\lim\limits_{x\to 1}\dfrac{1-x+\log x}{(x-1)^p}$. Now we are going to use the fact that

$$\log x = (x-1) - \frac{1}{2}(x-1)^2 + \frac{1}{3}(x-1)^3 + o((x-1)^3).$$

If $p = 2$,

$$\lim_{x\to 1}\frac{1-x+\log x}{(x-1)^2} = \lim_{x\to 1}\frac{-\frac{1}{2}(x-1)^2 + \frac{1}{3}(x-1)^3 + o((x-1)^3)}{(x-1)^2} = -\frac{1}{2}.$$

If $p = 3$,

$$\lim_{x\to 1}\frac{1-x+\log x}{(x-1)^3} = \lim_{x\to 1}\frac{-\frac{1}{2}(x-1)^2 + \frac{1}{3}(x-1)^3 + o((x-1)^3)}{(x-1)^3}$$

$$= \lim_{x\to 1}\left(-\frac{1}{2}\frac{1}{x-1} + \frac{1}{3}\right).$$

Since $\lim\limits_{x\to 1}\dfrac{1}{x-1}$ does not exist, $\lim\limits_{x\to 1}\dfrac{1-x+\log x}{(x-1)^3}$ does not exist either.

(IV). $\lim\limits_{x\to 0}\dfrac{1-\cos x}{x^p}$. We will use that $\cos x = 1 - \frac{x^2}{2} + \frac{x^4}{4!} + o(x^4)$. If $p = 2$,

$$\lim_{x\to 0}\frac{1-\cos x}{x^2} = \lim_{x\to 0}\frac{\frac{x^2}{2} - \frac{x^4}{4!} - o(x^4)}{x^2}$$

$$= \lim_{x\to 0}\left[\frac{1}{2} - \frac{x^2}{4!} + \frac{o(x^4)}{x^2}\right] = \frac{1}{2}.$$

If $p = 3$,

$$\lim_{x\to 0}\frac{1-\cos x}{x^3} = \lim_{x\to 0}\frac{\frac{x^2}{2} - \frac{x^4}{4!} - o(x^4)}{x^3}$$

$$= \lim_{x\to 0}\left[\frac{1}{2x} - \frac{x}{4!} + \frac{o(x^4)}{x^3}\right] = \lim_{x\to 0}\frac{1}{2x}.$$

Since $\lim\limits_{x\to 0}\dfrac{1}{x}$ does not exist, $\lim\limits_{x\to 0}\dfrac{1-\cos x}{x^3}$ does not exist either.

(V). $\lim\limits_{x\to 0}\dfrac{\tan x - \sin x}{x^p}$. We will use

$$\tan x = x + \frac{1}{3}x^3 + \frac{2}{15}x^5 + o(x^5),$$

$$\sin x = x - \frac{1}{3!}x^3 + \frac{1}{5!}x^5 + o(x^5).$$

If $p = 3$,

$$\lim_{x\to 0}\frac{\tan x - \sin x}{x^3} = \lim_{x\to 0}\frac{\frac{1}{2}x^3 + \frac{1}{8}x^5 + o(x^5)}{x^3}$$

$$= \lim_{x\to 0}\left[\frac{1}{2} + \frac{1}{8}x^2 + \frac{o(x^5)}{x^3}\right] = \frac{1}{2}.$$

If $p = 4$,

$$\lim_{x \to 0} \frac{\tan x - \sin x}{x^4} = \lim_{x \to 0} \frac{\frac{1}{2}x^3 + \frac{1}{8}x^5 + o(x^5)}{x^4}$$

$$= \lim_{x \to 0} \left[\frac{1}{2x} + \frac{1}{8}x + \frac{o(x^5)}{x^4} \right] = \lim_{x \to 0} \frac{1}{2x}.$$

Since $\lim_{x \to 0} \frac{1}{x}$ does not exist, $\lim_{x \to 0} \frac{\tan x - \sin x}{x^4}$ does not exist either.

(VI). $\lim_{x \to 4} \frac{\sqrt{x} - 2}{(x - 4)^p}$. We will use

$$\sqrt{x} - 2 = \frac{1}{4}(x - 4) - \frac{1}{64}(x - 4)^2 + o((x - 4)^2).$$

If $p = 1$,

$$\lim_{x \to 4} \frac{\sqrt{x} - 2}{x - 4} = \lim_{x \to 4} \frac{\frac{1}{4}(x - 4) - \frac{1}{64}(x - 4)^2 + o((x - 4)^2)}{x - 4}$$

$$= \lim_{x \to 4} \left[\frac{1}{4} - \frac{1}{64}(x - 4) + \frac{o((x - 4)^2)}{x - 4} \right] = \frac{1}{4}.$$

If $p = 2$,

$$\lim_{x \to 4} \frac{\sqrt{x} - 2}{(x - 4)^2} = \lim_{x \to 4} \frac{\frac{1}{4}\frac{1}{x-4} - \frac{1}{64}(x - 4)^2 + o(x^2)}{(x - 4)^2}$$

$$= \lim_{x \to 4} \left[\frac{1}{4}(x - 4) - \frac{1}{64} + \frac{o((x - 4)^2)}{(x - 4)^2} \right]$$

$$= \lim_{x \to 4} \frac{1}{4(x - 4)} - \frac{1}{64}.$$

Since $\lim_{x \to 4} \frac{1}{x - 4}$ does not exist, $\lim_{x \to 4} \frac{\sqrt{x} - 2}{(x - 4)^2}$ does not exist either.

(VII). $\lim_{x \to 0} \frac{e^x - 1}{x^p}$. We will use $e^x = 1 + x + \frac{1}{2}x^2 + o(x^2)$. If $p = 1$,

$$\lim_{x \to 0} \frac{e^x - 1}{x} = \lim_{x \to 0} \frac{x + \frac{1}{2}x^2 + o(x^2)}{x}$$

$$= \lim_{x \to 0} \left[1 + \frac{1}{2}x + \frac{o(x^2)}{x} \right] = 1.$$

If $p = 2$,

$$\lim_{x \to 0} \frac{e^x - 1}{x^2} = \lim_{x \to 0} \frac{x + \frac{1}{2}x^2 + o(x^2)}{x^2} = \lim_{x \to 0} \left[\frac{1}{x} + \frac{1}{2} + \frac{o(x^2)}{x^2} \right] = \lim_{x \to 0} \frac{1}{x} + \frac{1}{2}.$$

Since $\lim_{x \to 0} \frac{1}{x}$ does not exist, $\lim_{x \to 4} \frac{e^x - 1}{x^2}$ does not exist either.

(VIII). $\lim\limits_{x\to 0} \dfrac{\cos x - e^{-\frac{x^2}{2}}}{x^p}$. This time we will use

$$\cos x = 1 - \frac{1}{2}x^2 + \frac{1}{4!}x^4 - \frac{1}{6!}x^6 + o(x^6),$$

$$e^{-\frac{x^2}{2}} = 1 - \frac{1}{2}x^2 + \frac{3}{4!}x^4 - \frac{15}{6!}x^6 + o(x^6).$$

If $p = 4$,

$$\lim_{x\to 0} \frac{\cos x - e^{-\frac{x^2}{2}}}{x^4} = \lim_{x\to 0} \frac{-\frac{1}{12}x^4 + \frac{14}{6!}x^6 + o(x^6)}{x^4}$$

$$= \lim_{x\to 0}\left[-\frac{1}{12} + \frac{14}{6!}x^2 + \frac{o(x^6)}{x^2}\right] = -\frac{1}{12}.$$

If $p = 5$,

$$\lim_{x\to 0} \frac{\cos x - e^{-\frac{x^2}{2}}}{x^5} = \lim_{x\to 0} \frac{-\frac{1}{12}x^4 + \frac{14}{6!}x^6 + o(x^6)}{x^5}$$

$$= \lim_{x\to 0}\left(-\frac{1}{12x} + \frac{14}{6!}x + \frac{o(x^6)}{x^5}\right) = -\frac{1}{12}\lim_{x\to 0}\frac{1}{x}.$$

Since $\lim\limits_{x\to 0}\frac{1}{x}$ does not exist, $\lim\limits_{x\to 0}\dfrac{\cos x - e^{-\frac{x^2}{2}}}{x^5}$ does not exist either.

(IX). $\lim\limits_{x\to 0} \dfrac{\sin x^2 - \log(1+x^2)}{x^p}$. We will use

$$\sin x^2 = x^2 - \frac{1}{3!}x^6 + o(x^6),$$

$$\log(1+x^2) = x^2 - \frac{1}{2}x^4 + \frac{1}{3}x^6 + o(x^6).$$

If $p = 4$,

$$\lim_{x\to 0} \frac{\sin x^2 - \log(1+x^2)}{x^4} = \lim_{x\to 0} \frac{\frac{1}{2}x^4 - \frac{1}{2}x^6 + o(x^6)}{x^4}$$

$$= \lim_{x\to 0}\left[\frac{1}{2} - \frac{1}{2}x^2 + \frac{o(x^6)}{x^4}\right] = \frac{1}{2}.$$

If $p = 5$,

$$\lim_{x\to 0} \frac{\sin x^2 - \log(1+x^2)}{x^5} = \lim_{x\to 0} \frac{\frac{1}{2}x^4 - \frac{1}{2}x^6 + o(x^6)}{x^5}$$

$$= \lim_{x\to 0}\left[\frac{1}{2x} - \frac{1}{2}x + \frac{o(x^6)}{x^5}\right] = \lim_{x\to 0}\frac{1}{2x}.$$

Since $\lim\limits_{x\to 0}\frac{1}{x}$ does not exist, $\lim\limits_{x\to 0}\dfrac{\sin x^2 - \log(1+x^2)}{x^5}$ does not exist either.

$\square$

Exercise 5.29. *Calculate the following limits, using Young's formula:*

(I) $\lim\limits_{x\to 0} \dfrac{3\sin ax - 3ax - a^3x^3}{6bx - 6\sin bx + b^3x^3}$,

$(b \neq 0)$

(II) $\lim\limits_{x\to 1} \dfrac{1 - x + \log x}{1 - \sqrt{2x - x^2}}$,

(III) $\lim\limits_{x\to 0} \dfrac{\sin x - x}{\arctan x - x}$,

(IV) $\lim\limits_{x\to 0} \dfrac{\sin x - x\cos x}{(x^2 - \sin^2 x)^{\frac{1}{2}}}$,

(V) $\lim\limits_{x\to\infty} x^2\left[1 - \sqrt{x(1+x)}\log\dfrac{1+x}{x}\right]$,

(VI) $\lim\limits_{x\to 0} \dfrac{\sinh x - \tan x}{\sin x - \arcsin x}$,

(VII) $\lim\limits_{x\to 0} \dfrac{\sin x - \tan x}{\arcsin x - \arctan x}$,

(VIII) $\lim\limits_{x\to 0} \dfrac{\cos x - \sqrt{1 - x}}{\sin x}$,

(IX) $\lim\limits_{x\to 0} \dfrac{(\sin 3x - 3\sin x)^2}{(\cos 2x - \cos x)^3}$.

Solution to Exercise 5.29.

(I). We will use $\sin(ax) = ax - \dfrac{(ax)^3}{3!} + o(x^3)$:

$$\lim_{x\to 0} \frac{3\sin ax - 3ax - a^3x^3}{6bx - 6\sin bx + b^3x^3} = \lim_{x\to 0} \frac{3\left(ax - \frac{(ax)^3}{3!} + o(x^3)\right) - 3ax - a^3x^3}{6bx - 6\left(bx - \frac{(bx)^3}{3!} + o(x^3)\right) + b^3x^3}$$

$$= \lim_{x\to 0} \frac{-\frac{a^3x^3}{2} + 3o(x^3) - a^3x^3}{b^3x^3 + 6o(x^3) + b^3x^3}$$

$$= \lim_{x\to 0} \frac{\frac{-3a^3x^3}{2x^3} + \frac{o(x^3)}{x^3}}{\frac{2b^3x^3}{x^3} + \frac{o(x^3)}{x^3}} = -\frac{3a^3}{4b^3}.$$

(II). We will use

$$\log x = (x-1) - \frac{1}{2}(x-1)^2 + o((x-1)^2),$$

$$\sqrt{2x - x^2} = 1 - \frac{1}{2}(x-1)^2 + o((x-1)^2).$$

Then,

$$\lim_{x\to 1} \frac{1 - x + \log x}{1 - \sqrt{2x - x^2}} = \lim_{x\to 1} \frac{1 - x + \left[(x-1) - \frac{(x-1)^2}{2} + o((x-1)^2)\right]}{1 - \left[1 - \frac{(x-1)^2}{2} + o((x-1)^2)\right]}$$

$$= \lim_{x\to 1} \frac{-\frac{(x-1)^2}{2} + o((x-1)^2)}{\frac{(x-1)^2}{2} + o((x-1)^2)} = -1.$$

(III). We will use

$$\sin x = x - \frac{x^3}{3!} + o(x^3),$$

$$\arctan x = x - \frac{x^3}{3} + o(x^3).$$

Then,

$$\lim_{x \to 0} \frac{\sin x - x}{\arctan x - x} = \lim_{x \to 0} \frac{x - \frac{x^3}{3!} + o(x^3) - x}{x - \frac{x^3}{3} + o(x^3) - x}$$

$$= \lim_{x \to 0} \frac{-\frac{x^3}{6x^3} + \frac{o(x^3)}{x^3}}{-\frac{x^3}{3x^3} + \frac{o(x^3)}{x^3}} = \frac{1}{2}.$$

(IV). We will use

$$\sin x = x - \frac{x^3}{3!} + o(x^3),$$

$$\cos x = 1 - \frac{x^2}{2} + o(x^2)$$

and

$$\sin^2 x = \left(x - \frac{x^3}{3!} + o(x^3) \right) \left(x - \frac{x^3}{3!} + o(x^3) \right)$$

$$= x^2 - \frac{x^4}{6} + o(x^4) - \frac{x^4}{6} + \frac{x^6}{36} - \frac{x^3}{6}o(x^3) + o(x^4)$$

$$- o(x^3)\frac{x^3}{6} + o(x^3)o(x^3)$$

$$= \frac{36x^2 - 12x^4 + x^6}{36} + o(x^4) + o(x^6)$$

$$= \frac{36x^2 - 12x^4 + x^6}{36} + o(x^4).$$

Consequently,

$$\lim_{x \to 0} \frac{\sin x - x \cos x}{(x^2 - \sin^2 x)^{\frac{1}{2}}} = \lim_{x \to 0} \frac{\left[x - \frac{x^3}{3!} + o(x^3)\right] - x\left[1 - \frac{x^2}{2} + o(x^2)\right]}{\left[x^2 - \left(\dfrac{36x^2 - 12x^4 + x^6}{36} + o(x^4)\right)\right]^{\frac{1}{2}}}$$

$$= \lim_{x \to 0} \frac{\frac{x^3}{3} + o(x^3)}{\left[\frac{x^4}{3} + \frac{x^6}{36} - o(x^4)\right]^{\frac{1}{2}}}$$

$$= \lim_{x \to 0} \frac{\frac{x^3}{3} + o(x^3)}{\left[\frac{x^4}{3} + o(x^4)\right]^{\frac{1}{2}}}$$

$$= \lim_{x \to 0} x \cdot \frac{\frac{1}{3} + \frac{o(x^3)}{x^3}}{\left[\frac{1}{3} + \frac{o(x^4)}{x^4}\right]^{\frac{1}{2}}} = 0.$$

(v). $\lim\limits_{x \to \infty} x^2 \left(1 - \sqrt{x(1+x)} \log \dfrac{1+x}{x}\right)$. Let $x = \frac{1}{z}$. Then $z \to 0$ as $x \to \infty$ and

$$\lim_{x \to \infty} x^2 \left[1 - \sqrt{x(1+x)} \log \frac{1+x}{x}\right] = \lim_{z \to 0} \frac{1}{z^2}\left[1 - \sqrt{\frac{1}{z}\left(1 + \frac{1}{z}\right)} \log \frac{1 + \frac{1}{z}}{\frac{1}{z}}\right]$$

$$= \lim_{z \to 0} \frac{1}{z^2}\left[1 - \frac{1}{z}\sqrt{z+1} \log(1+z)\right]$$

$$= \lim_{z \to 0} \frac{z - \sqrt{z+1}\log(1+z)}{z^3}.$$

Now we will use

$$\sqrt{1+z} = 1 + \frac{z}{2} - \frac{z^2}{8} + \frac{z^3}{16} + o(z^3),$$

$$\log(1+z) = z - \frac{z^2}{2} + \frac{z^3}{3} + o(z^3).$$

Then,

$$\lim_{z \to 0} \frac{z - \sqrt{z+1} \log(1+z)}{z^3}$$

$$= \lim_{z \to 0} \frac{z - \left[1 + \frac{z}{2} - \frac{z^2}{8} + \frac{z^3}{16} + o(z^3)\right] \left[z - \frac{z^2}{2} + \frac{z^3}{3} + o(z^3)\right]}{z^3}$$

$$= \lim_{z \to 0} \frac{z - z + \frac{z^2}{2} - \frac{z^3}{3} - o(z^3) - \frac{z^2}{2} + \frac{z^3}{4} - \frac{z^4}{6} - o(z^4) + \frac{z^3}{8}}{z^3}$$

$$+ \frac{-\frac{z^4}{16} + \frac{z^5}{24} - o(z^5) - \frac{z^4}{16} + \frac{z^5}{32} - \frac{z^6}{48} - o(z^6) - o(z^4)}{z^3}$$

$$= \lim_{z \to 0} \frac{-\frac{z^3}{3} + \frac{z^3}{4} + \frac{z^3}{8} + o(z^3)}{z^3} = \lim_{z \to 0} \frac{\frac{z^3}{24} + o(z^3)}{z^3} = \frac{1}{24}.$$

(VI). We will use

$$\sin x = x - \frac{x^3}{3!} + o(x^3),$$

$$\tan x = x + \frac{x^3}{3} + o(x^3),$$

$$\sinh x = x + \frac{x^3}{3!} + o(x^3),$$

$$\arcsin x = x + \frac{x^3}{3!} + o(x^3).$$

Then,

$$\lim_{x \to 0} \frac{\sinh x - \tan x}{\sin x - \arcsin x} = \lim_{x \to 0} \frac{\left[x + \frac{x^3}{3!} + o(x^3)\right] - \left[x + \frac{x^3}{3} + o(x^3)\right]}{\left[x - \frac{x^3}{3!} + o(x^3)\right] - \left[x + \frac{x^3}{3!} + o(x^3)\right]}$$

$$= \lim_{x \to 0} \frac{-\frac{x^3}{6} + o(x^3)}{-\frac{x^3}{3} + o(x^3)} = \lim_{x \to 0} \frac{-\frac{1}{6} + \frac{o(x^3)}{x^3}}{-\frac{1}{3} + \frac{o(x^3)}{x^3}} = \frac{1}{2}.$$

(VII). We will use

$$\sin x = x - \frac{x^3}{3!} + o(x^3),$$

$$\tan x = x + \frac{x^3}{3} + o(x^3),$$

$$\arcsin x = x + \frac{x^3}{3!} + o(x^3),$$

$$\arctan x = x - \frac{x^3}{3} + o(x^3).$$

Then,

$$\lim_{x \to 0} \frac{\sin x - \tan x}{\arcsin x - \arctan x} = \lim_{x \to 0} \frac{\left[x - \frac{x^3}{3!} + o(x^3)\right] - \left[x + \frac{x^3}{3} + o(x^3)\right]}{\left[x + \frac{x^3}{6} + o(x^3)\right] - \left[x - \frac{x^3}{3} + o(x^3)\right]}$$

$$= \lim_{x \to 0} \frac{-\frac{x^3}{2} + o(x^3)}{\frac{x^3}{2} + o(x^3)} = \lim_{x \to 0} \frac{-\frac{1}{2} + \frac{o(x^3)}{x^3}}{\frac{1}{2} + \frac{o(x^3)}{x^3}} = -1.$$

(VIII). We will use

$$\sin x = x - \frac{x^3}{3!} + o(x^3),$$

$$\cos x = 1 - \frac{x^2}{2} + o(x^2),$$

$$\sqrt{1-x} = 1 - \frac{x}{2} - \frac{x^2}{8} + o(x^2).$$

Then,

$$\lim_{x \to 0} \frac{\cos x - \sqrt{1-x}}{\sin x} = \lim_{x \to 0} \frac{\left[1 - \frac{x^2}{2} + o(x^2)\right] - \left[1 - \frac{x}{2} - \frac{x^2}{8} + o(x^2)\right]}{x - \frac{x^3}{6} + o(x^3)}$$

$$= \lim_{x \to 0} \frac{\frac{x}{2} - \frac{3x^2}{8} + o(x^2)}{x + o(x^2)} = \lim_{x \to 0} \frac{\frac{1}{2} - \frac{3x}{8} + \frac{o(x^2)}{x}}{1 + \frac{o(x^2)}{x}} = \frac{1}{2}.$$

(IX). We will use

$$\sin x = x - \frac{x^3}{3!} + o(x^3),$$

$$\cos x = 1 - \frac{x^2}{2} + o(x^2),$$

$$\sin 3x = 3x - \frac{(3x)^3}{3!} + o(x^3),$$

$$\cos 2x = 1 - \frac{(2x)^2}{2} + o(x^2).$$

Then,

$$\lim_{x \to 0} \frac{(\sin 3x - 3\sin x)^2}{(\cos 2x - \cos x)^3} = \lim_{x \to 0} \frac{\left\{\left[3x - \frac{27x^3}{6} + o(x^3)\right] - 3\left[x - \frac{x^3}{6} + o(x^3)\right]\right\}^2}{\left\{\left[1 - \frac{(2x)^2}{2} + o(x^2)\right] - \left[1 - \frac{x^2}{2} + o(x^2)\right]\right\}^3}$$

$$= \lim_{x \to 0} \frac{\left[-4x^3 + o(x^3)\right]^2}{\left[-\frac{3x^2}{2} + o(x^2)\right]^3} = \lim_{x \to 0} \frac{\left[-4 + \frac{o(x^3)}{x^3}\right]^2}{\left[-\frac{3}{2} + \frac{o(x^2)}{x^2}\right]^3} = -\frac{128}{27}.$$

□

Exercise 5.30. *Study the monotonicity, asymptotes, local extremes and convexity of the following functions. Sketch their graphs:*

(I) $f(x) = \dfrac{x^3}{(x+1)^2}$,

(II) $f(x) = \dfrac{x^2+1}{x^2-4}$,

(III) $f(x) = \sqrt{4x^2-x}$,

(IV) $f(x) = \dfrac{1}{\log x}$,

(V) $f(x) = \dfrac{e^x}{x}$,

(VI) $f(x) = \sqrt[3]{x} - \sqrt[3]{x+1}$,

(VII) $f(x) = \tan^2 x$,

(VIII) $f(x) = x^6 - 3x^4 + 3x^2 - 5$.

Solution to Exercise 5.30.

(I). $f(x) = \dfrac{x^3}{(x+1)^2}$. The domain of f is clearly $D(f) = \mathbb{R} \setminus \{-1\}$ and

$$f'(x) = \frac{3x^2(x+1)^2 - 2x^3(x+1)}{(x+1)^4} = \frac{x^3 + 3x^2}{(x+1)^3} = \frac{x^2(x+3)}{(x+1)^3}.$$

Then the roots of f' are clearly $x = -3$ and $x = 0$. The sign of f' and monotonicity of f behave as shown in the following table.

$f'(x)$	$-\infty$	$+$	-3	$-$	-1	$+$	0	$+$	∞
$f(x)$		↗		↘		↗		↗	

Therefore, f is increasing in $(-\infty, -3) \cup (-1, \infty)$ and decreasing in $(-3, -1)$. Thus, f has a relative maximum at $x = -3$.

Let us study the asymptotes:

$$\lim_{x \to -1^+} f(x) = \infty \text{ and } \lim_{x \to -1^-} f(x) = -\infty.$$

Then, f has a vertical asymptote at $x = -1$.

Since

$$\lim_{x \to \infty} f(x) = \infty \text{ and } \lim_{x \to -\infty} f(x) = -\infty,$$

f has no horizontal asymptote. However,

$$m = \lim_{x \to \infty} \frac{f(x)}{x} = \lim_{x \to \infty} \frac{x^3}{x(x+1)^2} = 1$$

and

$$r = \lim_{x \to \infty} (f(x) - mx) = \lim_{x \to \infty} \left[\frac{x^3}{(x+1)^2} - x \right] = \lim_{x \to \infty} \frac{x^3 - x(x+1)^2}{(x+1)^2} = -2.$$

Hence f has an oblique asymptote $y = mx + r = x - 2$.

Finally, let us study the convexity of f, for which the second derivative is needed:

$$f''(x) = \frac{(3x^2 + 6x)(x+1)^3 - 3(x+1)^2(x^3 + 3x^2)}{(x+1)^6}$$

$$= \frac{(3x^2 + 6x)(x+1) - 3(x^3 + 3x^2)}{(x+1)^4}$$

$$= \frac{6x}{(x+1)^4}.$$

Hence the unique root of f'' is $x = 0$. The behavior of the sign of f'' and the convexity of f are shown in the next table.

$f''(x)$	$-\infty$		$-$		0	$+$		∞
$f(x)$			$\frown$			$\smile$		

The function f has clearly an inflexion point at $x = 0$. Taking into account the previous considerations, the graph of f can be drawn as in Figure 5.1.

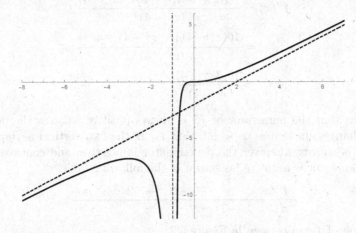

FIGURE 5.1: Graph of the function from Exercise 5.30 part (I) with the corresponding asymptotes (in dashed lines).

(II). $f(x) = \dfrac{x^2 + 1}{x^2 - 4}$. The domain of f is $D(f) = \mathbb{R} \setminus \{\pm 2\}$ and

$$f'(x) = \frac{2x(x^2 - 4) - (x^2 + 1)2x}{(x^2 - 4)^2} = \frac{-10x}{(x^2 - 4)^2}.$$

It is obvious that the unique root of f' is $x = 0$ and that the denominator is strictly positive in $D(f)$. The monotonicity of f is described in the following table.

$f'(x)$	$-\infty$		$+$		0		$-$		∞
$f(x)$			$\nearrow$				$\searrow$		

Therefore, f is decreasing in $(0, \infty)$ and increasing in $(-\infty, 0)$, from which f has a relative maximum at $x = 0$.

Let us study the asymptotes. Since

$$\lim_{x \to 2^+} f(x) = \infty \quad \text{and} \quad \lim_{x \to 2^-} f(x) = -\infty,$$

f has a vertical asymptote at $x = 2$. Similarly,

$$\lim_{x \to -2^+} f(x) = \infty \quad \text{and} \quad \lim_{x \to -2^-} f(x) = -\infty.$$

Then, f has another vertical asymptote at $x = -2$. On the other hand,

$$\lim_{x \to \infty} f(x) = 1 = \lim_{x \to -\infty} f(x),$$

so $y = 1$ is a horizontal asymptote of f.

Finally, we calculate f'' to study the convexity of f:

$$f''(x) = \frac{-10(x^2 - 4)^2 + 40x^2(x^2 - 4)}{(x^2 - 4)^4}$$

$$= \frac{10(x^2 - 4)\left[-(x^2 - 4) + 4x^2\right]}{(x^2 - 4)^4}$$

$$= \frac{10(3x^2 + 4)}{(x^2 - 4)^3}.$$

It is clear that the numerator of f'' is always positive whereas the denominator changes its sign at $x = \pm 2$, that is, at the two vertical asymptotes. Thus, f is convex wherever the denominator is positive, and concave when the denominator is negative, as shown in the following table.

$f''(x)$	$-\infty$	$+$	-2	$-$	2	$+$	∞
$f(x)$		$\smile$		$\frown$		$\smile$	

The graph of f can be seen in Figure 5.2.

(III). $f(x) = \sqrt{4x^2 - x}$. The domain of f is the set of real numbers x such that

$$4x^2 - x = x(4x - 1) \geq 0.$$

The latter inequality holds whenever $x \leq 0$ or $x \geq 1/4$. Hence $D(f) = (-\infty, 0] \cup \left[\frac{1}{4}, \infty\right)$. Also $f'(x) = \dfrac{8x - 1}{2\sqrt{4x^2 - x}}$. Observe that the sign of f' coincides with the sign of the numerator since the denominator is always positive. The numerator vanishes at $x = 1/8$, which is not in the domain. The monotonicity of f is shown in the table below.

$f'(x)$	$-\infty$	$-$	0	$-$	$\frac{1}{4}$	$+$	∞
$f(x)$		$\searrow$		$\curvearrowbotright$		$\nearrow$	

Therefore, f is increasing in $\left[\frac{1}{4}, \infty\right)$ and decreasing in $(-\infty, 0]$.

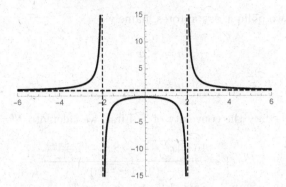

FIGURE 5.2: Graph of the function from Exercise 5.30 part (II) with the corresponding asymptotes (in dashed lines).

Now we study the asymptotes of f. Neither vertical nor horizontal asymptotes are observed. However there is oblique asymptotes whose slopes are given by

$$m_1 = \lim_{x \to \infty} \frac{f(x)}{x} = \lim_{x \to \infty} \frac{\sqrt{4x^2 - x}}{x} = \lim_{x \to \infty} \sqrt{\frac{4x^2}{x^2} - \frac{x}{x^2}} = 2$$

and

$$m_2 = \lim_{x \to -\infty} \frac{f(x)}{x} = \lim_{x \to -\infty} \frac{\sqrt{4x^2 - x}}{x} = \lim_{x \to -\infty} \left[-\sqrt{\frac{4x^2}{x^2} - \frac{x}{x^2}} \right] = -2.$$

The y-intercept corresponding to m_1 and m_2 are, respectively

$$r_1 = \lim_{x \to \infty} [f(x) - m_1 x] = \lim_{x \to \infty} \left[\sqrt{4x^2 - x} - 2x \right]$$

$$= \lim_{x \to \infty} \frac{\left[\sqrt{4x^2 - x} - 2x \right] \left[\sqrt{4x^2 - x} + 2x \right]}{\sqrt{4x^2 - x} + 2x}$$

$$= \lim_{x \to \infty} \frac{-x}{\sqrt{4x^2 - x} + 2x} = \lim_{x \to \infty} \frac{-1}{\sqrt{4 - \frac{1}{x}} + 2} = -\frac{1}{4}$$

and, similarly,

$$r_2 = \lim_{x \to -\infty} [f(x) - m_2 x] = \lim_{x \to -\infty} \left[\sqrt{4x^2 - x} + 2x \right]$$

$$= \lim_{x \to -\infty} \frac{\left[\sqrt{4x^2 - x} + 2x \right] \left[\sqrt{4x^2 - x} - 2x \right]}{\sqrt{4x^2 - x} - 2x}$$

$$= \lim_{x \to -\infty} \frac{-x}{\sqrt{4x^2 - x} - 2x} = \lim_{x \to -\infty} \frac{|x|}{\sqrt{4x^2 - x} + 2|x|}$$

$$= \lim_{x \to -\infty} \frac{1}{\sqrt{4 - \frac{1}{x}} + 2} = \frac{1}{4}.$$

Hence f has two oblique asymptotes, namely

$$y = -2x + \frac{1}{4},$$

$$y = 2x - \frac{1}{4}.$$

Finally, we study the convexity of f. First, we calculate f'':

$$
\begin{aligned}
f''(x) &= \frac{16\sqrt{4x^2 - x} - (8x - 1)\frac{8x-1}{\sqrt{4x^2-x}}}{(2\sqrt{4x^2 - x})^2} \\
&= \frac{16(4x^2 - x) - (8x - 1)^2}{4(4x^2 - x)\sqrt{4x^2 - x}} \\
&= \frac{64x^2 - 16x - 64x^2 + 16x - 1}{4x(4x - 1)\sqrt{4x^2 - x}} \\
&= \frac{-1}{4x(4x - 1)\sqrt{4x^2 - x}}.
\end{aligned}
$$

Since the denominator is positive in $D(f)$, the second derivative is negative in $D(f)$ and therefore f is concave as pointed out in the table below.

$f'(x)$	$-\infty$	$-$	0	$-$	$\frac{1}{4}$	$+$	∞
$f(x)$		$\frown$		$-$		$\frown$	

The graph of f has been represented in Figure 5.3.

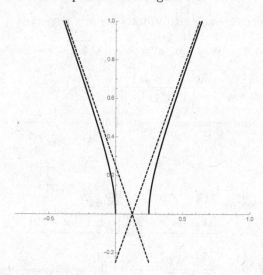

FIGURE 5.3: Graph of the function from Exercise 5.30 part (III) with the corresponding asymptotes (in dashed lines).

(IV). $f(x) = \frac{1}{\log x}$. Notice that $\log x$ is well-defined as long as $x > 0$ and that f is not defined for $x = 1$. Then $D(f) = (0,1) \cup (1,\infty)$. Also $f'(x) = -\dfrac{1}{x(\log x)^2}$, that is $f'(x) < 0$ for all $x \in D(f)$. Then f is decreasing in $(0,1) \cup (1,\infty)$.

Since
$$\lim_{x \to 1^+} f(x) = \infty \quad \text{and} \quad \lim_{x \to 1^-} f(x) = -\infty$$
clearly f has a vertical asymptote at $x = 1$.

Finally, let us study the convexity of f. The second derivative will be needed:
$$f''(x) = \frac{2 + \log x}{(\log x)^3 x^2}.$$

The unique solution of $f''(x) = 0$ is clearly $x = e^{-2}$. Having this in mind, the sign of f'' and the convexity of f are provided in the table below.

$f''(x)$	$-\infty$	$+$	e^{-2}	$-$	1	$+$	∞
$f(x)$		$\smile$		$\frown$		$\smile$	

It is straightforward from the table that f has an inflection point at $x = e^{-2}$. The graph of f can be seen in Figure 5.4.

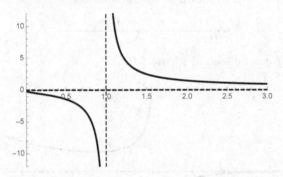

FIGURE 5.4: Graph of the function from Exercise 5.30 part (IV) with the corresponding asymptotes (in dashed lines).

(V). $f(x) = \frac{e^x}{x}$. The function f is well defined for all real numbers except for $x = 0$, so its domain is given by $D(f) = \mathbb{R} \setminus \{0\}$. Also
$$f'(x) = \frac{xe^x - e^x}{x^2} = \frac{e^x(x-1)}{x^2} = (x-1)\frac{e^x}{x^2}.$$

The sign of f' clearly coincides with the sign of $x - 1$, which changes from positive to negative exactly at $x = 1$. Then the sign of f' and the monotonicity of f behave as shown in the table below.

$f'(x)$	$-\infty$	$-$	1	$+$	∞
$f(x)$		$\searrow$		$\nearrow$	

Since f is increasing in $(1, \infty)$ and decreasing in $(-\infty, 1)$, f must have a relative minimum at $x = 0$.

Next we study the asymptotes of f. Since

$$\lim_{x \to 0^+} f(x) = \infty \quad \text{and} \quad \lim_{x \to 0^-} f(x) = -\infty$$

f has a vertical asymptote at $x = 0$. Also, $\lim_{x \to -\infty} f(x) = 0$, so $y = 0$ is a horizontal asymptote of f. No oblique asymptotes are observed.

Finally, we study the convexity of f by calculating first f'':

$$f''(x) = \frac{e^x(x-1)x^2 - e^x(x-1)2x}{x^4}$$

$$= \frac{e^x(x^3 - 2x^2 + 2x)}{x^4}$$

$$= \frac{e^x(x^2 - 2x + 2)}{x^3}.$$

Notice that the polynomial $x^2 - 2x + 2$ does not have real roots. As a matter of fact the numerator $e^x(x^2 - 2x + 2)$ is positive for all $x \in \mathbb{R}$. Then the sign of f'' coincides with the sign of x^3, which allows us to calculate the sign of f'' and the convexity of f as shown in the table below.

$f''(x)$	$-\infty$	$-$	0	$+$	∞
$f(x)$		$\frown$		$\smile$	

The graph of f can be seen in Figure 5.5.

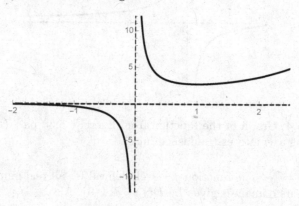

FIGURE 5.5: Graph of the function from Exercise 5.30 part (V) and asymptotes.

(VI). $f(x) = \sqrt[3]{x} - \sqrt[3]{x+1}$. This function is well-defined in the whole real line, so $D(f) = \mathbb{R}$. Also,

$$f'(x) = \frac{1}{3\sqrt[3]{x^2}} - \frac{1}{3\sqrt[3]{(x+1)^2}} = \frac{\sqrt[3]{(x+1)^2} - \sqrt[3]{x^2}}{3\sqrt[3]{x^2(x+1)^2}}$$

if $x \neq 0$ and $x \neq 1$. The denominator vanishes at $x = 0$ and $x = -1$. This means that the tangent at $x = 0$ and $x = 1$ are vertical. The denominator is positive in the rest of the real line, so the sign of f' coincides with the sign of the numerator. To determine the latter we calculate first the roots of the numerator:

$$\sqrt[3]{(x+1)^2} - \sqrt[3]{x^2} = 0 \Leftrightarrow \sqrt[3]{(x+1)^2} = \sqrt[3]{x^2} \Leftrightarrow (x+1)^2 = x^2 \Leftrightarrow x = -\frac{1}{2}.$$

Then the sign of f' and the monotonicity of f are as pointed out in the following table.

$f'(x)$	$-\infty$	$-$	$-\frac{1}{2}$	$+$	∞
$f(x)$		↘		↗	

Therefore, f is increasing in $\left(-\frac{1}{2}, \infty\right)$ and decreasing in $\left(-\infty, -\frac{1}{2}\right)$, so f has a relative minimum at $x = -\frac{1}{2}$.

Next we study the asymptotes. Since

$$\lim_{x \to \infty} f(x) = 0 = \lim_{x \to -\infty} f(x),$$

$y = 0$ is a horizontal asymptote of f. No oblique nor vertical asymptotes are observed.

Finally, we study the convexity of f via its second derivative. If $x \neq 0$ and $x \neq 1$ then

$$f''(x) = -\frac{2}{9\sqrt[3]{x^5}} + \frac{2}{9\sqrt[3]{(x+1)^5}} = \frac{2}{9} \cdot \frac{\sqrt[3]{x^5} - \sqrt[3]{(x+1)^5}}{\sqrt[3]{x^5(x+1)^5}}.$$

We have that f'' does not have roots since

$$f''(x) = 0 \Leftrightarrow \sqrt[3]{x^5} - \sqrt[3]{(x+1)^5} = 0 \Leftrightarrow \sqrt[3]{x^5} = \sqrt[3]{(x+1)^5} \Leftrightarrow x = x + 1$$

and the last equation does not have a solution. However the sign of f'' changes at $x = 0$ and $x = 1$. As a matter of fact, the sign of f'' and the convexity of f are shown in the table below.

$f''(x)$	$-\infty$	$-$	-1	$+$	0	$-$	∞
$f(x)$		⌢		⌣		⌢	

Although the tangents at $x = -1$ and $x = 0$ are vertical, f has inflection points at $x = -1$ and $x = 0$. The graph of f is shown in Figure 5.6.

(VII). $f(x) = \tan^2 x$. First, notice that f is well-defined except for $x = \frac{\pi}{2} + k\pi$ with $k \in \mathbb{Z}$. Also we have that $f'(x) = 2\tan x (1 + \tan^2 x)$. Then,

$$f'(x) = 0 \Leftrightarrow \tan x = 0 \Leftrightarrow x = \pi k, \, k \in \mathbb{Z}.$$

On the other hand

$$f''(x) = 2(1 + \tan^2 x)^2 + 4\tan^2 x (1 + \tan^2 x),$$

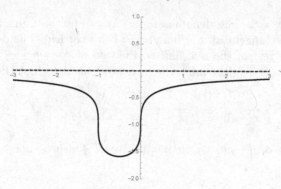

FIGURE 5.6: Graph of the function from Exercise 5.30 part (VI) and asymptote in dashed line.

so $f''(k\pi) = 2 > 0$ for all $k \in \mathbb{Z}$. Then the points $x = k\pi$ with $k \in \mathbb{Z}$ are minimums and therefore, for every $k \in \mathbb{Z}$, f is increasing in $\left(\pi k, \pi k + \frac{\pi}{2}\right)$ and decreasing in $\left(-\frac{\pi}{2} + \pi k, \pi k\right)$. Also, since $f''(x) > 0$ we have that f is convex.

Finally, we study the asymptotes. Since

$$\lim_{x \to (\frac{\pi}{2} + \pi k)^+} f(x) = \lim_{x \to (\frac{\pi}{2} + \pi k)^-} f(x) = \infty$$

then f has a vertical asymptote at $x = \frac{\pi}{2} + \pi k$ for all $k \in \mathbb{Z}$. The graph of f has been drawn in Figure 5.7.

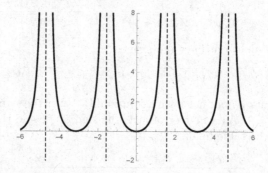

FIGURE 5.7: Graph of the function from Exercise 5.30 part (VII) and its infinitely many vertical asymptotes.

(VIII). $f(x) = x^6 - 3x^4 + 3x^2 - 5$. This function is well-defined in the whole real line, i.e., $D(f) = \mathbb{R}$. Also

$$f'(x) = 6x^5 - 12x^3 + 6x = 6x(x^4 - 2x^2 + 1) = 6x(x^2 - 1)^2,$$

then

$$f'(x) = 0 \Leftrightarrow 6x(x^2 - 1)^2 = 0 \Leftrightarrow x = -1, 0, 1.$$

It is straightforward to check now that the sign of f' and the monotonicity of f are as shown in the table below.

$f'(x)$	$-\infty$	$-$	-1	$-$	0	$+$	1	$+$	∞
$f(x)$		↘		↘		↗		↗	

Clearly f is increasing in $(0, \infty)$ and decreasing in $(-\infty, 0)$, so f has a relative minimum at $x = 0$.

To study the convexity of f we calculate first its second derivative.

$$f''(x) = 30x^4 - 36x^2 + 6 = 6(5x^4 - 6x^2 + 1).$$

Then

$$f''(x) = 0 \Leftrightarrow 5x^4 - 6x^2 + 1 = 0 \overset{x^2 = t}{\Leftrightarrow} 5t^2 - 6t + 1 = 0$$

$$\Leftrightarrow t = \frac{1}{5}, 1 \Leftrightarrow x = \pm 1, \pm \frac{1}{\sqrt{5}}.$$

Now it is easy to see that the sign of f'' and the convexity of f are as pointed out in the next table.

$f''(x)$	$-\infty$	$+$	-1	$-$	$-\frac{1}{\sqrt{5}}$	$+$	$\frac{1}{\sqrt{5}}$	$-$	1	$+$	∞
$f(x)$		⌣		⌢		⌣		⌢		⌣	

Then f has four inflexion points at $x = \pm 1$ and $x = \pm\frac{1}{\sqrt{5}}$. The graph of f can be seen in Figure 5.8.

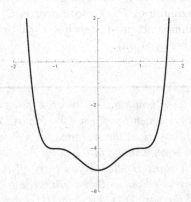

FIGURE 5.8: Graph of the function from Exercise 5.30 part (VIII).

□

Exercise 5.31. *The price of a diamond is proportional to the square of its weight. A diamond that weighs C Carats accidentally fell and broke into two pieces. What is the worst that could have happened? In other words, provide the weight of each piece so that the price, after the break, is the minimum possible.*

Solution to Exercise 5.31.

The function that models the price of our diamond will have the form $f(x) = \alpha \cdot x^2$, where α is some positive constant and x stands for the weight of the gem. After breaking, if the pieces weigh C_1 and C_2 carats respectively (with $C_1 + C_2 = C$), the final price we have will be

$$f(C_1) + f(C_2) = \alpha C_1^2 + \alpha C_2^2.$$

We need to find out the weight of each one of the pieces in order to maximize our loss (or minimize the final value after breaking). Let us call x to the weight of one of the pieces (with $x \in [0, C]$). Then the other one shall weigh $C - x$. Therefore, the final price we have will be

$$P(x) = f(x) + f(C - x) = \alpha x^2 + \alpha(C - x)^2 = \alpha\left(2x^2 - 2Cx + C^2\right).$$

From the fact that $P'(x) = \alpha\left(4x - 2C\right)$ we obtain that the only critical point of P in $(0, C)$ is at $x = C/2$, which is a minimum (since $P''(x) = 4 > 0$). Thus, by breaking the diamond into 2 pieces of identical weight each we will lower the value of the original gem to its minimum possible. On the other hand, if we aim to find the maximum of $P(x)$ above over $[0, C]$ it is straightforward to see that this maximum is attained at either $x = 0$ or $x = C$ (which means that the gem does not break at all).

$\square$

Exercise 5.32. *Imagine you want to reach a point P that is located across the woods from a nearby street (see Figure 5.9). Suppose that the street is straight, and H is the distance from P to the closest point C on the street. You begin walking at the point A, which is at a distance L from C. At which point B should you turn off the street and head across the woods to reach P in order to minimize the walking distance (that is, the sum of the segments AB and BP)?*

Solution to Exercise 5.32.

To find the minimum of the sum of the distances AB and BP we need to minimize the function given by

$$d(x) = L - x + \text{length}(BP) = L - x + \sqrt{H^2 + x^2},$$

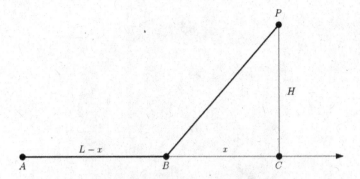

FIGURE 5.9: Representation for Problem 5.32.

with $x \in [0, L]$. Let us first check the critical points of d in $(0, L)$ by finding the zeros of $d'(x)$:

$$d'(x) = -1 + \frac{x}{\sqrt{H^2 + x^2}},$$

which is easy to check that is never 0. Thus, the extreme points must be located at the endpoints of the interval $[0, L]$. Since we have that

$$d(L) = \sqrt{L^2 + H^2} \leq \sqrt{L^2 + H^2 + 2LH} = \sqrt{(L + H)^2} = L + H = d(0).$$

The maximum is attained at $x = 0$ and the minimum at $x = L$. Thus, in order to maximize the distance we should got from A to C and, then, to cross over to P, and in order to minimize the distance we walk, we should go directly from A to P diagonally.

□

Exercise 5.33. *Suppose we have a right and equilateral cone (the slant height equals the diameter of its base) inscribed in a sphere of radius R. Consider the annuli that result in intersecting the sphere and the cone with planes parallel to the base of the cone (see Figure 5.10). Determine the height H (measured from the base of the cone) at which we must intersect the sphere and the cone to obtain an annulus with the maximum possible area.*

Solution to Exercise 5.33.

First, we clarify the notations that will be used in this problem. We denote by r the radius of the cone. Then the slant height must be $2r$. The height of the cone will be denoted by h. According to the Pythagorean theorem we have $h^2 + r^2 = (2r)^2$, from which $h = \sqrt{3}r$. Let us denote by a the distance from the base of the cone to the center of the sphere ($a = h - R$). Applying once more the Pythagorean theorem $R^2 = a^2 + r^2$. Since $h = \sqrt{3}r$, we obtain

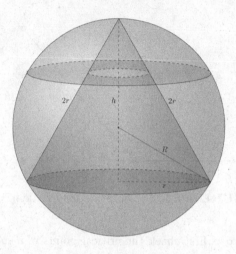

FIGURE 5.10: Situation presented in Exercise 5.33.

$r = \sqrt{3}R/2$ and $h = \sqrt{3}r = 3R/2$. Now denote the radii of the annulus by r_1 and r_2 assuming that $r_2 > r_1$. Then the area we want to optimize is given by $A = \pi(r_2^2 - r_1^2)$. The function A depends on two variables, namely r_1 and r_2. Next we will try to find a relationship between r_1 and r_2 in order to reduce the number of variables to just one. To do so we will express r_1 and r_2 in terms of the height H.

To write r_1 in terms of H we use the intercept theorem about similar triangles (also known as Thales theorem) to obtain $\dfrac{r}{h} = \dfrac{r_1}{h-H}$, from which

$$r_1 = r - r\frac{H}{h} = \frac{\sqrt{3}R}{2} - \frac{\sqrt{3}}{3}H.$$

Thus,

$$r_1^2 = \frac{3R^2}{4} + \frac{H^2}{3} - RH.$$

From Figure 5.11 we see that $x = H - R/2$. Then another use of the Pythagorean theorem shows that $x^2 + r_2^2 = R^2$, that is,

$$r_2^2 = R^2 - x^2 = R^2 - (H - R/2)^2 = \frac{3R^2}{4} + RH - H^2.$$

We have managed to write r_1 and r_2 as a function of the same variable, H, and therefore,

$$A(H) = \pi(r_2^2 - r_1^2)$$
$$= \pi\left[\left(\frac{3R^2}{4} + RH - H^2\right) - \left(\frac{3R^2}{4} + \frac{H^2}{3} - RH\right)\right]$$
$$= \pi\left(2RH - \frac{4H^2}{3}\right).$$

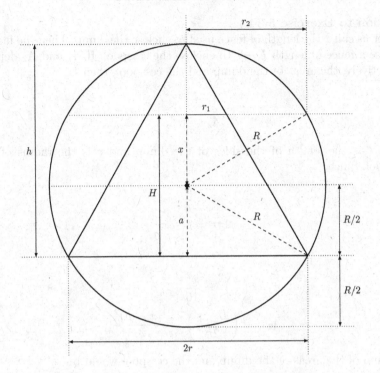

FIGURE 5.11: Geometric elements to tackle Exercise 5.33.

It is straightforward to check that

$$A'(H) = \pi\left(2R - \frac{8H}{3}\right) = 0$$

if and only if $H = \frac{3R}{4}$, and that $A''\left(\frac{3R}{4}\right) = -\frac{8\pi}{3} < 0$, so the annulus attains its maximum area for $H = \frac{3R}{4}$ or, equivalently, for $H = \frac{\sqrt{3}r}{2}$. $\qquad\square$

> **Exercise 5.34.** *A farmer owns enough materials to build a fence of length L with which he wants to enclose a squared dump and a circular cesspool. Find the dimensions of the dump and the cesspool so that the sum of their areas is minimized.*

Solution to Exercise 5.34.

Let us call x the length of fence used to enclose the dump. Then the farmer can use a fence of length $L - x$ to enclose the cesspool. If A_1 and A_2 denote, respectively, the areas of the dump and the cesspool, then

$$A_1 = \ell^2,$$
$$A_2 = \pi r^2,$$

where ℓ is the length of the sides of the dump and r is the radius of the cesspool. Since

$$\ell = \frac{x}{4},$$
$$2\pi r = L - x,$$

we have

$$A_1 = \left(\frac{x}{4}\right)^2 = \frac{x^2}{16},$$
$$A_2 = \pi \left(\frac{L-x}{2\pi}\right)^2 = \frac{(x-L)^2}{4\pi}.$$

The sum of the areas of the dump and the cesspool would be

$$A(x) = A_1(x) + A_2(x) = \frac{x^2}{16} + \frac{(x-L)^2}{4\pi}$$

with $x \in [0, L]$. The equation

$$A'(x) = \frac{x}{8} + \frac{x-L}{2\pi} = 0$$

has a unique root at $x = \frac{4L}{4+\pi}$. It is straightforward that $\frac{4L}{4+\pi} \in [0, L]$ and $A''\left(\frac{4L}{4+\pi}\right) = \frac{4+\pi}{8\pi} > 0$, so A attains its minimum at $x = \frac{4L}{4+\pi}$. Finally, the side ℓ of the dump and the radius r of the cesspool are given, respectively, by

$$\ell = \frac{x}{4} = \frac{L}{4+\pi},$$
$$r = \frac{L-x}{2\pi} = \frac{L}{8+2\pi}.$$

$\square$

Chapter 6

Riemann Integral

Exercise 6.1. *Find primitives of the following functions:*

(I) $\displaystyle\int \frac{1}{\sqrt{x-1}+\sqrt{x+1}}dx.$ (II) $\displaystyle\int \frac{1}{1+\sin x}dx.$

(III) $\displaystyle\int \tan^2 x\,dx.$ (IV) $\displaystyle\int x\sqrt{1-x^2}dx.$

(V) $\displaystyle\int \sin^2 x\,dx.$

Solution to Exercise 6.1.

(I) Rationalizing the integrand we obtain the following:

$$\frac{1}{\sqrt{x-1}+\sqrt{x+1}} = \frac{1}{\sqrt{x-1}+\sqrt{x+1}} \cdot \frac{\sqrt{x-1}-\sqrt{x+1}}{\sqrt{x-1}-\sqrt{x+1}}$$
$$= \frac{1}{2}(\sqrt{x+1}-\sqrt{x-1}).$$

Therefore, we have

$$\int \frac{1}{\sqrt{x-1}+\sqrt{x+1}}dx = \int \frac{1}{2}(\sqrt{x+1}-\sqrt{x-1})dx$$
$$= \frac{1}{2}\int \sqrt{x+1}dx - \frac{1}{2}\int \sqrt{x-1}dx.$$

For the integrand $\sqrt{x+1}$ we make the substitution $u = x+1$ (with $du = dx$). For the integrand $\sqrt{x-1}$ we make $s = x-1$ and $ds = dx$.

DOI: 10.1201/9781003400745-6

Therefore,

$$\int \frac{1}{\sqrt{x-1}+\sqrt{x+1}}dx = \frac{1}{2}\int \sqrt{x+1}dx - \frac{1}{2}\int \sqrt{x-1}dx$$

$$= \frac{1}{2}\int \sqrt{u}du - \frac{1}{2}\int \sqrt{s}ds$$

$$= \frac{1}{2}\cdot\frac{2u^{3/2}}{3} - \frac{1}{2}\cdot\frac{2s^{3/2}}{3} + C$$

$$= \frac{1}{3}u^{3/2} - \frac{1}{3}s^{3/2} + C$$

$$= \frac{1}{3}(x+1)^{3/2} - \frac{1}{3}(x-1)^{3/2} + C$$

$$= \frac{1}{3}((x+1)^{3/2} - (x-1)^{3/2}) + C.$$

(II) As before, let us rationalize the integrand:

$$\frac{1}{1+\sin x} = \frac{1}{1+\sin x}\cdot\frac{1-\sin x}{1-\sin x} = \frac{1-\sin x}{1-\sin^2 x} = \frac{1-\sin x}{\cos^2 x}.$$

Thus,

$$\int \frac{1}{1+\sin x}dx = \int \frac{1}{\cos^2 x}dx - \int \frac{\sin x}{\cos^2 x}dx = \int \sec^2 x dx - \int \frac{\sin x}{\cos^2 x}dx.$$

For the integrand $\frac{\sin x}{\cos^2 x}$, substitute $u = \cos x$ and $du = -\sin x dx$:

$$\int \frac{1}{1+\sin x}dx = \int \sec^2 x dx - \int \frac{\sin x}{\cos^2 x}dx$$

$$= \tan x + \int \frac{1}{u^2}du$$

$$= \tan x - \frac{1}{u} + C$$

$$= \tan x - \frac{1}{\cos x} + C$$

$$= \tan x - \sec x + C.$$

(III) Write $\tan^2 x$ as $\sec^2 x - 1$. Thus

$$\int \tan^2 x dx = \int \sec^2 x dx - \int dx = \tan x - x + C.$$

(IV) For the integrand $x\sqrt{1-x^2}$, substitute $u = 1 - x^2$, with $du = -2xdx$. We have

$$\int x\sqrt{1-x^2}dx = -\frac{1}{2}\int \sqrt{u}du = -\frac{1}{2}\cdot\frac{2u^{3/2}}{3} + C$$

$$= -\frac{1}{3}u^{3/2} + C = -\frac{1}{3}(1-x^2)^{3/2} + C.$$

(v) Write $\sin^2 x$ as $\frac{1}{2} - \frac{1}{2}\cos(2x)$. So,

$$\int \sin^2 x\, dx = \frac{1}{2}\int dx - \frac{1}{2}\int \cos(2x)\, dx.$$

For the integrand $\cos(2x)$, substitute $u = 2x$ and $du = 2dx$. We have

$$\int \sin^2 x\, dx = \frac{1}{2}\int dx - \frac{1}{2}\int \cos(2x)\, dx$$

$$= \frac{x}{2} - \frac{1}{4}\int \cos u\, du$$

$$= \frac{x}{2} - \frac{1}{4}\sin u + C$$

$$= \frac{x}{2} - \frac{1}{4}\sin(2x) + C$$

$$= \frac{x}{2} - \frac{1}{4}\cdot 2\sin x \cos x + C$$

$$= \frac{1}{2}(x - \sin x \cos x) + C.$$

□

Exercise 6.2. *Calculate the following indefinite integrals using integration by parts:*

(I) $\displaystyle\int x^2 e^x\, dx.$ (II) $\displaystyle\int x\cos x\, dx.$ (III) $\displaystyle\int \arctan x\, dx.$

(IV) $\displaystyle\int \log|x|\, dx.$ (V) $\displaystyle\int \arcsin x\, dx.$ (VI) $\displaystyle\int e^x \cos x\, dx.$

(VII) $\displaystyle\int \log^3 x\, dx.$ (VIII) $\displaystyle\int \frac{\log\log x}{x}\, dx.$ (IX) $\displaystyle\int x\log^2 x\, dx.$

(X) $\displaystyle\int \frac{x\arcsin x}{\sqrt{1-x^2}}\, dx.$

Solution to Exercise 6.2.

(I) For the integrand $x^2 e^x$ we integrate by parts ($\int f\, dg = fg - \int g\, df$) putting $f = x^2$, $dg = e^x dx$, $df = 2x dx$ and $g = e^x$. Thus,

$$\int x^2 e^x\, dx = x^2 e^x - 2\int x e^x\, dx.$$

For the integrand $e^x x$, we integrate by parts, $\int f\, dg = fg - \int g\, df$, where

$f = x$, $dg = e^x dx$, $df = dx$ and $g = e^x$. Therefore,

$$\int x^2 e^x dx = x^2 e^x - 2\left[xe^x - \int e^x dx\right]$$
$$= x^2 e^x - 2xe^x + 2e^x + C$$
$$= e^x(x^2 - 2x + 2) + C.$$

(II) For the integrand $x \cos x$, we integrate by parts, $\int f dg = fg - \int g df$, with

$$f = x, \ dg = \cos x \, dx, \ df = dx \text{ and } g = \sin x.$$

Thus

$$\int x \cos x dx = x \sin x - \int \sin x dx = x \sin x + \cos x + C.$$

(III) For the integrand $\arctan x$, integrating by parts, $\int f dg = fg - \int g df$, where

$$f = \arctan x, \ dg = dx, \ df = dx/(x^2 + 1) \text{ and } g = x.$$

Therefore,

$$\int \arctan x dx = x \arctan x - \int \frac{x}{x^2 + 1} dx.$$

For the integrand $x/(x^2 + 1)$, substitute $u = x^2 + 1$ and $du = 2x dx$. Thus

$$\int \arctan x dx = x \arctan x - \int \frac{x}{x^2 + 1} dx$$
$$= x \arctan x - \frac{1}{2} \int \frac{1}{u} du$$
$$= x \arctan x - \frac{\log|u|}{2} + C$$
$$= x \arctan x - \frac{1}{2} \log(x^2 + 1) + C.$$

(IV) We shall divide the solution of this exercise in two cases: $x \geq 0$ and $x < 0$.

(i) For $x \geq 0$ we have

$$\int \log|x| dx = \int \log x dx.$$

For the integrand $\log x$, we integrate by parts, $\int f dg = fg - \int g df$, where $f = \log x$, $dg = dx$, $df = 1/x$ and $g = x$. Thus,

$$\int \log x dx = x \log x - \int x \frac{1}{x} dx = x \log x - x + C.$$

(ii) For $x < 0$,

$$\int \log|x|dx = \int \log(-x)dx.$$

Consider $u = -x$ and $du = -dx$. Therefore, by item (i), it follows that

$$\int \log|x|dx = \int \log(-x)dx$$
$$= -\int \log(u)du$$
$$= -(u\log u - u) + C$$
$$= -(-x\log(-x) + x) + C$$
$$= x\log(-x) - x + C.$$

From cases (i) and (ii) above, we conclude that

$$\int \log|x|dx = x\log|x| - x + C.$$

(v) For the integrand $\arcsin x$, we integrate by parts ($\int fdg = fg - \int gdf$) where

$$f = \arcsin x, \; dg = dx, \; df = dx/\sqrt{1-x^2} \text{ and } g = x.$$

Thus,

$$\int \arcsin xdx = x\arcsin x - \int \frac{x}{\sqrt{1-x^2}}dx.$$

For the integrand $x/\sqrt{1-x^2}$, substitute $u = 1 - x^2$ (with $du = -2xdx$). We have

$$\int \arcsin xdx = x\arcsin x - \int \frac{x}{\sqrt{1-x^2}}dx$$
$$= x\arcsin x + \frac{1}{2}\int \frac{1}{\sqrt{u}}du$$
$$= x\arcsin x + \frac{1}{2}\cdot 2\sqrt{u} + C$$
$$= x\arcsin x + \sqrt{1-x^2} + C.$$

(VI) This is a typical example of integrating by parts twice (a "cyclic" integral). For the integrand $e^x\cos x$, integrate by parts, $\int fdg = fg - \int gdf$, where

$$f = \cos x, \; dg = e^xdx, \; df = -\sin xdx \text{ and } g = e^x.$$

Therefore,

$$\int e^x\cos xdx = e^x\cos x + \int e^x\sin xdx.$$

For the integrand $e^x \sin x$, we integrate by parts once more ($\int f dg = fg - \int g df$), where $f = \sin x$, $dg = e^x dx$, $df = \cos x dx$ and $g = e^x$, we have

$$\int e^x \cos x dx = e^x \cos x + \int e^x \sin x dx = e^x \cos x + e^x \sin x - \int e^x \cos x dx.$$

Adding $\int e^x \cos x dx$ to both sides and then dividing both sides by 2, we obtain

$$\int e^x \cos x dx = \frac{1}{2}(e^x \sin x + e^x \cos x) + C = \frac{e^x}{2}(\sin x + \cos x) + C.$$

(VII) For the integrand $\log^3 x$, we integrate by parts, $\int f dg = fg - \int g df$, where $f = \log^3 x$, $dg = dx$, $df = \frac{3\log^2 x}{x} dx$ and $g = x$. Thus

$$\int \log^3 x dx = x \log^3 x - 3 \int \log^2 x dx.$$

For the integrand $\log^2 x$, we integrate by parts, $\int f dg = fg - \int g df$, where $f = \log^2 x$, $dg = dx$, $df = \frac{2\log x}{x} dx$ and $g = x$. Then

$$\int \log^3 x dx = x \log^3 x - 3 \int \log^2 x dx = x \log^3 x - 3x \log^2 x + 6 \int \log x dx.$$

It follows that

$$\int \log^3 x dx = x \log^3 x - 3x \log^2 x + 6 \int \log x dx$$
$$= x \log^3 x - 3x \log^2 x + 6(x \log x - x) + C$$
$$= x(\log^3 x - 3 \log^2 x + 6 \log x - 6) + C.$$

(VIII) For the integrand $(\log \log x)/x$, substitute $u = \log x$ and $du = dx/x$. It follows that

$$\int \frac{\log \log x}{x} dx = \int \log u\, du = u \log u - u + C = \log x(\log \log x - 1) + C.$$

(IX) For the integrand $x \log^2 x$, we integrate by parts, $\int f dg = fg - \int g df$, where

$$f = \log^2 x, \; dg = x dx, \; df = \frac{2 \log x}{x} dx, \; g = x^2/2.$$

Therefore,

$$\int x \log^2 x dx = \frac{x^2 \log^2 x}{2} - \int x \log x dx.$$

For the integrand $x \log x$, we integrate by parts, $\int f dg = fg - \int g df$, making

$$f = \log x, \; dg = x dx, \; df = dx/x \text{ and } g = x^2/2.$$

Then

$$\int x \log^2 x\, dx = \frac{x^2 \log^2 x}{2} - \int x \log x\, dx$$

$$= \frac{x^2 \log^2 x}{2} - \frac{x^2 \log x}{2} + \frac{1}{2}\int x\, dx$$

$$= \frac{x^2 \log^2 x}{2} - \frac{x^2 \log x}{2} + \frac{x^2}{4} + C$$

$$= \frac{x^2}{2}\left(\log^2 x - \log x + \frac{1}{2}\right) + C.$$

(x) For the integrand $(x \arcsin x)/\sqrt{1-x^2}$, we integrate by parts, $\int f dg = fg - \int g df$, where

$$f = \arcsin x,\ dg = \frac{x}{\sqrt{1-x^2}} dx,\ df = dx/\sqrt{1-x^2} \text{ and } g = -\sqrt{1-x^2}.$$

Therefore,

$$\int \frac{x \arcsin x}{\sqrt{1-x^2}} dx = -\sqrt{1-x^2}\arcsin x + \int dx = -\sqrt{1-x^2}\arcsin x + x + C.$$

□

Exercise 6.3. *Calculate the primitives of the following rational functions:*

(I) $\int \dfrac{2x^2+7x-1}{x^3+x^2-x-1} dx.$ (II) $\int \dfrac{x+4}{x^2+1} dx.$

(III) $\int \dfrac{x^2+x+2}{x^4+2x^2+1} dx.$ (IV) $\int \dfrac{2x^2+x+1}{(x+3)(x-1)^2} dx.$

(V) $\int \dfrac{1}{x^4+1} dx.$ (VI) $\int \dfrac{2x-1}{(x^2+2)^2} dx.$

Solution to Exercise 6.3.

(I) Observe that

$$\int \frac{2x^2+7x-1}{x^3+x^2-x-1} dx = \int \frac{2x^2+7x-1}{(x-1)(x+1)^2} dx$$

and use simple fractions. The simple fraction expansion is of the form

$$\frac{2x^2+7x-1}{(x-1)(x+1)^2} = \frac{A}{x-1} + \frac{B}{x+1} + \frac{C}{(x+1)^2}.$$

Multiply both sides by $(x-1)(x+1)^2$:

$$2x^2 + 7x - 1 = A(x+1)^2 + B(x-1)(x+1) + C(x-1).$$

Expand and collect in terms of powers of x, that is,

$$2x^2 + 7x - 1 = (A+B)x^2 + (2A+C)x + A - B - C.$$

Equate coefficients on both sides, yielding 3 equations in 3 unknowns:

$$\begin{cases} -1 = A - B - C, \\ 7 = 2A + C, \\ 2 = A + B. \end{cases}$$

Solving the above system, we obtain

$$A = 2, B = 0 \text{ and } C = 3.$$

Therefore,

$$\int \frac{2x^2 + 7x - 1}{x^3 + x^2 - x - 1} dx = \int \frac{2x^2 + 7x - 1}{(x-1)(x+1)^2} dx$$

$$= \int \left(\frac{2}{x-1} + \frac{3}{(x+1)^2} \right) dx$$

$$= 2 \int \frac{1}{x-1} dx + 3 \int \frac{1}{(x+1)^2} dx.$$

For the integrand $1/(x-1)$, substitute $s = x - 1$ and $ds = dx$. For the integrand $1/(x+1)^2$ substitute $u = x + 1$ and $du = dx$. Thus

$$\int \frac{2x^2 + 7x - 1}{x^3 + x^2 - x - 1} dx = 2 \int \frac{1}{x-1} dx + 3 \int \frac{1}{(x+1)^2} dx$$

$$= 2 \int \frac{1}{s} ds + 3 \int \frac{1}{u^2} du$$

$$= 2 \log |s| - \frac{3}{u} + C$$

$$= 2 \log |x - 1| - \frac{3}{x+1} + C.$$

(II) Expanding the integrand $(x+4)/(x^2+1)$ we get

$$\int \frac{x+4}{x^2+1} dx = \int \left(\frac{x}{x^2+1} + \frac{4}{x^2+1} \right) dx = \int \frac{x}{x^2+1} dx + 4 \int \frac{1}{x^2+1} dx.$$

For the integrand $x/(x^2+1)$, substitute $u = x^2 + 1$ and $du = 2x dx$. Then

$$\int \frac{x+4}{x^2+1} dx = \frac{1}{2} \int \frac{1}{u} du + 4 \int \frac{1}{x^2+1} dx.$$

The integral of $1/u$ is $\log|u|$ and the integral of $1/(x^2+1)$ is $\arctan x$. Thus

$$\int \frac{x+4}{x^2+1}\,dx = \frac{1}{2}\log|u|+4\arctan x+C = \frac{1}{2}\log(x^2+1)+4\arctan x+C.$$

(III) Observe that

$$\int \frac{x^2+x+2}{x^4+2x^2+1}\,dx = \int \frac{x^2+x+2}{(x^2+1)^2}\,dx$$

and use simple fractions. The simple fraction expansion is of the form:

$$\frac{x^2+x+2}{(x^2+1)^2} = \frac{Ax+B}{x^2+1} + \frac{Cx+D}{(x^2+1)^2},$$

Multiply both sides by $(x^2+1)^2$, expand and collect in terms of powers of x:

$$x^2+x+2 = (Ax+B)(x^2+1)+(Cx+D) = Ax^3+Bx^2+(A+C)x+B+D.$$

Equate coefficients on both sides, yielding the system

$$\begin{cases} 0 = A, \\ 1 = B, \\ 1 = A+C, \\ 2 = B+D, \end{cases}$$

whose solution is

$$A = 0, B = C = D = 1.$$

Therefore,

$$\int \frac{x^2+x+2}{x^4+2x^2+1}\,dx = \int \frac{x^2+x+2}{(x^2+1)^2}\,dx$$

$$= \int \left(\frac{1}{x^2+1} + \frac{x+1}{(x^2+1)^2} \right) dx$$

$$= \int \frac{1}{x^2+1}\,dx + \int \frac{x+1}{(x^2+1)^2}\,dx.$$

The integral of $1/(x^2+1)$ is arc tan x. For the integrand $(x+1)/(x^2+1)^2$, substitute $u = \text{arc tan } x$, that is, $x = \tan u$ and $dx = \sec^2 u\, du$. Thus,

$$\int \frac{x^2 + x + 2}{x^4 + 2x^2 + 1}\, dx = \int \frac{1}{x^2 + 1}\, dx + \int \frac{x+1}{(x^2+1)^2}\, dx$$

$$= \text{arc tan } x + \int \frac{\tan u + 1}{(\tan^2 u + 1)^2} \sec^2 u\, du$$

$$= \text{arc tan } x + \int \frac{\tan u + 1}{\sec^4 u} \sec^2 u\, du$$

$$= \text{arc tan } x + \int \cos^2 u(\tan u + 1)\, du$$

$$= \text{arc tan } x + \int (1 - \sin^2 u)(\tan u + 1)\, du$$

$$= \text{arc tan } x + \int (1 - \sin^2 u + \tan u - \sin^2 u \tan u)\, du$$

$$= \text{arc tan } x + \int du - \int \sin^2 u\, du$$

$$+ \int \tan u\, du - \int \sin^2 u \tan u\, du$$

$$= \text{arc tan } x + \int du - \int \sin^2 u\, du$$

$$+ \int \tan u\, du - \int (1 - \cos^2 u) \tan u\, du$$

$$= \text{arc tan } x + \int du - \int \sin^2 u\, du + \int \tan u\, du$$

$$- \int \tan u\, du + \int \tan u \cos^2 u\, du$$

$$= \text{arc tan } x + \int du - \int \sin^2 u\, du + \int \sin u \cos u\, du.$$

The integral of 1 is u, and from previous exercises we know that a primitive of $\sin^2 u$ is $(u - \sin u \cos u)/2$. Therefore,

$$\int \frac{x^2 + x + 2}{x^4 + 2x^2 + 1}\, dx = \text{arc tan } x + u - \frac{1}{2}(u - \sin u \cos u) + \int \sin u \cos u\, du$$

$$= \frac{3}{2} \text{arc tan } x + \frac{1}{2} \sin(\text{arc tan } x) \cos(\text{arc tan } x)$$

$$+ \int \sin u \cos u\, du.$$

Consider the change $s = \cos u$ and $ds = -\sin u du$. So,

$$\int \frac{x^2 + x + 2}{x^4 + 2x^2 + 1} dx = \frac{3}{2} \arctan x + \frac{1}{2} \sin(\arctan x) \cos(\arctan x) - \int s\,ds$$

$$= \frac{3}{2} \arctan x + \frac{1}{2} \sin(\arctan x) \cos(\arctan x)$$

$$- \frac{s^2}{2} + C$$

$$= \frac{3}{2} \arctan x + \frac{1}{2} \sin(\arctan x) \cos(\arctan x)$$

$$- \frac{\cos^2(\arctan x)}{2} + C.$$

Recall that

$$\sin(\arctan x) = \frac{x}{\sqrt{1 + x^2}} \quad \text{and} \quad \cos(\arctan x) = \frac{1}{\sqrt{1 + x^2}}.$$

Thus

$$\int \frac{x^2 + x + 2}{x^4 + 2x^2 + 1} dx = \frac{3}{2} \arctan x + \frac{1}{2} \sin(\arctan x) \cos(\arctan x)$$

$$- \frac{\cos^2(\arctan x)}{2} + C$$

$$= \frac{3}{2} \arctan x + \frac{1}{2} \frac{x}{\sqrt{1 + x^2}} \frac{1}{\sqrt{1 + x^2}} - \frac{1}{2(1 + x^2)}$$

$$= \frac{3(x^2 + 1) \arctan x + x - 1}{2(x^2 + 1)}.$$

(IV) For the integrand $(2x^2 + x + 1)/((x - 1)^2(x + 3))$, use simple fraction decomposition. The simple fraction expansion is of the form

$$\frac{2x^2 + x + 1}{(x - 1)^2(x + 3)} = \frac{A}{x - 1} + \frac{B}{(x - 1)^2} + \frac{C}{x + 3}.$$

Multiply both sides by $(x - 1)^2(x + 3)$, expand and collect in terms of powers of x:

$$2x^2 + x + 1 = (A + C)x^2 + (2A + B - 2C)x - 3A + 3B + C.$$

Equate coefficients on both sides, yielding 3 equations in 3 unknowns:

$$\begin{cases} 2 = A + C, \\ 1 = 2A + B - 2C, \\ 1 = -3A + 3B + C. \end{cases}$$

The solution of the above system is

$$A = B = C = 1.$$

and therefore

$$\int \frac{2x^2 + x + 1}{(x-1)^2(x+3)}dx = \int \left(\frac{1}{x+3} + \frac{1}{x-1} + \frac{1}{(x-1)^2} \right) dx$$

$$= \int \frac{1}{x+3}dx + \int \frac{1}{x-1}dx + \int \frac{1}{(x-1)^2}dx.$$

For the integrand $1/(x+3)$, substitute $u = x + 3$ and $du = dx$. For the integrands $1/(x-1)$ and $1/(x-1)^2$, substitute $v = x - 1$ and $dv = dx$. Then

$$\int \frac{2x^2 + x + 1}{(x-1)^2(x+3)}dx = \int \frac{1}{x+3}dx + \int \frac{1}{x-1}dx + \int \frac{1}{(x-1)^2}dx$$

$$= \int \frac{1}{u}du + \int \frac{1}{v}dv + \int \frac{1}{v^2}dv$$

$$= \log|u| + \log|v| - 1/v + C$$

$$= \log|x+3| + \log|x-1| - \frac{1}{x-1} + C.$$

(v) First, notice that $x^4 + 1$ can be expressed as:

$$x^4 + 1 = x^4 + 2x^2 + 1 - 2x^2 = (x^2+1)^2 - 2x^2 = (x^2+1-\sqrt{2}x)(x^2+1+\sqrt{2}x),$$

where we have used in the last equality: $a^2 - b^2 = (a-b)(a+b)$. Hence,

$$\int \frac{1}{x^4+1}dx = \int \frac{1}{(x^2 - \sqrt{2}x + 1)(x^2 + \sqrt{2}x + 1)}dx$$

Observe that both $x^2 - \sqrt{2}x + 1$ and $x^2 + \sqrt{2}x + 1$ are irreducible polynomials of degree 2 and now use simple fractions. The simple fraction decomposition is as follows:

$$\frac{1}{(x^2 - \sqrt{2}x + 1)(x^2 + \sqrt{2}x + 1)} = \frac{Ax + B}{x^2 - \sqrt{2}x + 1} + \frac{Cx + D}{x^2 + \sqrt{2}x + 1}.$$

Multiplying both sides of the equality by $(x^2 - \sqrt{2}x + 1)(x^2 + \sqrt{2}x + 1)$ we obtain:

$$1 = (Ax + B)\left(x^2 + \sqrt{2}x + 1\right) + (Cx + D)\left(x^2 - \sqrt{2}x + 1\right)$$

$$= (A + C)x^3 + (\sqrt{2}A + B - \sqrt{2}C + D)x^2$$

$$+ (A + \sqrt{2}B + C - \sqrt{2}D)x + (B + D).$$

Identifying coefficients on both sides, yielding 4 equations in 4 unknowns:

$$\begin{cases} 0 = A + C, \\ 0 = \sqrt{2}A + B - \sqrt{2}C + D, \\ 0 = A + \sqrt{2}B + C - \sqrt{2}D, \\ 1 = B + D. \end{cases}$$

The solution of the above system is:

$$A = -\frac{1}{2\sqrt{2}}, \quad C = \frac{1}{2\sqrt{2}}, \quad B = D = \frac{1}{2},$$

Thus,

$$\int \frac{1}{x^4+1}\,dx = \int \left(\frac{-\frac{1}{2\sqrt{2}}x + \frac{1}{2}}{x^2 - \sqrt{2}x + 1} + \frac{\frac{1}{2\sqrt{2}}x + \frac{1}{2}}{x^2 + \sqrt{2}x + 1} \right) dx$$

$$= -\frac{1}{2\sqrt{2}} \int \frac{x - \sqrt{2}}{x^2 - \sqrt{2}x + 1}\,dx$$

$$+ \frac{1}{2\sqrt{2}} \int \frac{x + \sqrt{2}}{x^2 + \sqrt{2}x + 1}\,dx. \tag{6.1}$$

For the first integral in (6.1), observe that

$$x - \sqrt{2} = \frac{1}{2}\left(2x - \sqrt{2}\right) - \frac{\sqrt{2}}{2},$$

which implies

$$\int \frac{x + \sqrt{2}}{x^2 - \sqrt{2}x + 1}\,dx = \frac{1}{2} \int \frac{2x - \sqrt{2}}{x^2 - \sqrt{2}x + 1}\,dx - \frac{\sqrt{2}}{2} \int \frac{1}{x^2 - \sqrt{2}x + 1}\,dx.$$

Let us calculate now $\int \dfrac{2x - \sqrt{2}}{x^2 - \sqrt{2}x + 1}\,dx$. Apply the substitution

$$u = x^2 - \sqrt{2}x + 1 \text{ and } du = (2x - \sqrt{2})dx:$$

$$\int \frac{2x - \sqrt{2}}{x^2 - \sqrt{2}x + 1}\,dx = \int \frac{1}{u}\,du = \log|u| + C = \log\left|x^2 - \sqrt{2}x + 1\right| + C.$$

Next we calculate $\int \dfrac{1}{x^2 - \sqrt{2}x + 1}\,dx$:

$$\int \frac{1}{x^2 - \sqrt{2}x + 1}\,dx = \int \frac{1}{\left(x - \frac{\sqrt{2}}{2}\right)^2 - \frac{1}{2} + 1}\,dx$$

$$= \int \frac{1}{\left(x - \frac{\sqrt{2}}{2}\right)^2 + \frac{1}{2}}\,dx$$

$$= 2 \int \frac{1}{2\left(x - \frac{\sqrt{2}}{2}\right)^2 + 1}\,dx$$

$$= 2 \int \frac{1}{\left[\sqrt{2}\left(x - \frac{\sqrt{2}}{2}\right)\right]^2 + 1}\,dx$$

$$= 2 \int \frac{1}{\left(\sqrt{2}x - 1\right)^2 + 1}\,dx.$$

Applying the substitution $u = \sqrt{2}x - 1$ and $du = \sqrt{2}dx$ (so $dx = 1/\sqrt{2}du$) we obtain:

$$\int \frac{1}{x^2 - \sqrt{2}x + 1} dx = \frac{2}{\sqrt{2}} \int \frac{1}{u^2 + 1} dx$$

$$= \frac{2}{\sqrt{2}} \arctan(u) + C$$

$$= \frac{2}{\sqrt{2}} \arctan\left(\sqrt{2}x - 1\right) + C.$$

Hence,

$$\int \frac{x + \sqrt{2}}{x^2 - \sqrt{2}x + 1} dx = \frac{1}{2} \log\left|x^2 - \sqrt{2}x + 1\right|$$

$$- \frac{\sqrt{2}}{2} \cdot \sqrt{2} \arctan\left(\sqrt{2}x - 1\right) + C$$

$$= \frac{1}{2} \log\left|x^2 - \sqrt{2}x + 1\right|$$

$$- \arctan\left(\sqrt{2}x - 1\right) + C.$$

For the second integral in (6.1), observe that

$$x + \sqrt{2} = \frac{1}{2}\left(2x + \sqrt{2}\right) + \frac{\sqrt{2}}{2},$$

which implies

$$\int \frac{x + \sqrt{2}}{x^2 + \sqrt{2}x + 1} dx = \frac{1}{2} \int \frac{2x + \sqrt{2}}{x^2 + \sqrt{2}x + 1} dx + \frac{\sqrt{2}}{2} \int \frac{1}{x^2 + \sqrt{2}x + 1} dx.$$

Let us now calculate $\displaystyle\int \frac{2x + \sqrt{2}}{x^2 + \sqrt{2}x + 1} dx$. If we apply the substitution $u = x^2 + \sqrt{2}x + 1$ and $du = (2x + \sqrt{2})dx$, we have

$$\int \frac{2x + \sqrt{2}}{x^2 + \sqrt{2}x + 1} dx = \int \frac{1}{u} du = \log|u| + C = \log\left|x^2 + \sqrt{2}x + 1\right| + C.$$

Now we work on $\int \dfrac{1}{x^2 + \sqrt{2}x + 1}\,dx$:

$$\int \frac{1}{x^2 + \sqrt{2}x + 1}\,dx = \int \frac{1}{\left(x + \frac{\sqrt{2}}{2}\right)^2 - \frac{1}{2} + 1}\,dx$$

$$= \int \frac{1}{\left(x - \frac{\sqrt{2}}{2}\right)^2 + \frac{1}{2}}\,dx$$

$$= 2\int \frac{1}{2\left(x + \frac{\sqrt{2}}{2}\right)^2 + 1}\,dx$$

$$= 2\int \frac{1}{\left[\sqrt{2}\left(x + \frac{\sqrt{2}}{2}\right)\right]^2 + 1}\,dx$$

$$= 2\int \frac{1}{\left(\sqrt{2}x + 1\right)^2 + 1}\,dx.$$

Applying now the substitution

$$u = \sqrt{2}x + 1 \text{ and } du = \sqrt{2}dx \text{ (so } dx = 1/\sqrt{2}du),$$

we have:

$$\int \frac{1}{x^2 - \sqrt{2}x + 1}\,dx = \frac{2}{\sqrt{2}}\int \frac{1}{u^2 + 1}\,dx$$

$$= \frac{2}{\sqrt{2}}\arctan(u) + C$$

$$= \frac{2}{\sqrt{2}}\arctan\left(\sqrt{2}x + 1\right) + C.$$

Hence,

$$\int \frac{x + \sqrt{2}}{x^2 + \sqrt{2}x + 1}\,dx = \frac{1}{2}\log\left|x^2 + \sqrt{2}x + 1\right|$$

$$+ \frac{\sqrt{2}}{2}\cdot\frac{2}{\sqrt{2}}\arctan\left(\sqrt{2}x + 1\right) + C$$

$$= \frac{1}{2}\log\left|x^2 + \sqrt{2}x + 1\right|$$

$$+ \arctan\left(\sqrt{2}x + 1\right) + C.$$

Therefore,

$$\int \frac{1}{x^4+1}dx = -\frac{1}{2\sqrt{2}}\left(\frac{1}{2}\log\left|x^2-\sqrt{2}x+1\right| - \arctan\left(\sqrt{2}x-1\right)\right)$$

$$+ \frac{1}{2\sqrt{2}}\left(\frac{1}{2}\log\left|x^2+\sqrt{2}x+1\right| + \arctan\left(\sqrt{2}x+1\right)\right) + C$$

$$= \frac{1}{4\sqrt{2}}\log\left|\frac{x^2+\sqrt{2}x+1}{x^2-\sqrt{2}x+1}\right| + \frac{1}{2\sqrt{2}}\arctan\left(\sqrt{2}x+1\right)$$

$$+ \frac{1}{2\sqrt{2}}\arctan\left(\sqrt{2}x-1\right) + C.$$

(VI) For the integrand $(2x-1)/(x^2+2)^2$, substitute $u = \arctan(x/\sqrt{2})$, that is, $x = \sqrt{2}\tan u$ and $dx = \sqrt{2}\sec^2 u\,du$. Then

$$(x^2+2)^2 = (2\tan^2 u + 2)^2 = 4\sec^4 u$$

and

$$\int \frac{2x-1}{(x^2+2)^2}dx = \sqrt{2}\int \frac{1}{4}\cos^2 u(2\sqrt{2}\tan u - 1)du$$

$$= \frac{\sqrt{2}}{4}\int \cos^2 u(2\sqrt{2}\tan u - 1)du.$$

Write $\cos^2 u$ as $1 - \sin^2 u$ and expand the integrand:

$$\int \frac{2x-1}{(x^2+2)^2}dx = \frac{\sqrt{2}}{4}\int (1-\sin^2 u)(2\sqrt{2}\tan u - 1)du$$

$$= \int \tan u\,du - \frac{\sqrt{2}}{4}\int du - \int \sin^2 u \tan u\,du$$

$$+ \frac{\sqrt{2}}{4}\int \sin^2 u\,du$$

$$= \int \tan u\,du - \frac{\sqrt{2}}{4}\int du - \int (1-\cos^2 u)\tan u\,du$$

$$+ \frac{\sqrt{2}}{4}\int \sin^2 u\,du$$

$$= \int \tan u\,du - \frac{\sqrt{2}}{4}\int du - \int \tan u\,du$$

$$+ \int \cos^2 u \tan u\,du + \frac{\sqrt{2}}{4}\int \sin^2 u\,du$$

$$= -\frac{\sqrt{2}}{4}\int du + \int \sin u \cos u\,du + \frac{\sqrt{2}}{4}\int \sin^2 u\,du.$$

For the integrand $\sin u \cos u$, substitute $v = \cos u$ and $dv = -\sin u du$.
Then

$$\int \frac{2x-1}{(x^2+2)^2}dx = -\frac{\sqrt{2}}{4}\int du - \int v dv + \frac{\sqrt{2}}{4}\int \sin^2 u du.$$

We can now conclude that

$$\int \frac{2x-1}{(x^2+2)^2}dx = -\frac{\sqrt{2}}{4}u - \frac{v^2}{2} + \frac{\sqrt{2}}{4}\left(\frac{1}{2}(u - \sin u \cos u)\right) + C$$

$$= -\frac{\sqrt{2}}{4}u - \frac{\cos^2 u}{2} + \frac{\sqrt{2}}{8}(u - \sin u \cos u) + C$$

$$= -\frac{\sqrt{2}}{8}u - \frac{\cos^2 u}{2} - \frac{\sqrt{2}}{8}\sin u \cos u + C.$$

Recall that $u = \arctan(x/\sqrt{2})$ and that

$$\sin(\arctan z) = \frac{z}{\sqrt{1+z^2}} \quad \text{and} \quad \cos(\arctan z) = \frac{1}{\sqrt{1+z^2}}.$$

Therefore,

$$\int \frac{2x-1}{(x^2+2)^2}dx = -\frac{\sqrt{2}}{8}u - \frac{\cos^2 u}{2} - \frac{\sqrt{2}}{8}\sin u \cos u + C$$

$$= -\frac{\sqrt{2}}{8}\arctan\left(\frac{x}{\sqrt{2}}\right) - \frac{1}{2\left(1+\frac{x^2}{2}\right)} - \frac{x}{8\left(1+\frac{x^2}{2}\right)} + C$$

$$= -\frac{\sqrt{2}}{8}\arctan\left(\frac{x}{\sqrt{2}}\right) - \frac{1}{2+x^2} - \frac{x}{4(2+x^2)} + C$$

$$= -\frac{1}{4}\left(\frac{\sqrt{2}}{2}\arctan\left(\frac{x}{\sqrt{2}}\right) + \frac{4+x}{2+x^2}\right) + C.$$

$\square$

Exercise 6.4. *Calculate the following primitives by making an adequate change of variable:*

(I) $\int \sqrt{1-x^2}dx.$

(II) $\int \sqrt{x^2+1}dx.$

(III) $\int x\sqrt{x+1}dx.$

(IV) $\int \frac{e^x + 2e^{-x}}{e^{2x}+1}dx.$

(V) $\int \frac{\sin x}{\sin x + \cos x}dx.$

(VI) $\int \sin^3 x \cos^4 x dx.$

$$\text{(VII)}\ \int \frac{1+\cos^2 x}{\cos x(1+\sin^2 x)}dx. \qquad \text{(VIII)}\ \int \frac{1}{\sin^2 x \cos^2 x}dx.$$

$$\text{(IX)}\ \int \frac{\sqrt[4]{x}}{1+\sqrt{x}}dx. \qquad\qquad\quad \text{(X)}\ \int \frac{1}{\cos x}dx.$$

Solution to Exercise 6.4.

(I) For the integrand $\sqrt{1-x^2}$, substitute $u = \arcsin x$, that is, $x = \sin u$ and $dx = \cos u\, du$. Then $\sqrt{1-x^2} = \sqrt{1-\sin^2 u} = \cos u$ and

$$\int \sqrt{1-x^2}dx = \int \cos^2 u\, du.$$

Write $\cos^2 u$ as $\frac{1}{2}(\cos(2u)+1)$. Then

$$\int \sqrt{1-x^2}dx = \frac{1}{2}\int \cos(2u)du + \frac{1}{2}\int du.$$

For the integrand $\cos(2u)$, substitute $v = 2u$ and $dv = 2du$. So

$$\begin{aligned}
\int \sqrt{1-x^2}dx &= \frac{1}{2}\int \cos(2u)du + \frac{1}{2}\int du \\
&= \frac{1}{4}\int \cos v\, dv + \frac{1}{2}\int du \\
&= \frac{1}{4}\sin v + \frac{1}{2}u + C \\
&= \frac{1}{4}\sin(2u) + \frac{1}{2}u + C \\
&= \frac{1}{2}\sin u \cos u + \frac{1}{2}u + C \\
&= \frac{1}{2}\sin u \sqrt{1-\sin^2 u} + \frac{1}{2}u + C.
\end{aligned}$$

Substitute back for $u = \arcsin x$. Thus

$$\begin{aligned}
\int \sqrt{1-x^2}dx &= \frac{1}{2}\sin \arcsin x \sqrt{1-\sin^2 \arcsin x} + \frac{1}{2}\arcsin x \\
&= \frac{1}{2}(x\sqrt{1-x^2} + \arcsin x) + C.
\end{aligned}$$

(II) Substitute $u = \arctan x$, that is, $x = \tan u$ and $dx = \sec^2 u\, du$. Then $\sqrt{x^2+1} = \sqrt{\tan^2 u + 1} = \sec u$ and

$$\int \sqrt{x^2+1}dx = \int \sec^3 u\, du.$$

For the integrand $\sec^3 u$, we integrate by parts ($\int f\,dg = fg - \int g\,df$), where

$f = \sec u,\, dg = \sec^2 u\,du,\, df = \tan u \sec u\,du$ and $g = \tan u$.

Therefore,

$$\int \sec^3 u\,du = \tan u \sec u - \int \sec u \tan^2 u\,du$$

$$= \tan u \sec u - \int \sec u(\sec^2 u - 1)du$$

$$= \tan u \sec u - \int \sec^3 u\,du + \int \sec u\,du.$$

Adding $\int \sec^3 u\,du$ to both sides of the above equation, we obtain

$$2\int \sec^3 u\,du = \tan u \sec u + \int \sec u\,du = \tan u \sec u + \log(\tan u + \sec u).$$

Dividing both sides by 2 and substituting back $u = \arctan x$, we conclude that

$$\int \sqrt{x^2 + 1}\,dx = \int \sec^3 u\,du$$

$$= \frac{1}{2}(\tan u \sec u + \log(\tan u + \sec u)) + C$$

$$= \frac{1}{2}(\tan \arctan x \sec \arctan x$$

$$+ \log(\tan \arctan x + \sec \arctan x)) + C$$

$$= \frac{1}{2}\left(x\sqrt{x^2 + 1} + \log(x + \sqrt{x^2 + 1})\right) + C.$$

(III) Replace $u = x + 1$ and $du = dx$. Then

$$\int x\sqrt{x + 1}\,dx = \int (u - 1)\sqrt{u}\,du$$

$$= \int (u^{3/2} - \sqrt{u})du$$

$$= \int u^{3/2}\,du - \int \sqrt{u}\,du$$

$$= \frac{2}{5}u^{5/2} - \frac{2}{3}u^{3/2} + C$$

$$= \frac{2}{5}(x + 1)^{5/2} - \frac{2}{3}(x + 1)^{3/2} + C.$$

(IV) Observe that

$$\int \frac{e^x + 2e^{-x}}{e^{2x} + 1} dx = \int \left(\frac{e^x}{e^{2x} + 1} + \frac{2e^{-x}}{e^{2x} + 1} \right) dx$$

$$= \int \frac{e^x}{e^{2x} + 1} dx + 2 \int \frac{e^{-x}}{e^{2x} + 1} dx.$$

Consider $u = e^x$ and $du = e^x dx$. Thus

$$\int \frac{e^x + 2e^{-x}}{e^{2x} + 1} dx = \int \frac{e^x}{e^{2x} + 1} dx + 2 \int \frac{e^{-x}}{e^{2x} + 1} dx$$

$$= \int \frac{1}{u^2 + 1} du + 2 \int \frac{1}{u^2(u^2 + 1)} du$$

$$= \arctan u + 2 \int \frac{1}{u^2(u^2 + 1)} du.$$

For the integrand $1/(u^2(u^2 + 1))$, use simple fractions to achieve the following:

$$\int \frac{e^x + 2e^{-x}}{e^{2x} + 1} dx = \arctan u + 2 \int \frac{1}{u^2(u^2 + 1)} du$$

$$= \arctan u + 2 \int \left(\frac{1}{u^2} - \frac{1}{u^2 + 1} \right) du$$

$$= \arctan u + 2 \int \frac{1}{u^2} du - 2 \int \frac{1}{u^2 + 1} du$$

$$= \arctan u - \frac{2}{u} - 2\arctan u + C$$

$$= -\arctan e^x - \frac{2}{e^x} + C.$$

(V) Observe that

$$\int \frac{\sin x}{\sin x + \cos x} dx = \frac{1}{2} \int \frac{2\sin x}{\sin x + \cos x} dx$$

$$= \frac{1}{2} \int \frac{\sin x + \sin x + \cos x - \cos x}{\sin x + \cos x} dx$$

$$= \frac{1}{2} \left(\int \frac{\sin x + \cos x}{\sin x + \cos x} dx + \int \frac{\sin x - \cos x}{\sin x + \cos x} dx \right)$$

$$= \frac{1}{2} \left(\int dx - \int \frac{\cos x - \sin x}{\sin x + \cos x} dx \right).$$

A primitive for the function 1 is x. For the integrand $(\cos x - \sin x)/(\sin x + \cos x)$, substitute $u = \sin x + \cos x$ and $du = (\cos x - \sin x)dx$. Thus

$$\int \frac{\sin x}{\sin x + \cos x}dx = \frac{1}{2}\left(\int dx - \int \frac{\cos x - \sin x}{\sin x + \cos x}\right)dx$$

$$= \frac{1}{2}\left(x - \int \frac{1}{u}du\right)$$

$$= \frac{1}{2}(x - \log|u|) + C$$

$$= \frac{1}{2}(x - \log|\sin x + \cos x|) + C.$$

(VI) Using that $\sin^2 x + \cos^2 x = 1$, observe that

$$\int \sin^3 x \cos^4 x dx = \int \sin^2 x \cos^4 x \sin x dx = \int (1-\cos^2 x)\cos^4 x \sin x dx.$$

For the integrand $(1 - \cos^2 x)\cos^4 x \sin x$, substitute $u = \cos x$ and $du = -\sin x dx$. Thus

$$\int \sin^3 x \cos^4 x dx = \int (1 - \cos^2 x)\cos^4 x \sin x dx$$

$$= -\int (1 - u^2)u^4 du$$

$$= \int (u^6 - u^4)du$$

$$= \frac{u^7}{7} - \frac{u^5}{5} + C$$

$$= \frac{\cos^7 x}{7} - \frac{\cos^5 x}{5} + C.$$

(VII) Using that $\sin^2 x + \cos^2 x = 1$, we get

$$\int \frac{1 + \cos^2 x}{\cos x(1 + \sin^2 x)}dx = \int \frac{2 - \sin^2 x}{\cos x(1 + \sin^2 x)}dx$$

$$= \int \cos x \cdot \frac{2 - \sin^2 x}{\cos^2 x(1 + \sin^2 x)}dx$$

$$= \int \cos x \cdot \frac{2 - \sin^2 x}{(1 - \sin^2 x)(1 + \sin^2 x)}dx$$

Make the substitution $u = \sin x$ and $du = \cos x\, dx$:

$$\int \cos x \cdot \frac{2 - \sin^2 x}{(1 - \sin^2 x)(1 + \sin^2 x)}dx = \int \frac{2 - u^2}{(1 - u^2)(1 + u^2)}du$$

$$= \int \frac{2 - u^2}{(1 - u)(1 + u)(1 + u^2)}du.$$

Apply the technique of simple fraction decomposition to the expression

$$\frac{2 - u^2}{(1 - u)(1 + u)(1 + u^2)}.$$

This decomposition has the following form:

$$\frac{2 - u^2}{(1 - u)(1 + u)(1 + u^2)} = \frac{A}{1 - u} + \frac{B}{1 + u} + \frac{Cu + D}{1 + u^2}.$$

Multiplying both sides by $(1 - u)(1 + u)(1 + u^2)$, we have

$$
\begin{aligned}
2 - u^2 &= A(1 + u)(1 + u^2) + B(1 - u)(1 + u^2) \\
&\quad + (Cu + D)(1 - u)(1 + u) \\
&= (A - B - C)u^3 + (A + B - D)u^2 + (A - B + C)u \\
&\quad + (A + B + D).
\end{aligned}
$$

Equate coefficients on both sides, yielding the system with 4 equations and 4 unknowns

$$
\begin{cases}
0 = A - B - C, \\
-1 = A + B - D, \\
0 = A - B + C, \\
2 = A + B + D.
\end{cases}
$$

The solution of the system is:

$$A = B = \frac{1}{4}, \quad C = 0, \quad D = \frac{3}{2}.$$

Hence,

$$
\begin{aligned}
\int \frac{1 + \cos^2 x}{\cos x(1 + \sin^2 x)}\, dx &= \int \frac{2 - u^2}{(1 - u)(1 + u)(1 + u^2)}\, du \\
&= \frac{1}{4}\int \frac{1}{1 - u}\, du + \frac{1}{4}\int \frac{1}{1 + u}\, du + \frac{3}{2}\int \frac{1}{1 + u^2}\, du \\
&= -\frac{1}{4}\log|1 - u| + \frac{1}{4}\log|1 + u| + \frac{3}{2}\arctan(u) + C \\
&= -\frac{1}{4}\log|1 - \sin x| + \frac{1}{4}\log|1 + \sin x| \\
&\quad + \frac{3}{2}\arctan(\sin x) + C \\
&= \frac{1}{4}\log\left|\frac{1 + \sin x}{1 - \sin x}\right| + \frac{3}{2}\arctan(\sin x) + C.
\end{aligned}
$$

(VIII) For the integrand $1/(\sin^2 x \cos^2 x) = \csc^2 x \sec^2 x$, recall that $\csc^2 x \sec^2 x = 4\csc^2(2x)$. Then

$$\int \frac{1}{\sin^2 x \cos^2 x}\, dx = \int \csc^2 x \sec^2 x\, dx = 4\int \csc^2(2x)\, dx.$$

For the integrand $\csc^2(2x)$, substitute $u = 2x$ and $du = 2dx$. Thus

$$\int \frac{1}{\sin^2 x \cos^2 x} dx = 4 \int \csc^2(2x)dx$$

$$= 2 \int \csc^2 u\, du$$

$$= -2\cot u + C$$

$$= -2\cot(2x) + C.$$

(IX) For the integrand $\sqrt[4]{x}/(1 + \sqrt{x})$, substitute $u = x^{1/4}$ and $du = dx/(4x^{3/4})$. Then

$$\int \frac{\sqrt[4]{x}}{1 + \sqrt{x}} dx = 4 \int \frac{u^4}{u^2 + 1} du.$$

Observe that

$$\frac{u^4}{u^2 + 1} = u^2 + \frac{1}{u^2 + 1} - 1$$

and thus

$$\int \frac{\sqrt[4]{x}}{1 + \sqrt{x}} dx = 4 \int \left(u^2 + \frac{1}{u^2 + 1} - 1 \right) du$$

$$= 4 \left(\int u^2 du + \int \frac{1}{u^2 + 1} du - \int du \right).$$

The integral of u^2 is $u^3/3$, the integral of $1/(u^2 + 1)$ is $\arctan u$ and the integral of 1 is u. So

$$\int \frac{\sqrt[4]{x}}{1 + \sqrt{x}} dx = 4 \left(\frac{u^3}{3} + \arctan u - u \right) + C$$

$$= 4 \left(\frac{x^{3/4}}{3} - \sqrt[4]{x} + \arctan \sqrt[4]{x} \right) + C.$$

(X) Using that $\sin^2 x + \cos^2 x = 1$, observe that

$$\int \frac{1}{\cos x} dx = \int \frac{\cos x}{\cos^2 x} dx = \int \frac{\cos x}{1 - \sin^2 x} dx.$$

For the integrand $\cos x/(1 - \sin^2 x)$, substitute $u = \sin x$ and $du = \cos x dx$. Thus

$$\int \frac{1}{\cos x} dx = \int \frac{\cos x}{1 - \sin^2 x} dx$$

$$= \int \frac{1}{1 - u^2} du$$

$$= \int \frac{1}{2} \left(\frac{1}{1 - u} + \frac{1}{1 + u} \right) du$$

$$= \frac{1}{2} \left(\int \frac{1}{1 - u} du + \int \frac{1}{1 + u} du \right).$$

For the integrand $1/(1 - u)$, substitute $r = 1 - u$ and $dr = -du$. For the integrand $1/(1 + u)$, substitute $s = 1 + u$ and $ds = du$. Thus

$$\int \frac{1}{\cos x} dx = \frac{1}{2} \left(-\int \frac{1}{r} dr + \int \frac{1}{s} ds \right) = \frac{1}{2} (\log |s| - \log |r|) + C.$$

Substituting back for $r = 1 - u$, $s = 1 + u$ and $u = \sin x$, and noting that $r, s \geq 0$, we conclude that

$$\int \frac{1}{\cos x} dx = \frac{1}{2} (\log(1 + \sin x) - \log(1 - \sin x)) + C = \frac{1}{2} \log \frac{1 + \sin x}{1 - \sin x} + C.$$

Remark: It is usual to find

$$\int \frac{1}{\cos x} dx = \log (\tan x + \sec x) + C. \tag{6.2}$$

Both answers are correct. It is a simple task to show that

$$\frac{1}{2} \log \frac{1 + \sin x}{1 - \sin x} = \log (\tan x + \sec x).$$

In order to arrive at (6.2), we proceed as follows:

$$\int \frac{1}{\cos x} dx = \int \sec x dx = \int \sec x \frac{\tan x + \sec x}{\tan x + \sec x} dx$$

$$= \int \frac{\sec x \tan x + \sec^2 x}{\tan x + \sec x} dx$$

$$= \int \frac{1}{u} du$$

$$= \log |u| + C$$

$$= \log(\tan x + \sec x) + C,$$

where, above, we used the change $u = \tan x + \sec x$ and $du = (\sec x \tan x + \sec^2 x) dx$.

$\square$

Exercise 6.5. *Calculate the following primitives:*

(I) $\displaystyle\int \frac{2-\sin x}{2+\cos x}\,dx.$

(II) $\displaystyle\int \frac{1}{1+\sin^2 x}\,dx.$

(III) $\displaystyle\int \frac{1-r\cos x}{1-2r\cos x+r^2}\,dx.$

(IV) $\displaystyle\int \frac{1}{1+\sqrt{x+1}}\,dx.$

(V) $\displaystyle\int \frac{4^x+1}{2^x+1}\,dx.$

(VI) $\displaystyle\int \frac{\sqrt{1-x}}{1-\sqrt{x}}\,dx.$

(VII) $\displaystyle\int \cos 2x \sin^2 x\,dx.$

(VIII) $\displaystyle\int \frac{1}{\sin 3x \cos x}\,dx.$

(IX) $\displaystyle\int x\left(1+\frac{1}{\sqrt{x}}\right)^2\,dx.$

(X) $\displaystyle\int \frac{2+\sqrt{x+1}}{(x+1)^2-\sqrt{x+1}}\,dx.$

(XI) $\displaystyle\int \sqrt{2x-x^2}\,dx.$

(XII) $\displaystyle\int \frac{1}{\sqrt{x-x^2}}\,dx.$

(XIII) $\displaystyle\int \frac{\sin^3 x}{\sqrt{\cos x}}\,dx.$

(XIV) $\displaystyle\int \frac{1}{a^2 e^x + b^2 e^{-x}}\,dx.$

(XV) $\displaystyle\int x^5 \sqrt{1-x^3}\,dx.$

(XVI) $\displaystyle\int \frac{1}{x(x^7+1)}\,dx.$

(XVII) $\displaystyle\int \cos x \log(1+\cos x)\,dx.$

(XVIII) $\displaystyle\int \frac{x^2-3x+3}{x^2-3x+2}\,dx.$

(XIX) $\displaystyle\int \frac{3x+5}{(x^2-2x+2)^2}\,dx.$

(XX) $\displaystyle\int \frac{1}{x\sqrt{x^2-1}}\,dx.$

(XXI) $\displaystyle\int \frac{1}{x+1}\sqrt{\frac{x+3}{x-1}}\,dx.$

(XXII) $\displaystyle\int \frac{1}{2+3\tan x}\,dx.$

(XXIII) $\displaystyle\int \frac{x^2}{\sqrt{3x^2-x+1}}\,dx.$

Solution to Exercise 6.5.

(I) Observe that

$$\int \frac{2-\sin x}{2+\cos x}\,dx = 2\int \frac{1}{2+\cos x}\,dx + \int \frac{-\sin x}{2+\cos x}\,dx.$$

For the integrand $-\sin x/(2+\cos x)$, make the substitution $v = 2+\cos x$ and $dv = -\sin x\,dx$. Then

$$\int \frac{2-\sin x}{2+\cos x}dx = 2\int \frac{1}{2+\cos x}dx + \int \frac{-\sin x}{2+\cos x}dx$$

$$= 2\int \frac{1}{2+\cos x}dx + \int \frac{1}{v}dv$$

$$= 2\int \frac{1}{2+\cos x}dx + \log|v|$$

$$= 2\int \frac{1}{2+\cos x}dx + \log(2+\cos x).$$

For the integrand $1/(2+\cos x)$, we make the substitution

$$u = \tan(x/2) \quad \text{and} \quad du = \frac{\sec^2(x/2)}{2}dx.$$

Notice that

$$\cos x = \frac{1-u^2}{1+u^2} \quad \text{and} \quad \cos^2\frac{x}{2} = \frac{1}{1+u^2}.$$

Then

$$dx = 2\cos^2\frac{x}{2}du = \frac{2}{1+u^2}du$$

and

$$\int \frac{2-\sin x}{2+\cos x}dx = 2\int \frac{1}{2+\cos x}dx + \log(2+\cos x)$$

$$= 2\int \frac{1}{2+\frac{1-u^2}{1+u^2}}\frac{2}{1+u^2}du + \log(2+\cos x)$$

$$= 4\int \frac{1}{3+u^2}du + \log(2+\cos x)$$

$$= \frac{4}{3}\int \frac{1}{1+\frac{u^2}{3}}du + \log(2+\cos x)$$

$$= \frac{4}{3}\int \frac{1}{1+\left(\frac{u}{\sqrt{3}}\right)^2}du + \log(2+\cos x).$$

For the integrand $1/(1 + (u/\sqrt{3})^2)$, substitute $s = u/\sqrt{3}$ and $ds = du/\sqrt{3}$. Thus

$$\int \frac{2 - \sin x}{2 + \cos x} dx = \frac{4}{3} \int \frac{1}{1 + \left(\frac{u}{\sqrt{3}}\right)^2} du + \log(2 + \cos x)$$

$$= \frac{4\sqrt{3}}{3} \int \frac{1}{1 + s^2} ds + \log(2 + \cos x)$$

$$= \frac{4\sqrt{3}}{3} \arctan s + \log(2 + \cos x) + C$$

$$= \frac{4\sqrt{3}}{3} \arctan \frac{\tan \frac{x}{2}}{\sqrt{3}} + \log(2 + \cos x) + C.$$

(II) Multiply numerator and denominator of $1/(1 + \sin^2 x)$ by $\sec^2 x$. Then

$$\int \frac{1}{1 + \sin^2 x} dx = \int \frac{\sec^2 x}{\sec^2 x + \tan^2 x} dx$$

For the integrand $\sec^2 x/(\sec^2 x + \tan^2 x)$, substitute $u = \tan x$ and $du = \sec^2 x\, dx$. Note that

$$\cos^2 x = \frac{1}{1 + u^2}$$

or, equivalently,

$$\sec^2 x = 1 + u^2.$$

Then

$$\int \frac{1}{1 + \sin^2 x} dx = \int \frac{\sec^2 x}{\sec^2 x + \tan^2 x} dx$$

$$= \int \frac{1}{1 + u^2 + u^2} du$$

$$= \int \frac{1}{1 + (\sqrt{2}u)^2} du.$$

For the integrand $1/(1 + (\sqrt{2}u)^2)$, substitute $s = \sqrt{2}u$ and $ds = \sqrt{2}du$. Thus

$$\int \frac{1}{1 + \sin^2 x} dx = \int \frac{1}{1 + (\sqrt{2}u)^2} du$$

$$= \frac{1}{\sqrt{2}} \int \frac{1}{1 + s^2} ds$$

$$= \frac{\arctan s}{\sqrt{2}} + C$$

$$= \frac{\arctan(\sqrt{2}u)}{\sqrt{2}} + C$$

$$= \frac{\arctan(\sqrt{2}\tan x)}{\sqrt{2}} + C.$$

(III) For the integrand $(1-r\cos x)/(1-2r\cos x+r^2)$, substitute $u = \tan(x/2)$ and $du = \dfrac{\sec^2(x/2)}{2}dx$. Observe that

$$\cos x = \frac{1-u^2}{u^2+1} \quad \text{and} \quad dx = \frac{2}{u^2+1}du.$$

Therefore,

$$\int \frac{1-r\cos x}{1-2r\cos x+r^2}dx$$

$$= \int \frac{2\left(1 - \frac{r(1-u^2)}{u^2+1}\right)}{(u^2+1)\left(1 - \frac{2r(1-u^2)}{u^2+1} + r^2\right)}du$$

$$= 2\int \frac{u^2+1-r+ru^2}{(u^2+1)(u^2+1-2r+2ru^2+r^2u^2+r^2)}du$$

$$= 2\int \left(\frac{1}{2(u^2+1)} + \frac{1-r^2}{2(u^2+1-2r+2ru^2+r^2u^2+r^2)}\right)du,$$

where, in the last identity, we used simple fractions. Integrating the sum term by term and factoring out the constants we get

$$\int \frac{1-r\cos x}{1-2r\cos x+r^2}dx$$

$$= \int \frac{1}{u^2+1}du$$

$$+ (1-r^2)\int \frac{1}{u^2+1-2r+2ru^2+r^2u^2+r^2}du$$

$$= \arctan u + (1-r^2)\int \frac{1}{(1+2r+r^2)u^2+1-2r+r^2}du$$

$$= \arctan u + (1-r^2)\int \frac{1}{(1+r)^2u^2+(1-r)^2}du$$

$$= \arctan u + \frac{1-r^2}{(1-r)^2}\int \frac{1}{\frac{(1+r)^2u^2}{(1-r)^2}+1}du.$$

In the integral above, substitute

$$s = \frac{(1+r)u}{1-r} \quad \text{and} \quad ds = \frac{1+r}{1-r}du.$$

Thus

$$\int \frac{1 - r\cos x}{1 - 2r\cos x + r^2}dx = \arctan u + \frac{1 - r^2}{(1-r)^2}\int \frac{1}{\frac{(1+r)^2 u^2}{(1-r)^2} + 1}du$$

$$= \arctan u + \frac{(1-r^2)(1-r)}{(1-r)^2(1+r)}\int \frac{1}{s^2+1}ds$$

$$= \arctan u + \int \frac{1}{s^2+1}ds$$

$$= \arctan u + \arctan s + C.$$

Substitute back for $s = ((1+r)u)/(1-r)$ and $u = \tan(x/2)$:

$$\int \frac{1 - r\cos x}{1 - 2r\cos x + r^2}dx = \frac{x}{2} + \arctan \frac{(1+r)\tan \frac{x}{2}}{1-r} + C.$$

(IV) For the integrand $1/(1 + \sqrt{x+1})$, substitute $u = x + 1$ and $du = dx$. Thus

$$\int \frac{1}{1 + \sqrt{x+1}}dx = \int \frac{1}{1 + \sqrt{u}}du.$$

For the integrand $1/(1 + \sqrt{u})$, substitute $s = \sqrt{u}$ and $ds = du/(2\sqrt{u})$. Then

$$\int \frac{1}{1 + \sqrt{x+1}}dx = \int \frac{1}{1 + \sqrt{u}}du = 2\int \frac{s}{s+1}ds = 2\int \left(1 - \frac{1}{s+1}\right)ds.$$

Integrate the sum term by term, factor out constants and then, for the integrand $1/(s+1)$, substitute $p = s+1$ and $dp = ds$. Therefore,

$$\int \frac{1}{1 + \sqrt{x+1}}dx = 2\int ds - 2\int \frac{1}{p}dp$$

$$= 2s - 2\log p + C$$

$$= 2s - 2\log(s+1) + C$$

$$= 2\sqrt{u} - 2\log(\sqrt{u}+1) + C$$

$$= 2\sqrt{x+1} - 2\log(\sqrt{x+1}+1) + C.$$

(v) For the integrand $(4^x + 1)/(2^x + 1)$, substitute $u = 2^x$ and $du = 2^x \log 2\,dx$. Then

$$\int \frac{4^x+1}{2^x+1}dx = \frac{1}{\log 2}\int \frac{u^2+1}{u(u+1)}du = \frac{1}{\log 2}\int \left(1 + \frac{1}{u} - \frac{2}{u+1}\right)du.$$

Integrating the sum term by term and factoring out the constants, we get

$$\int \frac{4^x+1}{2^x+1}dx = \frac{1}{\log 2}\left(\int du + \int \frac{1}{u}du - 2\int \frac{1}{u+1}du\right).$$

For the integrand $1/(u+1)$, substitute $s = u+1$ and $ds = du$. Then

$$\int \frac{4^x + 1}{2^x + 1} dx = \frac{1}{\log 2} \left(\int du + \int \frac{1}{u} du - 2 \int \frac{1}{u+1} du \right)$$

$$= \frac{1}{\log 2} \left(\int du + \int \frac{1}{u} du - 2 \int \frac{1}{s} ds \right)$$

$$= \frac{1}{\log 2} \left(u + \log |u| - 2 \log |s| \right) + C$$

$$= \frac{1}{\log 2} \left(u + \log |u| - 2 \log |u + 1| \right) + C$$

$$= \frac{1}{\log 2} \left(2^x + \log 2^x - 2 \log(2^x + 1) \right) + C.$$

(VI) Rationalize the integrand:

$$\frac{\sqrt{1-x}}{1-\sqrt{x}} = \frac{\sqrt{1-x}(1+\sqrt{x})}{1-x}.$$

So

$$\int \frac{\sqrt{1-x}}{1-\sqrt{x}} dx = \int \frac{\sqrt{1-x}(1+\sqrt{x})}{1-x} dx$$

$$= \int \frac{\sqrt{1-x}}{1-x} dx + \int \frac{\sqrt{x}\sqrt{1-x}}{1-x} dx$$

$$= \int \frac{1}{\sqrt{1-x}} dx + \int \frac{\sqrt{x}}{\sqrt{1-x}} dx.$$

For the integrand $1/\sqrt{1-x}$, substitute $v = 1-x$ and $dv = -dx$. For the integrand $\sqrt{x}/\sqrt{1-x}$, substitute $u = \sqrt{x}$ and $du = dx/(2\sqrt{x})$. Thus

$$\int \frac{\sqrt{1-x}}{1-\sqrt{x}} dx = \int \frac{1}{\sqrt{1-x}} dx + \int \frac{\sqrt{x}}{\sqrt{1-x}} dx$$

$$= -\int \frac{1}{\sqrt{v}} dv + \int \frac{2u^2}{\sqrt{1-u^2}} du$$

$$= -2\sqrt{v} + 2 \int \frac{u^2}{\sqrt{1-u^2}} du.$$

For the integrand $u^2/\sqrt{1-u^2}$, substitute $u = \sin s$ and $du = \cos s\, ds$. Then $\sqrt{1-u^2} = \sqrt{1-\sin^2 s} = \sqrt{\cos^2 s} = \cos s$ and $s = \arcsin u$. So

$$\int \frac{\sqrt{1-x}}{1-\sqrt{x}} dx = -2\sqrt{v} + 2 \int \frac{u^2}{\sqrt{1-u^2}} du$$

$$= -2\sqrt{v} + 2 \int \sin^2 s\, ds.$$

For the integrand $\sin^2 s$ proceed as in Exercise 4.2(XI) to obtain

$$\int \frac{\sqrt{1-x}}{1-\sqrt{x}} dx = -2\sqrt{v} + 2 \int \sin^2 s \, ds$$

$$= -2\sqrt{v} + s - \sin s \cos s + C.$$

Substitute back for $v = 1 - x$, $s = \arcsin u$ and $u = \sqrt{x}$:

$$\int \frac{\sqrt{1-x}}{1-\sqrt{x}} dx = -2\sqrt{1-x} + \arcsin\sqrt{x} - \sqrt{x}\cos\arcsin\sqrt{x} + C.$$

Simplify using that $\cos\arcsin z = \sqrt{1-z^2}$. Therefore,

$$\int \frac{\sqrt{1-x}}{1-\sqrt{x}} dx = -2\sqrt{1-x} + \arcsin\sqrt{x} - \sqrt{x}\sqrt{1-x} + C$$

$$= \arcsin\sqrt{x} - (2+\sqrt{x})\sqrt{1-x} + C.$$

(VII) Write $\sin^2 x$ as $\frac{1}{2} - \frac{1}{2}\cos(2x)$. Thus

$$\int \cos 2x \sin^2 x \, dx = \int \cos 2x \left(\frac{1}{2} - \frac{1}{2}\cos(2x) \right) dx$$

$$= \frac{1}{2} \left(\int \cos(2x) dx - \int \cos^2(2x) dx \right).$$

In both integrals above, substitute $u = 2x$ and $du = 2dx$. Then

$$\int \cos 2x \sin^2 x \, dx = \frac{1}{4} \left(\int \cos u \, du - \int \cos^2 u \, du \right).$$

The integral of $\cos u$ is $\sin u$. Therefore,

$$\int \cos 2x \sin^2 x \, dx = \frac{1}{4} \left(\sin u - \frac{1}{2}\sin u \cos u - \frac{1}{2}u \right) + C$$

$$= \frac{1}{4} \left(\sin 2x - \frac{1}{2}\sin(2x)\cos(2x) - x \right) + C.$$

(VIII) Recall the following trigonometric identities:

$$\sin a \cos b = \frac{\sin(a+b) + \sin(a-b)}{2} \quad \text{and} \quad \sin(a+b) = \sin a \cos b + \sin b \cos a.$$

Thus

$$\int \frac{1}{\sin 3x \cos x} dx = \int \frac{2}{\sin(3x+x) + \sin(3x-x)} dx$$

$$= 2 \int \frac{1}{\sin 4x + \sin 2x} dx$$

$$= 2 \int \frac{1}{2 \sin 2x \cos 2x + \sin 2x} dx$$

$$= 2 \int \frac{1}{\sin 2x (2 \cos 2x + 1)} dx$$

$$= 2 \int \frac{\csc 2x}{2 \cos 2x + 1} dx.$$

For the integrand $\csc 2x / (2 \cos 2x + 1)$, substitute $u = 2x$ and $du = 2dx$. Then

$$\int \frac{1}{\sin 3x \cos x} dx = 2 \int \frac{\csc 2x}{2 \cos 2x + 1} dx$$

$$= \int \frac{\csc u}{2 \cos u + 1} du$$

$$= \int \frac{\csc u}{2 \cos u + 1} \cdot \frac{\tan^3 u}{\tan^3 u} du$$

$$= \int \frac{\sec^2 u \tan u}{2 \tan^2 u + \sec u \tan^2 u} du$$

$$= \int \frac{\sec u \sec u \tan u}{(2 + \sec u)(\sec^2 u - 1)} du.$$

Above we used that $\tan^2 u = \sec^2 u - 1$. For the integrand $(\sec u \sec u \tan u)/((2 + \sec u)(sec^2 u - 1))$, substitute $s = \sec u$ and $ds = \tan u \sec u \, du$. Therefore,

$$\int \frac{1}{\sin 3x \cos x} dx = \int \frac{\sec u \sec u \tan u}{(2 + \sec u)(\sec^2 u - 1)} du = \int \frac{s}{(s+2)(s^2-1)} ds.$$

For the integrand $s/((s+2)(s^2-1))$, use simple fractions. Factor the denominator into linear and irreducible quadratic terms and then the simple fraction decomposition is of the form

$$\frac{s}{(s-1)(s+1)(s+2)} = \frac{A}{s-1} + \frac{B}{s+1} + \frac{C}{s+2}.$$

Multiply both sides by $(s-1)(s+1)(s+2)$ and simplify to get

$$s = (A + B + C)s^2 + (3A + B)s + 2A - 2B - C.$$

Equate coefficients on both sides, yielding 3 equations in 3 unknowns:

$$\begin{cases} 0 = A + B + C \\ 1 = 3A + B \\ 0 = 2A - 2B - C. \end{cases}$$

The solution of the above system is

$$A = \frac{1}{6}, \quad B = \frac{1}{2} \quad \text{and} \quad C = -\frac{2}{3}.$$

Therefore,

$$\int \frac{1}{\sin 3x \cos x} dx = \int \frac{s}{(s+2)(s^2-1)} ds$$

$$= \int \left(\frac{1}{6(s-1)} + \frac{1}{2(s+1)} - \frac{2}{3(s+2)} \right) ds$$

$$= \frac{1}{6} \int \frac{1}{s-1} ds + \frac{1}{2} \int \frac{1}{s+1} ds - \frac{2}{3} \int \frac{1}{s+2} ds.$$

For the integrand $1/(s-1)$, substitute $v = s - 1$ and $dv = ds$, for the integrand $1/(s+1)$, substitute $w = s+1$ and $dw = ds$ and for the integrand $1/(s+2)$, substitute $p = s+2$ and $dp = ds$. Thus

$$\int \frac{1}{\sin 3x \cos x} dx = \frac{1}{6} \int \frac{1}{s-1} ds + \frac{1}{2} \int \frac{1}{s+1} ds - \frac{2}{3} \int \frac{1}{s+2} ds$$

$$= \frac{1}{6} \int \frac{1}{v} dv + \frac{1}{2} \int \frac{1}{w} dw - \frac{2}{3} \int \frac{1}{p} dp$$

$$= \frac{1}{6} \log|v| + \frac{1}{2} \log|w| - \frac{2}{3} \log|p| + C.$$

Substitute back for $v = s-1$, $w = s+1$, $p = s+2$, $s = \sec u$ and $u = 2x$. So

$$\int \frac{1}{\sin 3x \cos x} dx = \frac{1}{6} \log|v| + \frac{1}{2} \log|w| - \frac{2}{3} \log|p| + C$$

$$= \frac{1}{6} \log|\sec 2x - 1| + \frac{1}{2} \log|\sec 2x + 1|$$

$$- \frac{2}{3} \log|\sec 2x + 2| + C.$$

(IX) Substitute $u = 1/\sqrt{x}$ and $du = -dx/(2x^{3/2})$. Thus

$$\int x \left(1 + \frac{1}{\sqrt{x}} \right)^2 dx = -2 \int \frac{(u+1)^2}{u^5} du.$$

For the integrand $(u+1)^2/u^5$, substitute $s = u+1$ and $ds = du$. So

$$\int x \left(1 + \frac{1}{\sqrt{x}}\right)^2 dx = -2 \int \frac{(u+1)^2}{u^5} du = -2 \int \frac{s^2}{(s-1)^5} ds.$$

For the integrand $s^2/(s-1)^5$, use simple fractions to get

$$\int x \left(1 + \frac{1}{\sqrt{x}}\right)^2 dx$$

$$= -2 \int \frac{s^2}{(s-1)^5} ds$$

$$= -2 \int \left(\frac{1}{(s-1)^3} + \frac{2}{(s-1)^4} + \frac{1}{(s-1)^5}\right) ds$$

$$= -2 \int \frac{1}{(s-1)^3} ds - 4 \int \frac{1}{(s-1)^4} ds - 2 \int \frac{1}{(s-1)^5} ds.$$

Substitute $p = s - 1$ and $dp = ds$. Thus

$$\int x \left(1 + \frac{1}{\sqrt{x}}\right)^2 dx$$

$$= -2 \int \frac{1}{(s-1)^3} ds - 4 \int \frac{1}{(s-1)^4} ds - 2 \int \frac{1}{(s-1)^5} ds$$

$$= -2 \int \frac{1}{p^3} dp - 4 \int \frac{1}{p^4} dp - 2 \int \frac{1}{p^5} dp$$

$$= \frac{1}{p^2} + \frac{4}{3p^3} + \frac{1}{2p^4} + C.$$

Substitute back for $p = s - 1$, $s = u + 1$ and $u = 1/\sqrt{x}$. Then

$$\int x \left(1 + \frac{1}{\sqrt{x}}\right)^2 dx = \frac{1}{p^2} + \frac{4}{3p^3} + \frac{1}{2p^4} + C$$

$$= x + \frac{4x^{3/2}}{3} + \frac{x^2}{2} + C.$$

(x) Apply the substitution $u = \sqrt{x+1}$ and $du = \frac{1}{2\sqrt{x+1}} dx$. So $dx = 2u\,du$ and $(x+1)^2 = u^4$. Then

$$\int \frac{2 + \sqrt{x+1}}{(x+1)^2 - \sqrt{x+1}} dx = 2 \int \frac{2+u}{u^4 - u} \cdot u\,du$$

$$= 2 \int \frac{(2+u)u}{u(u^3 - 1)} du$$

$$= 2 \int \frac{2+u}{u^3 - 1} du$$

$$= 2 \int \frac{2+u}{(u-1)(u^2 + u + 1)} du.$$

Observe that $u^2 + u + 1$ is an irreducible polynomial of degree 2. Use simple fraction decomposition to $\frac{2+u}{(u-1)(u^2+u+1)}$. This decomposition is given by

$$\frac{2+u}{(u-1)(u^2+u+1)} = \frac{A}{u-1} + \frac{Bu+C}{u^2+u+1}.$$

Multiplying both sides by $(u-1)(u^2+u+1)$, we have

$$2 + u = A(u^2 + u + 1) + (Bu + C)(u-1)$$
$$= (A+B)u^2 + (A-B+C)u + (A-C).$$

Equate coefficients on both sides, yielding the system with 3 equations and 3 unknowns

$$\begin{cases} 0 = A + B, \\ 1 = A - B + C, \\ 2 = A - C. \end{cases}$$

The solution of the system is:

$$A = 1, \ B = C = -1.$$

Hence,

$$\int \frac{2+u}{(u-1)(u^2+u+1)}du = \int \frac{1}{u-1}du - \int \frac{u+1}{u^2+u+1}du. \qquad (6.3)$$

In (6.3), express $u+1$ as

$$u + 1 = \frac{1}{2}(2u+1) + \frac{1}{2},$$

which implies that

$$\int \frac{u+1}{u^2+u+1}du = \int \frac{\frac{1}{2}(2u+1) + \frac{1}{2}}{u^2+u+1}du$$
$$= \frac{1}{2}\int \frac{2u+1}{u^2+u+1}du + \frac{1}{2}\int \frac{1}{u^2+u+1}du. \qquad (6.4)$$

In the first integral of (6.4), use the substitution $v = u^2 + u + 1$ and $dv = (2u+1)du$:

$$\int \frac{2u+1}{u^2+u+1}du = \int \frac{1}{v}dv = \log|v| + C = \log|u^2 + u + 1| + C.$$

In the second integral of (6.4),

$$\int \frac{1}{u^2 + u + 1} du = \int \frac{1}{\left(u + \frac{1}{2}\right)^2 - \frac{1}{4} + 1} du$$

$$= \int \frac{1}{\left(u + \frac{1}{2}\right)^2 + \frac{3}{4}} du$$

$$= \frac{4}{3} \int \frac{1}{\frac{4}{3}\left(u + \frac{1}{2}\right)^2 + 1} du$$

$$= \frac{4}{3} \int \frac{1}{\left[\frac{2}{\sqrt{3}}\left(u + \frac{1}{2}\right)\right]^2 + 1} du.$$

Now apply the substitution $v = \frac{2}{\sqrt{3}}\left(u + \frac{1}{2}\right)$ and $dv = \frac{2}{\sqrt{3}} du$:

$$\int \frac{1}{u^2 + u + 1} du = \frac{4}{3} \int \frac{1}{\left[\frac{2}{\sqrt{3}}\left(u + \frac{1}{2}\right)\right]^2 + 1} du$$

$$= \frac{4}{3} \cdot \frac{\sqrt{3}}{2} \int \frac{1}{v^2 + 1} dv$$

$$= \frac{2}{\sqrt{3}} \arctan(v) + C$$

$$= \frac{2}{\sqrt{3}} \arctan\left[\frac{2}{\sqrt{3}}\left(u + \frac{1}{2}\right)\right] + C.$$

Thus,

$$\int \frac{u + 1}{u^2 + u + 1} du = \frac{1}{2} \log\left|u^2 + u + 1\right| + \frac{1}{2} \cdot \frac{2}{\sqrt{3}} \arctan\left[\frac{2}{\sqrt{3}}\left(u + \frac{1}{2}\right)\right] + C$$

$$= \frac{1}{2} \log\left|u^2 + u + 1\right| + \frac{1}{\sqrt{3}} \arctan\left[\frac{2}{\sqrt{3}}\left(u + \frac{1}{2}\right)\right] + C.$$

Therefore,

$$\int \frac{2 + u}{(u - 1)(u^2 + u + 1)} du = \log|u - 1| - \frac{1}{2} \log\left|u^2 + u + 1\right|$$

$$- \frac{1}{\sqrt{3}} \arctan\left[\frac{2}{\sqrt{3}}\left(u + \frac{1}{2}\right)\right] + C.$$

In conclusion,

$$\int \frac{2+\sqrt{x+1}}{(x+1)^2 - \sqrt{x+1}} dx$$

$$= 2\left(\log|u-1| - \frac{1}{2}\log|u^2+u+1| - \frac{1}{\sqrt{3}}\arctan\left[\frac{2}{\sqrt{3}}\left(u+\frac{1}{2}\right)\right]\right) + C$$

$$= 2\log|\sqrt{x+1}-1| - \log|\sqrt{x+1}+x+2|$$

$$- \frac{2}{\sqrt{3}}\arctan\left[\frac{2}{\sqrt{3}}\left(\sqrt{x+1}+\frac{1}{2}\right)\right] + C.$$

(XI) Observe that $\sqrt{2x-x^2} = \sqrt{1-(x-1)^2}$. For the integrand $\sqrt{1-(x-1)^2}$, substitute $u = x - 1$ and $du = dx$. So

$$\int \sqrt{2x-x^2}\,dx = \int \sqrt{1-(x-1)^2}\,dx = \int \sqrt{1-u^2}\,du.$$

For the integrand $\sqrt{1-u^2}$, substitute $u = \sin s$ (that is, $s = \arcsin u$) and $du = \cos s\,ds$. Then

$$\int \sqrt{2x-x^2}\,dx = \int \sqrt{1-u^2}\,du = \int \cos^2 s\,ds.$$

For the integrand $\cos^2 s$ proceed as in the solution of the Exercise 4.4(I). Therefore,

$$\int \sqrt{2x-x^2}\,dx = \int \cos^2 s\,ds = \frac{1}{4}\sin(2s) + \frac{s}{2} + C$$

$$= \frac{1}{2}\sin s \cos s + \frac{s}{2} + C$$

$$= \frac{1}{2}u\cos\arcsin u + \frac{\arcsin u}{2} + C$$

$$= \frac{x-1}{2}\cos\arcsin(x-1) + \frac{\arcsin(x-1)}{2} + C.$$

(XII) Observe that $1/\sqrt{x-x^2} = 1/\sqrt{1/4-(x-1/2)^2}$. For the integrand $1/\sqrt{1/4-(x-1/2)^2}$, substitute $u = x - 1/2$ and $du = dx$. Then

$$\int \frac{1}{\sqrt{x-x^2}}\,dx = \int \frac{1}{\sqrt{\frac{1}{4}-u^2}}\,du = 2\int \frac{1}{\sqrt{1-4u^2}}\,du.$$

Consider $s = 2u$ and $ds = 2du$. Thus

$$\int \frac{1}{\sqrt{x-x^2}}\,dx = 2\int \frac{1}{\sqrt{1-4u^2}}\,du = \int \frac{1}{\sqrt{1-s^2}}\,ds$$

$$= \arcsin s + C = \arcsin(2x-1) + C.$$

(XIII) Use the trigonometric identity $\sin^2 x = 1 - \cos^2 x$ to get

$$\int \frac{\sin^3 x}{\sqrt{\cos x}}\,dx = \int \frac{\sin x (1 - \cos^2 x)}{\sqrt{\cos x}}\,dx.$$

For the integrand $\sin x (1 - \cos^2 x)/\sqrt{\cos x}$, substitute $u = \cos x$ and $du = -\sin x\,dx$. So

$$\int \frac{\sin^3 x}{\sqrt{\cos x}}\,dx = \int \frac{\sin x (1 - \cos^2 x)}{\sqrt{\cos x}}\,dx = -\int \frac{1 - u^2}{\sqrt{u}}\,du.$$

For the integrand $(1 - u^2)/\sqrt{u}$, substitute $s = \sqrt{u}$ and $ds = du/2\sqrt{u}$. Then

$$\int \frac{\sin^3 x}{\sqrt{\cos x}}\,dx = -\int \frac{1 - u^2}{\sqrt{u}}\,du$$

$$= -2 \int (1 - s^4)\,ds$$

$$= -2 \int ds + 2 \int s^4\,ds$$

$$= -2s + \frac{2s^5}{5} + C.$$

Substitute back for $s = \sqrt{u}$ and $u = \cos x$:

$$\int \frac{\sin^3 x}{\sqrt{\cos x}}\,dx = -2\sqrt{\cos x} + \frac{2\cos^{5/2} x}{5} + C.$$

(XIV) Simplify the integrand $1/(a^2 e^x + b^2 e^{-x})$ to get $e^x/(a^2 e^{2x} + b^2)$. So

$$\int \frac{1}{a^2 e^x + b^2 e^{-x}}\,dx = \int \frac{e^x}{a^2 e^{2x} + b^2}\,dx.$$

For the integrand $e^x/(a^2 e^{2x} + b^2)$, substitute $u = e^x$ and $du = e^x dx$. Then

$$\int \frac{1}{a^2 e^x + b^2 e^{-x}}\,dx = \int \frac{e^x}{a^2 e^{2x} + b^2}\,dx = \int \frac{1}{a^2 u^2 + b^2}\,du$$

$$= \int \frac{1}{b^2\left(\frac{a^2 u^2}{b^2} + 1\right)}\,du = \frac{1}{b^2} \int \frac{1}{\frac{a^2 u^2}{b^2} + 1}\,du.$$

Consider $s = (au)/b$ and $ds = (a/b)du$. Thus

$$\int \frac{1}{a^2 e^x + b^2 e^{-x}}\,dx = \frac{1}{b^2} \int \frac{1}{\frac{a^2 u^2}{b^2} + 1}\,du = \frac{1}{ab} \int \frac{1}{s^2 + 1}\,ds$$

$$= \frac{1}{ab} \arctan s + C.$$

Substitute back for $s = (au)/b$ and $u = e^x$:

$$\int \frac{1}{a^2 e^x + b^2 e^{-x}} dx = \frac{1}{ab} \arctan\left(\frac{ae^x}{b}\right) + C.$$

(XV) For the integrand $x^5\sqrt{1-x^3}$, substitute $u = x^3$ and $du = 3x^2 dx$. Then

$$\int x^5\sqrt{1-x^3}\,dx = \frac{1}{3}\int u\sqrt{1-u}\,du.$$

Consider $s = 1 - u$ and $ds = -du$. Thus

$$\int x^5\sqrt{1-x^3}\,dx = \frac{1}{3}\int u\sqrt{1-u}\,du = \frac{1}{3}\int (s-1)\sqrt{s}\,ds$$

$$= \frac{1}{3}\int (s^{3/2} - \sqrt{s})\,ds = \frac{1}{3}\int s^{3/2}\,ds - \frac{1}{3}\int \sqrt{s}\,ds$$

$$= \frac{2}{15}s^{5/2} - \frac{2}{9}s^{3/2} + C.$$

Substitute back for $s = 1 - u$ and $u = x^3$:

$$\int x^5\sqrt{1-x^3}\,dx = \frac{2}{15}s^{5/2} - \frac{2}{9}s^{3/2} + C = \frac{2}{15}(1-x^3)^{5/2} - \frac{2}{9}(1-x^3)^{3/2} + C.$$

(XVI) Substitute $u = x^7$ and $du = 7x^6 dx$. Then

$$\int \frac{1}{x(x^7+1)} dx = \frac{1}{7}\int \frac{1}{u(u+1)} du.$$

For the integrand $1/(u(u+1))$, use simple fractions to arrive at

$$\int \frac{1}{x(x^7+1)} dx = \frac{1}{7}\int \frac{1}{u(u+1)} du$$

$$= \frac{1}{7}\int \left(\frac{1}{u} - \frac{1}{u+1}\right) du$$

$$= \frac{1}{7}\int \frac{1}{u} du - \frac{1}{7}\int \frac{1}{u+1} du.$$

For the integrand $1/(u+1)$, substitute $s = u+1$ and $ds = du$. Thus

$$\int \frac{1}{x(x^7+1)} dx = \frac{1}{7}\int \frac{1}{u} du - \frac{1}{7}\int \frac{1}{u+1} du$$

$$= \frac{1}{7}\int \frac{1}{u} du - \frac{1}{7}\int \frac{1}{s} ds$$

$$= \frac{1}{7}\log|u| - \frac{1}{7}\log|s| + C.$$

Substitute back for $s = u+1$ and $u = x^7$:

$$\int \frac{1}{x(x^7+1)} dx = \frac{1}{7}\log|x^7| - \frac{1}{7}\log|x^7+1| + C = \frac{1}{7}\log\left|\frac{x^7}{x^7+1}\right| + C.$$

(XVII) Integrate by parts, $\int f dg = fg - \int g df$, where $f = \log(1 + \cos x)$, $dg = \cos x dx$, $df = -\sin x dx/(1 + \cos x)$ and $g = \sin x$. Then

$$\int \cos x \log(1 + \cos x) dx = \sin x \log(1 + \cos x) + \int \frac{\sin^2 x}{1 + \cos x} dx$$

$$= \sin x \log(1 + \cos x) + \int \frac{1 - \cos^2 x}{1 + \cos x} dx$$

$$= \sin x \log(1 + \cos x) + \int \frac{(1 - \cos x)(1 + \cos x)}{1 + \cos x} dx$$

$$= \sin x \log(1 + \cos x) + \int (1 - \cos x) dx$$

$$= \sin x \log(1 + \cos x) + \int dx - \int \cos x dx$$

$$= \sin x \log(1 + \cos x) + x - \sin x + C$$

$$= x + \sin x (\log(1 + \cos x) - 1) + C.$$

(XVIII) Observe that

$$\frac{x^2 - 3x + 3}{x^2 - 3x + 2} = -\frac{1}{x - 1} + \frac{1}{x - 2} + 1.$$

Thus

$$\int \frac{x^2 - 3x + 3}{x^2 - 3x + 2} dx = \int \frac{1}{x - 2} dx - \int \frac{1}{x - 1} dx + \int dx.$$

For the integrand $1/(x - 1)$, substitute $u = x - 1$ and $du = dx$. For the integrand $1/(x - 2)$, substitute $s = x - 2$ and $ds = dx$. Then

$$\int \frac{x^2 - 3x + 3}{x^2 - 3x + 2} dx = \int \frac{1}{x - 2} dx - \int \frac{1}{x - 1} dx + \int dx$$

$$= \int \frac{1}{s} ds - \int \frac{1}{u} du + \int dx$$

$$= \log|s| - \log|u| + x + C$$

$$= x + \log|x - 2| - \log|x - 1| + C$$

$$= x + \log \left| \frac{x - 2}{x - 1} \right| + C.$$

(XIX) Observe that

$$\int \frac{3x + 5}{(x^2 - 2x + 2)^2} dx = \int \left(\frac{3(2x - 2)}{2(x^2 - 2x + 2)^2} + \frac{8}{(x^2 - 2x + 2)^2} \right) dx$$

$$= \frac{3}{2} \int \frac{2x - 2}{(x^2 - 2x + 2)^2} dx + 8 \int \frac{1}{(x^2 - 2x + 2)^2} dx.$$

For the integrand $(2x-2)/(x^2-2x+2)^2$, substitute $u = x^2 - 2x + 2$ and $du = (2x-2)dx$. Then

$$\int \frac{3x+5}{(x^2-2x+2)^2}dx = \frac{3}{2}\int \frac{2x-2}{(x^2-2x+2)^2}dx + 8\int \frac{1}{(x^2-2x+2)^2}dx$$

$$= \frac{3}{2}\int \frac{1}{u^2}du + 8\int \frac{1}{(x^2-2x+2)^2}dx$$

$$= -\frac{3}{2u} + 8\int \frac{1}{(x^2-2x+2)^2}dx$$

$$= -\frac{3}{2u} + 8\int \frac{1}{((x-1)^2+1)^2}dx.$$

Substitute $s = x - 1$ and $ds = dx$. Thus

$$\int \frac{3x+5}{(x^2-2x+2)^2}dx = -\frac{3}{2u} + 8\int \frac{1}{((x-1)^2+1)^2}dx$$

$$= -\frac{3}{2u} + 8\int \frac{1}{(s^2+1)^2}ds.$$

For the integrand $1/(s^2+1)^2$, substitute $s = \tan p$ and $ds = \sec^2 p\, dp$. Then $(s^2+1)^2 = (\tan^2 p + 1)^2 = \sec^4 p$ and $p = \arctan s$. Therefore,

$$\int \frac{3x+5}{(x^2-2x+2)^2}dx = -\frac{3}{2u} + 8\int \cos^2 p\, dp.$$

Therefore,

$$\int \frac{3x+5}{(x^2-2x+2)^2}dx = -\frac{3}{2u} + 8\left(\frac{1}{4}\sin(2p) + \frac{p}{2}\right) + C$$

$$= -\frac{3}{2u} + 2\sin(2p) + 4p + C.$$

Substitute back for $p = \arctan s$, $s = x - 1$ and $u = x^2 - 2x + 2$. Thus

$$\int \frac{3x+5}{(x^2-2x+2)^2}dx = -\frac{3}{2u} + 2\sin(2p) + 4p + C$$

$$= -\frac{3}{2(x^2-2x+2)} + 2\sin(2\arctan(x-1))$$

$$+ 4\arctan(x-1) + C.$$

(xx) Substitute $u = \sqrt{x^2-1}$ and

$$du = \frac{x}{\sqrt{x^2-1}}dx \quad \left(\text{from which } \frac{1}{x}du = \frac{1}{\sqrt{x^2-1}}dx\right)$$

Then

$$\int \frac{1}{x\sqrt{x^2-1}}dx = \int \frac{1}{x}\cdot\frac{1}{x}du = \int \frac{1}{x^2}du.$$

Since $u = \sqrt{x^2 - 1}$, it follows that $x^2 = u^2 + 1$. Thus

$$\int \frac{1}{x\sqrt{x^2 - 1}} dx = \int \frac{1}{u^2 + 1} du$$

$$= \arctan u + C$$

$$= \arctan \sqrt{x^2 - 1} + C.$$

(XXI) Apply the substitution $u^2 = \dfrac{x+3}{x-1}$. Observe that by solving x in terms of u in $u^2 = \dfrac{x+3}{x-1}$ we have $x = \dfrac{u^2+3}{u^2-1}$. Thus, $\dfrac{1}{x+1} = \dfrac{u^2-1}{2u^2+2}$ and $dx = -\dfrac{8u}{(u^2-1)^2} du$. Thus,

$$\int \frac{1}{x+1} \sqrt{\frac{x+3}{x-1}} dx = \int \frac{u^2-1}{2u^2+2} \cdot u \left(-\frac{8u}{(u^2-1)^2} \right) du$$

$$= -4 \int \frac{u^2}{(u^2+1)(u+1)(u-1)} du.$$

Now use simple fraction decomposition of $\dfrac{u^2}{(u^2+1)(u+1)(u-1)}$ which gives us the following expansion:

$$\frac{u^2}{(u^2+1)(u+1)(u-1)} = \frac{A}{u+1} + \frac{B}{u-1} + \frac{Cu+D}{u^2+1}.$$

Multiplying both sides by $(u^2+1)(u+1)(u-1)$ gives us the equation:

$$u^2 = A(u-1)(u^2+1) + B(u+1)(u^2+1) + (Cu+D)(u+1)(u-1)$$
$$= (A+B+C)u^3 + (-A+B+D)u^2 + (A+B-C)u$$
$$+ (-A+B-D).$$

Hence, we have the following system of equations:

$$\begin{cases} 0 = A+B-C, \\ 0 = A+B+C, \\ 1 = -A+B+D, \\ 0 = -A+B-D, \end{cases}$$

with solution

$$A = -\frac{1}{4}, \ B = \frac{1}{4}, \ C = 0, \ D = \frac{1}{2}.$$

Thus,

$$\int \frac{u^2}{(u^2+1)(u+1)(u-1)} du = -\frac{1}{4}\int \frac{1}{u+1}du + \frac{1}{4}\int \frac{1}{u-1}du + \frac{1}{2}\int \frac{1}{u^2+1}du$$

$$= -\frac{1}{4}\log|u+1| + \frac{1}{4}\log|u-1| + \frac{1}{2}\arctan(u) + C$$

$$= -\frac{1}{4}\log\left|\frac{u+1}{u-1}\right| + \frac{1}{2}\arctan(u) + C.$$

Therefore,

$$\int \frac{1}{x+1}\sqrt{\frac{x+3}{x-1}}dx = \log\left|\frac{u+1}{u-1}\right| - 2\arctan(u) + C$$

$$= \log\left|\frac{\sqrt{\frac{x+3}{x-1}}+1}{\sqrt{\frac{x+3}{x-1}}-1}\right| - 2\arctan\left(\sqrt{\frac{x+3}{x-1}}\right) + C$$

$$= \log\left|\frac{\sqrt{x+3}+\sqrt{x-1}}{\sqrt{x+3}-\sqrt{x-1}}\right| - 2\arctan\left(\sqrt{\frac{x+3}{x-1}}\right) + C.$$

(XXII) Substitute $u = \tan x$ and $du = \sec^2 x\, dx$. Then

$$\int \frac{1}{2+3\tan x}dx = \int \frac{1}{(3u+2)(u^2+1)}du.$$

For the integrand $1/((3u+2)(u^2+1))$, use simple fractions to obtain

$$\int \frac{1}{2+3\tan x}dx = \int \frac{1}{(3u+2)(u^2+1)}du$$

$$= \frac{1}{13}\int \frac{2-3u}{u^2+1}du + \frac{9}{13}\int \frac{1}{3u+2}du$$

$$= -\frac{3}{13}\int \frac{u}{u^2+1}du + \frac{2}{13}\int \frac{1}{u^2+1}du + \frac{9}{13}\int \frac{1}{3u+2}du.$$

For the integrand $u/(u^2+1)$, substitute $s = u^2+1$ and $ds = 2u\,du$. For the integrand $1/(3u+2)$, substitute $p = 3u+2$ and $dp = 3du$. Thus

$$\int \frac{1}{2+3\tan x}dx = -\frac{3}{13}\int \frac{u}{u^2+1}du + \frac{2}{13}\int \frac{1}{u^2+1}du + \frac{9}{13}\int \frac{1}{3u+2}du$$

$$= -\frac{3}{26}\int \frac{1}{s}ds + \frac{2}{13}\int \frac{1}{u^2+1}du + \frac{3}{13}\int \frac{1}{p}dp$$

$$= -\frac{3}{26}\log|s| + \frac{2}{13}\arctan u + \frac{3}{13}\log|p| + C.$$

Substitute back for $s = u^2+1$, $p = 3u+2$ and $u = \tan x$:

$$\int \frac{1}{2+3\tan x}dx = -\frac{3}{26}\log(\tan^2 x+1) + \frac{2x}{13} + \frac{3}{13}\log|3\tan x+2| + C.$$

(XXIII) We will apply the reduction method. Since x^2 is a polynomial of degree 2, the method states that the following equality must be satisfied:

$$\int \frac{x^2}{\sqrt{3x^2 - x + 1}}dx = (Ax + B)\sqrt{3x^2 - x + 1} + K \int \frac{1}{\sqrt{3x^2 - x + 1}}dx, \quad (6.5)$$

where K is a constant and $Ax + B$ is a polynomial of degree < 2. Our goal is to find constants A, B and K. To do so, differentiate (6.5):

$$\frac{x^2}{\sqrt{3x^2 - x + 1}} = A\sqrt{3x^2 - x + 1} + (Ax + B)\frac{6x - 1}{2\sqrt{3x^2 - x + 1}} + \frac{K}{\sqrt{3x^2 - x + 1}}.$$

Now multiply the latter equality by $2\sqrt{3x^2 - x + 1}$, which gives us:

$$2x^2 = 2A(3x^2 - x + 1) + (Ax + B)(6x - 1) + 2K$$
$$= 12Ax^2 + (-3A + 6B)x + (2A - B + 2K).$$

Hence, we have the following system of equations:

$$\begin{cases} 2 = 12A, \\ 0 = -3A + 6B, \\ 0 = 2A - B + 2K, \end{cases}$$

with solution

$$A = \frac{1}{6}, \ B = \frac{1}{12}, \ K = -\frac{1}{8}.$$

Now it is enough to calculate $\int \frac{1}{\sqrt{3x^2 - x + 1}}dx$:

$$\int \frac{1}{\sqrt{3x^2 - x + 1}}dx = \int \frac{1}{\sqrt{\left(\sqrt{3}x - \frac{1}{2\sqrt{3}}\right)^2 + \frac{11}{12}}}dx.$$

Apply the substitution $u = \sqrt{3}x - \frac{1}{2\sqrt{3}}$ and $du = \sqrt{3}dx$:

$$\int \frac{1}{\sqrt{\left(\sqrt{3}x - \frac{1}{2\sqrt{3}}\right)^2 + \frac{11}{12}}}dx = \frac{1}{\sqrt{3}}\int \frac{1}{\sqrt{u^2 + \frac{11}{12}}}du$$

$$= \frac{1}{\sqrt{3}} \cdot \frac{2\sqrt{3}}{\sqrt{11}}\int \frac{1}{\sqrt{\left(\sqrt{\frac{12}{11}}u\right)^2 + 1}}du$$

$$= \frac{2}{\sqrt{11}}\int \frac{1}{\sqrt{\left(2\sqrt{\frac{3}{11}}u\right)^2 + 1}}du.$$

Now apply the substitution $v = 2\sqrt{\frac{3}{11}}u$ and $dv = 2\sqrt{\frac{3}{11}}du$:

$$\int \frac{1}{\sqrt{\left(2\sqrt{\frac{11}{3}}u\right)^2 + 1}}du = \frac{1}{2}\sqrt{\frac{11}{3}}\int \frac{1}{\sqrt{v^2 + 1}}dv$$

$$= \frac{1}{2}\sqrt{\frac{11}{3}}\operatorname{arc\,sinh}(v) + C$$

$$= \frac{1}{2}\sqrt{\frac{11}{3}}\operatorname{arc\,sinh}\left(2\sqrt{\frac{3}{11}}u\right) + C.$$

Hence,

$$\int \frac{1}{\sqrt{3x^2 - x + 1}}dx = \frac{2}{\sqrt{11}}\cdot\frac{1}{2}\sqrt{\frac{11}{3}}\operatorname{arc\,sinh}\left(2\sqrt{\frac{3}{11}}u\right) + C$$

$$= \frac{1}{\sqrt{3}}\operatorname{arc\,sinh}\left[2\sqrt{\frac{3}{11}}\left(\sqrt{3}x - \frac{1}{2\sqrt{3}}\right)\right] + C.$$

In conclusion,

$$\int \frac{x^2}{\sqrt{3x^2 - x + 1}}dx = \left(\frac{1}{6}x + \frac{1}{12}\right)\sqrt{3x^2 - x + 1}$$

$$- \frac{1}{8\sqrt{3}}\operatorname{arc\,sinh}\left[2\sqrt{\frac{3}{11}}\left(\sqrt{3}x - \frac{1}{2\sqrt{3}}\right)\right] + C$$

$$= \frac{1}{12}(2x + 1)\sqrt{3x^2 - x + 1}$$

$$- \frac{1}{8\sqrt{3}}\operatorname{arc\,sinh}\left[\frac{1}{\sqrt{11}}(6x - 1)\right] + C.$$

$\square$

Exercise 6.6. *Prove that a monotonic function defined on $[a, b]$ is integrable.*

Solution to Exercise 6.6.

Without loss of generality, assume that f is increasing on $[a, b]$. Thus, $f(a) \le f(x) \le f(b)$, and f is bounded on $[a, b]$. Given $\varepsilon > 0$, there exists $\alpha > 0$ such that $\alpha[f(b) - f(a)] < \varepsilon$. Let $P = \{x_0, x_1, \ldots, x_n\}$ be a partition of $[a, b]$ such that $|x_i - x_{i-1}| \le \alpha$ for all $i = 1, \ldots n$. Since f is increasing, it follows that

$$\inf_{x\in[x_{i-1},x_i]} f(x) = f(x_{i-1}) \quad \text{and} \quad \sup_{x\in[x_{i-1},x_i]} f(x) = f(x_i), \quad i = 1, \ldots, n.$$

Thus

$$U(f,P) - L(f,P) = \sum_{i=1}^{n}[f(x_i) - f(x_{i-1})]|x_i - x_{i-1}|$$

$$\leq \alpha \sum_{i=1}^{n}[f(x_i) - f(x_{i-1})]$$

$$= \alpha[f(b) - f(a)] < \varepsilon.$$

Consequently, Riemann's condition is satisfied, so f is integrable.

□

Exercise 6.7. *Prove that the composition of an integrable and a continuous function is integrable.*

Solution to Exercise 6.7.
This a classical result in Riemann integration and the usual proof is a bit tedious due to the calculations involved in it. However, we can shorten it by using a very famous result by Lebesgue. Let $f : I \to \mathbb{R}$ be an integrable function and $\phi : f(I) \to \mathbb{R}$ a continuous function. By the Lebesgue criterion for integrability (a bounded function is Riemann integrable if and only if the set of all discontinuities has measure zero) f is continuous away from a set of measure zero. So there exists a subset A of I with measure 0 such that the restriction of f to $I \setminus A$ is continuous. So the composition map $\phi \circ f$ is continuous on $I \setminus A$. Hence again by the Lebesgue criterion for integrability one can conclude that $\phi \circ f$ is Riemann Integrable on I.   □

Exercise 6.8. *Prove that a bounded function whose set of discontinuity points is finite is integrable.*

Solution to Exercise 6.8.
This is an immediate consequence of the Lebesgue criterion for integrability because every finite set has measure zero.

□

Exercise 6.9. *Prove that the function defined on $[0,1]$ by*

$$f(x) = \begin{cases} 0 & \text{if } x \notin \mathbb{Q} \\ \frac{1}{q} & \text{if } x = \frac{p}{q} \in \mathbb{Q}, \text{ with } p \in \mathbb{Z} \text{ and } q \in \mathbb{N} \text{ coprime} \end{cases}$$

is integrable and obtain its integral.

Solution to Exercise 6.9.

Let us prove the following facts:

- f is discontinuous at all rational numbers.

Let $x_0 = \frac{p}{q}$ be an arbitrary rational number, with $p \in \mathbb{Z}$, $q \in \mathbb{N}$, and p and q coprime. So $f(x_0) = \frac{1}{q}$. Let α be any irrational number and, for all $n \in \mathbb{N}$, define $x_n = x_0 + \frac{\alpha}{n}$. Observe that $x_n \in \mathbb{R} \setminus \mathbb{Q}$, and so $f(x_n) = 0$ for all $n \in \mathbb{N}$. Let $\varepsilon = \frac{1}{q}$, and given $\delta > 0$ let $n = 1 + \left[\frac{\alpha}{\delta}\right]$, where $\left[\frac{\alpha}{\delta}\right]$ is the first natural greater than or equal to $\frac{\alpha}{\delta}$. Observe that

$$|f(x_0) - f(x_n)| = \frac{1}{q} = \varepsilon$$

and

$$|x_0 - x_n| = \frac{\alpha}{n} = \frac{\alpha}{1 + \left[\frac{\alpha}{\delta}\right]} < \frac{\alpha}{\left[\frac{\alpha}{\delta}\right]} \leq \delta,$$

which shows the discontinuity of f at x_0.

- f is continuous at all irrational numbers.

Let $\varepsilon > 0$ and $i \in \mathbb{N}$. Let also $x_0 \in [0,1]$ be any irrational number. According to the Archimedean property of the reals, there exists $r \in \mathbb{N} \setminus \{1\}$ with $\frac{1}{r} < \varepsilon$, and for each $i = 1, \ldots, r$ there exists $k_i \in \mathbb{N}$, such that

$$0 < \frac{k_i}{i} < x_0 < \frac{k_i + 1}{i}.$$

Let d_i the minimal distance of x_0 to its i-th lower and upper limits equals, that is,

$$d_i := \min\left\{ \left|x_0 - \frac{k_i}{i}\right|, \left|x_0 - \frac{k_i + 1}{i}\right| \right\}.$$

Define

$$\delta := \min_{2 \leq i \leq r} \{d_i\}.$$

So for all $i = 2, \ldots, r$,

$$\left|x_0 - \frac{k_i}{i}\right| \geq \delta \quad \text{and} \quad \left|x_0 - \frac{k_i + 1}{i}\right| \geq \delta,$$

or equivalently,

$$\frac{k_i}{i}, \frac{k_i + 1}{i} \notin (x_0 - \delta, x_0 + \delta)$$

for all $1, \ldots, r$.

Now let $x \in \mathbb{Q} \cap (x_0 - \delta, x_0 + \delta)$ with the unique representation $x = p/q$ where $p, q \in \mathbb{N}$ are coprime. Then, necessarily, $q > r$ and therefore,

$$f(x) = \frac{1}{q} < \frac{1}{r} < \varepsilon.$$

Likewise, for all irrational $x \in [0, 1]$, $f(x) = 0 = f(x_0)$, and thus, if $\varepsilon > 0$ then any choice of (sufficiently small) $\delta > 0$ gives

$$|x - x_0| < \delta \implies |f(x) - f(x_0)| = 0 < \varepsilon.$$

Therefore, f is continuous on $[0, 1] \cap (\mathbb{R} \setminus \mathbb{Q})$.

Now let us prove that f is integrable. It is well known that $[0, 1] \cap \mathbb{Q}$ has measure zero (more generally, every countable subset of the real numbers has measure zero). So from the Lebesgue criterion for integrability, which states that a bounded function is Riemann integrable if and only if the set of all discontinuities has measure zero, we can conclude that f is Riemann integrable.

Observe that

$$\int_0^1 f(x)dx = 0$$

because the function is equal to zero almost everywhere. □

Exercise 6.10. *Prove that the composition of two integrable functions may not be integrable.*

Solution to Exercise 6.10.
Consider $f, g : [0, 1] \to \mathbb{R}$ the functions defined by

$$f(x) = \begin{cases} 0 & \text{if } x \notin \mathbb{Q}, \\ \frac{1}{q} & \text{if } x = \frac{p}{q} \in \mathbb{Q}, \text{ with } p \in \mathbb{Z} \text{ and } q \in \mathbb{N} \text{ coprime}, \end{cases}$$

and

$$g(x) = \begin{cases} 0 & \text{if } x = 0, \\ 1 & \text{if } x \neq 0. \end{cases}$$

From previous exercises, we know that f and g are integrable. The composition of these functions is

$$g(f(x)) = \begin{cases} 0 & \text{if } x \notin \mathbb{Q}, \\ 1 & \text{if } x \in \mathbb{Q} \end{cases}$$

which is nowhere continuous, so not Riemann-integrable over any interval (recall that the Lebesgue criterion for integrability states that a bounded function is Riemann integrable if and only if the set of all discontinuities has measure zero). □

Exercise 6.11. *Prove that the product of two integrable functions is integrable.*

Solution to Exercise 6.11.
If f and g are integrable, then so is $f + g$. Hence, by Exercise 6.7, with $\phi(y) = y^2$, we have that f^2, g^2 and $(f+g)^2 = f^2 + g^2 + 2fg$ are all integrable. The integrability of

$$fg = \frac{1}{2}\left[(f+g)^2 - f^2 - g^2\right]$$

now follows by linearity.

□

Exercise 6.12. *Let f be a non-negative integrable function on $[a, b]$. Prove that if f is continuous at c and $f(c) > 0$, then $\int_a^b f > 0$.*

Solution to Exercise 6.12.
Let $M = f(c)/2$. Suppose $c \in (a, b)$. Thus, there exists $\delta > 0$ such that

$$x \in [c - \delta, c + \delta] \implies f(x) \geq M.$$

Consider the partition $P_1 = \{a = t_0, \ldots, c - \delta, c + \delta, \ldots, t_n = b\}$ of $[a, b]$. Let $m = \inf\{f(x) : x \in [c - \delta, c + \delta]\}$. We have $L(f, P_1) \geq 2m\delta \geq 2M\delta > 0$. Thus

$$\int_a^b f = \sup_P L(f, P) \geq L(f, P_1) > 0.$$

If $c = a$, then there exists $\delta > 0$ such that $x \in [c, c + \delta] \Rightarrow f(x) \geq M$. Considering $P_1 = \{c, c + \delta, \ldots, t_n = b\}$ and $m = \inf\{f(x) : x \in [c, c + \delta]\}$, we have $L(f, P_1) \geq m\delta \geq M\delta > 0$. Therefore,

$$\int_a^b f \geq L(f, P_1) > 0.$$

The case $c = b$ follows analogously.

□

Exercise 6.13. *Prove that if f has a continuous derivative of order n in $[a, b]$, then*

$$f(b) - \sum_{k=0}^{n-1} \frac{f^{(k)}(a)}{k!}(b - a)^k = \int_a^b \frac{f^{(n)}(x)}{(n-1)!}(b - x)^{n-1}dx.$$

Solution to Exercise 6.13.

We will prove it by induction on n.

- Base case ($n = 1$). Since f' is integrable on $[a, b]$ and f is a primitive of f', by the Fundamental Theorem of Calculus we have

$$f(b) - f(a) = \int_a^b f'(x)dx,$$

as needed.

- Induction step. Assume that

$$f(b) - \sum_{k=0}^{n-1} \frac{f^{(k)}(a)}{k!}(b-a)^k = \int_a^b \frac{f^{(n)}(x)}{(n-1)!}(b-x)^{n-1}dx,$$

for some $n \in \mathbb{N}$ (I.H.). Apply integration by parts to $\int_a^b \frac{f^{(n)}(x)}{(n-1)!}(b - x)^{n-1}dx$ by taking $u = f^{(n)}(x)$ and $dv = \frac{(b-x)^{n-1}}{(n-1)!}dx$. So $du = f^{(n+1)}(x)dx$ and $v = -\frac{(b-x)^n}{n(n-1)!} = -\frac{(b-x)^n}{n!}$. Hence,

$$\int_a^b \frac{f^{(n)}(x)}{(n-1)!}(b-x)^{n-1}dx$$

$$= -f^{(n)}(x)\frac{(b-x)^n}{n!}\Big|_a^b - \int_a^b \left(-\frac{(b-x)^n}{n!}\right)f^{(n+1)}(x)dx$$

$$= -f^{(n)}(b)\frac{(b-b)^n}{n!} + f^{(n)}(a)\frac{(b-a)^n}{n!} + \int_a^b \frac{f^{(n+1)}(x)}{n!}(b-x)^n dx$$

$$= f^{(n)}(a)\frac{(b-a)^n}{n!} + \int_a^b \frac{f^{(n+1)}(x)}{n!}(b-x)^n dx,$$

that is,

$$\int_a^b \frac{f^{(n+1)}(x)}{n!}(b-x)^n dx = f(b) - \sum_{k=0}^{n-1} \frac{f^{(k)}(a)}{k!}(b-a)^k - f^{(n)}(a)\frac{(b-a)^n}{n!}$$

$$= f(b) - \sum_{k=0}^{n} \frac{f^{(k)}(a)}{k!}(b-a)^k.$$

$\square$

Exercise 6.14. *Express the following integrals as the limit of a sum, and find the following limits by evaluating an integral using Barrow's rule:*

(I) $\int_1^2 \log x\, dx;\ \lim_{n\to\infty} \sqrt[n]{\left(1+\frac{1}{n}\right)\left(1+\frac{2}{n}\right)\cdots\left(1+\frac{n}{n}\right)}.$

(II) $\int_0^1 x^p\, dx;\ \lim_{n\to\infty}\frac{1^p+2^p+\cdots+n^p}{n^{p+1}}\ \text{with}\ p\in\mathbb{N}.$

(III) $\int_1^2 \frac{dx}{x};\ \lim_{n\to\infty}\left(\frac{1}{n+1}+\frac{1}{n+2}+\cdots+\frac{1}{n+n}\right).$

(IV) $\int_0^1 \frac{dx}{x^2+1};\ \lim_{n\to\infty}\left[n\left(\frac{1}{n^2+1^2}+\frac{1}{n^2+2^2}+\cdots+\frac{1}{n^2+n^2}\right)\right].$

(V) $\int_0^1 \frac{dx}{\sqrt{x^2+1}};\ \lim_{n\to\infty}\sum_{k=1}^{n}\frac{1}{\sqrt{n^2+k^2}}.$

(VI) $\int_0^1 \sqrt{x}\, dx;\ \lim_{n\to\infty}\frac{1+\sqrt{2}+\sqrt{3}+\cdots+\sqrt{n}}{n\sqrt{n}}.$

(VII) $\int_0^1 \sqrt{1-x^2}\, dx;\ \lim_{n\to\infty}\sum_{k=1}^{n}\frac{\sqrt{n^2-k^2}}{n^2}.$

Solution to Exercise 6.14.

(I) Define $f(x)=\log(x)$ and observe that

$$\lim_{n\to\infty}\sqrt[n]{\left(1+\frac{1}{n}\right)\left(1+\frac{2}{n}\right)\cdots\left(1+\frac{n}{n}\right)} = \lim_{n\to\infty} e^{\log\left(\sqrt[n]{\left(1+\frac{1}{n}\right)\left(1+\frac{2}{n}\right)\cdots\left(1+\frac{n}{n}\right)}\right)}$$

$$= \lim_{n\to\infty} e^{\log\left[\left(1+\frac{1}{n}\right)\left(1+\frac{2}{n}\right)\cdots\left(1+\frac{n}{n}\right)\right]^{1/n}}$$

$$= e^{\lim_{n\to\infty}\left(\frac{1}{n}\log\left[\left(1+\frac{1}{n}\right)\left(1+\frac{2}{n}\right)\cdots\left(1+\frac{n}{n}\right)\right]\right)}$$

$$= e^{\lim_{n\to\infty}\left(\frac{1}{n}\sum_{k=1}^{n}\log\left(1+\frac{k}{n}\right)\right)}$$

$$= e^{\lim_{n\to\infty}\left(\frac{1}{n}\sum_{k=1}^{n}f\left(1+\frac{k}{n}\right)\right)}.$$

This limit exists and it is equal to

$$e^{\int_0^1 f(1+y)\, dy} = e^{\int_0^1 \log(1+y)\, dy} = e^{\int_1^2 \log x\, dx} = e^{(x\log x - x)\big|_1^2} = e^{2\log 2 - 1} = \frac{4}{e}.$$

(II) Define $f(x) = x^p$ and observe that

$$\lim_{n \to \infty} \frac{1^p + 2^p + \cdots + n^p}{n^{p+1}} = \lim_{n \to \infty} \sum_{k=1}^{n} \frac{k^p}{n^{p+1}}$$

$$= \lim_{n \to \infty} \frac{1}{n} \sum_{k=1}^{n} \frac{k^p}{n^p}$$

$$= \lim_{n \to \infty} \frac{1}{n} \sum_{k=1}^{n} f\left(\frac{k}{n}\right).$$

This limit exists and it is equal to

$$\int_0^1 f(x)dx = \int_0^1 x^p dx = \frac{x^{p+1}}{p+1}\bigg|_0^1 = \frac{1}{p+1}.$$

(III) Define $f(y) = \frac{1}{1+y}$ and observe that

$$\lim_{n \to \infty} \left(\frac{1}{n+1} + \frac{1}{n+2} + \cdots + \frac{1}{n+n}\right) = \lim_{n \to \infty} \sum_{k=1}^{n} \frac{1}{n+k}$$

$$= \lim_{n \to \infty} \sum_{k=1}^{n} \frac{1}{n\left(1 + \frac{k}{n}\right)}$$

$$= \lim_{n \to \infty} \frac{1}{n} \sum_{k=1}^{n} \frac{1}{1 + \frac{k}{n}}$$

$$= \lim_{n \to \infty} \frac{1}{n} \sum_{k=1}^{n} f\left(\frac{k}{n}\right).$$

This limit exists and it is equal to

$$\int_0^1 f(y)dy = \int_0^1 \frac{dy}{1+y} = \int_1^2 \frac{dx}{x} = \log 2,$$

where $x = 1 + y$ and $dx = dy$.

(IV) Define $f(x) = \dfrac{1}{x^2 + 1}$ and observe that

$$\lim_{n\to\infty} \left[n \left(\frac{1}{n^2 + 1^2} + \frac{1}{n^2 + 2^2} + \cdots + \frac{1}{n^2 + n^2} \right) \right]$$

$$= \lim_{n\to\infty} \sum_{k=1}^{n} \frac{n}{n^2 + k^2}$$

$$= \lim_{n\to\infty} \sum_{k=1}^{n} \frac{n}{n^2 \left[1 + \left(\frac{k}{n}\right)^2 \right]}$$

$$= \lim_{n\to\infty} \frac{1}{n} \sum_{k=1}^{n} \frac{1}{1 + \left(\frac{k}{n}\right)^2}$$

$$= \lim_{n\to\infty} \frac{1}{n} \sum_{k=1}^{n} f\left(\frac{k}{n}\right).$$

This limit exists and it is equal to

$$\int_0^1 f(x)\,dx = \int_0^1 \frac{dx}{x^2 + 1} = \arctan x \big|_0^1 = \frac{\pi}{4}.$$

(V) Define $f(x) = \frac{1}{x^2+1}$ and observe that

$$\lim_{n\to\infty} \sum_{k=1}^{n} \frac{1}{\sqrt{n^2 + k^2}} = \lim_{n\to\infty} \sum_{k=1}^{n} \frac{1}{\sqrt{n^2 \left[1 + \left(\frac{k}{n}\right)^2 \right]}}$$

$$= \lim_{n\to\infty} \frac{1}{n} \sum_{k=1}^{n} \frac{1}{\sqrt{1 + \left(\frac{k}{n}\right)^2}}$$

$$= \lim_{n\to\infty} \frac{1}{n} \sum_{k=1}^{n} f\left(\frac{k}{n}\right).$$

This limit exists and it is equal to

$$\int_0^1 f(x)\,dx = \int_0^1 \frac{dx}{\sqrt{x^2 + 1}} = \operatorname{arcsinh} x \big|_0^1 = \operatorname{arcsinh} 1.$$

(VI) Define $f(x) = \sqrt{x}$ and observe that

$$\lim_{n\to\infty} \frac{1 + \sqrt{2} + \sqrt{3} + \cdots + \sqrt{n}}{n\sqrt{n}} = \lim_{n\to\infty} \sum_{k=1}^{n} \frac{\sqrt{k}}{n\sqrt{n}}$$

$$= \lim_{n\to\infty} \frac{1}{n} \sum_{k=1}^{n} \frac{\sqrt{k}}{\sqrt{n}}$$

$$= \lim_{n\to\infty} \frac{1}{n} \sum_{k=1}^{n} f\left(\frac{k}{n}\right).$$

This limit exists and it is equal to

$$\int_0^1 f(x)dx = \int_0^1 \sqrt{x}dx = \frac{2x^{3/2}}{3}\Big|_0^1 = \frac{2}{3}.$$

(VII) Define $f(x) = \sqrt{1 - x^2}$ and observe that

$$\lim_{n\to\infty} \sum_{k=1}^n \frac{\sqrt{n^2 - k^2}}{n^2} = \lim_{n\to\infty} \sum_{k=1}^n \frac{\sqrt{n^2\left(1 - \frac{k^2}{n^2}\right)}}{n^2}$$

$$= \lim_{n\to\infty} \sum_{k=1}^n \frac{\sqrt{1 - \frac{k^2}{n^2}}}{n}$$

$$= \lim_{n\to\infty} \frac{1}{n} \sum_{k=1}^n \sqrt{1 - \frac{k^2}{n^2}}$$

$$= \lim_{n\to\infty} \frac{1}{n} \sum_{k=1}^n f\left(\frac{k}{n}\right).$$

This limit exists and it is equal to

$$\int_0^1 f(x)dx = \int_0^1 \sqrt{1 - x^2}dx = \frac{1}{2}\left(x\sqrt{1 - x^2} + \arcsin x\right)\Big|_0^1 = \frac{\pi}{4}.$$

□

Exercise 6.15. *Let $f : [0,1] \to \mathbb{R}$ be a continuous function. Prove that the following functions are differentiable and find their derivatives:*

(I) $F(t) = \displaystyle\int_t^1 f(x)dx.$

(II) $F(t) = \displaystyle\int_0^{t^2} f(x)dx.$

Solution to Exercise 6.15.

(I) Fixed $t \in [0,1]$, by the continuity of f, given $\varepsilon > 0$, there exists $\delta > 0$ such that

$$x \in [0,1], |x - t| < \delta \implies |f(x) - f(t)| < \varepsilon.$$

So, if $0 < h < \delta$ and $t + h \in [0, 1]$, we have

$$\left| \frac{F(t+h) - F(t)}{h} - (-f(t)) \right| = \left| \frac{1}{h} \left(\int_{t+h}^{1} f(x)dx - \int_{t}^{1} f(x)dx \right) + f(t) \right|$$

$$= \frac{1}{h} \left| -\int_{t}^{t+h} f(x)dx + f(t) \right|$$

$$= \frac{1}{h} \left| -\int_{t}^{t+h} [f(x) - f(t)]\, dx \right|$$

$$\leq \frac{1}{h} \int_{t}^{t+h} |f(x) - f(t)|dx$$

$$\leq \frac{1}{h} \varepsilon \int_{t}^{t+h} dx = \varepsilon.$$

Thus

$$\lim_{h \to 0^+} \frac{F(t+h) - F(t)}{h} = -f(t).$$

Similarly, we can conclude that

$$\lim_{h \to 0^-} \frac{F(t+h) - F(t)}{h} = -f(t),$$

from which it follows that F is derivable and that $F'(t) = -f(t)$.

(II) Consider $G(s) = \int_0^s f(x)dx$ and $g(t) = t^2$. Observe that

$$(G \circ g)(t) = G(g(t)) = \int_0^{g(t)} f(x)dx = \int_0^{t^2} f(x)dx = F(t).$$

Analogously to the previous item, we can conclude that G is differentiable and that

$$G'(s) = f(s).$$

Since g is also differentiable, we conclude that $F = G \circ g$ is differentiable and

$$F'(t) = (G \circ g)'(t) = G'(g(t))g'(t) = f(g(t))2t = 2tf(t^2).$$

$\square$

Exercise 6.16. *Calculate the following integrals:*

(I) $\displaystyle\int_0^{2\pi} \max\{\sin x, \cos x\}dx.$

(II) $\displaystyle\int_0^{\pi/2} \frac{\cos^3 x \sin^3 x}{1 + \sin^2 x}dx.$

(III) $\displaystyle\int_{-3}^{3} |x(x-1)(x+1)(x-2)|dx.$

Solution to Exercise 6.16.

(I) In $(0, 2\pi)$ the solutions of $\sin x = \cos x$ are $x = \frac{\pi}{4}$ and $x = \frac{5\pi}{4}$. Moreover, recall that

$$\begin{cases} \cos x > \sin x & \text{for } x \in \left(0, \frac{\pi}{4}\right), \\ \sin x > \cos x & \text{for } x \in \left(\frac{\pi}{4}, \frac{5\pi}{4}\right), \\ \cos x > \sin x & \text{for } x \in \left(\frac{5\pi}{4}, 2\pi\right). \end{cases}$$

Thus

$$\int_0^{2\pi} \max\{\sin x, \cos x\}dx = \int_0^{\pi/4} \cos x\, dx + \int_{\pi/4}^{5\pi/4} \sin x\, dx + \int_{5\pi/4}^{2\pi} \cos x\, dx$$

$$= \sin x \Big|_0^{\pi/4} - \cos x \Big|_{\pi/4}^{5\pi/4} + \sin x \Big|_{5\pi/4}^{2\pi}$$

$$= \frac{\sqrt{2}}{2} - \left(-\frac{\sqrt{2}}{2} - \frac{\sqrt{2}}{2}\right) + \frac{\sqrt{2}}{2}$$

$$= 2\sqrt{2}.$$

(II) Using that $\cos^2 x = 1 - \sin^2 x$, observe that

$$\int_0^{\pi/2} \frac{\cos^3 x \sin^3 x}{1 + \sin^2 x}dx = \int_0^{\pi/2} \frac{(\sin^3 x - \sin^5 x)\cos x}{1 + \sin^2 x}dx.$$

For the integrand $((\sin^3 x - \sin^5 x)\cos x)/(1 + \sin^2 x)$, substitute $u = \sin x$ and $du = \cos x\, dx$. This gives us a new lower limit $u = \sin 0 = 0$ and upper limit $u = \sin(\pi/2) = 1$. Then

$$\int_0^{\pi/2} \frac{\cos^3 x \sin^3 x}{1 + \sin^2 x} dx = \int_0^{\pi/2} \frac{(\sin^3 x - \sin^5 x) \cos x}{1 + \sin^2 x} dx$$

$$= \int_0^1 \frac{u^3 - u^5}{1 + u^2} du$$

$$= \int_0^1 \left(-u^3 - \frac{2u}{1 + u^2} + 2u \right) du$$

$$= -\int_0^1 u^3 du - 2 \int_0^1 \frac{u}{1 + u^2} du + 2 \int_0^1 u du.$$

For the integrand $u/(1+u^2)$, substitute $s = 1 + u^2$ and $ds = 2u du$. This gives a new lower limit $s = 1 + 0^2 = 1$ and upper limit $s = 1 + 1^2 = 2$. So

$$\int_0^{\pi/2} \frac{\cos^3 x \sin^3 x}{1 + \sin^2 x} dx = -\int_0^1 u^3 du - 2 \int_0^1 \frac{u}{1 + u^2} du + 2 \int_0^1 u du$$

$$= -\int_0^1 u^3 du - \int_1^2 \frac{1}{s} ds + 2 \int_0^1 u du$$

$$= -\frac{u^4}{4} \Big|_0^1 - \log |s| \Big|_1^2 + u^2 \Big|_0^1$$

$$= -\frac{1}{4} - \log 2 + 1$$

$$= \frac{3}{4} - \log 2.$$

(III) Let $f(x) = x(x-1)(x+1)(x-2) = x^4 - 2x^3 - x^2 + 2x$. Observe that the solutions of $f(x) = 0$ are $x = -1, 0, 1, 2$ and that

$$\begin{cases} f(x) > 0 & \text{for } x \in (-3, -1), \\ f(x) < 0 & \text{for } x \in (-1, 0), \\ f(x) > 0 & \text{for } x \in (0, 1), \\ f(x) < 0 & \text{for } x \in (1, 2), \\ f(x) > 0 & \text{for } x \in (2, 3). \end{cases}$$

Then

$$\int_{-3}^{3} |x(x-1)(x+1)(x-2)|dx$$

$$= \int_{-3}^{3} |f(x)|dx$$

$$= \int_{-3}^{-1} f(x)dx - \int_{-1}^{0} f(x)dx + \int_{0}^{1} f(x)dx - \int_{1}^{2} f(x)dx + \int_{2}^{3} f(x)dx$$

$$= \int_{-3}^{-1} (x^4 - 2x^3 - x^2 + 2x)dx - \int_{-1}^{0} (x^4 - 2x^3 - x^2 + 2x)dx$$

$$+ \int_{0}^{1} (x^4 - 2x^3 - x^2 + 2x)dx - \int_{1}^{2} (x^4 - 2x^3 - x^2 + 2x)dx$$

$$+ \int_{2}^{3} (x^4 - 2x^3 - x^2 + 2x)dx$$

$$= \left(\frac{x^5}{5} - \frac{x^4}{2} - \frac{x^3}{3} + x^2 \right)\Bigg|_{-3}^{-1} - \left(\frac{x^5}{5} - \frac{x^4}{2} - \frac{x^3}{3} + x^2 \right)\Bigg|_{-1}^{0}$$

$$+ \left(\frac{x^5}{5} - \frac{x^4}{2} - \frac{x^3}{3} + x^2 \right)\Bigg|_{0}^{1} - \left(\frac{x^5}{5} - \frac{x^4}{2} - \frac{x^3}{3} + x^2 \right)\Bigg|_{1}^{2}$$

$$+ \left(\frac{x^5}{5} - \frac{x^4}{2} - \frac{x^3}{3} + x^2 \right)\Bigg|_{2}^{3}$$

$$= \frac{1076}{15} + \frac{19}{30} + \frac{11}{30} + \frac{19}{30} + \frac{251}{30} = \frac{1226}{15}.$$

□

Exercise 6.17. *Let f be a continuous function on $\mathbb{R}$ and let u and v be two differentiable functions on $\mathbb{R}$. The function g on $\mathbb{R}$ is defined as*

$$g(x) = \int_{u(x)}^{v(x)} f(t)dt.$$

Prove that g is differentiable on $\mathbb{R}$ and that

$$g'(x) = f(v(x))v'(x) - f(u(x))u'(x).$$

Solution to Exercise 6.17.

Write g as

$$g(x) = \int_{u(x)}^{0} f(t)dt + \int_{0}^{v(x)} f(t)dt.$$

Consider

$$K(s) = \int_{s}^{0} f(t)dt$$

and

$$L(s) = \int_{0}^{s} f(t)dt.$$

Observe that

$$(K \circ u)(t) + (L \circ v)(t) = K(u(t)) + L(v(t)) = \int_{u(x)}^{0} f(t)dt + \int_{0}^{v(x)} f(t)dt = g(x).$$

We can conclude that K and L are differentiable and that

$$K'(s) = -f(s) \quad \text{and} \quad L'(s) = f(s).$$

Since u and v are also differentiable, we conclude that $g = K \circ u + L \circ v$ is differentiable and

$$
\begin{aligned}
g'(x) &= (K \circ u + L \circ v)'(x) \\
&= (K \circ u)'(x) + (L \circ v)'(x) \\
&= K'(u(x))u'(x) + L'(v(x))v'(x) \\
&= -f(u(x))u'(x) + f(v(x))v'(x) \\
&= f(v(x))v'(x) - f(u(x))u'(x).
\end{aligned}
$$

□

Exercise 6.18. *Find the derivative of the following functions:*

(I) $f(x) = \displaystyle\int_{0}^{x} \frac{e^t}{t^4 + t^2 + 2} dt.$

(II) $f(x) = \displaystyle\int_{x}^{x^2} \frac{dt}{\sqrt{t^2 + 1}}.$

Solution to Exercise 6.18.

By Exercise 6.17, we can conclude that:

(I) $f'(x) = \frac{e^x}{x^4 + x^2 + 2}.$

(II) $f'(x) = \frac{1}{\sqrt{(x^2)^2 + 1}} \cdot 2x - \frac{1}{\sqrt{x^2 + 1}} = \frac{2x}{\sqrt{x^4 + 1}} - \frac{1}{\sqrt{x^2 + 1}}.$

□

Exercise 6.19. *Calculate the following limits:*

(I) $\displaystyle\lim_{x\to 0^+} \frac{\int_0^{x^2} \sin\sqrt{t}\,dt}{x^3}$.

(II) $\displaystyle\lim_{x\to +\infty} \frac{\left(\int_0^x e^{t^2}\,dt\right)^2}{\int_0^x e^{2t^2}\,dt}$.

Solution to Exercise 6.19.

(I) Notice that we can use L'Hôpital's Rule. Since

$$\frac{d}{dx}\int_0^{x^2}\sin\sqrt{t}\,dt = 2x\sin\sqrt{x^2},$$

it follows that

$$\lim_{x\to 0^+}\frac{\int_0^{x^2}\sin\sqrt{t}\,dt}{x^3} = \lim_{x\to 0^+}\frac{2x\sin\sqrt{x^2}}{3x^2} = \frac{2}{3}\lim_{x\to 0^+}\frac{\sin x}{x}.$$

Applying again L'Hôpital's Rule, we obtain that

$$\lim_{x\to 0^+}\frac{\int_0^{x^2}\sin\sqrt{t}\,dt}{x^3} = \frac{4}{3}\lim_{x\to 0^+}\cos x = \frac{4}{3}.$$

(II) It is not difficult to see that

$$\lim_{x\to +\infty}\int_0^x e^{t^2}\,dt = \infty \quad\text{and}\quad \lim_{x\to +\infty}\int_0^x e^{2t^2}\,dt = \infty.$$

Thus, applying L'Hôpital's Rule twice, we obtain that

$$\lim_{x\to +\infty}\frac{\left(\int_0^x e^{t^2}\,dt\right)^2}{\int_0^x e^{2t^2}\,dt} = \lim_{x\to +\infty}\frac{2\left(\int_0^x e^{t^2}\,dt\right)e^{x^2}}{e^{2x^2}}$$

$$= 2\lim_{x\to +\infty}\frac{\int_0^x e^{t^2}\,dt}{e^{x^2}}$$

$$= 2\lim_{x\to +\infty}\frac{e^{x^2}}{2xe^{x^2}}$$

$$= \lim_{x\to +\infty}\frac{1}{x}$$

$$= 0.$$

□

Exercise 6.20. *Calculate the area of the region bounded by the curve* $y = x^3 - x$ *and its tangent at the point of abscissa* $x = -1$.

Solution to Exercise 6.20.
Let $f(x) = x^3 - x$. Thus, the curve in question is the graph of f. The ordinate of the point of abscissa $x = -1$ is $y = f(-1) = 0$. The tangent line to the curve $y = x^3 - x$ at the point $(-1, 0)$ is the line r that passes through $(-1, 0)$ and whose slope is $f'(-1) = 3(-1)^2 - 1 = 2$, that is, $r(x) = 2x + 2$.

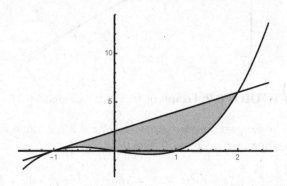

FIGURE 6.1: Graph of the Exercise 6.20.

Let us now find the abscissa of the other point of intersection between the graph of f and the line r. By equating f and r, we obtain $x^3 - 3x - 2 = 0$, whose solutions are $x = -1$ and $x = 2$. Since $r(x) > f(x)$ for $x \in (-1, 2)$ (by continuity, this can be seen by noting, for example, that $r(0) = 2 > 0 = f(0)$), the area of the region bounded by the curve $y = x^3 - x$ and its tangent at the point of abscissa $x = -1$ will be given by

$$\int_{-1}^{2} (r(x) - f(z)) \, dx = \int_{-1}^{2} \left(2x + 2 - x^3 + x\right) dx$$

$$= \left(\frac{3x^2}{2} + 2x - \frac{x^4}{4} \right) \Bigg|_{-1}^{2}$$

$$= \frac{27}{4}.$$

The region related to this problem can be seen in Figure 6.1. □

Exercise 6.21. *Find the area of the region bounded by the parabola* $y = -x^2 - 2x + 3$, *its tangent at the point* $(2, -5)$ *and the axis* y.

Solution to Exercise 6.21.
The region bounded by the parabola $y = -x^2 - 2x + 3$, its tangent at the point $(2, -5)$ and the axis y can be seen graphically in Figure 6.2

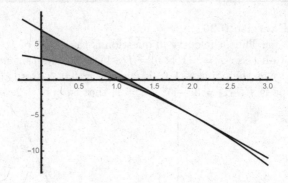

FIGURE 6.2: Graph of region for Exercise 6.21.

At $x = 2$, the tangent line to $y = -x^2 - 2x + 3$ has slope $m = -6$. The equation of this line is $y = -6x + 7$. Then

$$\text{Area} = \int_0^2 [(-6x + 7) - (-x^2 - 2x + 3)]dx = \int_0^2 (x^2 - 4x + 4)dx = \frac{8}{3}.$$

$\square$

Exercise 6.22 (Euler's Gamma function). *We ask the following questions regarding the famous Gamma function, due to Euler.*

(I) *Show that, given $x \in (0, \infty)$, the following integral is convergent:*

$$\Gamma(x) = \int_0^\infty t^{x-1} e^{-t}\, dt.$$

(II) *Prove that $\Gamma(x + 1) = x\Gamma(x)$ for all $x > 0$.*

(III) *To finish, show that $\Gamma(n + 1) = n!$ for all $n \in \mathbb{N} \cup \{0\}$.*

Solution to Exercise 6.22.
(I). Let

$$\Gamma(x) = \int_0^\infty t^{x-1} e^{-t} dt = \underbrace{\int_0^1 t^{x-1} e^{-t} dt}_{(1)} + \underbrace{\int_1^\infty t^{x-1} e^{-t} dt}_{(2)},$$

and let us show that both integrals above are convergent.

For the first integral, it suffices to notice that, since $t \geq 0$ it is $e^{-t} \leq 1$ on the interval $[0, 1]$, and
$$t^{x-1}e^{-t} \leq t^{x-1} \cdot 1,$$
which (by the comparison test), entails
$$\int_0^1 t^{x-1}e^{-t}dt \leq \int_0^1 t^{x-1}dt = \frac{t^x}{x}\Big|_{t=0}^{t=1} = 1/x,$$
obtaining convergence for the first integral.

Regarding the second integral, we have that
$$\lim_{t\to\infty} \frac{t^{x-1}e^{-t}}{\frac{1}{t^2}} = \lim_{t\to\infty} t^{x+1}e^{-t} = 0,$$
which (by the asymptotic comparison test) also converges, since
$$\int_1^\infty \frac{1}{t^2}\,dt = 1.$$

(II). Let $x > 0$, then, integrating by parts, we have
$$\Gamma(x+1) = \int_0^\infty t^x e^{-t}dt$$
$$= -t^x e^{-t}\Big|_0^\infty + \int_0^\infty xt^{x-1}e^{-t}dt$$
$$= 0 + x\int_0^\infty t^{x-1}e^{-t}dt$$
$$= x\Gamma(x).$$

(III). This last part is a direct consequence of the previous one, since, for $n \in \mathbb{N} \cup \{0\}$, we have
$$\Gamma(n+1) = n\Gamma(n) = n \cdot (n-1) \cdot \Gamma(n-2) = \cdots = n!\,\Gamma(1) = n!,$$
because
$$\Gamma(1) = \int_0^\infty t^{1-1}e^{-t}\,dt = \int_0^\infty e^{-t}\,dt = -e^{-t}\Big|_0^\infty = 0 - (-1) = 1.$$

$\square$

Exercise 6.23. *Find, when possible (that is, if they converge), the following improper integrals:*

(I) $\displaystyle\int_{0+}^{1} \log x\, dx.$ (II) $\displaystyle\int_{0}^{+\infty} e^{-x}\, dx.$

(III) $\displaystyle\int_{0}^{+\infty} x^n e^{-x}\, dx$ *with* $n \in \mathbb{N}.$ (IV) $\displaystyle\int_{0}^{+\infty} e^{-\sqrt{x}}\, dx.$

Solution to Exercise 6.23.

(I) Integrate by parts, $\int f\, dg = fg - \int g\, df$, where $f = \log x$, $dg = dx$, $df = dx/x$ and $g = x$. Then

$$\int_{0+}^{1} \log x\, dx = \lim_{a \to 0^+} x \log x \Big|_a^1 - \int_0^1 dx$$

$$= -\lim_{a \to 0^+} a \log a - 1$$

$$= -\lim_{a \to 0^+} \frac{\log a}{1/a} - 1$$

$$= -\lim_{a \to 0^+} \frac{1/a}{-1/a^2} - 1$$

$$= \lim_{a \to 0^+} a - 1$$

$$= -1,$$

where above we have used L'Hôpital's Rule.

(II) Substitute $u = -x$ and $du = -dx$. Then

$$\int_0^{+\infty} e^{-x}\, dx = -\int_0^{-\infty} e^u\, du$$

$$= \int_{-\infty}^0 e^u\, du$$

$$= \lim_{a \to -\infty} e^u \Big|_a^0$$

$$= 1 - \lim_{a \to -\infty} e^a$$

$$= 1.$$

(III) Use integration by parts with $u = x^n$ and $dv = e^{-x}dx$, so $du = nx^{n-1}$ and $v = -e^{-x}$:

$$\int_0^{+\infty} x^n e^{-x}dx = -\lim_{b\to+\infty} x^n e^{-x}\Big|_0^b + n\int_0^{+\infty} x^{n-1}e^{-x}dx$$

$$= -\lim_{b\to+\infty} b^n e^{-b} + n\int_0^{+\infty} x^{n-1}e^{-x}dx.$$

Recall that

$$\lim_{b\to+\infty} b^n e^{-b} = \lim_{b\to+\infty} \frac{b^n}{e^b} = 0,$$

since b^n is a polynomial and e^b the exponential function (or simply apply L'Hôpital's Rule n times). Therefore (using also Exercise 6.22),

$$\int_0^{+\infty} x^n e^{-x}dx = n\int_0^{+\infty} x^{n-1}e^{-x}dx$$

$$= n\Gamma(n)$$

$$= n(n-1)!$$

$$= n!.$$

(IV) Substitute $u = \sqrt{x}$ and $du = dx/(2\sqrt{x})$. Then

$$\int_0^{+\infty} e^{-\sqrt{x}}dx = 2\int_0^{+\infty} e^{-u}u\,du.$$

For the integrand $e^{-u}u$, integrate by parts, $\int f\,dg = fg - \int g\,df$, where $f = u$, $dg = e^{-u}du$, $df = du$, $g = -e^{-u}$. Then

$$\int_0^{+\infty} e^{-\sqrt{x}}dx = 2\int_0^{+\infty} e^{-u}u\,du$$

$$= -2\lim_{b\to+\infty} e^{-u}u\Big|_0^b + 2\int_0^{+\infty} e^{-u}du$$

$$= -2\lim_{b\to+\infty} e^{-u}u\Big|_0^b - 2\lim_{b\to+\infty} e^{-u}\Big|_0^b$$

$$= -2\lim_{b\to+\infty} e^{-b}b - 2\lim_{b\to+\infty} e^{-b} + 2$$

$$= 2.$$

□

Exercise 6.24. *Prove that:*

(I) $\displaystyle\int_0^{+\infty} 2xe^{-x^2}\,dx = 1.$

(II) $\displaystyle\int_1^{+\infty} \frac{\arctan x}{x^2}\,dx = \frac{\pi}{4} + \log\sqrt{2}.$

(III) $\displaystyle\int_{0+}^{1} x^{-2}e^{-1/x}\,dx = \frac{1}{e}.$

(IV) $\displaystyle\int_{0+}^{1} x^2 \log x\,dx = -\frac{1}{9}.$

Solution to Exercise 6.24.

(I) Substitute $u = -x^2$ and $du = -2x\,dx$. This gives a new lower limit $u = -0^2 = 0$ and upper limit $u = -\infty$. Then

$$\int_0^{+\infty} 2xe^{-x^2}\,dx = -\int_0^{-\infty} e^u\,du$$
$$= \int_{-\infty}^{0} e^u\,du$$
$$= \lim_{a\to-\infty} e^u\Big|_a^0$$
$$= 1 - \lim_{a\to-\infty} e^a$$
$$= 1.$$

(II) Integrate by parts, $\int f\,dg = fg - \int g\,df$, with

$$f = \arctan x, \; dg = dx/x^2, \; df = dx/(x^2+1), \; g = -1/x.$$

Then

$$\int_1^{+\infty} \frac{\arctan x}{x^2}\,dx = -\lim_{b\to+\infty} \frac{\arctan x}{x}\Big|_1^b + \int_1^{+\infty} \frac{1}{x(x^2+1)}\,dx$$
$$= -\lim_{b\to+\infty} \frac{\arctan b}{b} + \arctan 1 + \int_1^{+\infty} \frac{1}{x(x^2+1)}\,dx$$
$$= \frac{\pi}{4} + \int_1^{+\infty} \frac{1}{x(x^2+1)}\,dx.$$

For the integrand $1/(x(x^2+1))$, substitute $u = x^2$ and $du = 2xdx$. This gives a new lower limit $u = 1^2 = 1$ and upper limit $u = \infty$. Then

$$\int_1^{+\infty} \frac{\arctan x}{x^2} dx = \frac{\pi}{4} + \int_1^{+\infty} \frac{1}{x(x^2+1)} dx$$

$$= \frac{\pi}{4} + \frac{1}{2} \int_1^{+\infty} \frac{1}{u(u+1)} du$$

$$= \frac{\pi}{4} + \frac{1}{2} \int_1^{+\infty} \left(\frac{1}{u} - \frac{1}{u+1} \right) du$$

$$= \frac{\pi}{4} + \frac{1}{2} \lim_{b \to +\infty} (\log|u| - \log|u+1|) \Big|_1^b$$

$$= \frac{\pi}{4} + \frac{1}{2} \lim_{b \to +\infty} (\log b - \log(b+1)) + \frac{\log 2}{2}$$

$$= \frac{\pi}{4} + \log\sqrt{2},$$

where above we have used simple fraction decomposition.

(III) Substitute $u = -1/x$ and $du = dx/x^2$. This gives a new lower limit $u = -\infty$ and upper limit $u = -1/1 = -1$ and so

$$\int_{0+}^1 x^{-2} e^{-1/x} dx = \int_{-\infty}^{-1} e^u du$$

$$= \lim_{a \to -\infty} e^u \Big|_a^{-1}$$

$$= \frac{1}{e} - \lim_{a \to -\infty} e^a$$

$$= \frac{1}{e}.$$

(IV) Integrate by parts, $\int f\,dg = fg - \int g\,df$, where $f = \log x$, $dg = x^2 dx$, $df = dx/x$ and $g = x^3/3$. Then

$$\int_{0+}^1 x^2 \log x\,dx = \frac{1}{3} \lim_{a \to 0+} x^3 \log x \Big|_a^1 - \frac{1}{3} \int_0^1 x^2 dx$$

$$= -\frac{1}{3} \lim_{a \to 0+} a^3 \log a - \frac{1}{3} \frac{x^3}{3} \Big|_0^1$$

$$= -\frac{1}{9}.$$

$\square$

Exercise 6.25. *Determine, for each of the following improper integrals, whether it is finite or not, and calculate it when possible:*

(I) $\displaystyle\int_{1+}^{2}\frac{dx}{x\log x}.$ (II) $\displaystyle\int_{0+}^{1}x\log x\,dx.$

(III) $\displaystyle\int_{2}^{+\infty}\frac{\log x}{x}dx.$ (IV) $\displaystyle\int_{-1+}^{1^-}\frac{dx}{\sqrt{1-x^2}}.$

(V) $\displaystyle\int_{1+}^{2}\frac{dx}{\sqrt{x-1}}.$ (VI) $\displaystyle\int_{2}^{+\infty}\frac{dx}{x\log^2 x}.$

(VII) $\displaystyle\int_{0+}^{+\infty}\frac{dx}{x(x+4)}.$ (VIII) $\displaystyle\int_{1}^{+\infty}\frac{x}{x^4+1}dx.$

(IX) $\displaystyle\int_{2}^{+\infty}\frac{dx}{x^2+x-2}.$ (X) $\displaystyle\int_{0}^{\pi/2^-}\frac{dx}{\cos x}.$

(XI) $\displaystyle\int_{0+}^{+\infty}\frac{dx}{x^2+\sqrt{x}}.$ (XII) $\displaystyle\int_{1}^{+\infty}\frac{dx}{x(x^2+1)}.$

(XIII) $\displaystyle\int_{-1}^{1^-}\sqrt{\frac{1+x}{1-x}}dx.$ (XIV) $\displaystyle\int_{-\infty}^{\infty}\frac{\sin x}{1+x^2}dx.$

Solution to Exercise 6.25.

(I) This integral has a single point of conflict, which is 1, since $\lim_{x\to1}x\log x=0$. Consider the change of variable $u=\log x$. Then $x=e^u$ and $dx=e^u du$. So

$$\int_{1+}^{2}\frac{dx}{x\log x}=\int_{0}^{\log 2}\frac{e^u du}{e^u u}=\int_{0}^{\log 2}\frac{du}{u}=-\infty.$$

(II) Integrate by parts, $\int f dg=fg-\int g df$, where $f=\log x$, $dg=x dx$, $df=dx/x$ and $g=x^2/2$. Then

$$\int_{0+}^{1}x\log x\,dx=\frac{1}{2}\lim_{a\to0+}x^2\log x\Big|_{a}^{1}-\frac{1}{2}\int_{0}^{1}x dx$$
$$=-\frac{1}{2}\lim_{a\to0+}a^2\log a-\frac{x^2}{4}\Big|_{0}^{1}$$
$$=-\frac{1}{4}.$$

(III) Considering $u=\log x$ and $du=dx/x$, observe that

$$\int_{2}^{+\infty}\frac{\log x}{x}dx=\int_{\log 2}^{+\infty}u du=\lim_{b\to+\infty}\frac{u^2}{2}\Big|_{\log 2}^{b}=\infty.$$

(IV) Observe that

$$\int_{-1+}^{1^-} \frac{dx}{\sqrt{1-x^2}} = \lim_{\substack{a\to-1^+\\b\to1^-}} \arcsin x|_a^b = \lim_{b\to1^-}\arcsin b - \lim_{a\to-1^+}\arcsin a = \pi.$$

(V) Substitute $u = x - 1$ and $du = dx$. Then

$$\int_{1+}^{2}\frac{dx}{\sqrt{x-1}} = \int_{0+}^{1}\frac{1}{\sqrt{u}}du = 2\lim_{a\to0^+}\sqrt{u}\Big|_a^1 = 2.$$

(VI) Substitute $u = \log x$ and $du = dx/x$. Then

$$\int_{2}^{+\infty}\frac{dx}{x\log^2 x} = \int_{\log 2}^{+\infty}\frac{1}{u^2}du = -\lim_{b\to+\infty}\frac{1}{u}\Big|_{\log 2}^{b} = \frac{1}{\log 2}.$$

(VII) By simple fraction decomposition, we obtain

$$\int_{0+}^{+\infty}\frac{dx}{x(x+4)} = \int_{0+}^{+\infty}\left(\frac{1}{4x} - \frac{1}{4(x+4)}\right)dx$$

$$= \frac{1}{4}\int_{0+}^{+\infty}\left(\frac{1}{x} - \frac{1}{x+4}\right)dx$$

$$= \frac{1}{4}\lim_{\substack{a\to0^+\\b\to+\infty}}(\log x - \log(x+4))\Big|_a^b$$

$$= \frac{1}{4}\lim_{b\to+\infty}(\log b - \log(b+4))$$

$$- \frac{1}{4}\lim_{a\to0^+}(\log a - \log(a+4))$$

$$= \infty.$$

This last integral is not convergent, because in order to converge both limits must exist, which is not the case here.

(VIII) Substitute $u = x^2$ and $du = 2xdx$. Then

$$\int_{1}^{+\infty}\frac{x}{x^4+1}dx = \frac{1}{2}\int_{1}^{+\infty}\frac{1}{u^2+1}du = \frac{1}{2}\lim_{b\to+\infty}\arctan u\Big|_1^b = \frac{\pi}{8}.$$

(IX) Observe that

$$\int_{2}^{+\infty}\frac{dx}{x^2+x-2} = \int_{2}^{+\infty}\frac{dx}{(x+1/2)^2 - 9/4}.$$

Substitute $u = x + 1/2$ and $du = dx$. Then

$$\int_{2}^{+\infty}\frac{dx}{x^2+x-2} = \int_{2}^{+\infty}\frac{dx}{(x+1/2)^2 - 9/4}$$

$$= \int_{5/2}^{+\infty}\frac{1}{u^2 - 9/4}du$$

$$= -\frac{4}{9}\int_{5/2}^{+\infty}\frac{1}{1-(4u^2)/9}du.$$

For the integrand $1/(1 - (4u^2)/9)$, substitute $s = (2u)/3$ and $ds = \frac{2}{3}du$. Thus

$$\int_2^{+\infty} \frac{dx}{x^2 + x - 2} = -\frac{4}{9} \int_{5/2}^{+\infty} \frac{1}{1 - (4u^2)/9} du$$

$$= -\frac{2}{3} \int_{5/3}^{+\infty} \frac{1}{1 - s^2} ds$$

$$= -\frac{2}{3} \lim_{b \to +\infty} \operatorname{arctanh} s \Big|_{5/3}^b$$

$$= -\frac{1}{3} \lim_{b \to +\infty} \log \left| \frac{1+s}{1-s} \right| \Big|_{5/3}^b$$

$$= \frac{2 \log 2}{3}.$$

(x) From Exercise 6.4 (x) we know that

$$\int_0^{\pi/2^-} \frac{dx}{\cos x} = \lim_{b \to \pi/2^-} \log(\tan x + \sec x) \Big|_0^b = \lim_{b \to \pi/2^-} \log(\tan b + \sec b) = \infty.$$

(xi) Making the substitution $u = \sqrt{x}$ and $du = dx/(2\sqrt{x})$ gives a new lower limit $u = 0$ and upper limit $u = +\infty$. Then

$$\int_{0+}^{+\infty} \frac{dx}{x^2 + \sqrt{x}}$$

$$= 2 \int_{0+}^{\infty} \frac{u}{u^4 + u} du$$

$$= 2 \int_{0+}^{\infty} \frac{1}{u^3 + 1} du$$

$$= 2 \int_{0+}^{\infty} \left(\frac{2 - u}{3(u^2 - u + 1)} + \frac{1}{3(u + 1)} \right) du$$

$$= 2 \lim_{\substack{a \to 0^+ \\ b \to +\infty}} \left(\frac{1}{\sqrt{3}} \arctan \frac{2u - 1}{\sqrt{3}} - \frac{1}{6} \log(u^2 - u + 1) + \frac{1}{3} \log(u + 1) \right) \Big|_a^b$$

$$= \frac{4\pi}{3\sqrt{3}}.$$

The calculation of the antiderivative above was obtained as follows:

$$\int \left(\frac{2-u}{3(u^2-u+1)} + \frac{1}{3(u+1)} \right) du$$

$$= \frac{1}{3} \int \frac{2-u}{u^2-u+1} du + \frac{1}{3} \int \frac{1}{u+1} du$$

$$= \frac{1}{3} \int \left(\frac{3}{2(u^2-u+1)} - \frac{2u-1}{2(u^2-u+1)} \right) du + \frac{1}{3} \log(u+1)$$

$$= \frac{1}{2} \int \frac{1}{u^2-u+1} du - \frac{1}{6} \int \frac{2u-1}{u^2-u+1} du + \frac{1}{3} \log(u+1)$$

$$= \frac{1}{2} \int \frac{1}{u^2-u+1} du - \frac{1}{6} \log(u^2-u+1) + \frac{1}{3} \log(u+1)$$

$$= \frac{1}{2} \int \frac{1}{(u-1/2)^2 + 3/4} du - \frac{1}{6} \log(u^2-u+1) + \frac{1}{3} \log(u+1)$$

$$= \frac{1}{\sqrt{3}} \arctan \frac{2u-1}{\sqrt{3}} - \frac{1}{6} \log(u^2-u+1) + \frac{1}{3} \log(u+1) + C.$$

(XII) Substitute $u = x^2$ and $du = 2x dx$. Then

$$\int_1^{+\infty} \frac{dx}{x(x^2+1)} = \frac{1}{2} \int_1^\infty \frac{1}{u(u+1)} du.$$

By simple fractions, we obtain

$$\int_1^{+\infty} \frac{dx}{x(x^2+1)} = \frac{1}{2} \int_1^\infty \frac{1}{u(u+1)} du$$

$$= \frac{1}{2} \int_1^{+\infty} \left(\frac{1}{u} - \frac{1}{u+1} \right) du$$

$$= \frac{1}{2} \lim_{b \to +\infty} \left. (\log u - \log(u+1)) \right|_1^b$$

$$= \frac{1}{2} \lim_{b \to +\infty} (\log b - \log(b+1)) + \frac{\log 2}{2}$$

$$= \frac{\log 2}{2}.$$

(XIII)

$$\int_{-1}^{1^-} \sqrt{\frac{1+x}{1-x}} dx = \lim_{b \to 1^-} \int_{-1}^b \sqrt{\frac{1+x}{1-x}} dx.$$

Use the substitution $u^2 = \frac{1+x}{1-x}$ to solve the indefinite integral $\sqrt{\frac{1+x}{1-x}}dx$. Solving x in terms of u in the latter equation gives us $x = \frac{u^2-1}{u^2+1}$ and $dx = \frac{4u}{(u^2+1)^2}du$. So,

$$\int \sqrt{\frac{1+x}{1-x}}dx = \int \frac{4u^2}{u^2+1}du.$$

Apply simple fraction decomposition to $\frac{4u^2}{(u^2+1)^2}$, which gives us the expansion

$$\frac{4u^2}{(u^2+1)^2} = \frac{Au+B}{u^2+1} + \frac{Cu+D}{(u^2+1)^2}.$$

Multiplying both sides by $(u^2+1)^2$ gives the equation:

$$4u^2 = (Au+B)(u^2+1) + Cu + D$$
$$= Au^3 + Bu^2 + (A+C)u + (B+D).$$

Thus, we have the following system of equations

$$\begin{cases} 0 = A, \\ 4 = B, \\ 0 = A+C, \\ 0 = B+D, \end{cases}$$

with solution

$$A = C = 0, \ B = 4, \ D = -4.$$

Therefore,

$$\int \frac{4u^2}{u^2+1}du = 4\int \frac{1}{u^2+1}du - 4\int \frac{1}{(u^2+1)^2}du.$$

Let us calculate the integral $\int \frac{1}{(u^2+1)^2}du$. Now use the reduction formula

$$\int \frac{1}{(1+x^2)^n}dx = \frac{1}{2n-2}\cdot\frac{x}{(1+x^2)^{n-1}} + \frac{2n-3}{2n-2}\int \frac{x}{(1+x^2)^{n-1}}dx,$$

where in this case $n = 2$. So,

$$\int \frac{1}{(u^2+1)^2}du = \frac{1}{2}\cdot\frac{u}{1+u^2} + \frac{1}{2}\int \frac{u}{1+u^2}du$$
$$= \frac{1}{2}\cdot\frac{u}{1+u^2} + \frac{1}{2}\arctan(u) + C.$$

Hence,

$$\int \sqrt{\frac{1+x}{1-x}}\,dx = 4\arctan(u) - 2\cdot\frac{u}{1+u^2} - 2\arctan(u) + C$$

$$= 2\arctan\left(\sqrt{\frac{1+x}{1-x}}\right) - 2\cdot\frac{\sqrt{\frac{1+x}{1-x}}}{1+\frac{1+x}{1-x}} + C$$

$$= 2\arctan\left(\sqrt{\frac{1+x}{1-x}}\right) - 2\cdot\frac{\sqrt{\frac{1+x}{1-x}}}{\frac{2}{1-x}} + C$$

$$= 2\arctan\left(\sqrt{\frac{1+x}{1-x}}\right) - 2\cdot\frac{\sqrt{1+x}(1-x)}{2\sqrt{1-x}} + C$$

$$= 2\arctan\left(\sqrt{\frac{1+x}{1-x}}\right) - \sqrt{1+x}\sqrt{1-x} + C$$

$$= 2\arctan\left(\sqrt{\frac{1+x}{1-x}}\right) - \sqrt{1-x^2} + C.$$

Thus,

$$\int_{-1}^{1^-}\sqrt{\frac{1+x}{1-x}}\,dx = \lim_{b\to 1^-}\left(2\arctan\left(\sqrt{\frac{1+x}{1-x}}\right) - \sqrt{1-x^2}\right)\Big|_{-1}^{b}$$

$$= \lim_{b\to 1^-}\left(2\arctan\left(\sqrt{\frac{1+b}{1-b}}\right) - \sqrt{1-b^2}\right)$$

$$= 2\cdot\frac{\pi}{2} = \pi.$$

(XIV) This integral has an integrand that is not always positive, it alternates its sign. However, we can prove that it is convergent as follows: If we can prove that the value $|\int_{-\infty}^{\infty}\frac{\sin x}{1+x^2}\,dx|$ is finite, then the original integral will converge.

Since

$$\left|\int_{-\infty}^{\infty}\frac{\sin x}{1+x^2}\,dx\right| \le \int_{-\infty}^{\infty}\left|\frac{\sin x}{1+x^2}\right|\,dx$$

$$= \int_{-\infty}^{\infty}\frac{|\sin x|}{1+x^2}\,dx \le \int_{-\infty}^{\infty}\frac{1}{1+x^2}\,dx = \pi,$$

we then conclude that the original integral converges (absolutely!). Thus, since the integrand is an odd function, it follows that the value of this integral on a symmetric interval is actually 0.

□

Exercise 6.26. *Let f and g be two functions defined on (a, b) such that the two improper integrals $\int_a^b f$ and $\int_a^b g$ are finite. Show with an example that, in general, the improper integral $\int_a^b fg$ does not have to be finite.*

Solution to Exercise 6.26.
Let $f, g : (1, 2) \to \mathbb{R}$ be defined as

$$f(x) = g(x) = \frac{1}{\sqrt{x - .1}}.$$

Let us prove that

$$\int_1^2 f(x)dx = \int_1^2 g(x)dx = 2 \quad \text{and} \quad \int_1^2 f(x)g(x)dx = \infty.$$

Considering $u = x - 1$ and $du = dx$ we get

$$\int_1^2 \frac{1}{\sqrt{x - 1}}dx = \int_0^1 \frac{1}{\sqrt{u}}du$$

$$= 2 \lim_{a \to 0^+} \sqrt{u} \Big|_a^1$$

$$= 2.$$

However

$$\int_1^2 f(x)g(x)dx = \int_1^2 \frac{1}{x - 1}dx = \lim_{a \to 1^+} \log(x - 1) \Big|_a^2$$

$$= \log(2 - 1) - \lim_{a \to 1^+} \log(a - 1) = \infty.$$

□

Exercise 6.27. *Determine whether the following improper integrals are finite or not:*

(I) $\displaystyle\int_0^{+\infty} \frac{e^{-x} \sin x}{\sqrt{x + 1}}dx.$ (II) $\displaystyle\int_{0^+}^{+\infty} \frac{dx}{\sqrt{x}(e^x + 1)}.$

(III) $\displaystyle\int_{0^+}^{+\infty} e^{-(x^2 + 1/x^2)}dx.$ (IV) $\displaystyle\int_{0^+}^{+\infty} \frac{\cos x}{\sqrt{x}}dx.$

(V) $\displaystyle\int_0^{+\infty} \frac{\sin^3 x}{e^x + \cos x + 1}dx.$

Solution to Exercise 6.27.

(I) First, let us show that the improper integral $\int_0^{+\infty} e^{-x} \sin x dx$ converges. Notice that

$$0 \leq \left| e^{-x} \sin x \right| \leq e^{-x}.$$

Let us show that $\int_0^{+\infty} e^{-x}$ converges. We have

$$\int_0^{+\infty} e^{-x} = \lim_{b \to \infty} \int_0^b e^{-x} dx.$$

Use the substitution $u = e^{-x}$ and $du = -e^{-x} dx$. Now the lower and upper limits of integration are 1 and $-e^{-b}$, respectively:

$$\lim_{b \to \infty} \int_0^b e^{-x} dx = - \lim_{b \to \infty} \int_1^{-e^{-b}} du$$

$$= \lim_{b \to \infty} \int_{-e^{-b}}^1 du$$

$$= \lim_{b \to \infty} u \Big|_{-e^{-b}}^1$$

$$= \lim_{b \to \infty} \left(1 + e^{-b} \right)$$

$$= 1.$$

Thus, $\int_0^{+\infty} e^{-x} \sin x dx$ is absolutely convergent and, in particular, converges. Second, observe that $\frac{1}{\sqrt{x+1}}$ is bounded between 0 and 1 on $[0, +\infty)$ and it is an strictly decreasing function on $[0, +\infty)$. Hence, by Abel's Criterion, the improper integral

$$\int_0^{+\infty} \frac{e^{-x} \sin x}{\sqrt{x+1}} dx$$

converges.

(II)

$$\int_{0+}^{+\infty} \frac{dx}{\sqrt{x(e^x + 1)}} = \int_{0+}^1 \frac{dx}{\sqrt{x(e^x + 1)}} + \int_1^{+\infty} \frac{dx}{\sqrt{x(e^x + 1)}}.$$

For the first integral, we have that

$$\lim_{x \to 0+} \frac{\frac{1}{\sqrt{x(e^x+1)}}}{\frac{1}{\sqrt{x}}} = \lim_{x \to 0+} \frac{\sqrt{x}}{\sqrt{x}\sqrt{e^x + 1}} = \lim_{x \to 0+} \frac{1}{\sqrt{e^x + 1}} = \frac{1}{\sqrt{2}}.$$

Since $\int_{0+}^1 \frac{dx}{\sqrt{x}}$ is convergent, we have that $\int_{0+}^1 \frac{dx}{\sqrt{x(e^x+1)}}$ is convergent by the Asymptotic Comparison Criterion. For second integral, we have that

$$\lim_{x \to +\infty} \frac{\frac{1}{\sqrt{x(e^x+1)}}}{\frac{1}{\sqrt{e^x}}} = \lim_{x \to +\infty} \frac{\sqrt{e^x}}{\sqrt{x}\sqrt{e^x + 1}} = \lim_{x \to +\infty} \frac{1}{\sqrt{x}} = 0.$$

Now let us show that $\int_1^{+\infty} e^{-x/2}dx$ is convergent. To do so, first observe that

$$\int_1^{+\infty} e^{-x/2}dx = \lim_{b \to +\infty} \int_0^b e^{-x/2}dx.$$

Now apply the substitution $u = e^{-x/2}$, so $du = -\frac{1}{2}e^{-x/2}dx$. Hence,

$$\int_1^{+\infty} e^{-x/2}dx = -2 \lim_{b \to +\infty} \int_{e^{-1/2}}^{e^{-b/2}} du$$

$$= 2 \lim_{b \to +\infty} \int_{e^{-b/2}}^{e^{-1/2}} du$$

$$= 2 \lim_{b \to +\infty} u \Big|_{e^{-b/2}}^{e^{-1/2}}$$

$$= 2 \lim_{b \to +\infty} \left(e^{-1/2} - e^{-b/2} \right)$$

$$= 2e^{-1/2}.$$

Hence, by the Asymptotic Comparison Criterion, $\int_1^{+\infty} \frac{dx}{\sqrt{x(e^x+1)}}$ converges. Thus, $\int_{0+}^{+\infty} \frac{dx}{\sqrt{x(e^x+1)}}$ converges.

(III)

$$\int_{0+}^{+\infty} e^{-(x^2+1/x^2)}dx = \int_{0+}^1 e^{-(x^2+1/x^2)}dx + \int_1^{+\infty} e^{-(x^2+1/x^2)}dx.$$

Notice that

$$e^{-(x^2+1/x^2)} \le e^{-x^2} \le e^{-x}$$

for $x > 0$. Hence, by the Comparison Criterion, both improper integrals $\int_{0+}^1 e^{-(x^2+1/x^2)}dx$ and $\int_1^{+\infty} e^{-(x^2+1/x^2)}dx$ converge. So, $\int_{0+}^{+\infty} e^{-(x^2+1/x^2)}dx$ converges.

(IV)

$$\int_{0+}^{+\infty} \frac{\cos x}{\sqrt{x}}dx = \int_{0+}^1 \frac{\cos x}{\sqrt{x}}dx + \int_1^{+\infty} \frac{\cos x}{\sqrt{x}}dx.$$

For the first integral, observe that

$$0 \le \left| \frac{\cos x}{\sqrt{x}} \right| \le \frac{1}{\sqrt{x}}.$$

Since $\int_{0+}^{1} \frac{1}{\sqrt{x}} dx$ converges, we have by the Comparison Criterion that $\int_{0+}^{1} \frac{\cos x}{\sqrt{x}} dx$ converges. For the second integral, observe that f is locally integrable on $[0, +\infty)$ and consider the function

$$F(x) := \int_{1}^{x} \cos y \, dy = \sin y \Big|_{1}^{x} = \sin x - 1.$$

It is clear that F is bounded between -2 and 0. Now observe that $\frac{1}{\sqrt{x}}$ is an strictly decreasing function with $\lim_{x \to +\infty} \frac{1}{\sqrt{x}} = 0$. Thus, by the Dirichlet Criterion, we have that $\int_{1}^{+\infty} \frac{\cos x}{\sqrt{x}} dx$ converges. Therefore, $\int_{0+}^{+\infty} \frac{\cos x}{\sqrt{x}} dx$ converges.

(v) First, notice that by the reverse triangular inequality we have

$$|e^x + \cos x + 1| \geq |e^x + 1| - |\cos x| \geq e^x + 1 - 1 = e^x.$$

So,

$$\frac{1}{|e^x + \cos x + 1|} \leq e^{-x}.$$

Thus,

$$0 \leq \left| \frac{\sin^3 x}{e^x + \cos x + 1} \right| = \frac{|\sin^3 x|}{|e^x + \cos x + 1|} \leq e^{-x}.$$

Since the improper integral $\int_{0}^{+\infty} e^{-x} dx$ converges, we have by the Comparison Criterion that $\int_{0}^{+\infty} \frac{\sin^3 x}{e^x + \cos x + 1} dx$ converges.

$\square$

Exercise 6.28. *Prove that if $p > 1$ the following improper integrals are absolutely convergent:*

(I) $\displaystyle\int_{1}^{+\infty} \frac{\sin x}{x^p} dx.$

(II) $\displaystyle\int_{1}^{+\infty} \sin^p \frac{1}{x} dx.$

Solution to Exercise 6.28.

(I) Observe that

$$\int_1^{+\infty} \left| \frac{\sin x}{x^p} \right| dx \le \int_1^{+\infty} \frac{1}{x^p} dx$$

$$= \lim_{b \to +\infty} \frac{x^{1-p}}{1-p} \Big|_1^b$$

$$= \lim_{b \to +\infty} \frac{b^{1-p}}{1-p} - \frac{1}{1-p}$$

$$= \frac{1}{p-1}$$

and thus by the comparison criterion we conclude that the original integral is absolutely convergent.

(II) Recall that $\sin x \sim x$ when $x \to 0$, so $\sin \frac{1}{x} \sim \frac{1}{x}$ when $x \to \infty$. Hence,

$$\sin^p \frac{1}{x} \sim \frac{1}{x^p},$$

when $x \to \infty$. Thus, by the Asymptotic Comparison Criterion, since $\int_1^{+\infty} \frac{1}{x^p} dx$ converges for $p > 1$ we have that $\int_1^{+\infty} \frac{\sin x}{x^p} dx$ converges.

□

Exercise 6.29. *Prove that the improper integral given by*

$$\int_{\pi/2}^{+\infty} \frac{\cos x}{\sqrt{x}} dx$$

is conditionally convergent.

Solution to Exercise 6.29.
Let us show first that the improper integral converges:

$$\int_{\pi/2}^{+\infty} \frac{\cos x}{\sqrt{x}} dx = \lim_{b \to +\infty} \int_{\pi/2}^b \frac{\cos x}{\sqrt{x}} dx.$$

To do so, we will use integration by parts with $u = \frac{1}{\sqrt{x}}$ and $dv = \cos x$, so $du = -\frac{1}{2x^{3/2}}$ and $v = \sin x$. Hence,

$$\int_{\pi/2}^b \frac{\cos x}{\sqrt{x}} dx = -\frac{\sin x}{2x^{3/2}} \Big|_{\pi/2}^b + \frac{1}{2} \int_{\pi/2}^b \frac{\sin x}{x^{3/2}}$$

$$= -\frac{\sin b}{2b^{3/2}} + \frac{1}{2(\pi/2)^{3/2}} + \frac{1}{2} \int_{\pi/2}^b \frac{\sin x}{x^{3/2}}.$$

Thus,

$$\int_{\pi/2}^{+\infty} \frac{\cos x}{\sqrt{x}} dx = \lim_{b\to+\infty} \left(-\frac{\sin b}{2b^{3/2}} + \frac{1}{2(\pi/2)^{3/2}} + \frac{1}{2}\int_{\pi/2}^{b} \frac{\sin x}{x^{3/2}} \right)$$

$$= \frac{1}{2(\pi/2)^{3/2}} + \frac{1}{2}\int_{\pi/2}^{+\infty} \frac{\sin x}{x^{3/2}}.$$

Now, since

$$0 \le \left| \frac{\sin x}{x^{3/2}} \right| \le \frac{1}{x^{3/2}}$$

for $x \ge \pi/2$ and $\int_{\pi/2}^{+\infty} \frac{1}{x^{3/2}}$ is convergent, we have that $\int_{\pi/2}^{+\infty} \frac{\sin x}{x^{3/2}}$ converges. Therefore, the improper integral $\int_{\pi/2}^{+\infty} \frac{\cos x}{\sqrt{x}} dx$ converges.

Finally, we will show that $\int_{\pi/2}^{+\infty} \frac{\cos x}{\sqrt{x}} dx$ is not absolutely convergent. Observe that

$$\left| \frac{\cos x}{\sqrt{x}} \right| \ge \frac{\cos^2 x}{\sqrt{x}} = \frac{1+\cos(2x)}{2\sqrt{x}} = \frac{1}{2\sqrt{x}} + \frac{\cos(2x)}{2\sqrt{x}}.$$

On the one hand, it is clear that $\int_{\pi/2}^{+\infty} \frac{1}{2\sqrt{x}} dx$ diverges. On the other hand, $\int_{\pi/2}^{+\infty} \frac{\cos(2x)}{2\sqrt{x}} dx$ converges. Indeed, notice that $\cos(2x)$ is locally integrable on $[\pi/2, +\infty)$ and the function

$$F(x) := \int_{\pi/2}^{x} \cos y\, dy = \sin y\big|_{\pi/2}^{x} = \sin x - 1$$

is bounded between -2 and 0. Moreover, the function $\frac{1}{\sqrt{x}}$ is strictly decreasing and $\lim_{x\to+\infty} \frac{1}{\sqrt{x}} = 0$. So, by the Dirichlet Criterion, the improper integral $\int_{\pi/2}^{+\infty} \frac{\cos(2x)}{2\sqrt{x}} dx$ converges. Hence, $\int_{\pi/2}^{+\infty} \frac{\cos x}{\sqrt{x}} dx$ is not absolutely convergent.

Remark: Another way of studying the absolute of convergence is the following: Observe that for each $n \in \mathbb{N}$ we have

$$\int_{\frac{n\pi}{2}}^{\frac{(n+1)\pi}{2}} \left| \frac{\cos x}{\sqrt{x}} \right| dx \ge \frac{1}{\sqrt{\frac{n\pi}{2}}} \int_{\frac{n\pi}{2}}^{\frac{(n+1)\pi}{2}} |\cos x| dx = \frac{2}{\sqrt{\frac{n\pi}{2}}} = 2\sqrt{\frac{2}{\pi}} \cdot \frac{1}{\sqrt{n}}.$$

Thus, for any $n \in \mathbb{N}$,

$$\int_{\pi/2}^{\frac{(n+1)\pi}{2}} \left| \frac{\cos x}{\sqrt{x}} \right| dx \ge 2\sqrt{\frac{2}{\pi}} \left(1 + \frac{1}{\sqrt{2}} + \frac{1}{\sqrt{3}} + \cdots + \frac{1}{\sqrt{n}} \right)$$

$$\ge 2\sqrt{\frac{2}{\pi}} \left(1 + \frac{1}{2} + \frac{1}{3} + \cdots + \frac{1}{n} \right).$$

So,

$$\int_{\pi/2}^{+\infty} \left| \frac{\cos x}{\sqrt{x}} \right| dx \ge \lim_{n\to\infty} 2\sqrt{\frac{2}{\pi}} \left(1 + \frac{1}{2} + \frac{1}{3} + \cdots + \frac{1}{n} \right) = \infty.$$

$\square$

Exercise 6.30. *Find the arc length of the graphs of the following functions:*

(I) $f(x) = \sqrt{2x}$ *for* $x \in [1,3]$.

(II) $f(x) = e^x$ *for* $x \in [-1, 2]$.

(III) $f(x) = \log\cos x$ *for* $x \in \left[0, \frac{\pi}{4}\right]$.

Solution to Exercise 6.30.

(I) The arc length of the graph of f with $x \in [1,3]$ will be given by

$$\text{arclength}(f, 1, 3) = \int_1^3 \sqrt{1 + [f'(x)]^2}\,dx$$

$$= \int_1^3 \sqrt{1 + \frac{1}{2x}}\,dx$$

$$= \frac{1}{\sqrt{2}} \int_1^3 \frac{\sqrt{2x+1}}{\sqrt{x}}\,dx.$$

Substitute $u = \sqrt{x}$ and $du = \frac{dx}{2\sqrt{x}}$. Then

$$\text{arclength}(f, 1, 3) = \frac{1}{\sqrt{2}} \int_1^3 \frac{\sqrt{2x+1}}{\sqrt{x}}\,dx = \sqrt{2} \int_1^{\sqrt{3}} \sqrt{2u^2 + 1}\,du.$$

Substitute $u = \tan s / \sqrt{2}$ and $du = \frac{\sec^2 s}{\sqrt{2}}\,ds$. Thus $\sqrt{2u^2 + 1} = \sec s$ and $s = \arctan(\sqrt{2}u)$ and therefore

$$\text{arclength}(f, 1, 3) = \sqrt{2} \int_1^{\sqrt{3}} \sqrt{2u^2 + 1}\,du$$

$$= \int_{\arctan\sqrt{2}}^{\arctan(3\sqrt{2})} \sec^3 s\,ds$$

$$= \frac{1}{2}\tan s \sec s + \frac{1}{2}\log(\tan s + \sec s)\Big|_{\arctan\sqrt{2}}^{\arctan(3\sqrt{2})}$$

$$= \frac{3\sqrt{2}}{2}\sec\arctan(3\sqrt{2}) + \frac{1}{2}\log(3\sqrt{2} + \sec\arctan(3\sqrt{2}))$$

$$- \frac{\sqrt{2}}{2}\sec\arctan\sqrt{2} - \frac{1}{2}\log(\sqrt{2} + \sec\arctan\sqrt{2}).$$

(II) The arc length of the graph of f with $x \in [-1, 2]$ will be given by

$$\text{arclength}(f, -1, 2) = \int_{-1}^{2} \sqrt{1 + [f'(x)]^2}\,dx$$

$$= \int_{-1}^{2} \sqrt{1 + e^{2x}}\,dx.$$

Apply the substitution $u = \sqrt{e^{2x} + 1}$ and $du = \frac{e^{2x}}{\sqrt{e^{2x}+1}}dx$. In particular, $dx = \frac{u}{u^2-1}du$, and now the lower and upper limits of integration are, respectively, $\sqrt{e^{-2} + 1}$ and $\sqrt{e^4 + 1}$. Hence,

$$\int_{-1}^{2} \sqrt{1 + e^{2x}}\,dx = \int_{\sqrt{e^{-2}+1}}^{\sqrt{e^4+1}} \frac{u^2}{u^2 - 1}\,du$$

$$= \int_{\sqrt{e^{-2}+1}}^{\sqrt{e^4+1}} \frac{u^2 - 1 + 1}{u^2 - 1}\,du$$

$$= \int_{\sqrt{e^{-2}+1}}^{\sqrt{e^4+1}} \frac{u^2 - 1}{u^2 - 1}\,du + \int_{\sqrt{e^{-2}+1}}^{\sqrt{e^4+1}} \frac{1}{u^2 - 1}\,du$$

$$= \int_{\sqrt{e^{-2}+1}}^{\sqrt{e^4+1}} du + \int_{\sqrt{e^{-2}+1}}^{\sqrt{e^4+1}} \frac{1}{u^2 - 1}\,du.$$

Use simple fraction decomposition to $\frac{1}{u^2-1}$ which gives us the expansion:

$$\frac{1}{u^2 - 1} = \frac{A}{u - 1} + \frac{B}{u + 1}.$$

By multiplying both sides of the latter equality by $u^2 - 1$ we obtain the equation:

$$1 = A(u + 1) + B(u - 1) = (A + B)u + (A - B),$$

and therefore we have the system of equations

$$\begin{cases} 0 = A + B, \\ 1 = A - B, \end{cases}$$

with solution

$$A = \frac{1}{2}, \ B = -\frac{1}{2}.$$

Hence,

$$\int_{\sqrt{e^{-2}+1}}^{\sqrt{e^4+1}} \frac{1}{u^2-1}du = \frac{1}{2}\int_{\sqrt{e^{-2}+1}}^{\sqrt{e^4+1}} \frac{1}{u-1}du - \frac{1}{2}\int_{\sqrt{e^{-2}+1}}^{\sqrt{e^4+1}} \frac{1}{u+1}du$$

$$= \frac{1}{2}\left(\log|u-1| - \log|u+1|\right)\Big|_{\sqrt{e^{-2}+1}}^{\sqrt{e^4+1}}$$

$$= \frac{1}{2}\left(\log\left|\frac{\sqrt{e^4+1}-1}{\sqrt{e^{-2}+1}-1}\right| - \log\left|\frac{\sqrt{e^4+1}+1}{\sqrt{e^{-2}+1}+1}\right|\right)$$

$$= \frac{1}{2}\log\left|\frac{\left(\sqrt{e^4+1}-1\right)\left(\sqrt{e^{-2}+1}+1\right)}{\left(\sqrt{e^4+1}+1\right)\left(\sqrt{e^{-2}+1}-1\right)}\right|.$$

Thus,

arclength$(f,-1,2)$

$$= \frac{1}{2}\log\left|\frac{\left(\sqrt{e^4+1}-1\right)\left(\sqrt{e^{-2}+1}+1\right)}{\left(\sqrt{e^4+1}+1\right)\left(\sqrt{e^{-2}+1}-1\right)}\right| + \sqrt{e^4+1} - \sqrt{e^{-2}+1}.$$

(III) The arc length of the graph of f with $x \in \left[0,\frac{\pi}{4}\right]$ will be given by

$$\text{arclength}\left(f,0,\frac{\pi}{4}\right) = \int_0^{\frac{\pi}{4}} \sqrt{1+[f'(x)]^2}dx$$

$$= \int_0^{\frac{\pi}{4}} \sqrt{1+\tan^2 x}dx$$

$$= \int_0^{\frac{\pi}{4}} \sqrt{\sec^2 x}dx$$

$$= \int_0^{\frac{\pi}{4}} \frac{1}{\sqrt{\cos^2 x}}dx.$$

Since $\cos x > 0$ on $[0,\pi/4]$ we have that $\sqrt{\cos^2 x} = |\cos x| = \cos x$ on $[0,\pi/4]$. Hence,

$$\int_0^{\frac{\pi}{4}} \frac{1}{\sqrt{\cos^2 x}}dx = \int_0^{\frac{\pi}{4}} \frac{1}{\cos x}dx$$

$$= \int_0^{\frac{\pi}{4}} \sec x\,dx$$

$$= \log|\tan x + \sec x|\Big|_0^{\pi/4}$$

$$= \log|\tan(\pi/4) + \sec(\pi/4)| - \log|\tan(0) + \sec(0)|$$

$$= \log\left|1+\sqrt{2}\right| - \log|1|$$

$$= \log\left|1+\sqrt{2}\right|.$$

□

Exercise 6.31. *Using integrals to obtain the arc length, show that the perimeter of a circumference of radius r is given by $2\pi r$.*

Solution to Exercise 6.31.
Recall that the formula of a circumference of radius r is given by $x^2 + y^2 = r^2$. By solving y as a function of x in the previous equation, we obtain two curves $y = \pm\sqrt{r^2 - x^2}$. Observe that the curve $y = \sqrt{r^2 - x^2}$ corresponds to the upper half semicircle whereas $y = -\sqrt{r^2 - x^2}$ is the lower half semicircle. Therefore, the calculation of the arc length can be reduced to calculating the arc length of the upper half semicircle and then multiply by 2. Hence, the arc length is equal to

$$
\begin{aligned}
2\int_{-r}^{r} \sqrt{1 + \frac{x^2}{r^2 - x^2}}\, dx &= 2\int_{-r}^{r} \sqrt{\frac{r^2}{r^2 - x^2}}\, dx \\
&= 2\int_{-r}^{r} \frac{|r|}{\sqrt{r^2 - x^2}}\, dx \\
&= 2r\int_{-r}^{r} \frac{1}{\sqrt{r^2 - x^2}}\, dx \\
&= 2r\int_{-r}^{r} \frac{1}{|r|\sqrt{1 - \left(\frac{x}{r}\right)^2}}\, dx \\
&= 2\int_{-r}^{r} \frac{1}{\sqrt{1 - \left(\frac{x}{r}\right)^2}}\, dx,
\end{aligned}
$$

where $\frac{x^2}{r^2 - x^2}$ is the square of the derivative of $\sqrt{r^2 - x^2}$ with respect to x, and $|r| = r$ since $r > 0$. Apply the substitution $u = \frac{x}{r}$ and $du = \frac{1}{r}dx$, and now the lower and upper limits of integration are -1 and 1, respectively:

$$
\begin{aligned}
2\int_{-r}^{r} \frac{1}{\sqrt{1 - \left(\frac{x}{r}\right)^2}}\, dx &= 2r\int_{-1}^{1} \frac{1}{\sqrt{1 - u^2}}\, du \\
&= 2r\, \arcsin u \Big|_{-1}^{1} \\
&= 2r\, (\arcsin(1) - \arcsin(-1)) \\
&= 2r\left(\frac{\pi}{2} + \frac{\pi}{2}\right) \\
&= 2\pi r.
\end{aligned}
$$

$\square$

> **Exercise 6.32.** *Calculate volume and surface area of a sphere of a given radius r.*

Solution to Exercise 6.32.
Without loss of generality, let us consider the sphere centered at the origin, that is, $x^2 + y^2 = r^2$. Let f be the positive part of this sphere, that is, $f(x) = \sqrt{r^2 - x^2}$. Then the volume and area we are looking for will be given by

$$V = \pi \int_{-r}^{r} [f(x)]^2 dx$$

$$= \pi \int_{-r}^{r} (r^2 - x^2) dx$$

$$= 2\pi \int_{0}^{r} (r^2 - x^2) dx$$

$$= 2\pi \left(r^2 x - \frac{x^3}{3} \right) \Big|_{0}^{r}$$

$$= \frac{4}{3}\pi r^3$$

and

$$A = 2\pi \int_{-r}^{r} f(x)\sqrt{1 + [f'(x)]^2} dx$$

$$= 4\pi \int_{0}^{r} \sqrt{r^2 - x^2}\sqrt{1 + \left(-\frac{x}{\sqrt{r^2 - x^2}} \right)^2} dx$$

$$= 4\pi \int_{0}^{r} \sqrt{r^2 - x^2 + (r^2 - x^2) \cdot \frac{x^2}{r^2 - x^2}} dx$$

$$= 4\pi \int_{0}^{r} r\, dx$$

$$= 4\pi r x \Big|_{0}^{r}$$

$$= 4\pi r^2.$$

Exercise 6.33. *Find the volume of the solid generated by the rotation around the x-axis of the graphs of the following functions:*

(I) $f(x) = \sin x$ *with* $x \in [0, \pi]$.

(II) $f(x) = e^{-x}$ *with* $x \in [0, a]$.

(III) $f(x) = x\sqrt{1 - x^2}$ *with* $x \in [0, 1]$.

(IV) $f(x) = a \cosh \frac{x}{a}$ *with* $x \in [-c, c]$ *and* $a, c > 0$.

Solution to Exercise 6.33.

(I) Let V be the volume we are looking for. Writing $\sin^2 x$ as $\frac{1}{2} - \frac{\cos 2x}{2}$, we get

$$
\begin{aligned}
V &= \pi \int_0^\pi [f(x)]^2 dx \\
&= \pi \int_0^\pi \sin^2 x dx \\
&= \pi \int_0^\pi \left(\frac{1}{2} - \frac{\cos 2x}{2} \right) dx \\
&= \pi \left(\frac{1}{2} \int_0^\pi dx - \frac{1}{2} \int_0^\pi \cos 2x dx \right) \\
&= \pi \left(\frac{\pi}{2} - \frac{1}{2} \int_0^\pi \cos 2x dx \right).
\end{aligned}
$$

For the integrand $\cos 2x$, substitute $u = 2x$ and $du = 2dx$. Then

$$
\begin{aligned}
V &= \pi \left(\frac{\pi}{2} - \frac{1}{2} \int_0^\pi \cos 2x dx \right) \\
&= \pi \left(\frac{\pi}{2} - \frac{1}{4} \int_0^{2\pi} \cos u du \right) \\
&= \pi \left(\frac{\pi}{2} - \frac{1}{4} \sin u \Big|_0^{2\pi} \right) \\
&= \frac{\pi^2}{2}.
\end{aligned}
$$

(II) Let V be the volume we are looking for. Then

$$
V = \pi \int_0^a [f(x)]^2 dx = \pi \int_0^a \frac{1}{e^{2x}} dx.
$$

Let $u = -2x$ and $du = -2dx$. Thus

$$V = \pi \int_0^a \frac{1}{e^{2x}} dx$$

$$= -\frac{\pi}{2} \int_0^{-2a} e^u du$$

$$= -\frac{\pi}{2} e^u \Big|_0^{-2a}$$

$$= -\frac{\pi}{2} \left(\frac{1}{e^{2a}} - 1 \right)$$

$$= \frac{\pi}{2} \left(1 - \frac{1}{e^{2a}} \right).$$

(III) Let V be the volume we are looking for. Then

$$V = \pi \int_0^1 [f(x)]^2 dx$$

$$= \pi \int_0^1 x^2(1 - x^2) dx$$

$$= \pi \int_0^1 (x^2 - x^4) dx$$

$$= \pi \left(\frac{x^3}{3} - \frac{x^5}{5} \right) \Big|_0^1$$

$$= \frac{2\pi}{15}.$$

(IV) Let V be the volume we are looking for. Then

$$V = \pi \int_{-c}^c [f(x)]^2 dx = a^2 \pi \int_{-c}^c \cosh^2 \frac{x}{a} dx.$$

For the integrand $\cosh^2(x/a)$, substitute $u = x/a$ and $du = dx/a$. This gives a new lower limit $u = -c/a$ and upper limit $u = c/a$. Thus

$$V = a^2 \pi \int_{-c}^c \cosh^2 \frac{x}{a} dx = a^3 \pi \int_{-c/a}^{c/a} \cosh^2 u \, du.$$

Writing $\cosh^2 u$ as $\frac{1}{2} + \frac{1}{2}\cosh 2u$, we get

$$V = a^3\pi \int_{-c/a}^{c/a} \cosh^2 u\, du$$

$$= a^3\pi \int_{-c/a}^{c/a} \left(\frac{1}{2} + \frac{1}{2}\cosh 2u\right) du$$

$$= a^3\pi \left(\frac{u}{2} + \frac{1}{4}\sinh 2u\right)\Big|_{-c/a}^{c/a}$$

$$= a^3\pi \left(\frac{c}{a} + \frac{1}{2}\sinh\frac{2c}{a}\right).$$

□

Exercise 6.34. *Calculate the volume of the solid generated by the rotation around the x-axis of the curve $a^2y^2 = ax^3 - x^4$ with $a > 0$.*

Solution to Exercise 6.34.
Consider f the positive part of the curve $a^2y^2 = ax^3 - x^4$, that is, $f(x) = \frac{x}{a}\sqrt{x(a-x)}$. Observe that $0 \le x \le a$. So the volume we are looking for will be given by

$$V = \pi \int_0^a [f(x)]^2 dx$$

$$= \frac{\pi}{a^2} \int_0^a (ax^3 - x^4) dx$$

$$= \frac{\pi}{a^2} \left(\frac{ax^4}{4} - \frac{x^5}{5}\right)\Big|_0^a$$

$$= \frac{\pi}{a^2} \left(\frac{a^5}{4} - \frac{a^5}{5}\right)$$

$$= \frac{\pi a^3}{20}.$$

□

Exercise 6.35. *Find the volume of the solid generated by the rotation around the y-axis of the figure bounded by the curve $y = \sin x$, with $0 \le x \le \frac{\pi}{2}$, the y-axis and the line $y = 1$.*

Solution to Exercise 6.35.

From $y = \sin x$, $0 \leq x \leq \frac{\pi}{2}$, we get $x = \arcsin y$, $0 \leq y \leq 1$. The volume of the solid in question (see Figure 6.3) will be given by

$$V = \pi \int_0^1 x^2 dy = \pi \int_0^1 \arcsin^2 y \, dy.$$

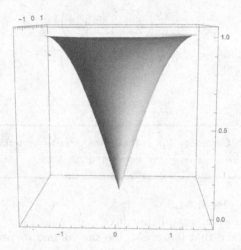

FIGURE 6.3: Sketch of the solid from Exercise 6.35.

For the integrand $\arcsin^2 y$, integrate by parts, $\int f \, dg = fg - \int g \, df$, where $f = \arcsin^2 y$, $dg = dy$, $df = \frac{2 \arcsin y}{\sqrt{1-y^2}} dy$ and $g = y$. Then

$$V = \pi \int_0^1 \arcsin^2 y \, dy$$

$$= \pi \left(y \arcsin^2 y \Big|_0^1 - 2 \int_0^1 \frac{y \arcsin y}{\sqrt{1-y^2}} dy \right)$$

$$= \pi \left(\frac{\pi^2}{4} - 2 \int_0^1 \frac{y \arcsin y}{\sqrt{1-y^2}} dy \right).$$

For the integrand $y \arcsin y / \sqrt{1-y^2}$, integrate by parts, $\int f \, dg = fg - \int g \, df$, with $f = \arcsin y$, $dg = y \, dy / \sqrt{1-y^2}$, $df = dy / \sqrt{1-y^2}$ and $g = -\sqrt{1-y^2}$.

Thus

$$V = \pi \left(\frac{\pi^2}{4} - 2 \int_0^1 \frac{y \arcsin y}{\sqrt{1-y^2}} dy \right)$$

$$= \pi \left(\frac{\pi^2}{4} + 2\sqrt{1-y^2} \arcsin y \Big|_0^1 - 2 \int_0^1 dy \right)$$

$$= \pi \left(\frac{\pi^2}{4} - 2 \int_0^1 dy \right)$$

$$= \pi \left(\frac{\pi^2}{4} - 2y \Big|_0^1 \right)$$

$$= \frac{\pi^3}{4} - 2\pi.$$

□

Exercise 6.36. *Calculate the area of the surface generated by revolving around the x-axis the graphs of the following functions:*

(I) $f(x) = \tan x$ *with* $x \in \left[0, \frac{\pi}{4}\right]$.

(II) $f(x) = \frac{x^2}{2}$ *with* $x \in [1,2]$.

Solution to Exercise 6.36.

(I) The area we are looking for is given by

$$A = 2\pi \int_0^{\pi/4} f(x)\sqrt{1 + [f'(x)]^2} dx$$

$$= 2\pi \int_0^{\pi/4} \tan x \sqrt{1 + [\sec^2 x]^2} dx$$

$$= 2\pi \int_0^{\pi/4} \tan x \sqrt{1 + \sec^4 x} dx.$$

For the integrand $\tan x \sqrt{1 + \sec^4 x}$, substitute $u = \sec^4 x$ and $du = 4 \tan x \sec^4 x dx$. So

$$A = 2\pi \int_0^{\pi/4} \tan x \sqrt{1 + \sec^4 x} dx = \frac{\pi}{2} \int_1^4 \frac{\sqrt{1+u}}{u} du.$$

Substitute $s = \sqrt{u+1}$ and $ds = du/(2\sqrt{u+1})$. Then

$$A = \frac{\pi}{2} \int_1^4 \frac{\sqrt{u+1}}{u} du$$

$$= \pi \int_{\sqrt{2}}^{\sqrt{5}} \frac{s^2}{s^2-1} ds$$

$$= \frac{\pi}{2} \left(\int_{\sqrt{2}}^{\sqrt{5}} \left(2 + \frac{1}{s-1} - \frac{1}{s+1} \right) ds \right)$$

$$= \frac{\pi}{2} \left(2s + \log|s-1| - \log|s+1| \right) \Big|_{\sqrt{2}}^{\sqrt{5}}$$

$$= \frac{\pi}{2} \left(2s + \log \frac{|s-1|}{|s+1|} \right) \Big|_{\sqrt{2}}^{\sqrt{5}}$$

$$= \pi \left(\sqrt{5} - \sqrt{2} + \frac{1}{2} \log \frac{(\sqrt{5}-1)(\sqrt{2}+1)}{(\sqrt{5}+1)(\sqrt{2}-1)} \right).$$

(II) The area we are asked to find is given by

$$A = 2\pi \int_1^2 f(x)\sqrt{1 + [f'(x)]^2} dx = \pi \int_1^2 x^2 \sqrt{1+x^2} dx.$$

Let $x = \tan u$ and $dx = \sec^2 u\, du$. Then

$$A = \pi \int_1^2 x^2 \sqrt{1+x^2} dx$$

$$= \pi \int_{\pi/4}^{\arctan 2} \tan^2 u \sqrt{1 + \tan^2 u} \sec^2 u\, du$$

$$= \pi \int_{\pi/4}^{\arctan 2} \tan^2 u \sec^3 u\, du$$

$$= \pi \int_{\pi/4}^{\arctan 2} (\sec^2 u - 1) \sec^3 u\, du$$

$$= \pi \int_{\pi/4}^{\arctan 2} (\sec^5 u - \sec^3 u)\, du.$$

Let us calculate below the antiderivatives of $\sec^5$ and $\sec^3$. For the integrand $\sec^5 u$, integrate by parts, $\int f\, dg = fg - \int g\, df$, where $f = \sec^3 u$, $dg = \sec^2 u\, du$, $df = 3\sec^3 u \tan u\, du$ and $g = \tan u$. Therefore,

$$\int \sec^5 u\, du = \sec^3 u \tan u - 3 \int \tan^2 u \sec^3 u\, du$$

$$= \sec^3 u \tan u - 3 \int (\sec^2 u - 1) \sec^3 u\, du$$

$$= \sec^3 u \tan u - 3 \int (\sec^5 u - \sec^3 u)\, du,$$

that is,

$$\int \sec^5 u\, du = \frac{1}{4}\left(\sec^3 u \tan u + 3 \int \sec^3 u\, du\right).$$

From the solution of the Exercise 4.4(II) we know that

$$\int \sec^3 u\, du = \frac{1}{2}\left(\tan u \sec u + \log(\tan u + \sec u)\right) + C$$

and consequently

$$\int \sec^5 u\, du = \frac{1}{4}\left(\sec^3 u \tan u + \frac{3}{2}\left(\tan u \sec u + \log(\tan u + \sec u)\right)\right) + C.$$

Therefore,

$$A = \pi \int_{\pi/4}^{\arctan 2} (\sec^5 u - \sec^3 u)\, du$$

$$= \pi \left[\frac{1}{4}\left(\sec^3 u \tan u + \frac{3}{2}\left(\tan u \sec u + \log(\tan u + \sec u)\right)\right)\right.$$

$$\left.- \frac{1}{2}\left(\tan u \sec u + \log(\tan u + \sec u)\right)\right]\Bigg|_{\pi/4}^{\arctan 2}$$

$$= \frac{\pi}{4}\left[\sec^3 u \tan u - \frac{1}{2}\left(\tan u \sec u + \log(\tan u + \sec u)\right)\right]\Bigg|_{\pi/4}^{\arctan 2}$$

$$= \frac{\pi}{4}\left[\left(2\sec^3 \arctan 2 - \frac{1}{2}\left(2 \sec \arctan 2 + \log(2 + \sec \arctan 2)\right)\right)\right.$$

$$\left.- \left(\frac{3\sqrt{2}}{2} - \frac{1}{2}\log(1 + \sqrt{2})\right)\right]$$

$$= \frac{\pi}{8}\left[18\sqrt{5} - 3\sqrt{2} + \log\frac{1 + \sqrt{2}}{2 + \sqrt{5}}\right],$$

where above we used that $\sec \arctan z = \sqrt{1 + z^2}$.

$\square$

Exercise 6.37. *Find the area of the ellipsoid formed by rotating the ellipse $\frac{x^2}{a^2} + \frac{y^2}{b^2} = 1$, $a > b > 0$, around the x-axis.*

Solution to Exercise 6.37.

Consider f the positive part of the ellipsoid $y^2 = b^2 - \frac{b^2x^2}{a^2}$, that is, $f(x) = \sqrt{b^2 - \frac{b^2x^2}{a^2}}$. So the area we are looking for will be given by

$$A = 2\pi \int_{-a}^{a} f(x)\sqrt{1 + [f'(x)]^2}dx$$

or, equivalently,

$$A = 4\pi \int_0^a f(x)\sqrt{1 + [f'(x)]^2}dx$$

$$= 4\pi \int_0^a \sqrt{b^2 - \frac{b^2x^2}{a^2}}\sqrt{1 + \left[-\frac{2b^2x}{2a^2\sqrt{b^2 - \frac{b^2x^2}{a^2}}}\right]^2}dx$$

$$= 4\pi \int_0^a \sqrt{b^2 - \frac{b^2x^2}{a^2} + \frac{b^4x^2}{a^4}}dx$$

$$= 4\pi \int_0^a \sqrt{b^2 - \left(\frac{b^2}{a^2} - \frac{b^4}{a^4}\right)x^2}dx$$

$$= 4\pi \frac{b}{a} \int_0^a \sqrt{a^2 - \left(\frac{a^2 - b^2}{a^2}\right)x^2}dx.$$

Let $\varepsilon^2 = \frac{a^2 - b^2}{a^2}$ (the number ε is called the eccentricity of the ellipse) and use the parametrization $x = a\cos\theta$. Then

$$A = 4\pi\frac{b}{a} \int_0^a \sqrt{a^2 - \varepsilon^2 x^2}dx$$

$$= 4\pi\frac{b}{a} \int_{\pi/2}^0 \sqrt{a^2 - \varepsilon^2 a^2 \cos^2\theta}(-a\sin\theta)d\theta$$

$$= -4\pi ab \int_{\pi/2}^0 \sqrt{1 - \varepsilon^2\cos^2\theta}\sin\theta d\theta.$$

Next set $\sin\varphi = \varepsilon\cos\theta$. Note that $\theta = \pi/2$ implies $\sin\varphi = 0$ and thus $\varphi = 0$, and $\theta = 0$ implies $\sin\varphi = \varepsilon$, that is, $\varphi = \arcsin\varepsilon$. Since $a > b > 0$, note also that $0 < \varepsilon < 1$. So $\cos\varphi\,d\varphi = -\varepsilon\sin\theta\,d\theta$ and

$$A = -4\pi ab \int_{\pi/2}^{0} \sqrt{1 - \varepsilon^2 \cos^2\theta}\,\sin\theta\,d\theta$$

$$= 4\pi \frac{ab}{\varepsilon} \int_{0}^{\arcsin\varepsilon} \sqrt{1 - \sin^2\varphi}\,\cos\varphi\,d\varphi$$

$$= 4\pi \frac{ab}{\varepsilon} \int_{0}^{\arcsin\varepsilon} \cos^2\varphi\,d\varphi.$$

Since

$$\int \cos^2\varphi\,d\varphi = \frac{1}{2}\sin\varphi\cos\varphi + \frac{1}{2}\varphi + C,$$

we have that

$$A = 4\pi \frac{ab}{\varepsilon} \int_{0}^{\arcsin\varepsilon} \cos^2\varphi\,d\varphi$$

$$= 2\pi \frac{ab}{\varepsilon} (\sin\varphi\cos\varphi + \varphi)\Big|_{0}^{\arcsin\varepsilon}$$

$$= 2\pi \frac{ab}{\varepsilon} \left(\varepsilon\sqrt{1 - \varepsilon^2} + \arcsin\varepsilon\right)$$

$$= 2\pi \frac{ab}{\varepsilon} \left(\varepsilon\frac{b}{a} + \arcsin\varepsilon\right)$$

$$= 2\pi b^2 + 2\pi \frac{ab}{\varepsilon}\arcsin\varepsilon,$$

that is,

$$A = 2\pi b^2 \left(1 + \frac{a}{b}\cdot\frac{\arcsin\varepsilon}{\varepsilon}\right),$$

where $\varepsilon = \sqrt{1 - (b/a)^2}$.

□

Exercise 6.38. *Find the value of the area of the surface generated by revolving around the x-axis the portion of the parabola $y^2 = x + 4$ determined by the line $x = 2$.*

Solution to Exercise 6.38.
Consider f the positive part of the parabola $y^2 = x+4$, that is, $f(x) = \sqrt{x+4}$ (see Figure 6.4).

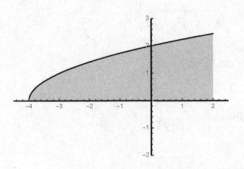
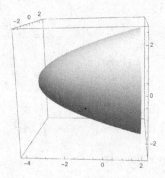

FIGURE 6.4: Images for Exercise 6.38.

Then the area sought will be given by

$$A = 2\pi \int_{-4}^{2} f(x)\sqrt{1 + [f'(x)]^2}\,dx$$

$$= 2\pi \int_{-4}^{2} \sqrt{x+4}\sqrt{1 + \left[\frac{1}{2\sqrt{x+4}}\right]^2}\,dx$$

$$= 2\pi \int_{-4}^{2} \sqrt{x+4}\sqrt{1 + \frac{1}{4(x+4)}}\,dx$$

$$= 2\pi \int_{-4}^{2} \sqrt{x + \frac{17}{4}}\,dx$$

$$= \frac{4\pi}{3}\left(x + \frac{17}{4}\right)^{3/2}\Bigg|_{-4}^{2}$$

$$= \frac{4\pi}{3}\left[\left(\frac{25}{4}\right)^{3/2} - \left(\frac{1}{4}\right)^{3/2}\right]$$

$$= \frac{62\pi}{3}.$$

$\square$

Exercise 6.39. *Find primitives for the following functions:*

(I) $x^5\sqrt{1-x^3}$.

(II) $\dfrac{1}{\sin x + \cos x}$.

(III) $\dfrac{x \arcsin x}{\sqrt{1-x^2}}$.

(IV) $x\tan^2 x$.

(V) $\dfrac{xe^x}{(1+x)^2}$.

(VI) $\dfrac{3x+5}{(x^2-2x+2)^2}$.

(VII) $\dfrac{\sqrt{x-x^2}}{x^4}$.

(VIII) $\dfrac{3x^{2/3}-7}{x-7x^{1/3}+6}$.

(IX) $\dfrac{3}{x+3(x+4)^{2/3}}$.

(X) $\dfrac{1}{x+1}\sqrt{\dfrac{3+x}{x-1}}$.

(XI) $\dfrac{1}{2+3\tan x}$.

(XII) $\dfrac{x^2}{\sqrt{3x^2-x+1}}$.

Solution to Exercise 6.39.

(I) We first make the change of variable $u = x^3$, that is, $x = u^{1/3}$ and $dx = \frac{1}{3}u^{-2/3}du$. Then

$$\int x^5\sqrt{1-x^3}dx = \int u^{5/3}\sqrt{1-u}\,\frac{du}{3u^{2/3}} = \frac{1}{3}\int u\sqrt{1-u}\,du.$$

Now consider $1-u = v^2$, that is, $u = 1-v^2$ and $du = -2vdv$. Thus

$$\frac{1}{3}\int u\sqrt{1-u}\,du = \frac{1}{3}\int (1-v^2)\cdot v(-2v)dv$$

$$= -\frac{2}{3}\int (v^2-v^4)dv$$

$$= -\frac{2}{9}v^3 + \frac{2}{15}v^5 + C.$$

Substituting back for $1-u = v^2$ and $u = x^3$, we obtain

$$\int x^5\sqrt{1-x^3}dx = -\frac{2}{9}v^3 + \frac{2}{15}v^5 + C$$

$$= -\frac{2}{9}(1-u)^{3/2} + \frac{2}{15}(1-u)^{5/2} + C$$

$$= -\frac{2}{45}(2+3u)(1-u)\sqrt{1-u} + C$$

$$= -\frac{2}{45}(2+3x^3)(1-x^3)\sqrt{1-x^3} + C.$$

(II) Consider $u = \tan(x/2)$. Recall that

$$\sin x = \frac{2u}{1+u^2}, \qquad \cos x = \frac{1-u^2}{1+u^2}, \qquad dx = \frac{2du}{1+u^2}.$$

Then

$$\int \frac{1}{\sin x + \cos x}dx = \int \frac{\frac{2}{1+u^2}}{\frac{2u}{1+u^2} + \frac{1-u^2}{1+u^2}}du$$

$$= -2 \int \frac{du}{u^2 - 2u - 1}$$

$$= -2 \int \frac{du}{(u-1-\sqrt{2})(u-1+\sqrt{2})}.$$

For the integrand $1/((u-1-\sqrt{2})(u-1+\sqrt{2}))$, observe that

$$\frac{1}{(u-1-\sqrt{2})(u-1+\sqrt{2})} = \frac{\sqrt{2}}{4}\left(\frac{1}{u-1-\sqrt{2}} - \frac{1}{u-1+\sqrt{2}}\right)$$

and thus

$$-2 \int \frac{du}{(u-1-\sqrt{2})(u-1+\sqrt{2})}$$

$$= \frac{\sqrt{2}}{2}\left(\int \frac{1}{u-1+\sqrt{2}}du - \int \frac{1}{u-1-\sqrt{2}}du\right)$$

$$= \frac{\sqrt{2}}{2}\left(\log|u-1+\sqrt{2}| - \log|u-1-\sqrt{2}|\right) + C$$

$$= \frac{\sqrt{2}}{2}\log\left|\frac{u-1+\sqrt{2}}{u-1-\sqrt{2}}\right| + C.$$

Substituting back for $u = \tan(x/2)$, we obtain

$$\int \frac{1}{\sin x + \cos x}dx = \frac{\sqrt{2}}{2}\log\left|\frac{u-1+\sqrt{2}}{u-1-\sqrt{2}}\right| + C$$

$$= \frac{\sqrt{2}}{2}\log\left|\frac{\tan\frac{x}{2}-1+\sqrt{2}}{\tan\frac{x}{2}-1-\sqrt{2}}\right| + C.$$

(III) Integrate by parts, $\int f dg = fg - \int g df$, where

$$f = \arcsin x, \qquad dg = \frac{x}{\sqrt{1-x^2}}dx$$

and

$$df = \frac{1}{\sqrt{1-x^2}}dx, \qquad g = \int \frac{x dx}{\sqrt{1-x^2}}.$$

Let $u = 1 - x^2$ and $du = -2x dx$. Then

$$g = \int \frac{x dx}{\sqrt{1-x^2}} = -\frac{1}{2}\int u^{-1/2}du = -u^{1/2} = -\frac{1}{\sqrt{1-x^2}}$$

(we do not write any constant in the primitive, because any primitive works for us). So, by the formula of integration by parts, it follows that

$$\int \frac{x \arcsin x}{\sqrt{1-x^2}}\, dx = -\frac{\arcsin x}{\sqrt{1-x^2}} + \int \frac{1}{1-x^2}\, dx.$$

Now we only have to calculate this last integral:

$$\int \frac{dx}{1-x^2} = \frac{1}{2}\int \frac{dx}{1+x} + \frac{1}{2}\int \frac{dx}{1-x}$$

$$= \frac{1}{2}\log|1+x| + \frac{1}{2}\log|1-x| + C$$

$$= \log\sqrt{|1-x^2|} + C.$$

Thus

$$\int \frac{x \arcsin x}{\sqrt{1-x^2}}\, dx = -\frac{\arcsin x}{\sqrt{1-x^2}} + \log\sqrt{|1-x^2|} + C.$$

(IV) Integrate by parts, $\int f\,dg = fg - \int g\,df$, where

$$f = x, \qquad dg = \tan^2 x\,dx$$

and

$$df = dx, \qquad g = \int \tan^2 x\,dx.$$

Consider $u = \tan x$, that is, $x = \arctan u$ and $dx = du/(1+u^2)$. Then

$$\int \tan^2 x\,dx = \int \frac{u^2}{1+u^2}\,du$$

$$= \int \left(1 - \frac{1}{1+u^2}\right) du$$

$$= u - \arctan u$$

$$= \tan x - x$$

(we do not write any constant in the primitive, because any primitive works for us). So, by the formula of integration by parts, it follows that

$$\int x \tan^2 x\,dx = x(\tan x - x) - \int (\tan x - x)\,dx$$

$$= x \tan x - x^2 + \log|\cos x| + \frac{x^2}{2} + C$$

$$= x \tan x + \log|\cos x| - \frac{x^2}{2} + C.$$

(V) Observe that the derivative of xe^x is

$$(xe^x)' = e^x + xe^x = (1+x)e^x.$$

So, by the formula of integration by parts, $\int f\,dg = fg - \int g\,df$, where

$$f = xe^x, \qquad dg = \frac{dx}{(1+x)^2}$$

and

$$df = (1+x)e^x\,dx, \qquad g = -\frac{1}{1+x},$$

it follows that

$$\int \frac{xe^x}{(1+x)^2}\,dx = -\frac{1}{1+x} \cdot xe^x + \int \frac{1}{1+x} \cdot (1+x)e^x\,dx$$

$$= -\frac{x}{1+x}e^x + \int e^x\,dx$$

$$= -\frac{x}{1+x}e^x + e^x + C$$

$$= \frac{e^x}{1+x} + C.$$

(VI) Observe that the polynomial $x^2 - 2x + 2$ has no real roots, so the denominator has multiple imaginary roots. The most indicated method, then, to calculate this primitive is the Hermite method. According to this method, there are four coefficients A, B, C and D such that

$$\frac{3x+5}{(x^2 - 2x + 2)^2} = \frac{Ax + B}{x^2 - 2x + 2} + \left(\frac{Cx + D}{x^2 - 2x + 2} \right)'.$$

Calculating its derivative, we arrive at

$$\frac{3x+5}{(x^2 - 2x + 2)^2}$$

$$= \frac{Ax + B}{x^2 - 2x + 2} + \frac{C(x^2 - 2x + 2) - (Cx + D)(2x - 2)}{(x^2 - 2x + 2)^2}$$

$$= \frac{Ax^3 + (-2A + B - C)x^2 + (2A - 2B - 2D)x + (2B + 2C + 2D)}{(x^2 - 2x + 2)^2}.$$

Equating the coefficients of the numerators, we obtain the linear system

$$\begin{cases} A = 0, \\ -2A + B - C = 0, \\ 2A - 2B - 2D = 3, \\ 2B + 2C + 2D = 5, \end{cases}$$

whose solutions are

$$A = 0, \qquad B = 4, \qquad C = 4, \qquad D = -\frac{11}{2}.$$

Then

$$\frac{3x+5}{(x^2-2x+2)^2} = \frac{4}{x^2-2x+2} + \left(\frac{8x-11}{2(x^2-2x+2)}\right)'.$$

Next, integrating on both sides,

$$\int \frac{3x+5}{(x^2-2x+2)^2}dx = 4\int \frac{dx}{x^2-2x+2} + \frac{8x-11}{2(x^2-2x+2)}$$

$$= 4\int \frac{dx}{1+(x-1)^2} + \frac{8x-11}{2(x^2-2x+2)}$$

$$= 4\arctan(x-1) + \frac{8x-11}{2(x^2-2x+2)} + C.$$

(VII) Consider the change of variable $x = 1/u$, $dx = -du/u^2$. So

$$\int \frac{\sqrt{x-x^2}}{x^4}dx = \int \frac{\sqrt{\frac{1}{u}-\frac{1}{u^2}}}{\frac{1}{u^4}}\left(-\frac{dx}{u^2}\right) = -\int u\sqrt{u-1}du.$$

We now make the change of variable $u = v^2+1$, $du = 2vdv$, and thus

$$\int \frac{\sqrt{x-x^2}}{x^4}dx = -\int u\sqrt{u-1}du$$

$$= -\int (v^2+1)\sqrt{(v^2+1)-1}2vdv$$

$$= -2\int (v^4+v^2)dv$$

$$= -\frac{2}{5}v^5 - \frac{2}{3}v^3 + C.$$

Returning to the original variables, we get

$$\int \frac{\sqrt{x-x^2}}{x^4}dx = -\frac{2}{5}v^5 - \frac{2}{3}v^3 + C$$

$$= -\frac{2}{5}(u-1)^{\frac{5}{2}} - \frac{2}{3}(u-1)^{\frac{3}{2}} + C$$

$$= -\left(\frac{2}{5}u^2 - \frac{2}{15}u - \frac{4}{15}\right)\sqrt{u-1} + C$$

$$= -\left(\frac{2}{5x^2} - \frac{2}{15x} - \frac{4}{15}\right)\sqrt{\frac{1}{x}-1} + C$$

$$= \frac{(4x^2+2x-6)\sqrt{1-x}}{15x^{5/2}} + C.$$

(VIII) Consider $x = u^3$ and $dx = 3u^2 du$. Then

$$\int \frac{3x^{2/3} - 7}{x - 7x^{1/3} + 6} dx = \int \frac{3u^2 - 7}{u^3 - 7u + 6} 3u^2 du$$

$$= \int \frac{9u^4 - 21u^2}{u^3 - 7u + 6} du$$

$$= \int \left(9u + \frac{42u^2 - 54u}{u^3 - 7u + 6} \right) du$$

$$= \frac{9}{2} u^2 + \int \frac{42u^2 - 54u}{(u-1)(u-2)(u-3)} du.$$

We have to calculate this last integral. We know that there exist three coefficients A, B, C, such that

$$\frac{42u^2 - 54u}{(u-1)(u-2)(u-3)}$$

$$= \frac{A}{u-1} + \frac{B}{u-2} + \frac{C}{u-3}$$

$$= \frac{A(u-2)(u-3) + B(u-1)(u-3) + C(u-1)(u-2)}{(u-1)(u-2)(u-3)}.$$

To calculate A, B, and C, let us equal the numerators, obtaining

$$42u^2 - 54u = A(u-2)(u-3) + B(u-1)(u-3) + C(u-1)(u-2).$$

If $u = 1$, we obtain that $2A = -12$, that is, $A = -6$. If $u = 2$, we get $-B = 60$, that is, $B = -60$. If $u = 3$, we finally obtain that $2C = 216$, that is, $C = 108$. Thus

$$\int \frac{42u^2 - 54u}{(u-1)(u-2)(u-3)} du$$

$$= -6 \int \frac{du}{u-1} - 60 \int \frac{du}{u-2} + 108 \int \frac{du}{u-3}$$

$$= -6 \log|u-1| - 60 \log|u-2| + 108 \log|u-3| + C$$

$$= \log \left| \frac{(u-3)^{108}}{(u-1)^6 (u-2)^{60}} \right| + C.$$

Therefore,

$$\int \frac{3x^{2/3} - 7}{x - 7x^{1/3} + 6} dx = \frac{9}{2} u^2 + \log \left| \frac{(u-3)^{108}}{(u-1)^6 (u-2)^{60}} \right| + C$$

$$= \frac{9}{2} x^{2/3} + \log \left| \frac{(x^{1/3} - 3)^{108}}{(x^{1/3} - 1)^6 (x^{1/3} - 2)^{60}} \right| + C.$$

(IX) Consider $x + 4 = u^3$, that is, $x = u^3 - 4$ and $dx = 3u^2 du$. So

$$\int \frac{3}{x + 3(x+4)^{2/3}} dx = \int \frac{3}{u^3 - 4 + 3u^2} 3u^2 du$$

$$= \int \frac{9u^2}{u^3 + 3u^2 - 4} du$$

$$= \int \frac{9u^2}{(u-1)(u-2)^2} du.$$

We know that there exist three constants A, B, C, such that

$$\frac{9u^2}{(u-1)(u-2)^2} = \frac{A}{u-1} + \frac{B}{u-2} + \frac{C}{(u-2)^2}$$

$$= \frac{A(u-2)^2 + B(u-1)(u-2) + C(u-1)}{(u-1)(u-2)^2},$$

that is,

$$9u^2 = A(u-2)^2 + B(u-1)(u-2) + C(u-1).$$

If $u = 1$, we get $A = 9$. If $u = 2$, we obtain $C = 36$. If $u = 0$, we get $4A + 2B - C = 0$, that is, $B = C/2 - 2A = 0$. Therefore,

$$\int \frac{3}{x + 3(x+4)^{2/3}} dx = \int \frac{3}{u^3 - 4 + 3u^2} 3u^2 du$$

$$= 9 \int \frac{du}{u-1} + 36 \int \frac{du}{(u-2)^2}$$

$$= 9 \log|u - 1| - \frac{36}{u-2} + C$$

$$= 9 \log|\sqrt[3]{x+4} - 1| - \frac{36}{\sqrt[3]{x+4} - 2} + C.$$

(X) Consider

$$\frac{3+x}{x-1} = u^2,$$

that is,

$$x = \frac{u^2 + 3}{u^2 - 1} \quad \text{and} \quad dx = -\frac{8u}{(u^2 - 1)^2} du.$$

Then

$$\int \frac{1}{x+1} \sqrt{\frac{3+x}{x-1}} dx = \int \frac{1}{\frac{u^2+3}{u^2-1} + 1} u \left(-\frac{8u}{(u^2-1)^2} \right) du$$

$$= -4 \int \frac{u^2}{u^4 - 1} du$$

$$= -4 \int \frac{u^2}{(u-1)(u+1)(u^2+1)} du.$$

We know that there exist four constants A, B, C, D, such that

$$\frac{u^2}{(u-1)(u+1)(u^2+1)}$$

$$= \frac{A}{u-1} + \frac{B}{u+1} + \frac{Cu+D}{u^2+1}$$

$$= \frac{A(u+1)(u^2+1) + B(u-1)(u^2+1) + (Cu+D)(u-1)(u+1)}{(u-1)(u+1)(u^2+1)}.$$

If $u = 1$, we obtain that $4A = 1$, that is, $A = 1/4$. If $u = -1$, we get $-4B = 1$, that is, $B = -1/4$. If $u = 0$, we can conclude that $A - B - D = 0$, that is, $D = 1/2$. Finally, comparing the terms in u^3 in the two denominators, we get $A + B + C = 0$. Then $C = 0$ and thus

$$-4\int \frac{u^2}{(u-1)(u+1)(u^2+1)}\,du = -\int \frac{du}{u-1} + \int \frac{du}{u+1} - 2\int \frac{du}{1+u^2}$$

$$= -\log|u-1| + \log|u+1| - 2\arctan u + C$$

$$= \log\left|\frac{u+1}{u-1}\right| + 2\arctan u + C.$$

Returning to the original variables, we get

$$\int \frac{1}{x+1}\sqrt{\frac{3+x}{x-1}}\,dx = \log\left|\frac{\sqrt{\frac{3+x}{x-1}}+1}{\sqrt{\frac{3+x}{x-1}}-1}\right| + 2\arctan\sqrt{\frac{3+x}{x-1}} + C$$

$$= \log\left|\frac{\sqrt{3+x}+\sqrt{x-1}}{\sqrt{3+x}-\sqrt{x-1}}\right| + 2\arctan\sqrt{\frac{3+x}{x-1}} + C.$$

(XI) Let $u = \tan x$, that is, $x = \arctan u$ and $dx = du/(1+u^2)$. Then

$$\int \frac{dx}{2+3\tan x} = \int \frac{\frac{du}{1+u^2}}{2+3u} = \int \frac{du}{(2+3u)(1+u^2)}.$$

We can find three constants A, B, C, such that

$$\frac{1}{(2+3u)(1+u^2)} = \frac{A}{3u+2} + \frac{Bu+C}{u^2+1}$$

$$= \frac{A(u^2 * 1) + (Bu+C)(3u+2)}{(3u+2)(u^2+1)}$$

$$= \frac{(A+3B)u^2 + (2B+3C)u + (A+2C)}{(3u+2)(u^2+1)}.$$

We get the linear system

$$\begin{cases} A + 3B = 0, \\ 2B + 3C = 0, \\ A + 2C = 1, \end{cases}$$

with solution

$$A = \frac{9}{13}, \qquad B = -\frac{3}{13}, \qquad C = \frac{2}{13}.$$

Consequently,

$$
\begin{aligned}
\int \frac{dx}{2 + 3\tan x} &= \frac{9}{13} \int \frac{du}{3u + 2} - \frac{1}{13} \int \frac{3u - 2}{u^2 + 1} du \\
&= \frac{9}{13} \int \frac{du}{3u + 2} - \frac{3}{26} \int \frac{2u\,du}{u^2 + 1} + \frac{2}{13} \int \frac{du}{1 + u^2} \\
&= \frac{9}{13} \log|3u + 2| - \frac{3}{26} \log(u^2 + 1) \\
&\quad + \frac{2}{13} \arctan u + C \\
&= \frac{9}{13} \log|3\tan x + 2| + \frac{3}{13} \log|\sec x| + \frac{2}{13}x + C.
\end{aligned}
$$

(XII) We know that there exist three constants A, B, and C, such that

$$\int \frac{x^2\,dx}{\sqrt{3x^2 - x + 1}} = (Ax + B)\sqrt{3x^2 - x + 1} + C \int \frac{dx}{\sqrt{3x^2 - x + 1}}.$$

Deriving both members, we obtain that

$$\frac{x^2}{\sqrt{3x^2 - x + 1}} = A\sqrt{3x^2 - x + 1} + (Ax + B)\frac{6x - 1}{2\sqrt{3x^3 - x + 1}} + \frac{C}{\sqrt{3x^2 - x + 1}}.$$

Then

$$
\begin{aligned}
x^2 &= A(3x^2 - x + 1) + (Ax + B)\frac{6x - 1}{2} + C \\
&= 6Ax^2 + \frac{-3A + 6B}{2}x + \left(A - \frac{B}{2} + C\right),
\end{aligned}
$$

from where we get the linear system

$$\begin{cases} 6A = 1, \\ -3A + 6B = 0, \\ 2A - B + 2C = 0, \end{cases}$$

with solution

$$A = \frac{1}{6}, \qquad B = \frac{1}{12}, \qquad C = -\frac{1}{8}.$$

Therefore,

$$\int \frac{x^2\,dx}{\sqrt{3x^3 - x + 1}} = \frac{2x + 1}{12}\sqrt{3x^2 - x + 1} - \frac{1}{8}\int \frac{dx}{\sqrt{3x^2 - x + 1}}.$$

Thus, we see that it is enough to calculate the integral of the last term. Completing the square, we have

$$\int \frac{dx}{\sqrt{3x^2 - x + 1}} = \int \frac{dx}{\sqrt{(\sqrt{3}x - \frac{1}{2\sqrt{3}})^2 + \frac{11}{12}}}$$

$$= \frac{\sqrt{3}}{3} \int \frac{\frac{6}{\sqrt{11}}dx}{\sqrt{(\frac{6x-1}{\sqrt{11}})^2 + 1}}.$$

Making the change of variable $u = (6x - 1)/\sqrt{11}$ and $du = 6dx/\sqrt{11}$, we obtain that

$$\int \frac{dx}{\sqrt{3x^2 - x + 1}} = \frac{\sqrt{3}}{3} \int \frac{du}{\sqrt{u^2 + 1}}$$

$$= \frac{\sqrt{3}}{3} \operatorname{arcsinh} u + C$$

$$= \frac{\sqrt{3}}{3} \log(u + \sqrt{u^2 + 1}) + C$$

$$= \frac{\sqrt{3}}{3} \log\left(\frac{6x - 1 + 2\sqrt{3x^2 - x + 1}}{\sqrt{11}}\right) + C.$$

Finally

$$\int \frac{x^2 dx}{\sqrt{3x^2 - x + 1}} = \frac{2x + 1}{12}\sqrt{3x^2 - x + 1}$$

$$- \frac{\sqrt{3}}{24} \log\left(\frac{6x - 1 + 2\sqrt{3x^2 - x + 1}}{\sqrt{11}}\right) + C.$$

□

Exercise 6.40. *Calculate the following limit using definite integrals:*

$$\lim_{n \to \infty} \frac{1}{n}\left(\cos\frac{x}{n} + \cos\frac{2x}{n} + \cdots + \cos\frac{nx}{n}\right).$$

Solution to Exercise 6.40.

If $x = 0$, the limit is clearly 1. If $x \neq 0$, consider x as a fixed number, define $f(t) = \cos tx$, and observe that

$$\lim_{n \to \infty} \frac{1}{n}\left(\cos\frac{x}{n} + \cos\frac{2x}{n} + \cdots + \cos\frac{nx}{n}\right)$$

$$= \lim_{n \to \infty} \frac{1}{n}\sum_{k=1}^{n} \cos\frac{kx}{n} = \lim_{n \to \infty} \frac{1}{n}\sum_{k=1}^{n} f\left(\frac{k}{n}\right).$$

This limit exists and it is equal to

$$\int_0^1 f(t)dt = \int_0^1 \cos(tx)dt.$$

Consider $u = tx$. Then $du = xdt$ and thus

$$\int_0^1 f(t)dt = \int_0^1 \cos(tx)dt = \frac{1}{x}\int_0^x \cos u\, du = \frac{1}{x}\sin u \Big|_{u=0}^{u=x} = \frac{\sin x}{x}.$$

Therefore,

$$\lim_{n\to\infty} \frac{1}{n}\left(\cos\frac{x}{n} + \cos\frac{2x}{n} + \cdots + \cos\frac{nx}{n}\right) = \frac{\sin x}{x}.$$

□

Exercise 6.41. *Let f be a continuous function on $[0,a]$. Prove that*

$$\int_0^a f(x)dx = \int_0^a f(a-x)dx$$

and, for $n = 1$ and $n = 3$, calculate

$$\int_0^a \frac{x\sin^n x}{1+\cos^2 x}dx.$$

Solution to Exercise 6.41.
Considering the change of variable $x = a - u$ and $dx = -du$, we obtain

$$\int_0^a f(x)dx = -\int_a^0 f(a-u)du = \int_0^a f(a-u)du.$$

This equality will help us calculate the two integrals that we are asked to do. In fact, the function $f(x) = x\sin^n x/(1+\cos^2 x)$ is continuous on $[0,\pi]$, so for f the previous equality is satisfied, concluding that

$$\int_0^\pi \frac{x\sin^n x}{1+\cos^2 x}dx = \int_0^\pi \frac{(\pi - x)\sin^n(\pi - x)}{1+\cos^2(\pi - x)}dx$$
$$= \int_0^\pi \frac{(\pi - x)\sin^n x}{1+\cos^2 x}dx$$
$$= \pi \int_0^\pi \frac{\sin^n x}{1+\cos^2 x}dx - \int_0^\pi \frac{x\sin^n x}{1+\cos^2 x}dx,$$

that is,

$$\int_0^\pi \frac{x\sin^n x}{1+\cos^2 x}dx = \frac{\pi}{2}\int_0^\pi \frac{\sin^n x}{1+\cos^2 x}dx.$$

388   *Real Analysis*

Now let us calculate

$$\int_0^\pi \frac{x\sin x}{1+\cos^2 x}dx \quad \text{and} \quad \int_0^\pi \frac{x\sin^3 x}{1+\cos^2 x}dx.$$

Considering $u = -\cos x$ and $du = \sin x\,dx$, we obtain that

$$\int_0^\pi \frac{x\sin x}{1+\cos^2 x}dx = \frac{\pi}{2}\int_0^\pi \frac{\sin x}{1+\cos^2 x}dx$$

$$= \frac{\pi}{2}\int_{-1}^1 \frac{du}{1+u^2}$$

$$= \frac{\pi}{2}\arctan u\Big|_{-1}^1$$

$$= \frac{\pi}{2}\left[\frac{\pi}{4} - \left(-\frac{\pi}{4}\right)\right] = \frac{\pi^2}{4}.$$

Now let $n = 3$. Making the same change of variable as before, we get

$$\int_0^\pi \frac{x\sin^3 x}{1+\cos^2 x}dx = \frac{\pi}{2}\int_0^\pi \frac{\sin^3 x}{1+\cos^2 x}dx$$

$$= \frac{\pi}{2}\int_0^\pi \frac{(1-\cos^2 x)\sin x}{1+\cos^2 x}dx$$

$$= \frac{\pi}{2}\int_{-1}^1 \frac{1-u^2}{1+u^2}du$$

$$= \frac{\pi}{2}\int_{-1}^1 \left(-1 + \frac{2}{1+u^2}\right)du$$

$$= -\pi + \pi\int_{-1}^1 \frac{du}{1+u^2}$$

$$= -\pi + \pi\arctan u\Big|_{-1}^1$$

$$= -\pi + \pi\left(\frac{\pi}{4} + \frac{\pi}{4}\right) = \frac{\pi}{2}(\pi - 2).$$

□

Exercise 6.42. *Find the value of the following definite integrals:*

(I) $\int_{-\sqrt3}^{\sqrt3} \sqrt{4-x^2}dx.$ (II) $\int_2^4 \frac{\sqrt{x^2-4}}{x^4}dx.$

(III) $\int_0^{\pi/2} \frac{\sin x}{3+\sin^2 x}dx.$ (IV) $\int_0^1 \sqrt{2x-x^2}dx.$

(V) $\int_0^1 \frac{\log(1+x)}{(1+x)^2}dx.$ (VI) $\int_{-\pi/2}^{\pi/2} \sqrt{\cos x - \cos^3 x}dx.$

Solution to Exercise 6.42.

(I) Consider $x = 2\sin u$ and $dx = 2\cos u \, du$. Observe that if $x = \sqrt{3}$, then $\sin u = \sqrt{3}/2$, i.e., $u = \pi/3$. Similarly, when $x = -\sqrt{3}$, we get $u = -\pi/3$. Thus

$$\int_{-\sqrt{3}}^{\sqrt{3}} \sqrt{4 - x^2} \, dx = \int_{-\pi/3}^{\pi/3} \sqrt{4 - 4\sin^2 u} \cdot 2\cos u \, du$$

$$= \int_{-\pi/3}^{\pi/3} 2\sqrt{\cos^2 x} \cdot 2\cos u \, du$$

$$= 4 \int_{-\pi/3}^{\pi/3} \cos^2 u \, du.$$

For the integrand $\cos^2 u$ proceed as in the solution of the Exercise 4.4(I). Therefore,

$$\int_{-\sqrt{3}}^{\sqrt{3}} \sqrt{4 - x^2} \, dx = 4 \int_{-\pi/3}^{\pi/3} \cos^2 u \, du$$

$$= 4 \left(\frac{1}{4} \sin 2u + \frac{1}{2} u \right) \Big|_{-\pi/3}^{\pi/3}$$

$$= \sin 2u \Big|_{-\pi/3}^{\pi/3} + 2u \Big|_{-\pi/3}^{\pi/3}$$

$$= \left(\frac{\sqrt{3}}{2} + \frac{\sqrt{3}}{2} \right) + 2 \left(\frac{\pi}{3} + \frac{\pi}{3} \right)$$

$$= \sqrt{3} + \frac{4\pi}{3}.$$

(II) Consider $x = 2\cosh u$ and $dx = 2\sinh u \, du$. So

$$\int_{2}^{4} \frac{\sqrt{x^2 - 4}}{x^4} \, dx = \int_{0}^{\text{arc cosh } 2} \frac{\sqrt{4\cosh^2 u - 4}}{16\cosh^4 u} 2\sinh u \, du$$

$$= \frac{1}{4} \int_{0}^{\text{arc cosh } 2} \frac{\sinh^2 u}{\cosh^4 u} \, du$$

$$= \frac{1}{4} \int_{0}^{\text{arc cosh } 2} \frac{\tanh^2 u}{\cosh^2 u} \, du.$$

Now let $v = \tanh u$ and $dv = du/\cosh^2 u$. Thus

$$\int_{2}^{4} \frac{\sqrt{x^2 - 4}}{x^4} \, dx = \frac{1}{4} \int_{0}^{\tanh(\text{arc cosh } 2)} v^2 \, dv$$

$$= \frac{1}{12} v^3 \Big|_{0}^{\tanh(\text{arc cosh } 2)}$$

$$= \frac{1}{12} \tanh^3(\text{arc cosh } 2).$$

We know that

$$1 = \cosh^2(\text{arc}\cosh 2) - \sinh^2(\text{arc}\cosh 2) = 4 - \sinh^2(\text{arc}\cosh 2),$$

and thus $\sinh(\text{arc}\cosh 2) = \pm\sqrt{3}$. Since $\text{arc}\cosh 2 > 0$, and the hyperbolic sine is positive in the positives, we obtain that $\sinh(\text{arc}\cosh 2) = \sqrt{3}$. So $\tanh(\text{arc}\cosh u) = \sqrt{3}/2$ and so we conclude that

$$\int_2^4 \frac{x^2 - 4}{x^4}\,dx = \frac{1}{12}\left(\frac{\sqrt{3}}{2}\right)^3 = \frac{\sqrt{3}}{32}.$$

(III) Observe that

$$\int_0^{\pi/2} \frac{\sin x}{3 + \sin^2 x}\,dx = \int_0^{\pi/2} \frac{\sin x}{4 - \cos^2 x}\,dx.$$

Consider $u = \cos x$ and $du = -\sin x\,dx$. Then

$$\int_0^{\pi/2} \frac{\sin x}{3 + \sin^2 x}\,dx = \int_0^{\pi/2} \frac{\sin x}{4 - \cos^2 x}\,dx$$

$$= -\int_1^0 \frac{du}{4 - u^2}$$

$$= \int_0^1 \frac{du}{4 - u^2}$$

$$= \int_0^1 \frac{du}{(2 + u)(2 - u)}$$

$$= \frac{1}{4}\int_0^1 \frac{du}{2 + u} + \frac{1}{4}\int_0^1 \frac{du}{2 - u}$$

$$= \frac{1}{4}\log(2 + u)\Big|_0^1 - \frac{1}{4}\log(2 - u)\Big|_0^1$$

$$= \frac{1}{4}\log 3 - \frac{1}{4}\log 2$$

$$\quad - \frac{1}{4}\log 1 + \frac{1}{4}\log 2$$

$$= \frac{1}{4}\log 3.$$

(IV) Observe that

$$\int_0^1 \sqrt{2x - x^2}\,dx = \int_0^1 \sqrt{1 - (x - 1)^2}\,dx.$$

Consider $x - 1 = \sin u$ and $dx = \cos u\, du$. Then

$$\int_0^1 \sqrt{2x - x^2}\, dx = \int_0^1 \sqrt{1 - (x-1)^2}\, dx$$

$$= \int_{-\pi/2}^0 \sqrt{1 - \sin^2 u}\, \cos u\, du$$

$$= \int_{-\pi/2}^0 \cos^2 u\, du$$

$$= \frac{\pi}{4}.$$

(V) Integrating by parts, we obtain

$$\int_0^1 \frac{\log(1 + x)}{(1 + x)^2}\, dx = \int_0^1 \log(1 + x) \cdot \frac{1}{(1 + x)^2}\, dx$$

$$= -\frac{\log(1 + x)}{1 + x}\Big|_0^1 + \int_0^1 \frac{dx}{(1 + x)^2}$$

$$= -\frac{\log 2}{2} - \frac{1}{1 + x}\Big|_0^1 = \frac{1 - \log 2}{2}.$$

(VI) Observe that

$$\sqrt{\cos x - \cos^3 x}\, dx = \sqrt{\cos x(1 - \cos^2 x)} = \sqrt{\cos x \sin^2 x} = \sqrt{\cos x} \cdot |\sin x|.$$

Then

$$\int_{-\pi/2}^{\pi/2} \sqrt{\cos x - \cos^3 x}\, dx$$

$$= \int_{-\pi/2}^{\pi/2} \sqrt{\cos x} \cdot |\sin x|\, dx$$

$$= -\int_{-\pi/2}^0 \sqrt{\cos x} \cdot \sin x\, dx + \int_0^{\pi/2} \sqrt{\cos x} \cdot \sin x\, dx.$$

Let $u = \cos x$ and $du = -\sin x\, dx$. Thus

$$\int_{-\pi/2}^{\pi/2} \sqrt{\cos x - \cos^3 x}\, dx = \int_0^1 \sqrt{u}\, du - \int_1^0 \sqrt{u}\, du$$

$$= 2\int_0^1 \sqrt{u}\, du$$

$$= 2 \cdot \frac{2}{3} u^{3/2}\Big|_0^1 = \frac{4}{3}.$$

$\square$

Exercise 6.43. *Prove that the following functions are differentiable and find their corresponding derivatives:*

(I) $F(x) = \displaystyle\int_0^{x^3} \sin^3 t\, dt.$

(II) $F(x) = \displaystyle\int_a^b f(x+t)\, dt,$ *where f is a continuous function.*

(III) $F(x) = \displaystyle\int_0^x x f(t)\, dt,$ *where f is a continuous function.*

(IV) $F(x) = \displaystyle\int_{f(x)}^{g(x)} h(t)\, dt,$ *where f and g are differentiable functions.*

Solution to Exercise 6.43.

(I) Let

$$G(v) = \int_0^v \sin^3 x\, dx.$$

It is well known that G is differentiable in $\mathbb{R}$ and its derivative is

$$G'(v) = \sin^3 v.$$

On the other hand, notice that

$$F(x) = G(x^3).$$

Thus, using the Chain Rule it follows that

$$F'(x) = G'(x^3) \cdot 3x^2 = 3x^2 \sin^3 x^3.$$

(II) Let us first consider x fixed, and make the change of variable $u = x + t$ and $dx = dt$. So, we have

$$F(x) = \int_{a+x}^{b+x} f(u)\, du$$
$$= \int_p^{b+x} f(u)\, du - \int_p^{a+x} f(u)\, du,$$

for some p. So if we call

$$G(v) = \int_p^v f(u)\, du,$$

we have that G is differentiable in the entire domain of f. Also, we have that

$$F(x) = G(b+x) - G(a+x),$$

from where we conclude that F is also differentiable in all its domain and also

$$F'(x) = G'(b+x) - G'(a+x) = f(b+x) - f(a+x).$$

(III) Let

$$G(x) = \int_0^x f(t)dt.$$

Then

$$F(x) = \int_0^x xf(t)dt = x\int_0^x f(t)dt = xG(x)$$

and thus

$$F'(x) = G(x) + xG'(x) = \int_0^x f(t)dt + xf(x).$$

(IV) Let $p \in \mathbb{R}$ and

$$H(u) = \int_p^u h(t)dt.$$

Then

$$F(x) = \int_{f(x)}^{g(x)} h(t)dt$$
$$= \int_p^{g(x)} h(t)dt - \int_p^{f(x)} h(t)dt$$
$$= H(g(x)) - H(f(x)).$$

Thus

$$F'(x) = H'(g(x))g'(x) - H'(f(x))f'(x) = h(g(x))g'(x) - h(f(x))f'(x).$$

$\square$

Exercise 6.44. *Prove that, if f is continuous, then*

$$\int_0^x f(u)(x-u)du = \int_0^x \int_0^u f(t)dtdu.$$

Solution to Exercise 6.44.
Consider the functions

$$F(x) = \int_0^x f(u)(x-u)du, \qquad G(x) = \int_0^x \int_0^u f(t)dtdu.$$

We will prove that, for all x, $F'(x) = G'(x)$ holds. This will imply that $F(x) = G(x) + C$ for some $C \in \mathbb{R}$. Since $F(0) = G(0) = 0$, we get $C = 0$, and therefore $F(x) = G(x)$, which is exactly what we want to prove.

Let us then calculate $F'(x)$ and $G'(x)$. It directly follows that

$$G'(x) = \int_0^x f(t)dt.$$

To compute $F'(x)$, we decompose the integral into two. We obtain

$$F(x) = \int_0^x f(u)(x-u)du = x\int_0^x f(u)du - \int_0^x uf(u)du,$$

and thus we see that

$$F'(x) = \left(\int_0^x f(u)du + xf(x)\right) - xf(x) = \int_0^x f(u)du = G'(x).$$

$\square$

Exercise 6.45. *Calculate the following limit:*

$$\lim_{x \to 0^+} \frac{\displaystyle\int_{x^2}^x \frac{e^t - 1}{\sin t^2} dt}{\log x}.$$

Solution to Exercise 6.45.
Let

$$F(x) = \int_{x^2}^x \frac{e^t - 1}{\sin t^2} dt.$$

The requested limit can be written as

$$\lim_{x \to 0^+} \frac{F(x)}{\log x}$$

and we will calculate it using L'Hôpital's Rule. Indeed, $\log x \to -\infty$ as $x \to 0^+$, and it is not difficult to prove that $F(x) \to \infty$ as $x \to 0^+$ (use that the integrand is not less than $1/t$, which follows from the facts $e^t - 1 \geq t$ and $\sin t^2 \leq t^2$). Considering

$$G(x) = \int_1^x \frac{e^t - 1}{\sin t^2} dt,$$

we know that
$$G'(x) = \frac{e^x - 1}{\sin x^2}.$$

On the other hand, since
$$F(x) = G(x) - G(x^2),$$

we obtain that
$$F'(x) = G'(x) - 2xG'(x^2) = \frac{e^x - 1}{\sin x^2} - 2x\frac{e^{x^2} - 1}{\sin x^4}.$$

Therefore, the requested limit is
$$\lim_{x\to 0^+} \frac{F(x)}{\log x} = \lim_{x\to 0^+} \frac{F'(x)}{1/x}$$

$$= \lim_{x\to 0^+} \left(x\frac{e^x - 1}{\sin x^2} - 2x^2\frac{e^{x^2} - 1}{\sin x^4} \right)$$

$$= \lim_{x\to 0^+} \left(\frac{x \cdot x}{x^2} - \frac{2x^2 \cdot x^2}{x^4} \right)$$

$$= 1 - 2 = -1.$$

□

Exercise 6.46. *For every $n \geq 2$, show that*
$$\int_0^{\pi/2} \sin^n x\, dx = \frac{n-1}{n} \int_0^{\pi/2} \sin^{n-2} x\, dx.$$

Prove that for every $n \geq 1$ we have

(I) $\displaystyle\int_0^{\pi/2} \sin^{2n+1} x\, dx = \frac{2 \cdot 4 \cdot 6 \cdots (2n)}{3 \cdot 5 \cdot 7 \cdots (2n+1)},$

(II) $\displaystyle\int_0^{\pi/2} \sin^{2n} x\, dx = \frac{\pi}{2} \cdot \frac{1 \cdot 3 \cdot 5 \cdots (2n-1)}{2 \cdot 4 \cdot 6 \cdots (2n)}.$

Solution to Exercise 6.46.
Let $n \geq 2$. Integrating by parts, we have
$$\int_0^{\pi/2} \sin^n x\, dx = \int_0^{\pi/2} \sin^{n-1} x \cdot \sin x\, dx$$

$$= -\sin^{n-1} x \cos x \Big|_0^{\pi/2} + (n-1) \int_0^{\pi/2} \sin^{n-2} x \cos^2 x\, dx$$

$$= (n-1) \int_0^{\pi/2} \sin^{n-2} x\, dx - (n-1) \int_0^{\pi/2} \sin^n x\, dx.$$

Passing the last integral to the first member, we get

$$n \int_0^{\pi/2} \sin^n x\,dx = (n-1) \int_0^{\pi/2} \sin^{n-2} x\,dx,$$

that is,

$$\int_0^{\pi/2} \sin^n x\,dx = \frac{n-1}{n} \int_0^{\pi/2} \sin^{n-2} x\,dx.$$

Now we will prove each of the two final formulas by induction.

(I) For $n = 1$, using the previously proven formula, we have that

$$\int_0^{\pi/2} \sin^3 x\,dx = \frac{2}{3} \int_0^{\pi/2} \sin x\,dx = -\frac{2}{3} \cos x \Big|_0^{\pi/2} = \frac{2}{3}.$$

Suppose now that for a certain n we have

$$\int_0^{\pi/2} \sin^{2n+1} x\,dx = \frac{2 \cdot 4 \cdot 6 \cdots (2n)}{3 \cdot 5 \cdot 7 \cdots (2n+1)}.$$

Again using the formula from before, we conclude that

$$\int_0^{\pi/2} \sin^{2n+3} x\,dx = \frac{2n+2}{2n+3} \int_0^{\pi/2} \sin^{2n+1} x\,dx$$

$$= \frac{2 \cdot 4 \cdot 6 \cdots (2n)(2n+2)}{3 \cdot 5 \cdot 7 \cdots (2n+1)(2n+3)}.$$

(II) For $n = 1$, we have

$$\int_0^{\pi/2} \sin^2 x\,dx = \frac{1}{2} \int_0^{\pi/2} (1 - \cos 2x)\,dx$$

$$= \frac{1}{2} \left(x - \frac{1}{2} \sin 2x \right) \Big|_0^{\pi/2} = \frac{\pi}{2} \cdot \frac{1}{2}.$$

Supposing now that for a certain n we have

$$\int_0^{\pi/2} \sin^{2n} x\,dx = \frac{\pi}{2} \cdot \frac{1 \cdot 3 \cdot 5 \cdots (2n-1)}{2 \cdot 4 \cdot 6 \cdots (2n)},$$

according to the formula proved at the beginning, we obtain that

$$\int_0^{\pi/2} \sin^{2n+2} x\,dx = \frac{2n+1}{2n+2} \int_0^{\pi/2} \sin^{2n} x\,dx$$

$$= \frac{\pi}{2} \cdot \frac{1 \cdot 3 \cdot 5 \cdots (2n-1)(2n+1)}{2 \cdot 4 \cdot 6 \cdots (2n)(2n+2)}.$$

$\square$

Exercise 6.47. *Calculate the area of the figure bounded by the curve* $y^2 = x(x-1)^2$.

Solution to Exercise 6.47.
We can write this curve as $y = \pm\sqrt{x(x-1)^2} = \pm\sqrt{x}(x-1)$, so we see that this is the union of the graphics of two functions:

$$f_1(x) = \sqrt{x}(x-1), \qquad \text{and} \qquad f_2(x) = -\sqrt{x}(x-1).$$

These two functions are defined on $[0, \infty)$. Besides. $f_1(x) = f_2(x)$ if, and only if, $x = 0$ or $x = 1$. It is also seen that $f_1(x) > f_2(x)$ if, and only if, $x - 1 > 1 - x$, that is, $x > 1$. Finally, observe that

$$\lim_{x \to \infty} f_1(x) = +\infty \qquad \text{and} \qquad \lim_{x \to \infty} f_2(x) = -\infty.$$

Therefore, this curve will enclose two zones, one limited and the other unlimited (see Figure 6.5).

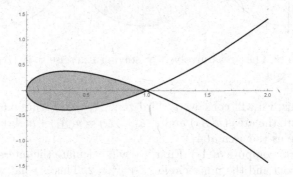

FIGURE 6.5: Area to be calculated in Exercise 6.47 and the curve that confines it.

Consequently, the required area is

$$A = \int_0^1 (f_2(x) - f_1(x))dx$$

$$= 2\int_0^1 \sqrt{x}(1-x)dx$$

$$= 2\int_0^1 (x^{1/2} - x^{3/2})dx$$

$$= \left(\frac{4}{3}x^{3/2} - \frac{4}{5}x^{5/2}\right)\Big|_0^1 = \frac{8}{15}.$$

$\square$

Exercise 6.48. *The annulus centered at the origin and with inner radius $\sqrt{2}$ and outer radius $\sqrt{6}$ is cut by the parabola of equation $x = y^2$. Find the area of any one of the two regions that are enclosed.*

Solution to Exercise 6.48.

The parabola divides the annulus into two zones, namely: Zone A_1 and Zone A_2 (see Figure 6.6). We will calculate the area of A_1, although, as we will see, this also allows us to know the area of A_2.

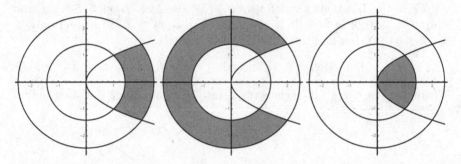

FIGURE 6.6: The areas involved in solving Exercise 6.48 (A_1, A_2, and A_3 from left to right).

To do this, we will consider a third zone A_3 (see Figure 6.6 as well). We shall calculate the areas $a(A_3)$ and $a(A_1 \cup A_3) = a(A_1) + a(A_3)$. Subtracting, this will give us the area $a(A_1)$.

Let us first compute $a(A_3)$. First, we will calculate the intersection points of the parabola and the inner circle $x^2 + y^2 = 2$. This is what we will get by analyzing the system

$$\begin{cases} x^2 + y^2 = 2, \\ x = y^2, \end{cases}$$

whose solutions are

$$(x, y) = (1, -1) \qquad \text{and} \qquad (x, y) = (1, 1).$$

The second coordinates of these vectors give us the limits of integration of the integral to perform. So

$$a(A_3) = \int_{-1}^{1} (\sqrt{2 - y^2} - y^2)\,dy$$

$$= \int_{-1}^{1} \sqrt{2 - y^2}\,dy - \frac{y^3}{3}\Big|_{-1}^{1}$$

$$= \int_{-1}^{1} \sqrt{2 - y^2}\,dy - \frac{2}{3}.$$

Considering $y = \sqrt{2}\sin u$ and $dy = \sqrt{2}\cos u\,du$, we get

$$\int_{-1}^{1} \sqrt{2 - y^2}\,dy = \int_{-\pi/4}^{\pi/4} \sqrt{2 - 2\sin^2 u} \cdot \sqrt{2}\cos u\,du$$

$$= 2\int_{-\pi/4}^{\pi/4} \cos^2 u\,du$$

$$= \int_{-\pi/4}^{\pi/4} (1 + \cos 2u)\,du$$

$$= \left(u + \frac{1}{2}\sin 2u\right)\Big|_{-\pi/4}^{\pi/4} = 1 + \frac{\pi}{2}.$$

Thus

$$a(A_3) = \left(1 + \frac{\pi}{2}\right) - \frac{2}{3} = \frac{1}{3} + \frac{\pi}{2}.$$

Now we will calculate the area $a(A_1 \cup A_3)$. We first calculate the intersection points of the parabola and the outer circumference $x^2 + y^2 = 6$, by solving the system

$$\begin{cases} x^2 + y^2 = 6, \\ x = y^2, \end{cases}$$

that has the solutions

$$(x, y) = (2, \sqrt{2}) \qquad \text{and} \qquad (x, y) = (2, -\sqrt{2}).$$

Therefore,

$$a(A_1 \cup A_3) = \int_{-\sqrt{2}}^{\sqrt{2}} (\sqrt{6 - y^2} - y^2)\,dy$$

$$= \int_{-\sqrt{2}}^{\sqrt{2}} \sqrt{6 - y^2}\,dy - \frac{y^3}{3}\Big|_{-\sqrt{2}}^{\sqrt{2}}$$

$$= \int_{-\sqrt{2}}^{\sqrt{2}} \sqrt{6 - y^2}\,dy - \frac{4\sqrt{2}}{3}.$$

Consider $y = \sqrt{6}\sin u$ and $dy = \sqrt{6}\cos u\,du$. Then

$$\int_{-\sqrt{2}}^{\sqrt{2}} \sqrt{6 - y^2}\,dy = \int_{-\arcsin\frac{\sqrt{3}}{3}}^{\arcsin\frac{\sqrt{3}}{3}} \sqrt{6 - 6\sin^2 u} \cdot \sqrt{6}\cos u\,du$$

$$= 6\int_{-\arcsin\frac{\sqrt{3}}{3}}^{\arcsin\frac{\sqrt{3}}{3}} \cos^2 u\,du$$

$$= 3\int_{-\arcsin\frac{\sqrt{3}}{3}}^{\arcsin\frac{\sqrt{3}}{3}} (1 + \cos 2u)\,du$$

$$= 3\left(u + \frac{1}{2}\sin 2u\right)\Big|_{-\arcsin\frac{\sqrt{3}}{3}}^{\arcsin\frac{\sqrt{3}}{3}}.$$

If $u = \arcsin \frac{\sqrt{3}}{3}$, then we have that

$$\sin 2u = 2\cos u \sin u$$

$$= 2\sin u\sqrt{1 - \sin^2 u}$$

$$= 2 \cdot \frac{\sqrt{3}}{3}\sqrt{1 - \frac{1}{3}} = \frac{2\sqrt{2}}{3}.$$

Therefore,

$$\int_{-\sqrt{2}}^{\sqrt{2}} \sqrt{6 - y^2}\,dy = 2\sqrt{2} + 6\arcsin\frac{\sqrt{3}}{3}$$

and consequently

$$a(A_1 \cup A_3) = \frac{2\sqrt{2}}{3} + 6\arcsin\frac{\sqrt{3}}{3}.$$

Finally,

$$a(A_1) = a(A_1 \cup A_3) - a(A_3)$$

$$= \left(\frac{2\sqrt{2}}{3} + 6\arcsin\frac{\sqrt{3}}{3}\right) - \left(\frac{1}{3} + \frac{\pi}{2}\right)$$

$$= \frac{2\sqrt{2} - 1}{3} + 6\arcsin\frac{\sqrt{3}}{3} - \frac{\pi}{2} \simeq 2.73.$$

In relation to area A_2, recall that the annulus has area

$$a(A_1 \cup A_2) = \pi((\sqrt{6})^2 - (\sqrt{2})^2) = 4\pi,$$

and so

$$a(A_2) = a(A_1 \cup A_2) - a(A_1)$$

$$= \frac{1 - 2\sqrt{2}}{3} - 6\arcsin\frac{\sqrt{3}}{3} + \frac{9\pi}{2} \simeq 9.83.$$

$\square$

Exercise 6.49. *Find the value of the parameter λ for which the curve $y = \lambda\cos x$ divides, into two parts of equal area, the region bounded by the x-axis, the curve $y = \sin x$ and the line $x = \pi/2$.*

Solution to Exercise 6.49.

If we represent $y = \lambda \cos x$ and $y = \sin x$, we obtain a graph as in Figure 6.7.

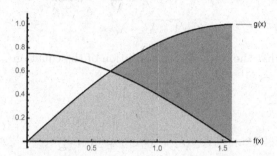

FIGURE 6.7: The two regions that appear in Exercise 6.49, where $f(x) = \lambda \cos x$ and $g(x) = \sin x$. Also, x_0 is x-intersection of both functions.

Note that zone A, bounded by $y = \sin x$, the x-axis and the line $x = \frac{\pi}{2}$, is divided by the graph of $y = \lambda \cos x$ into two other zones. Let us call these two zones A_1 (in light gray) and A_2 (in dark gray).

First, we calculate the area of A:

$$a(A) = \int_0^{\pi/2} \sin x dx = -\cos x \Big|_0^{\pi/2} = 1.$$

Therefore, we are looking for the λ for which the area of A_2 is equal to $1/2$. In order to find it, we must evaluate what is the area of A_2 for a generic λ. To do this, we must first calculate the intersection point of both graphs. If (x_0, y_0) is the point of intersection, it must satisfy

$$\sin x_0 = y_0 = \lambda \cos x_0,$$

and the area of A_2 will be

$$a(A_2) = \int_{x_0}^{\pi/2} (\sin x - \lambda \cos x) dx$$

$$= (-\cos x - \lambda \sin x) \Big|_{x_0}^{\pi/2}$$

$$= -\lambda + \cos x_0 + \lambda \sin x_0.$$

Observe that

$$1 = \sin^2 x_0 + \cos^2 x_0 = \lambda^2 \cos^2 x_0 + \cos^2 x_0 = (\lambda^2 + 1) \cos^2 x_0,$$

and thus

$$\cos x_0 = \frac{1}{\sqrt{1 + \lambda^2}} \qquad \text{and} \qquad \sin x_0 = \frac{\lambda}{\sqrt{1 + \lambda^2}}.$$

Therefore,

$$a(A_2) = -\lambda + \frac{1}{\sqrt{1+\lambda^2}} + \frac{\lambda^2}{\sqrt{1+\lambda^2}}$$
$$= \sqrt{1+\lambda^2} - \lambda,$$

and it follows that the λ we seek is a solution of the equation

$$\sqrt{1+\lambda^2} - \lambda = \frac{1}{2},$$

that is,

$$2\sqrt{1+\lambda^2} = 1 + 2\lambda.$$

By squaring,

$$4(1+\lambda^2) = (1+2\lambda)^2$$

whose only real solution is $\lambda = 3/4$.   □

Exercise 6.50. *Calculate the arc length of the curve $x = \frac{1}{4}y^2 - \frac{1}{2}\log y$ between the points $y = 1$ and $y = 2$.*

Solution to Exercise 6.50.

The problem asks to calculate the length of the graph of the function $f\colon [1,2] \to \mathbb{R}$, $f(y) = \frac{1}{4}y^2 - \frac{1}{2}\log y$, which is obtained as follows:

$$\text{arclength}(f,1,2) = \int_1^2 \sqrt{1 + f'(y)^2}\,dy$$

$$= \int_1^2 \sqrt{1 + \left(\frac{y}{2} - \frac{1}{2y}\right)^2}\,dy$$

$$= \frac{1}{2}\int_1^2 \frac{y^2+1}{y}\,dy$$

$$= \frac{1}{2}\left(\frac{y^2}{2} + \log y\right)\Big|_1^2 = \frac{3}{4} + \log\sqrt{2}.$$

□

Exercise 6.51. *Find the area of the surface generated by rotating around the y-axis the portion of the curve $y = x^2/2$ cut by the line $y = 3/2$.*

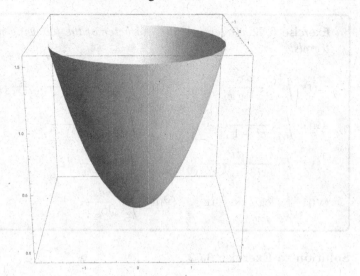

FIGURE 6.8: The surface of revolution generated by $y = \frac{x^2}{2}$ when we rotate it around the y-axis (Exercise 6.51).

Solution to Exercise 6.51.
Note that in this case we will have to consider the curve obtained as a function of y. We have two possible candidates: $x = \sqrt{2y}$ and $x = -\sqrt{2y}$, but when rotated they both generate the same surface, so we will consider only the function $f(y) = \sqrt{2y}$. The figure will be cut by the plane $y = 3/2$ and the limits of integration of the corresponding integral will be $y = 0$ and $y = 3/2$.

Then the searched area is

$$A = 2\pi \int_0^{3/2} f(y)\sqrt{1 + f'(y)^2}dy$$

$$= 2\pi \int_0^{3/2} \sqrt{2y}\sqrt{1 + \frac{1}{2y}}dy$$

$$= 2\pi \int_0^{3/2} \sqrt{2y + 1}dy.$$

Considering the change of variable $2y + 1 = u^2$, that is, $y = (u^2 - 1)/2$ and $dy = udu$, we obtain

$$A = 2\pi \int_1^2 u^2 du = \frac{2\pi}{3}u^3\Big|_1^2 = \frac{14}{3}\pi.$$

$\square$

Real Analysis

Exercise 6.52. *Determine the character of the following improper integrals:*

(I) $\displaystyle\int_0^{1/2} \frac{dx}{x \log x}.$ (II) $\displaystyle\int_1^2 \frac{dx}{(x^3 - 4x^2 + 4x)^{1/3}}.$

(III) $\displaystyle\int_0^\infty \frac{x^2}{x^4+1}dx.$ (IV) $\displaystyle\int_0^\infty \frac{dx}{(1+x^5)^{1/6}}.$

(V) $\displaystyle\int_0^3 \frac{dx}{(x(3-x))^{1/3}}.$ (VI) $\displaystyle\int_0^\infty \frac{x^2 e^{-x}}{1+x^2}dx.$

(VII) $\displaystyle\int_0^1 \log x \sin \frac{1}{x}dx.$ (VIII) $\displaystyle\int_2^\infty \frac{dx}{\log x}.$

Solution to Exercise 6.52.

(I) This integral has a single point of conflict, which is 0, since $\lim_{x\to 0+} x \log x = 0$. Consider the change of variable $u = -\log x$. Then $x = e^{-u}$ and $dx = -e^{-u}du$. So

$$\int_0^{1/2} \frac{dx}{x \log x} = \int_\infty^{\log 2} \frac{-e^{-u}du}{e^{-u}(-u)} = -\int_{\log 2}^\infty \frac{du}{u}.$$

This integral has the same character as the integral

$$\int_1^\infty \frac{dx}{x},$$

which diverges to ∞. Therefore,

$$\int_0^{1/2} \frac{dx}{x \log x} = -\infty.$$

(II) Observe that

$$\int_1^2 \frac{dx}{(x^3 - 4x^2 + 4x)^{1/3}} = \int_1^2 \frac{dx}{(x(x-2)^2)^{1/3}}.$$

Therefore, this integral has only one conflicting point in 2. When $x \to 2$, the integrand is equivalent to $1/(2(x-2))^{1/3}$. Therefore, this integral has the same character as

$$\int_1^2 \frac{dx}{(x-2)^{2/3}}.$$

Considering $x = 2 - u$ and $dx = -du$, we obtain that

$$\int_1^2 \frac{dx}{(x-2)^{2/3}} = \int_1^0 \frac{-du}{u^{2/3}} = \int_0^1 \frac{du}{u^{2/3}},$$

which is a convergent integral.

(III) The only conflicting point in this integral is ∞. This integral has the same character as the integral

$$\int_1^\infty \frac{x^2}{x^4+1}\,dx.$$

When $x \to \infty$, the integrand is equivalent to $x^2/x^4 = 1/x^2$. Therefore, the studied integral has the same character as the integral

$$\int_1^\infty \frac{dx}{x^2},$$

which is convergent.

Note that if we had dealt directly with the integral as given in the statement, the equivalence would have led us to the integral

$$\int_0^\infty \frac{dx}{x^2},$$

which diverges, because an additional conflicting point appears in it, the 0.

(IV) This integral has ∞ as its only conflicting point. Furthermore, it has the same character as the integral

$$\int_1^\infty \frac{dx}{(1+x^5)^{1/6}}.$$

On the other hand, when $x \to \infty$, the integrand is equivalent to $1/(x^5)^{1/6} = 1/x^{5/6}$. It follows that this last integral has the same character as

$$\int_1^\infty \frac{dx}{x^{5/6}},$$

which is divergent.

(V) This improper integral has two conflicting points, 0 and 3, since $\lim_{x\to 0+}(x(3-x))^{1/3} = 0$ and $\lim_{x\to 3-}(x(3-x))^{1/3} = 0$. This indicates that we should study this integral by dividing it in two, for example,

$$\int_0^1 \frac{dx}{(x(3-x))^{1/3}} \quad \text{and} \quad \int_1^3 \frac{dx}{(x(3-x))^{1/3}}.$$

In order for the integral of the statement to be convergent, the two integrals above must also be convergent.

Let us see the first one. This has 0 as the only conflicting point. It is easy to see that, when $x \to 0$, the integrand is equivalent to the function

$1/(3x)^{1/3}$. Therefore, the first of the two integrals has the same character as the integral

$$\int_0^1 \frac{dx}{x^{1/3}},$$

which is convergent.

Now let us study the other integral. It is also easy to see that when $x \to 3$, the integrand is equivalent to the function $1/(3(3-x))^{1/3}$. Therefore, the second integral has the same character as

$$\int_1^3 \frac{dx}{(3-x)^{1/3}}.$$

By changing the variable $x = 3 - u$ and $dx = -du$, we get

$$\int_1^3 \frac{dx}{(3-x)^{1/3}} = \int_2^0 \frac{-du}{u^{1/3}} = \int_0^2 \frac{du}{u^{1/3}},$$

which we know converges. So, the second integral also converges.

Combining the convergence of both integrals, we obtain that the integral of the statement is convergent.

(VI) The improper integral $\int_0^\infty e^{-x}dx$ is convergent. On the other hand, the function

$$\frac{x^2}{1+x^2} = 1 - \frac{1}{1+x^2}$$

is increasing and bounded on $[0, \infty)$. Then Abel's Criterion tells us that the integral of the statement converges.

(VII) This integral has 0 as its only conflicting point, because $\lim_{x \to 0+} \sin \frac{1}{x}$ does not exist and $\lim_{x \to 0+} \log x = -\infty$.

Note that we cannot use the Abel's Criterion: although $f(x) = \sin \frac{1}{x}$ is integrable in $[0, 1]$, and $g(x) = \log x$ is monotonic (increasing), this result is not applicable, as $g(x) = \log x$ is not bounded. Nor can we apply the Dirichlet's Criterion: one of the hypotheses (that the indefinite integral of $f(x) = \sin \frac{1}{x}$ is bounded) is fulfilled, but again we need $g(x) = \log x$ to be monotonous (which it is) and in addition that $\lim_{x \to 0} \log x = 0$ (which is false).

To solve this, let us change the variable $x = 1/u$ and $dx = -du/u^2$. Then

$$\int_0^1 \log x \sin \frac{1}{x} dx = -\int_\infty^1 (-\log u \sin u) \frac{du}{u^2}$$

$$= -\int_1^\infty \frac{\log u}{u^2} \sin u\, du.$$

This last integral has ∞ as the only conflicting point and let us see that we can apply Dirichlet Criterion. In fact, if $f(u) = \sin u$, then f is clearly locally integrable in $[1, \infty)$, and its indefinite integral is

$$F(t) = \int_1^t \sin u\, du = -\cos u \Big|_1^t = \cos 1 - \cos t,$$

which is bounded. On the other hand, let

$$g(u) = \frac{\log u}{u^2}.$$

Thus

$$g'(u) = \frac{\frac{1}{u} \cdot u^2 - 2u \log u}{u^4} = \frac{1 - 2\log u}{u^3}.$$

which takes a negative sign on $(\sqrt{e}, \infty)$. Therefore, g is decreasing on $[\sqrt{e}, \infty)$. Besides,

$$\lim_{u \to \infty} g(u) = \lim_{u \to \infty} \frac{\log u}{u^2} = 0.$$

Therefore, all the necessary hypotheses are fulfilled and, applying the Dirichlet's Criterion, the improper integral

$$\int_{\sqrt{e}}^{\infty} \frac{\log u}{u^2} \sin u\, du$$

converges and, therefore,

$$\int_1^{\infty} \frac{\log u}{u^2} \sin u\, du$$

also converges. Thus, we conclude that

$$\int_0^1 \log x \sin \frac{1}{x}\, dx$$

is convergent.

(VIII) This integral has ∞ as its only conflicting point. Let us make the change of variable $u = \log x$, that is, $x = e^u$ and $dx = e^u du$. So

$$\int_2^{\infty} \frac{dx}{\log x} = \int_{\log 2}^{\infty} \frac{e^u}{u}\, du,$$

which is clearly divergent, since $\lim_{u \to \infty} e^u/u = \infty$.

$\square$

Exercise 6.53. *Study the convergence of the integral*

$$\int_{-1}^{\infty} \frac{dx}{\sqrt{|x(1-x^2)|}}.$$

Solution to Exercise 6.53.

We can write this integral as

$$\int_{-1}^{\infty} \frac{dx}{\sqrt{|x(1+x)(1-x)|}}.$$

We note that this integral has four conflicting points, which are -1, 0, 1 and ∞. This tells us that we should study this integral by dividing it into six integrals:

$$\int_{-1}^{-\frac{1}{2}} \frac{dx}{\sqrt{|x(1+x)(1-x)|}}, \qquad \int_{-\frac{1}{2}}^{0} \frac{dx}{\sqrt{|x(1+x)(1-x)|}},$$

$$\int_{0}^{\frac{1}{2}} \frac{dx}{\sqrt{|x(1+x)(1-x)|}}, \qquad \int_{\frac{1}{2}}^{1} \frac{dx}{\sqrt{|x(1+x)(1-x)|}},$$

$$\int_{1}^{2} \frac{dx}{\sqrt{|x(1+x)(1-x)|}}, \qquad \int_{2}^{\infty} \frac{dx}{\sqrt{|x(1+x)(1-x)|}}.$$

We study each one of them separately.

(I) $\displaystyle \int_{-1}^{-\frac{1}{2}} \frac{dx}{\sqrt{|x(1+x)(1-x)|}}$

When $x \to -1$, the integrand is equivalent to $\frac{1}{\sqrt{2}\sqrt{1+x}}$. Consequently, the convergence of this integral is equivalent to the convergence of

$$\int_{-1}^{-\frac{1}{2}} \frac{dx}{\sqrt{1+x}}.$$

Let $x = u - 1$ and $dx = du$. Then

$$\int_{-1}^{-\frac{1}{2}} \frac{dx}{\sqrt{1+x}} = \int_{0}^{\frac{1}{2}} \frac{du}{\sqrt{u}} = 2\sqrt{u}\Big|_{0}^{1/2} = \sqrt{2}$$

and thus

$$\int_{-1}^{-1/2} \frac{dx}{\sqrt{|x(1+x)(1-x)|}} < \infty.$$

(II) $\displaystyle\int_{-\frac{1}{2}}^{0} \frac{dx}{\sqrt{|x(1+x)(1-x)|}}$

When $x \to 0$, the integrand is equivalent to $\frac{1}{\sqrt{-x}}$. Consequently, the convergence of this integral is equivalent to the convergence of

$$\int_{-\frac{1}{2}}^{0} \frac{dx}{\sqrt{-x}}.$$

Let $x = -u$ and $dx = -du$. Then

$$\int_{-\frac{1}{2}}^{0} \frac{dx}{\sqrt{-x}} = -\int_{\frac{1}{2}}^{0} \frac{du}{\sqrt{u}} = \int_{0}^{\frac{1}{2}} \frac{du}{\sqrt{u}} = \sqrt{2} \ .$$

and thus

$$\int_{-\frac{1}{2}}^{0} \frac{dx}{\sqrt{|x(1+x)(1-x)|}} < \infty.$$

(III) $\displaystyle\int_{0}^{1/2} \frac{dx}{\sqrt{|x(1+x)(1-x)|}}$

When $x \to 0$, the integrand is equivalent to $\frac{1}{\sqrt{x}}$. Consequently, the convergence of this integral is equivalent to the convergence of

$$\int_{0}^{\frac{1}{2}} \frac{dx}{\sqrt{x}}.$$

The above integral is $\sqrt{2}$ and thus

$$\int_{-\frac{1}{2}}^{0} \frac{dx}{\sqrt{|x(1+x)(1-x)|}} < \infty.$$

(IV) $\displaystyle\int_{\frac{1}{2}}^{1} \frac{dx}{\sqrt{|x(1+x)(1-x)|}}$

When $x \to 1$, the integrand is equivalent to $\frac{1}{\sqrt{2}\sqrt{1-x}}$. Consequently, the convergence of this integral is equivalent to the convergence of

$$\int_{\frac{1}{2}}^{1} \frac{dx}{\sqrt{1-x}}.$$

Let $x = 1 - u$ and $dx = -du$. Then

$$\int_{\frac{1}{2}}^{1} \frac{dx}{\sqrt{1-x}} = -\int_{\frac{1}{2}}^{0} \frac{du}{\sqrt{u}} = \int_{0}^{\frac{1}{2}} \frac{du}{\sqrt{u}} = \sqrt{2}$$

and thus

$$\int_{1/2}^{1} \frac{dx}{\sqrt{|x(1+x)(1-x)|}} < \infty.$$

(v) $\displaystyle\int_1^2 \frac{dx}{\sqrt{|x(1+x)(1-x)|}}$

When $x \to 1$, the integrand is equivalent to $\frac{1}{\sqrt{2}\sqrt{1-x}}$. Consequently, the convergence of this integral is equivalent to the convergence of

$$\int_1^2 \frac{dx}{\sqrt{1-x}}.$$

Let $x = 1 - u$ and $dx = -du$. Then

$$\int_1^2 \frac{dx}{\sqrt{1-x}} = -\int_1^2 \frac{du}{\sqrt{u}} = -2\sqrt{u}\big|_1^2 = -2\sqrt{2} + 2$$

and thus

$$\int_1^2 \frac{dx}{\sqrt{|x(1+x)(1-x)|}} < \infty.$$

(VI) $\displaystyle\int_2^\infty \frac{dx}{\sqrt{|x(1+x)(1-x)|}}$

When $x \to \infty$, the integrand is equivalent to $1/x^{3/2}$. Consequently, the convergence of this integral is equivalent to the convergence of

$$\int_2^\infty \frac{dx}{x^{3/2}}.$$

Observe that

$$\int_2^\infty \frac{dx}{x^{3/2}} = -\frac{2}{\sqrt{x}}\Big|_2^\infty = \sqrt{2}$$

and thus

$$\int_2^\infty \frac{dx}{\sqrt{|x(1+x)(1-x)|}} < \infty.$$

Since the above six integrals are convergent, the integral of the statement is also convergent.

□

Exercise 6.54. *Keeping in mind that the value of the Gamma function at $1/2$ is $\Gamma(1/2) = \sqrt{\pi}$, find the value of the following integrals:*

(I) $\displaystyle\int_0^\infty e^{-x^2}\,dx.$ (II) $\displaystyle\int_0^\infty x^2 e^{-x^2}\,dx.$

(III) $\displaystyle\int_0^\infty 3^{-4x^2}\,dx.$ (IV) $\displaystyle\int_0^\infty (x^2+1)e^{-\sqrt{x}}\,dx.$

(V) $\displaystyle\int_{-\infty}^\infty x^3 e^{-x^2}\,dx.$ (VI) $\displaystyle\int_0^\infty (x-3)e^{-x^2}\,dx.$

(VII) $\displaystyle\int_{-\infty}^\infty x^2 e^{-|x-1|}\,dx.$ (VIII) $\displaystyle\int_0^1 x^2 \log^4 x\,dx.$

Solution to Exercise 6.54.

Here, we shall use Exercise 6.22 together with the information provided in the statement of this question.

(I) We observe that if $f(x) = e^{-x^2}$, we have that $f(x) = f(-x)$, so we have symmetry with respect to the axis $x = 0$. Therefore,

$$\int_0^\infty e^{-x^2}\,dx = \frac{1}{2}\int_{-\infty}^\infty e^{-x^2}\,dx = \frac{1}{2}\sqrt{\pi}.$$

(II) Consider $x = \sqrt{t}$ and $dx = \frac{1}{2\sqrt{t}}dt$. Thus

$$\int_0^\infty x^2 e^{-x^2}\,dx = \int_0^\infty t e^{-t}\frac{1}{2\sqrt{t}}dt$$

$$= \frac{1}{2}\int_0^\infty t^{\frac{1}{2}} e^{-t}\,dt$$

$$= \frac{1}{2}\Gamma(3/2).$$

We know that $\Gamma(x+1) = x\Gamma(x)$, therefore $\Gamma\left(\frac{3}{2}\right) = \frac{1}{2}\Gamma\left(\frac{1}{2}\right) = \frac{\sqrt{\pi}}{2}$. So

$$\int_0^\infty x^2 e^{-x^2}\,dx = \frac{\sqrt{\pi}}{4}.$$

(III) Let $t = 4x^2\log 3$, that is, $x = \sqrt{\frac{t}{4\log 3}}$. Then $dx = \frac{1}{4\sqrt{t\log 3}}dt$ and consequently,

$$\int_0^\infty 3^{-4x^2}\,dx = \int_0^\infty e^{\log 3^{-4x^2}}\,dx = \int_0^\infty e^{-4x^2\log 3}\,dx$$

$$= \int_0^\infty e^{-t}\cdot t^{-1/2}\cdot\frac{1}{4\sqrt{\log 3}}dt = \frac{\sqrt{\pi}}{4\sqrt{\log 3}}.$$

(IV) Consider $x = t^2$ and $dx = 2tdt$. Then

$$\int_0^\infty (x^2 + 1)e^{-\sqrt{x}}dx = \int_0^\infty (t^4 + 1)e^{-t}2tdt = 2\int_0^\infty (t^5 e^{-t} + t \cdot e^{-t})dt$$

$$= 2\int_0^\infty t^5 e^{-t}dt + 2\int_0^\infty te^{-t}dt = 2(\Gamma(6) + \Gamma(2))$$

$$= 2(5! + 1!) = 2 \cdot 121$$

$$= 242.$$

(V) Let $f(x) = x^3 e^{-x^2}$. Then $f(-x) = -f(x)$, so f is an odd function. Therefore,

$$\int_{-\infty}^\infty x^3 e^{-x^2}\,dx = \int_{-\infty}^0 x^3 e^{-x^2}\,dx + \int_0^\infty x^3 e^{-x^2}\,dx$$

$$= -\int_0^\infty x^3 e^{-x^2}\,dx + \int_0^\infty x^3 e^{-x^2}\,dx$$

$$= 0.$$

(VI) Consider $x = \sqrt{t}$ and $dx = \frac{dt}{2\sqrt{t}}$. Thus

$$\int_0^\infty (x - 3)e^{-x^2}\,dx = \frac{1}{2}\int_0^\infty (t^{\frac{1}{2}} - 3)e^{-t} \cdot t^{-\frac{1}{2}}dt$$

$$= \frac{1}{2}\int_0^\infty e^{-t} \cdot dt - \frac{3}{2}\int_0^\infty e^{-t} \cdot t^{-\frac{1}{2}}dt$$

$$= \frac{1}{2}\Gamma(1) - \frac{3}{2}\Gamma(1/2) = \frac{1}{2} - \frac{3}{2}\sqrt{\pi}$$

$$= \frac{1 - 3\sqrt{\pi}}{2}.$$

(VII) Let $f(x) = x^2 e^{-|x-1|}$. Then

$$f(x) = \begin{cases} x^2 e^{1-x} & \text{if } x > 1, \\ x^2 e^{x-1} & \text{if } x < 1. \end{cases}$$

Thus we have

$$\int_{-\infty}^\infty x^2 e^{-|x-1|}dx = \int_{-\infty}^1 x^2 e^{x-1}dx + \int_1^\infty x^2 e^{1-x}dx.$$

Consider $t = x - 1$ and $dt = dx$. So

$$\int_{-\infty}^{1} x^2 e^{x-1} dx + \int_{1}^{\infty} x^2 e^{1-x} dx$$

$$= \int_{-\infty}^{0} (t+1)^2 e^t dt + \int_{0}^{\infty} (t+1)^2 e^{-t} dt$$

$$= \int_{-\infty}^{0} (t+1)^2 e^t dt + \int_{0}^{\infty} t^2 e^{-t} dt + \int_{0}^{\infty} 2t e^{-t} dt + \int_{0}^{\infty} e^{-t} dt$$

$$= \int_{-\infty}^{0} (t+1)^2 e^t dt + \Gamma(3) + 2\Gamma(2) + \Gamma(1). \tag{6.6}$$

Let us calculate the missing integral in (6.6). Integrating by parts twice, we get

$$\int_{-\infty}^{0} (t+1)^2 e^t dt = (t+1)^2 e^t \Big|_{-\infty}^{0} - 2 \int_{-\infty}^{0} (t+1) e^t dt$$

$$= (t+1)^2 e^t \Big|_{-\infty}^{0} - 2 \left((t+1) e^t \Big|_{-\infty}^{0} - \int_{-\infty}^{0} e^t dt \right)$$

$$= (t+1)^2 e^t \Big|_{-\infty}^{0} - 2 \left((t+1) e^t \Big|_{-\infty}^{0} - e^t \Big|_{-\infty}^{0} \right)$$

$$= e^t (t^2 + 1) \Big|_{-\infty}^{0}$$

$$= 1 - \lim_{t \to -\infty} e^t (t^2 + 1) = 1 - \lim_{t \to \infty} \frac{t^2}{e^t}$$

$$= 1 - 2 \lim_{t \to \infty} \frac{t}{e^t} = 1 - 2 \lim_{t \to \infty} \frac{1}{e^t}$$

$$= 1.$$

Above we applied L'Hôpital's Rule twice. Therefore,

$$\int_{-\infty}^{\infty} x^2 e^{-|x-1|} dx = 1 + \Gamma(3) + 2\Gamma(2) + \Gamma(1) = 1 + 2 + 2 + 1 = 6.$$

(VIII) Let $t = \log x$. Then $e^t = x$ and $e^t dt = dx$. Observe that $0 = e^t \Rightarrow t = -\infty$, and $1 = e^t \Rightarrow t = 0$. Then

$$\int_0^1 x^2 \log^4 x\, dx = \int_{-\infty}^0 e^{2t} t^4 e^t dt = \int_{-\infty}^0 e^{3t} t^4 dt$$

$$= \frac{1}{3} t^4 e^{3t} \Big|_{-\infty}^0 - \frac{4}{3} \int_{-\infty}^0 t^3 e^{3t} dt$$

$$= \frac{1}{3} t^4 e^{3t} \Big|_{-\infty}^0 - \frac{4}{3} \left(\frac{1}{3} t^3 e^{3t} \Big|_{-\infty}^0 - \int_{-\infty}^0 t^2 e^{3t} dt \right)$$

$$= \frac{1}{3} t^4 e^{3t} \Big|_{-\infty}^0 - \frac{4}{9} t^3 e^{3t} \Big|_{-\infty}^0$$

$$+ \frac{4}{3} \left(t^2 \frac{1}{3} e^{3t} \Big|_{-\infty}^0 - \frac{2}{3} \int_{-\infty}^0 t e^{3t} dt \right)$$

$$= \frac{1}{3} t^4 e^{3t} \Big|_{-\infty}^0 - \frac{4}{9} t^3 e^{3t} \Big|_{-\infty}^0 + \frac{4}{3} t^2 e^{3t} \Big|_{-\infty}^0$$

$$- \frac{8}{9} \left(\frac{1}{3} t e^{3t} \Big|_{-\infty}^0 - \frac{1}{3} \int_{-\infty}^0 e^{3t} dt \right)$$

$$= \frac{1}{3} t^4 e^{3t} \Big|_{-\infty}^0 - \frac{4}{9} t^3 e^{3t} \Big|_{-\infty}^0 + \frac{4}{3} t^2 e^{3t} \Big|_{-\infty}^0$$

$$- \frac{8}{27} t e^{3t} \Big|_{-\infty}^0 + \frac{8}{81} e^{3t} \Big|_{-\infty}^0$$

$$= \frac{8}{81} - \lim_{t \to -\infty} \left(\frac{1}{3} t^4 e^{3t} - \frac{4}{9} t^3 e^{3t} + \frac{4}{3} t^2 e^{3t} - \frac{8}{27} t e^{3t} + \frac{8}{81} e^{3t} \right).$$

Since $\lim_{t \to -\infty} t^k e^{mt} = 0$ for every $m, k \in \mathbb{N}$ (this can be proved by using, for instance, L'Hôpital's Rule repeatedly on $\lim_{t \to -\infty} \dfrac{t^k}{e^{-mt}}$) we obtain that the above limit is zero. Thus

$$\int_0^1 x^2 \log^4 x\, dx = \frac{8}{81}.$$

$\square$

Exercise 6.55. *Find an expression for the derivative of*

$$\int_{5x^3}^{7x^9} \sin t\, dt.$$

Solution to Exercise 6.55.

Let $f(x) = \int_{5x^3}^{7x^9} \sin t \, dt$. Then

$$f'(x) = \sin(7x^9) \cdot 63x^8 - \sin(5x^3) \cdot 15x^2.$$

$\square$

Exercise 6.56. *Let*

$$f(x) = \int_{e^x}^{e^{\frac{x}{5}}} t^4 dt.$$

Solve the equation $f'(x) = 0$.

Solution to Exercise 6.56.
Observe that

$$f'(x) = \left(e^{\frac{x}{5}}\right)^4 \cdot \frac{1}{5}e^{\frac{x}{5}} - (e^x)^4 \cdot e^x = \frac{1}{5}e^x - e^{5x}.$$

Thus

$$f'(x) = 0 \Rightarrow \frac{1}{5}e^x - e^{5x} = 0 \Rightarrow e^x \cdot \left(\frac{1}{5} - e^{4x}\right) = 0.$$

Since $e^x \neq 0$, it follows that

$$\frac{1}{5} - e^{4x} = 0 \Rightarrow e^{4x} = \frac{1}{5} \Rightarrow x = \frac{1}{4}\log\frac{1}{5}.$$

$\square$

Exercise 6.57. *Calculate*

$$\lim_{x \to 0} \frac{\int_x^{2x} t^2 \cos t \, dt}{x^3}.$$

Solution to Exercise 6.57.
It is an indeterminate expression of the type $\frac{0}{0}$, so we can use L'Hôpital's Rule. We first calculate the derivative of the numerator:

$$\left(\int_x^{2x} t^2 \cos t \, dt\right)' = (2x)^2 \cos(2x) \cdot 2 - x^2 \cos x = 8x^2 \cos(2x) - x^2 \cdot \cos x.$$

Therefore, applying L'Hôpital's Rule we have

$$\lim_{x \to 0} \frac{\int_x^{2x} t^2 \cos t \, dt}{x^3} = \lim_{x \to 0} \frac{8x^2 \cos(2x) - x^2 \cdot \cos x}{3x^2} = \lim_{x \to 0} \frac{8\cos(2x) - \cos x}{3} = \frac{7}{3}.$$

$\square$

Exercise 6.58. *Calculate:*

$$\lim_{n\to\infty}\left(\frac{1}{\sqrt{n^4+1}}+\frac{2}{\sqrt{n^4+16}}+\cdots+\frac{n}{\sqrt{2n^4}}\right).$$

Solution to Exercise 6.58.

Let us write the limit in the following way:

$$\lim_{n\to\infty}\left(\frac{1}{\sqrt{n^4+1}}+\frac{2}{\sqrt{n^4+16}}+\cdots+\frac{n}{\sqrt{2n^4}}\right)$$

$$=\lim_{n\to\infty}\sum_{k=1}^{n}\frac{k}{\sqrt{n^4+k^4}}$$

$$=\lim_{n\to\infty}\sum_{k=1}^{n}\frac{k}{\sqrt{n^4\left(1+\left(\frac{k}{n}\right)^4\right)}}$$

$$=\lim_{n\to\infty}\sum_{k=1}^{n}\frac{k}{n^2\sqrt{1+\left(\frac{k}{n}\right)^4}}$$

$$=\lim_{n\to\infty}\frac{1}{n}\sum_{k=1}^{n}\frac{k/n}{\sqrt{1+\left(\frac{k}{n}\right)^4}}.$$

The summation above is a Riemann sum and therefore

$$\lim_{n\to\infty}\left(\frac{1}{\sqrt{n^4+1}}+\frac{2}{\sqrt{n^4+16}}+\cdots+\frac{n}{\sqrt{2n^4}}\right)=\lim_{n\to\infty}\frac{1}{n}\sum_{k=1}^{n}\frac{k/n}{\sqrt{1+\left(\frac{k}{n}\right)^4}}$$

$$=\int_0^1\frac{x}{\sqrt{1+x^4}}dx.$$

For the integrand $x/\sqrt{1+x^4}$ substitute $u=x^2$ and $du=2xdx$. Thus

$$\lim_{n\to\infty}\left(\frac{1}{\sqrt{n^4+1}}+\frac{2}{\sqrt{n^4+16}}+\cdots+\frac{n}{\sqrt{2n^4}}\right)=\int_0^1\frac{x}{\sqrt{1+x^4}}dx$$

$$=\int_0^1\frac{1}{2}\frac{du}{\sqrt{1+u^2}}$$

$$=\frac{1}{2}(\operatorname{arcsinh}u)\Big|_0^1$$

$$=\frac{\operatorname{arcsinh}1}{2}.$$

□

Exercise 6.59. *Study the convergence of the following integrals and, if they converge, calculate their value:*

$$\int_1^\infty \frac{e^{-x}}{1+e^x}dx.$$

Solution to Exercise 6.59.

We will use the comparison criterion since on $[1,\infty)$ we have

$$\frac{e^{-x}}{1+e^x} < e^{-x}.$$

Since

$$\int_1^\infty e^{-x}dx = \lim_{t\to\infty}\left(-e^{-x}\right)\Big|_1^t$$

$$= \lim_{t\to\infty}\left(-e^{-t}+e^{-1}\right) = \frac{1}{e} < \infty,$$

we can conclude that $\int_1^\infty \frac{e^{-x}}{1+e^x}dx$ is convergent.

Let us then calculate the value of the integral. For the integrand $e^{-x}/(e^x+1)$, substitute $u = e^x$ and $du = e^x dx$. This gives a new lower limit $u = e$ and upper limit $u = \infty$. Therefore,

$$\int_1^\infty \frac{e^{-x}}{1+e^x}dx = \int_e^\infty \frac{1}{u^2(u+1)}du.$$

For the integrand $1/(u^2(u+1))$, use simple fractions to get

$$\int_1^\infty \frac{e^{-x}}{1+e^x}dx = \int_e^\infty \frac{1}{u^2(u+1)}du$$

$$= \int_e^\infty \left(\frac{1}{u^2} + \frac{1}{u+1} - \frac{1}{u}\right)du$$

$$= \lim_{b\to\infty}\left(-\frac{1}{u} + \log|u+1| - \log|u|\right)\Big|_e^b$$

$$= \lim_{b\to\infty}\left(-\frac{1}{b} + \log|b+1| - \log|b|\right) + \frac{1}{e} - \log(1+e) + 1$$

$$= \frac{1}{e} - \log(1+e) + 1.$$

$\square$

Chapter 7

Numerical Series

Exercise 7.1. *Study the following series by expressing them as telescopic series:*

(I) $\displaystyle\sum_{n=1}^{\infty} \frac{1}{2^n} \cdot \frac{n+2}{n(n+1)}$. *Hint: separate $\frac{n+2}{n(n+1)}$ into simple fractions.*

(II) $\displaystyle\sum_{n=1}^{\infty} 3^n \sin^3 \frac{a}{3^n}$ *with $a \in \mathbb{R}$. Hint: take into account the trigonometric formula $\sin x = 3\sin \dfrac{x}{3} - 4\sin^3 \dfrac{x}{3}$.*

(III) $\displaystyle\sum_{n=1}^{\infty} 2^{n-1} \tan^2 \frac{a}{2^n} \tan \frac{a}{2^{n-1}}$ *with $0 < a < \frac{\pi}{2}$. Hint: use the trigonometric formula $\tan x = \dfrac{2\tan\frac{x}{2}}{1-\tan^2\frac{x}{2}}$.*

(IV) $\displaystyle\sum_{n=1}^{\infty} \sin \frac{1}{2^n} \cos \frac{3}{2^n}$. *Hint: take into account the trigonometric formula $\cos x \sin y = \dfrac{1}{2}(\sin(x+y) - \sin(x-y))$.*

(V) $\displaystyle\sum_{n=1}^{\infty} \frac{1}{4^n \cos^2 \frac{x}{2^n}}$ *with $0 < x < \frac{\pi}{2}$. Hint: use the trigonometric formula $\dfrac{1}{4\cos^2 a} = \dfrac{1}{\sin^2 2a} - \dfrac{1}{4\sin^2 a}$.*

Solution to Exercise 7.1.

(I). Let us separate $\frac{n+2}{n(n+1)}$ into simple fractions:

$$\frac{n+2}{n(n+1)} = \frac{A}{n} + \frac{B}{n+1} \Rightarrow n+2 = A(n+1) + Bn$$

$$\Rightarrow \begin{cases} A+B = 1 \\ A = 2 \end{cases} \Rightarrow \begin{cases} A = 2, \\ B = -1. \end{cases}$$

DOI: 10.1201/9781003400745-7

Therefore,

$$\sum_{n=1}^{\infty} \frac{1}{2^n} \cdot \frac{n+2}{n(n+1)} = \sum_{n=1}^{\infty} \frac{1}{2^n} \left(\frac{2}{n} - \frac{1}{n+1} \right)$$

$$= \sum_{n=1}^{\infty} \left(\frac{1}{2^{n-1}n} - \frac{1}{2^n(n+1)} \right).$$

By taking $b_n = \frac{1}{2^{n-1}n}$, we have a telescopic series. Hence,

$$\sum_{n=1}^{\infty} \frac{1}{2^n} \cdot \frac{n+2}{n(n+1)} = b_1 - \lim_{n\to\infty} b_n = 1 - \lim_{n\to\infty} \frac{1}{2^{n-1}n} = 1.$$

(II). Taking into account that $\sin x = 3\sin\frac{x}{3} - 4\sin^3\frac{x}{3}$, we have that

$$\sin^3\frac{x}{3} = \frac{\sin x - 3\sin\frac{x}{3}}{-4}.$$

Denoting $x = a3^{1-n}$, we have

$$\sum_{n=1}^{\infty} 3^n \sin^3\frac{a}{3^n} = \sum_{n=1}^{\infty} 3^n \left(\frac{\sin\frac{a}{3^{n-1}} - 3\sin\frac{a}{3^n}}{-4} \right)$$

$$= -\frac{1}{4} \sum_{n=1}^{\infty} \left(3^n \sin\frac{a}{3^{n-1}} - 3^{n+1}\sin\frac{a}{3^n} \right).$$

If we take $b_n = 3^n \sin\frac{a}{3^{n-1}}$ we have a telescopic series. Thus,

$$\sum_{n=1}^{\infty} 3^n \sin^3\frac{a}{3^n} = -\frac{1}{4}\left(b_1 - \lim_{n\to\infty} 3^n \sin\frac{a}{3^{n-1}} \right).$$

Recall that if a sequence a_n converges to 0, then $\sin a_n \sim a_n$. Hence, we have

$$\sum_{n=1}^{\infty} 3^n \sin^3\frac{a}{3^n} = -\frac{1}{4}\left(3\sin a - \lim_{n\to\infty} 3^n \frac{a}{3^{n-1}} \right)$$

$$= -\frac{1}{4}(3\sin a - 3a)$$

$$= \frac{3}{4}(a - \sin a).$$

(III). Taking into account that $\tan x = \frac{2\tan\frac{x}{2}}{1 - \tan^2\frac{x}{2}}$, we have that

$$\tan x - \tan x \tan^2\frac{x}{2} = 2\tan\frac{x}{2} \Leftrightarrow \tan x - 2\tan\frac{x}{2} = \tan x \tan^2\frac{x}{2}.$$

Denoting $x = \dfrac{a}{2^{n-1}}$ we obtain:

$$\sum_{n=1}^{\infty} 2^{n-1} \tan^2 \frac{a}{2^n} \tan \frac{a}{2^{n-1}} = \sum_{n=1}^{\infty} 2^{n-1} \left(\tan \frac{a}{2^{n-1}} - 2 \tan \frac{a}{2^n} \right)$$

$$= \sum_{n=1}^{\infty} \left(2^{n-1} \tan \frac{a}{2^{n-1}} - 2^n \tan \frac{a}{2^n} \right).$$

By taking $b_n = 2^{n-1} \tan \frac{a}{2^{n-1}}$, we notice that we have a telescopic series. So

$$\sum_{n=1}^{\infty} 2^{n-1} \tan^2 \frac{a}{2^n} \tan \frac{a}{2^{n-1}} = b_1 - \lim_{n \to \infty} b_n$$

$$= \tan a - \lim_{n \to \infty} 2^{n-1} \tan \frac{a}{2^{n-1}}.$$

Recall that $\tan a_n \sim a_n$ as a_n converges to 0. Then,

$$\sum_{n=1}^{\infty} 2^{n-1} \tan^2 \frac{a}{2^n} \tan \frac{a}{2^{n-1}} = \tan a - \lim_{n \to \infty} 2^{n-1} \frac{a}{2^{n-1}} = \tan a - a.$$

(IV). Taking into account that $\cos x \sin y = \frac{1}{2}(\sin(x+y) - \sin(x-y))$, we have

$$\sum_{n=1}^{\infty} \sin \frac{1}{2^n} \cos \frac{3}{2^n} = \sum_{n=1}^{\infty} \frac{1}{2} \left(\sin \left(\frac{3}{2^n} + \frac{1}{2^n} \right) - \sin \left(\frac{3}{2^n} - \frac{1}{2^n} \right) \right)$$

$$= \sum_{n=1}^{\infty} \frac{1}{2} \left(\sin \frac{1}{2^{n-2}} - \sin \frac{1}{2^{n-1}} \right).$$

By making $b_n = \frac{1}{2} \sin \frac{1}{2^{n-2}}$, we have (using that $\sin a_n \sim a_n$ as a_n converges to 0)

$$\sum_{n=1}^{\infty} (b_n - b_{n+1}) = b_1 - \lim_{n \to \infty} b_n$$

$$= \frac{\sin 2}{2} - \lim_{n \to \infty} \frac{1}{2} \sin \frac{1}{2^{n-2}}$$

$$= \frac{\sin 2}{2} - \lim_{n \to \infty} \frac{1}{2 \cdot 2^{n-2}}$$

$$= \frac{\sin 2}{2}.$$

(v). Taking into account that $\frac{1}{4 \cos^2 a} = \frac{1}{\sin^2 2a} - \frac{1}{4 \sin^2 a}$, we have

$$\frac{1}{\cos^2 \frac{x}{2^n}} = \frac{4}{\sin^2 \frac{x}{2^{n-1}}} - \frac{1}{\sin^2 \frac{x}{2^n}}.$$

$$\sum_{n=1}^{\infty} \frac{1}{4^n \cos^2 \frac{x}{2^n}} = \sum_{n=1}^{\infty} \frac{1}{4^n} \left(\frac{4}{\sin^2 \frac{x}{2^{n-1}}} - \frac{1}{\sin^2 \frac{x}{2^n}} \right)$$

$$= \sum_{n=1}^{\infty} \left(\frac{1}{2^{2(n-1)} \sin^2 \frac{x}{2^{n-1}}} - \frac{1}{2^{2n} \sin^2 \frac{x}{2^n}} \right).$$

Letting $b_n = \dfrac{1}{2^{2(n-1)} \sin^2 \frac{x}{2^{n-1}}}$, we have (since $\sin a_n \sim a_n$ if $a_n \sim 0$)

$$\sum_{n=1}^{\infty} (b_n - b_{n+1}) = b_1 - \lim_{n \to \infty} b_n = \frac{1}{\sin^2 x} - \lim_{n \to \infty} \frac{1}{2^{2(n-1)} \sin^2 \frac{x}{2^{n-1}}}$$

$$= \frac{1}{\sin^2 x} - \lim_{n \to \infty} \frac{1}{2^{2(n-1)} \left(\frac{x}{2^{n-1}} \right)^2}$$

$$= \frac{1}{\sin^2 x} - \frac{1}{x^2}.$$

$\square$

Exercise 7.2. *Study the convergence of the following series:*

(I) $\displaystyle\sum_{n=1}^{\infty} (\sqrt{n+1} - \sqrt{n})$,

(II) $\displaystyle\sum_{n=1}^{\infty} \frac{\sin^4 n}{n^2}$,

(III) $\displaystyle\sum_{n=1}^{\infty} \frac{1}{\sqrt{n} - \frac{2}{3}}$,

(IV) $\displaystyle\sum_{n=1}^{\infty} \frac{1+n^2}{n!}$,

(V) $\displaystyle\sum_{n=1}^{\infty} \cos^n \left(a + \frac{b}{n} \right)$ *with* $b \in \mathbb{R}$ *and* $0 < a < \frac{\pi}{2}$,

(VI) $\displaystyle\sum_{n=1}^{\infty} \frac{n^2+1}{na^n}$ *with* $a \neq 0$,

(VII) $\displaystyle\sum_{n=1}^{\infty} \frac{n!}{n^n}$,

(VIII) $\displaystyle\sum_{n=1}^{\infty} \frac{3^n}{n^2+1}$,

(IX) $\displaystyle\sum_{n=1}^{\infty} \left(\frac{n+1}{n} \right)^{-n^3}$,

(X) $\displaystyle\sum_{n=1}^{\infty} (\sqrt[n]{n} - 1)^n$,

(XI) $\displaystyle\sum_{n=1}^{\infty} \frac{1}{\log n}$,

(XII) $\displaystyle\sum_{n=1}^{\infty} \frac{1}{na+b}$ *with* $(a,b) \neq (0,0)$,

(XIII) $\displaystyle\sum_{n=1}^{\infty} \frac{\sin nx}{n^2}$.

Solution to Exercise 7.2.

(I). By taking $a_n = -\sqrt{n}$, we have a telescopic series. So

$$\sum_{n=1}^{\infty}(\sqrt{n+1}-\sqrt{n}) = a_1 - \lim_{n\to\infty} a_n = -1 + \lim_{n\to\infty} \sqrt{n} = \infty,$$

that is, the series is divergent.

(II). Denoting $a_n = \frac{\sin^4 n}{n^2}$ and $b_n = \frac{1}{n^2}$, we have that $0 \le a_n \le b_n$ for every $n \in \mathbb{N}$. Therefore, by the Comparison Test, as $\sum_{n=1}^{\infty} \frac{1}{n^2}$ converges (p-series with $p = 2 > 1$) and $0 \le a_n \le b_n$ for all $n \in \mathbb{N}$, we have that $\sum_{n=1}^{\infty} \frac{\sin^4 n}{n^2}$ converges.

(III). As $a_n = \frac{1}{\sqrt{n}-\frac{2}{3}} > \frac{1}{\sqrt{n}} > \frac{1}{n} = b_n$ and $0 \le b_n \le a_n$ for every $n \in \mathbb{N}$. By the Comparison Test, since $\sum_{n=1}^{\infty} \frac{1}{n}$ diverges, we have that $\sum_{n=1}^{\infty} \frac{1}{\sqrt{n}-\frac{2}{3}}$ is also divergent.

(IV). In order to study the convergence, let us apply the Ratio Test:

$$\lim_{n\to\infty} \frac{\frac{1+(n+1)^2}{(n+1)!}}{\frac{1+n^2}{n!}} = \lim_{n\to\infty} \frac{1+(n+1)^2}{(1+n^2)(n+1)} = \lim_{n\to\infty} \frac{n^2+2n+2}{n^3+n^2+n+1} = 0 < 1.$$

Therefore, by the Ratio Test, the series converges.

(V). In this case, we will make use of the Root Test:

$$\lim_{n\to\infty} \sqrt[n]{\left|\cos^n\left(a+\frac{b}{n}\right)\right|} = \lim_{n\to\infty} \left|\cos\left(a+\frac{b}{n}\right)\right| = |\cos a| < 1.$$

Therefore, by the Root Test, the series converges (in particular, it is absolutely convergent).

(VI). Let us apply the Ratio Test:

$$\lim_{n\to\infty} \frac{\frac{(n+1)^2+1}{(n+1)a^{n+1}}}{\frac{n^2+1}{na^n}} = \lim_{n\to\infty} \frac{(n^2+2n+2)n}{(n+1)a(n^2+1)}$$

$$= \lim_{n\to\infty} \frac{n^3+2n^2+2n}{an^3+an^2+an+a} = \frac{1}{a}.$$

So that, if $\frac{1}{a} > 1$ by the Ratio Test the series diverges, and if $\frac{1}{a} < 1$ by the Ratio Test the series converges. However, if $a = 1$, the Ratio Test is inconclusive, so we must try another method in this latter case.

Observe that in the particular that $a = 1$, we have the series

$$\sum_{n=1}^{\infty} \frac{n^2+1}{n}$$

which satisfies

$$\lim_{n\to\infty} \frac{n^2+1}{n} = \infty.$$

Hence, by the Divergence Test, the series diverges.

(VII). Let us apply the Root Test:

$$\lim_{n\to\infty} \sqrt[n]{\frac{n!}{n^n}} = \lim_{n\to\infty} \frac{\sqrt[n]{n!}}{n}.$$

By the weak form of Stirling's Formula (that is, $\sqrt[n]{n!} \sim n/e$):

$$\lim_{n\to\infty} \sqrt[n]{\frac{n!}{n^n}} = \lim_{n\to\infty} \frac{\frac{n}{e}}{n} = \frac{1}{e} < 1.$$

Therefore, by the Root Test, the series converges.

(VIII). We will apply the Ratio Test:

$$\lim_{n\to\infty} \frac{\frac{3^{n+1}}{(n+1)^2+1}}{\frac{3^n}{n^2+1}} = \lim_{n\to\infty} \frac{3(n^2+1)}{n^2+2n+2} = 3 > 1.$$

Therefore, by the Ratio Test, the series diverges.

(IX). Let us apply the Root Test:

$$\lim_{n\to\infty} \sqrt[n]{\left(\frac{n+1}{n}\right)^{-n^3}} = \lim_{n\to\infty} \left(1+\frac{1}{n}\right)^{-n^2}$$

$$= \lim_{n\to\infty} \left[\left(1+\frac{1}{n}\right)^n\right]^{-n}$$

$$= \lim_{n\to\infty} e^{-n} = 0 < 1.$$

Therefore, by the Root Test, the series converges.

(X). We will apply the Root Test:

$$\lim_{n\to\infty} \sqrt[n]{(\sqrt[n]{n}-1)^n} = \lim_{n\to\infty} (\sqrt[n]{n}-1) = 0 < 1.$$

Therefore, by the Root Test, the series converges.

(XI). As $0 \le b_n = \frac{1}{n} \le \frac{1}{\log n} = a_n$ for all $n \in \mathbb{N}$ with $n > 1$ and $\sum_{n=1}^{\infty} \frac{1}{n}$.

diverges, we have by the Direct the Comparison Test that $\sum_{n=1}^{\infty} \frac{1}{\log n}$ also diverges.

(XII). If $a = 0$ and $b \neq 0$ the series diverges. So, assume that $a \neq 0$. Then,

$$\sum_{n=1}^{\infty} \frac{1}{na+b} = \frac{1}{a} \sum_{n=1}^{\infty} \frac{1}{n+\frac{b}{a}}.$$

So that, for some $N \in \mathbb{N}$ we have that $\frac{1}{n+\frac{b}{a}} > 0$ for every integer $n \geq N$. Moreover,

$$\lim_{n \to \infty} \frac{\frac{1}{n}}{\frac{1}{n+\frac{b}{a}}} = 1.$$

Therefore, by the Limit Comparison Test, as $\sum_{n=1}^{\infty} \frac{1}{n}$ diverges, we have that

$\sum_{n=1}^{\infty} \frac{1}{na+b}$ also diverges.

(XIII). As $0 \leq |a_n| = \left| \frac{\sin nx}{n^2} \right| \leq b_n = \frac{1}{n^2}$ for every $n \in \mathbb{N}$ and $\sum_{n=1}^{\infty} \frac{1}{n^2}$ converges (p-series with $p = 2 > 1$), we have by the Comparison Test that $\sum_{n=1}^{\infty} \frac{\sin nx}{n^2}$ converges.

$\square$

Exercise 7.3. *Study the convergence of the following series:*

(I) $\displaystyle\sum_{n=1}^{\infty} \frac{1}{n - \frac{3}{2}},$

(VII) $\displaystyle\sum_{n=1}^{\infty} \left(\frac{1}{n} \right)^{\frac{n+1}{n}},$

(II) $\displaystyle\sum_{n=1}^{\infty} \frac{1}{n(n+1)(n+2)},$

(VIII) $\displaystyle\sum_{n=1}^{\infty} \frac{1}{1 + a^n}$ *with* $a \neq -1,$

(III) $\displaystyle\sum_{n=1}^{\infty} \frac{1 + \sin^2 nx}{n^2}$ *with* $x \in \mathbb{R},$

(IX) $\displaystyle\sum_{n=1}^{\infty} \frac{\log(n+1) - 1}{(1+n)^2},$

(IV) $\displaystyle\sum_{n=1}^{\infty} \frac{1}{n} \sin \frac{1}{n},$

(X) $\displaystyle\sum_{n=1}^{\infty} \frac{1}{3 - \cos \frac{1}{n}},$

(V) $\displaystyle\sum_{n=1}^{\infty} \frac{\sqrt{n+1} - \sqrt{n}}{n},$

(XI) $\displaystyle\sum_{n=1}^{\infty} \left(\frac{x}{n} \right)^n n!$ *with* $x \in \mathbb{R},$

(VI) $\displaystyle\sum_{n=1}^{\infty} \frac{n(n+1)}{n^2 + 2n},$

(XII) $\displaystyle\sum_{n=1}^{\infty} \frac{1}{n(1 + \frac{1}{2} + \cdots + \frac{1}{n})}.$

Solution to Exercise 7.3.

(I). By the Limit Comparison Test, since

$$\lim_{n \to \infty} \frac{\frac{1}{n}}{\frac{1}{n-3/2}} = 1,$$

we have that $\displaystyle\sum_{n=1}^{\infty} \frac{1}{n - \frac{3}{2}}$ diverges as $\displaystyle\sum_{n=1}^{\infty} \frac{1}{n}$ also diverges (harmonic series).

(II). As

$$\lim_{n \to \infty} \frac{\frac{1}{n^3}}{\frac{1}{n(n+1)(n+2)}} = 1,$$

we have by the Limit Comparison Test that $\displaystyle\sum_{n=1}^{\infty} \frac{1}{n(n+1)(n+2)}$ converges since $\displaystyle\sum_{n=1}^{\infty} \frac{1}{n^3}$ converges (p-series with $p = 3 > 1$).

(III). Since

$$0 \le \frac{1 + \sin^2 nx}{n^2} \le \frac{2}{n^2},$$

we have by the Comparison Test that $\displaystyle\sum_{n=1}^{\infty} \frac{1 + \sin^2 nx}{n^2}$ converges as $\displaystyle\sum_{n=1}^{\infty} \frac{1}{n^2}$ converges (p-series with $p = 2 > 1$).

(IV). Observe that

$$\lim_{n \to \infty} \frac{\frac{1}{n} \sin \frac{1}{n}}{\frac{1}{n^2}} = \lim_{n \to \infty} \frac{\sin \frac{1}{n}}{\frac{1}{n}}.$$

Then, since $\sin a_n \sim a_n$ provided that (a_n) converges to 0, we have that

$$\lim_{n \to \infty} \frac{\frac{1}{n} \sin \frac{1}{n}}{\frac{1}{n^2}} = \lim_{n \to \infty} \frac{\frac{1}{n}}{\frac{1}{n}} = 1.$$

Hence, by the Limit Comparison Test, as $\displaystyle\sum_{n=1}^{\infty} \frac{1}{n^2}$ converges (p-series with $p = 2 > 1$), we obtain that $\displaystyle\sum_{n=1}^{\infty} \frac{1}{n} \sin \frac{1}{n}$ converges.

(V). Let us denote

$$b_n = \frac{\sqrt{n+1} - \sqrt{n}}{n} = \frac{\sqrt{n+1} - \sqrt{n}}{n} \cdot \frac{\sqrt{n+1} + \sqrt{n}}{\sqrt{n+1} + \sqrt{n}} = \frac{1}{n(\sqrt{n+1} + \sqrt{n})}$$

and $a_n = \frac{1}{n^{3/2}}$. Since

$$\lim_{n\to\infty} \frac{a_n}{b_n} = \lim_{n\to\infty} \frac{\frac{1}{n^{\frac{3}{2}}}}{\frac{1}{n(\sqrt{n+1}+\sqrt{n})}} = \lim_{n\to\infty} \frac{\sqrt{n+1}+\sqrt{n}}{\sqrt{n}}$$

$$= \lim_{n\to\infty} \left(1 + \sqrt{\frac{n+1}{n}}\right) = 2,$$

we have by the Limit Comparison Test that $\displaystyle\sum_{n=1}^{\infty} \frac{\sqrt{n+1}-\sqrt{n}}{n}$ converges as

$\displaystyle\sum_{n=1}^{\infty} \frac{1}{n^{3/2}}$ converges (it is a p-series with $p = \frac{3}{2} > 1$).

(VI). Observe that

$$\lim_{n\to\infty} \frac{n(n+1)}{n^2+2n} = \lim_{n\to\infty} \frac{n(n+1)}{n(n+2)} = \lim_{n\to\infty} \frac{n+1}{n+2} = 1 \neq 0.$$

Then, by Divergence Test, the series diverges.

(VII). Since

$$\lim_{n\to\infty} \frac{\left(\frac{1}{n}\right)^{\frac{n+1}{n}}}{\frac{1}{n}} = \lim_{n\to\infty} \frac{n}{n^{\frac{n+1}{n}}} = \lim_{n\to\infty} \frac{1}{n^{\frac{n+1}{n}-1}} = \lim_{n\to\infty} \frac{1}{n^{\frac{1}{n}}} = \lim_{n\to\infty} \frac{1}{\sqrt[n]{n}} = 1,$$

we have by the Limit Comparison Test that $\displaystyle\sum_{n=1}^{\infty} \left(\frac{1}{n}\right)^{\frac{n+1}{n}}$ diverges as $\displaystyle\sum_{n=1}^{\infty} \frac{1}{n}$

also diverges (harmonic series).

(VIII). If $|a| < 1$, we obtain that

$$\lim_{n\to\infty} \frac{1}{1+a^n} = 1;$$

and if $a = 1$, we have

$$\lim_{n\to\infty} \frac{1}{1+a^n} = \frac{1}{2}.$$

So that, by the Divergence Test, the series diverges when $a \in (-1, 1]$.
If $|a| > 1$, notice that

$$\lim_{n\to\infty} \frac{\left|\frac{1}{1+a^{n+1}}\right|}{\left|\frac{1}{1+a^n}\right|} = \lim_{n\to\infty} \left|\frac{1+a^n}{1+a^{n+1}}\right| = \lim_{n\to\infty} \left|\frac{a^n\left(\frac{1}{a^n}+1\right)}{a^n\left(\frac{1}{a^n}+a\right)}\right| = \left|\frac{1}{a}\right| = \frac{1}{|a|} < 1.$$

So, by the Ratio Test, the series absolutely converges (in particular, it converges) if $|a| > 1$, or equivalently, when $a \in (-\infty, -1) \cup (1, \infty)$.

(IX). Note that $f(x) = \frac{\log(x+1)-1}{(1+x)^2}$ is continuous on $[1,\infty)$. Let us analyze when $f(x) \geq 0$ on $[1,\infty)$. Take $x \in [1,\infty)$, then

$$\frac{\log(x+1)-1}{(1+x)^2} \geq 0 \iff \log(x+1)-1 \geq 0 \iff \log(x+1) \geq 1$$

$$\iff x+1 \geq e \iff x \geq e-1.$$

Thus, $f(x) \geq 0$ on $[e-1,\infty)$. Now let us study when f is decreasing on $[1,\infty)$. To do so, let us show when $f'(x) \geq 0$. We begin by calculating the derivative of f:

$$f'(x) = \frac{\frac{1}{(1+x)}(1+x)^2 - 2(\log(x+1)-1)(1+x)}{(1+x)^4}$$

$$= \frac{1+x - 2(\log(x+1)-1)(1+x)}{(1+x)^4}$$

$$= \frac{(1+x)(3 - 2\log(x+1))}{(1+x)^4}$$

$$= \frac{3 - 2\log(x+1)}{(1+x)^3};$$

then (recall that $x \geq 1$),

$$\frac{3 - 2\log(x+1)}{(1+x)^3} \leq 0 \iff 3 - 2\log(x+1) \leq 0 \iff \log(x+1) \geq \frac{3}{2}$$

$$\iff x+1 \geq e^{3/2} \iff x \geq e^{3/2} - 1.$$

Thus, $f'(x) \leq 0$ on $[e^{3/2}-1,\infty)$. Hence, we can apply the Integral Test to the series

$$\sum_{n=4}^{\infty} \frac{\log(n+1)-1}{(1+n)^2}$$

(notice that $4 > \max\{e-1, e^{3/2}-1\} > 1$).

Let us find the value of the improper integral

$$\int_4^{\infty} \frac{\log(x+1)-1}{(1+x)^2} dx = \lim_{b\to\infty} \int_4^b \frac{\log(x+1)-1}{(1+x)^2} dx.$$

Now apply the change of variable $u = x+1$ so that $du = dx$ to obtain

$$\int_4^b \frac{\log(x+1)-1}{(1+x)^2} dx = \int_5^{b+1} (\log u - 1)\frac{1}{u^2} du.$$

Observe that we can apply integration by parts with

$$\begin{cases} v = \log u - 1 \implies dv = \frac{1}{u}, \\ dw = \frac{1}{u^2}du \implies w = -\frac{1}{u}du. \end{cases}$$

Then,

$$\int_5^{b+1} (\log u - 1)\frac{1}{u^2} du = \left((\log u - 1) \left(-\frac{1}{u} \right) \right) \Big|_5^{b+1} + \int_5^{b+1} \frac{1}{u^2} du$$

$$= \left(-\frac{\log u}{u} + \frac{1}{u} \right) \Big|_5^{b+1} - \left(\frac{1}{u} \right) \Big|_5^{b+1}$$

$$= \frac{\log 5}{5} - \frac{\log(b+1)}{b+1} + \frac{1}{b+1} - \frac{1}{5} - \frac{1}{b+1} + \frac{1}{5}$$

$$= \frac{\log 5}{5} - \frac{\log(b+1)}{b+1}.$$

Since by L'Hôpital's Rule applied to $\displaystyle\lim_{b\to\infty} \frac{\log(b+1)}{b+1}$ we have

$$\lim_{b\to\infty} \frac{\log(b+1)}{b+1} = \lim_{b\to\infty} \frac{1}{b+1} = 0,$$

it yields

$$\int_4^\infty \frac{\log(x+1)-1}{(1+x)^2} dx = \lim_{b\to\infty} \int_4^b \frac{\log(x+1)-1}{(1+x)^2} dx$$

$$= \lim_{b\to\infty} \left(\frac{\log 5}{5} - \frac{\log(b+1)}{b+1} \right) = \frac{\log 5}{5}.$$

By the Integral Test, as $\displaystyle\int_4^\infty \frac{\log(x+1)-1}{(1+x)^2} dx$ converges, we obtain that $\displaystyle\sum_{n=4}^\infty \frac{\log(n+1)-1}{(1+n)^2}$ also converges. In particular, $\displaystyle\sum_{n=1}^\infty \frac{\log(n+1)-1}{(1+n)^2}$ converges.

(X). As

$$0 < \frac{1}{4} \le \frac{1}{3 - \cos\frac{1}{n}}$$

for all $n \in \mathbb{N}$, we obtain by the Comparison Test that $\displaystyle\sum_{n=1}^\infty \frac{1}{3 - \cos\frac{1}{n}}$ diverges since $\displaystyle\sum_{n=1}^\infty \frac{1}{4}$ diverges.

(XI). Let us apply the Root Test:

$$\lim_{n\to\infty} \sqrt[n]{\left| \left(\frac{x}{n} \right)^n n! \right|} = \lim_{n\to\infty} \frac{|x|}{n} \sqrt[n]{n!}.$$

By the weak form of Stirling's Formula, it yields that

$$\lim_{n\to\infty} \sqrt[n]{\left| \left(\frac{x}{n} \right)^n n! \right|} = \lim_{n\to\infty} \frac{|x|}{n} \cdot \frac{n}{e} = \frac{|x|}{e}.$$

Therefore, if $|x| > e$ the series diverges; whereas if $|x| < e$ the series converges.

Let us analyze separately the case when $|x| = e$, that is, either $x = e$ or $x = -e$. Observe that

$$\lim_{n \to \infty} \left(\frac{x}{2n}\right)^{2n} (2n)! = \lim_{n \to \infty} \frac{(e^2)^n}{(2n)^{2n}} \cdot \frac{(2n)^{2n}}{e^{2n}} \sqrt{4\pi n}$$

$$= \lim_{n \to \infty} \frac{(e^2)^n}{(2n)^{2n}} \frac{(2n)^{2n}}{(e^2)^n} \sqrt{4\pi n}$$

$$= \lim_{n \to \infty} \sqrt{2\pi n} = \infty,$$

where we have applied Stirling's Formula. Hence, since the subsequence of even indices $\left(\left(\frac{x}{2n}\right)^{2n} (2n)!\right)_{n=1}^{\infty}$ (with $|x| = e$) diverges, we have that $\left(\left(\frac{x}{n}\right)^n n!\right)_{n=1}^{\infty}$ also diverges. Therefore, by the Divergence Test, the series diverges when $|x| = e$.

(XII). Take $b_n = \frac{1}{n \log n}$ and $a_n = \frac{1}{n(1 + \frac{1}{2} + \cdots + \frac{1}{n})}$, and apply the Limit Comparison Test:

$$\lim_{n \to \infty} \frac{b_n}{a_n} = \lim_{n \to \infty} \frac{n(1 + \frac{1}{2} + \cdots + \frac{1}{n})}{n \log n} = \lim_{n \to \infty} \frac{\sum_{k=1}^{n} \frac{1}{k}}{\log n}.$$

By the Stolz Criterion, we have

$$\lim_{n \to \infty} \frac{b_n}{a_n} = \lim_{n \to \infty} \frac{\sum_{k=1}^{n} \frac{1}{k}}{\log n} = \lim_{n \to \infty} \frac{\sum_{k=1}^{n+1} \frac{1}{k} - \sum_{k=1}^{n} \frac{1}{k}}{\log(n+1) - \log n}$$

$$= \lim_{n \to \infty} \frac{\frac{1}{n+1}}{\log \frac{n+1}{n}} = \lim_{n \to \infty} \frac{\frac{1}{n+1}}{\log\left(1 + \frac{1}{n}\right)}.$$

Since $\log(1 + s_n) \sim s_n$ provided that s_n converges to 0, we arrive at

$$\lim_{n \to \infty} \frac{b_n}{a_n} = \lim_{n \to \infty} \frac{\frac{1}{n+1}}{\frac{1}{n}} = \lim_{n \to \infty} \frac{n}{n+1} = 1.$$

Thus, $\sum_{n=1}^{\infty} a_n$ converges if, and only if, $\sum_{n=1}^{\infty} b_n$ converges.

Let us see that $\sum_{n=2}^{\infty} b_n$ diverges. First, note that $f(x) = \frac{1}{x \log x}$ is continuous and positive on $[2, \infty)$. But also decreasing on $[2, \infty)$. Indeed,

$$f'(x) = -\frac{\log x + x \cdot \frac{1}{x}}{(x \log x)^2} = -\frac{\log x + 1}{(x \log x)^2} \leq 0,$$

for any $x \in [2, \infty)$. Hence, we can apply the Integral Test as follows:

$$\int_{2}^{\infty} \frac{1}{x \log x} dx = \lim_{c \to \infty} \int_{2}^{c} \frac{1}{x \log x} dx.$$

Now apply the following substitution:

$$u = \log x \implies du = \frac{1}{x}dx,$$

which implies

$$\int_2^c \frac{1}{x\log x}dx = \int_{\log 2}^{\log c} \frac{1}{u}du = \log|u|\Big|_{\log 2}^{\log c} = \log(\log c) - \log(\log 2).$$

Thus,

$$\int_2^\infty \frac{1}{x\log x}dx = \lim_{c\to\infty}(\log(\log c) - \log(\log 2)) = \infty.$$

By the Integral Test, we obtain that $\displaystyle\sum_{n=2}^\infty b_n$ diverges and, as an immediate consequence, we have that $\displaystyle\sum_{n=1}^\infty \frac{1}{n\left(1 + \frac{1}{2} + \cdots + \frac{1}{n}\right)}$ also diverges.

$\square$

Exercise 7.4. *Study the convergence of the following series:*

(I) $\displaystyle\sum_{n=1}^\infty \frac{1 + \frac{1}{2} + \cdots + \frac{1}{n}}{n^3\log n}$,

(II) $\displaystyle\sum_{n=1}^\infty \frac{1}{(\log n)^{2n}}$,

(III) $\displaystyle\sum_{n=1}^\infty \log\frac{n+1}{n}$,

(IV) $\displaystyle\sum_{n=1}^\infty e^{-\sqrt{n^2+1}}$,

(V) $\displaystyle\sum_{n=2}^\infty \frac{1}{(\log n)^p}$ *with* $p \in \mathbb{R}$,

(VI) $\displaystyle\sum_{n=1}^\infty \frac{x^n}{\sqrt{n}}$ *with* $x \in \mathbb{R}$,

(VII) $\displaystyle\sum_{n=1}^\infty \frac{\log n}{n^p}$ *with* $p \in \mathbb{R}$,

(VIII) $\displaystyle\sum_{n=1}^\infty \log\left(1 + \frac{x}{n}\right)$ *with* $x \geq 0$,

(IX) $\displaystyle\sum_{n=1}^\infty \frac{(-1)^n}{1 + \frac{1}{2} + \cdots + \frac{1}{n}}$,

(X) $\displaystyle\sum_{n=1}^\infty \frac{(-1)^n(n+1)}{n!}$,

(XI) $\displaystyle\sum_{n=1}^\infty \frac{(n^2+1)x^n}{(n+1)!}$,

(XII) $\displaystyle\sum_{n=1}^\infty \left(e^{1/n^2} - e^{1/(n^2+1)}\right)$,

(XIII) $\displaystyle\sum_{n=1}^\infty (-1)^{n+1}\frac{n}{n^2+1}$,

(XIV) $\displaystyle\sum_{n=1}^\infty \frac{(n!)^2}{(2n)!}x^{2n}$ *with* $x \in \mathbb{R}$.

Solution to Exercise 7.4.

(I). Let us denote $b_n = \frac{1}{n^3}$ and $a_n = \frac{1+\frac{1}{2}+\cdots+\frac{1}{n}}{n^3 \log n}$, and apply the Limit Comparison Test:

$$\lim_{n\to\infty} \frac{a_n}{b_n} = \lim_{n\to\infty} \frac{n^3\left(1+\frac{1}{2}+\cdots+\frac{1}{n}\right)}{n^3 \log n} = \lim_{n\to\infty} \frac{1+\frac{1}{2}+\cdots+\frac{1}{n}}{\log n} = 1,$$

where in the last equality we have applied the Stolz Criterion. Thus, by the Limit Comparison Test, as $\sum_{n=1}^{\infty} \frac{1}{n^3}$ converges (p-series with $p = 3 > 1$), we have that $\sum_{n=1}^{\infty} \frac{1+\frac{1}{2}+\cdots+\frac{1}{n}}{n^3 \log n}$ converges.

(II). Let us apply the Root Test:

$$\lim_{n\to\infty} \sqrt[n]{\frac{1}{(\log n)^{2n}}} = \lim_{n\to\infty} \frac{1}{(\log n)^2} = 0 < 1.$$

So, the series converges.

(III). Let us denote $b_n = \frac{1}{n}$ and $a_n = \log \frac{n+1}{n}$, and apply the Limit Comparison Test:

$$\lim_{n\to\infty} \frac{a_n}{b_n} = \lim_{n\to\infty} n \log \frac{n+1}{n} = \lim_{n\to\infty} \log \left(\frac{n+1}{n}\right)^n$$

$$= \lim_{n\to\infty} \log \left(1+\frac{1}{n}\right)^n = \log e = 1.$$

As $\sum_{n=1}^{\infty} \frac{1}{n}$ diverges (harmonic series), we have that $\sum_{n=1}^{\infty} \log \frac{n+1}{n}$ also diverges.

(IV). Note that

$$0 \le e^{-\sqrt{n^2+1}} \le e^{-n}.$$

Hence, by the Comparison Test, since $\sum_{n=1}^{\infty} e^{-n}$ converges (geometric series with common ratio $r = e^{-1}$, so that $|r| < 1$), we have that $\sum_{n=1}^{\infty} e^{-\sqrt{n^2+1}}$ converges.

(V). Let us apply Cauchy's Condensation Criterion. To do so, first note that $a_n = \frac{1}{(\log n)^p}$ is a decreasing sequence for any integer $n \ge 2$. Now,

$$\lim_{n\to\infty} \frac{2^n}{(\log 2^n)^p} = \lim_{n\to\infty} \frac{2^n}{(n \log 2)^p} = \lim_{n\to\infty} \frac{2^n}{n^p (\log 2)^p} = \infty.$$

So, by Cauchy's Condensation Criterion, the series diverges.

(VI). Let us apply the Ratio Test:

$$\lim_{n\to\infty}\left|\frac{a_{n+1}}{a_n}\right| = \lim_{n\to\infty}\frac{|x|^{n+1}}{\sqrt{n+1}}\cdot\frac{\sqrt{n}}{|x|} = |x|.$$

If $|x| > 1$, the series is divergent; and if $|x| < 1$, the series converges. If $x = 1$, then the series is divergent (p-series with $p = \frac{1}{2} < 1$). If $x = -1$, then the series converges by the Leibniz Alternating Series Test (also known as Leibniz Criterion).

(VII). For $p \leq 0$ we have

$$\lim_{n\to\infty}\frac{\log n}{n^p} = \infty.$$

So, by the Divergence Test, the series diverges when $p \leq 0$.

Assume for the rest of this part that $p > 0$. Let us try to apply Cauchy's Condensation Criterion. To do so, we will first study the monotonicity of the sequence $a_n = \frac{\log n}{n^p}$. Take $f(x) = \frac{\log x}{x^p}$ which is continuous and differentiable on $[1, \infty)$, then

$$f'(x) = \frac{\frac{1}{x}\cdot x^p - \log(x)px^{p-1}}{x^{2p}} = \frac{x^{p-1}(1 - p\log x)}{x^{2p}} = \frac{1 - p\log x}{x^{p+1}}.$$

Then, for $x > 1$,

$$\frac{1 - p\log x}{x^{p+1}} \leq 0 \iff 1 - p\log x \leq 0 \iff \log x \geq \frac{1}{p} \iff x \geq e^{1/p}.$$

Thus, there exists $n_p \in \mathbb{N}$ such that a_n is decreasing for any $n \geq n_p$.

Now, in order to apply Cauchy's Condensation Criterion, we must study the convergence of the series $\sum_{n=n_p}^{\infty} 2^n a_{2^n} = \sum_{n=n_p}^{\infty} 2^n\frac{\log(2^n)}{(2^n)^p} = \sum_{n=n_p}^{\infty} 2^n\frac{n\log 2}{(2^n)^p}.$

In order to do this, we will make use of the Root Test:

$$\lim_{n\to\infty}\sqrt[n]{\frac{2^n n\log 2}{(2^n)^p}} = \lim_{n\to\infty}\frac{2\sqrt[n]{n}\sqrt[n]{\log 2}}{2^p} = \frac{1}{2^{p-1}}.$$

As $\frac{1}{2^{p-1}} < 1 \iff p > 1$, the series converges if $p > 1$; and as $\frac{1}{2^{p-1}} > 1 \iff p < 1$, the series diverges if $p < 1$.

If $p = 1$, we have that

$$\frac{\log n}{n} > \frac{1}{n} > 0$$

for any integer $n \geq 2$. Then, by the Comparison test, the series diverges since $\sum_{n=2}^{\infty}\frac{1}{n}$ diverges (harmonic series).

(VIII). For $x = 0$ we have

$$\sum_{n=1}^{\infty}\log\left(1 + \frac{x}{n}\right) = \sum_{n=1}^{\infty}\log 1 = 0.$$

So, assume for the rest of this part that $x > 0$. Let us denote $b_n = \frac{1}{n}$ and $a_n = \log \frac{n+x}{n}$, and apply the Limit Comparison Test:

$$\lim_{n \to \infty} \frac{a_n}{b_n} = \lim_{n \to \infty} n \log \frac{n+x}{n} = \lim_{n \to \infty} \log \left(\frac{n+x}{n} \right)^n = \lim_{n \to \infty} \log \left(1 + \frac{1}{n/x} \right)^n$$

$$= \lim_{n \to \infty} \left[\log \left(1 + \frac{1}{n/x} \right)^{n/x} \right]^x = e^x > 0$$

for any $x > 0$. As $\sum_{n=1}^{\infty} \frac{1}{n}$ diverges (harmonic series), we have that $\sum_{n=1}^{\infty} \log \frac{n+x}{n}$ also diverges.

(IX). Observe that $a_n = \frac{(-1)^n}{1 + \frac{1}{2} + \cdots + \frac{1}{n}}$ is an alternating sequence which satisfies:

$$\lim_{n \to \infty} |a_n| = \lim_{n \to \infty} \frac{1}{1 + \frac{1}{2} + \cdots + \frac{1}{n}} = \frac{1}{\sum_{n=1}^{\infty} \frac{1}{n}} = 0.$$

Now let show that a_n is decreasing:

$$a_n > a_{n+1} \iff \frac{1}{1 + \frac{1}{2} + \cdots + \frac{1}{n}} > \frac{1}{1 + \frac{1}{2} + \cdots + \frac{1}{n} + \frac{1}{n+1}} \iff \frac{1}{n+1} > 0.$$

Therefore, by the Leibniz Alternating Series Test, the series is convergent.

(X). Let us denote $a_n = \frac{(-1)^n (n+1)}{n!}$, then $|a_n| = \frac{n+1}{n!}$. Let us apply now the Ratio Test:

$$\lim_{n \to \infty} \frac{|a_{n+1}|}{|a_n|} = \lim_{n \to \infty} \frac{\frac{n+2}{(n+1)!}}{\frac{n+1}{n!}} = \lim_{n \to \infty} \frac{(n+2)n!}{(n+1)(n+1)!}$$

$$= \lim_{n \to \infty} \frac{(n+2)}{(n+1)(n+1)} = 0 < 1.$$

Hence, by the Ratio Test, the series $\sum_{n=1}^{\infty} \frac{(-1)^n (n+1)}{n!}$ is absolutely convergent (so, in particular, convergent).

(XI). If $x = 0$, then the series is clearly convergent. So, assume that $x \neq 0$. Let $a_n = \frac{(n^2+1)x^n}{(n+1)!}$ and apply the Ratio Test:

$$\lim_{n \to \infty} \frac{|a_{n+1}|}{|a_n|} = \lim_{n \to \infty} \frac{\frac{((n+1)^2+1)|x|^{n+1}}{(n+2)!}}{\frac{(n^2+1)|x|^n}{(n+1)!}}$$

$$= \lim_{n \to \infty} \frac{((n+1)^2 + 1)|x|^{n+1}}{(n+2)!} \cdot \frac{(n+1)!}{(n^2+1)|x|^n}$$

$$= \lim_{n \to \infty} \frac{|x|(n^2 + 2n + 2)}{n^3 + 2n^2 + n + 2} = 0 < 1.$$

So, by the Ratio Test, the series converges.

(XII). Let us denote $a_n = e^{\frac{1}{n^2}} - e^{\frac{1}{n^2+1}}$ and $b_n = e^{\frac{1}{n^2}} - e^{\frac{1}{(n+1)^2}}$ and observe that $0 < a_n$. Let us prove that $a_n < b_n$:

$$a_n < b_n \iff e^{\frac{1}{n^2}} - e^{\frac{1}{n^2+1}} < e^{\frac{1}{n^2}} - e^{\frac{1}{(n+1)^2}}$$
$$\iff e^{\frac{1}{(n+1)^2}} < e^{\frac{1}{n^2+1}} \iff \frac{1}{(n+1)^2} < \frac{1}{n^2+1}$$
$$\iff n^2 + 1 < (n+1)^2 = n^2 + 2n + 1$$
$$\iff 0 < 2n.$$

Now, notice that $\sum_{n=1}^{\infty} b_n$ is a telescopic series that is convergent with

$$\sum_{n=1}^{\infty} \left(e^{\frac{1}{n^2}} - e^{\frac{1}{(n+1)^2}} \right) = e - \lim_{n\to\infty} e^{\frac{1}{n^2}} = e - 1.$$ Hence, by the Direct Comparison

Test, the series $\sum_{n=1}^{\infty} \left(e^{\frac{1}{n^2}} - e^{\frac{1}{n^2+1}} \right)$ is convergent.

(XIII). Let $a_n = (-1)^{n+1} \frac{n}{n^2+1}$. Then (a_n) is an alternating sequence and satisfies

$$\lim_{n\to\infty} |a_n| = \lim_{n\to\infty} \frac{n}{n^2+1} = 0.$$

Now let us show that $|a_n|$ decreases. To do so, define $f(x) = \frac{x}{x^2+1}$. Then,

$$f'(x) = \frac{(x^2+1) - x \cdot 2x}{(x^2+1)^2} = \frac{-x^2+1}{(x^2+1)^2}.$$

As $f'(x) < 0$ for all $x > 1$, we have that $|a_n|$ decreases. Therefore, by Leibniz Criterion, the series is convergent.

(XIV). Let us apply the Ratio Test:

$$\lim_{n\to\infty} \frac{\left| \frac{((n+1)!)^2}{(2n+2)!} x^{2n+2} \right|}{\left| \frac{(n!)^2}{(2n)!} x^{2n} \right|} = \lim_{n\to\infty} \frac{((n+1)!)^2 \cdot |x|^{2n+2} \cdot (2n)!}{(2n+2)! \cdot (n!)^2 \cdot |x|^{2n}}$$
$$= \lim_{n\to\infty} \frac{(n+1)!(n+1)!x^2}{n!n!(2n+1)(2n+2)}$$
$$= x^2 \lim_{n\to\infty} \frac{n^2+2n+1}{4n^2+4n+2} = \frac{x^2}{4}.$$

So that, by the Ratio Test, if $\frac{x^2}{4} < 1$, the series converges; and if $\frac{x^2}{4} > 1$, the series diverges. That is, the series converges if $|x| < 2$, and diverges if $|x| > 2$.

Let $|x| = 2$ (that is, either $x = 2$ or $x = -2$), then the sequence of general terms of the series satisfies $b_n = \frac{(n!)^2}{(2n)!} 4^n$. We can observe that $b_1 = 2$. Let us see that b_n is increasing:

$$\frac{(n!)^2 4^n}{(2n)!} < \frac{((n+1)!)^2 4^{n+1}}{(2n+2)!} \iff (2n+1)(2n+2) < (n+1)^2 4$$

$$\iff 4n^2 + 6n + 2 < 4n^2 + 8n + 4$$

$$\iff 0 < 2n + 2.$$

So $\lim_{n\to\infty} b_n \geq 2$ and therefore, by the Divergence Test, the series is divergent. $\square$

Exercise 7.5. *Prove that the convergence of $\sum_{n=1}^{\infty} a_n$ implies the convergence of $\sum_{n=1}^{\infty} \frac{\sqrt{a_n}}{n}$ if $a_n \geq 0$ for all $n \in \mathbb{N}$.*

Solution to Exercise 7.5.

Let us define the following sets:

$$S = \left\{ n \in \mathbb{N} \colon \frac{\sqrt{a_n}}{n} \leq a_n \right\}$$

and

$$T = \left\{ n \in \mathbb{N} \colon \frac{\sqrt{a_n}}{n} > a_n \right\}.$$

We can observe that, if $\frac{\sqrt{a_n}}{n} > a_n$ and $a_n \geq 0$ for all $n \in \mathbb{N}$, it yields that $\frac{a_n}{n^2} > a_n^2$. Hence, $a_n < \frac{1}{n^2}$ which implies that $\frac{\sqrt{a_n}}{n} < \frac{1}{n^2}$. Therefore, we have that

$$\sum_{n=1}^{\infty} \frac{\sqrt{a_n}}{n} = \sum_{n \in S} \frac{\sqrt{a_n}}{n} + \sum_{n \in T} \frac{\sqrt{a_n}}{n} \leq \sum_{n \in S} a_n + \sum_{n \in T} \frac{1}{n^2} \leq \sum_{n \in \mathbb{N}} a_n + \frac{\pi^2}{6}.$$

Thus, the convergence of $\sum_{n=1}^{\infty} a_n$ implies the convergence of $\sum_{n=1}^{\infty} \frac{\sqrt{a_n}}{n}$. $\square$

Exercise 7.6. *Assume that $a_n > 0$ for every $n \in \mathbb{N}$, $\displaystyle\sum_{n=1}^{\infty} a_n$ diverges and let $s_n = a_1 + a_2 + \cdots + a_n$.*

(I) *Prove that $\displaystyle\sum_{n=1}^{\infty} \frac{a_n}{1+a_n}$ diverges.*

(II) *Prove that*

$$\frac{a_{n+1}}{s_{n+1}} + \cdots + \frac{a_{n+k}}{s_{n+k}} \geq 1 - \frac{s_n}{s_{n+k}}$$

for every $n, k \in \mathbb{N}$, and deduce that $\displaystyle\sum_{n=1}^{\infty} \frac{a_n}{s_n}$ diverges.

(III) *Prove that*

$$\frac{a_n}{s_n^2} < \frac{1}{s_{n-1}} + \frac{1}{s_n}$$

for every $n \in \mathbb{N}$, and deduce that $\displaystyle\sum_{n=1}^{\infty} \frac{a_n}{s_n^2}$ converges.

(IV) *What can we say about $\displaystyle\sum_{n=1}^{\infty} \frac{a_n}{1+na_n}$ and $\displaystyle\sum_{n=1}^{\infty} \frac{a_n}{1+n^2 a_n}$?*

Solution to Exercise 7.6.

(I). First of all, since the series is a series of positive terms we have that $\displaystyle\sum_{n=1}^{\infty} a_n = \infty$. Now let us consider two cases.

Case 1. Assume that there are infinitely many n's such that $a_n > 1$. Let $A := \{n \in \mathbb{N}: a_n > 1\}$, then A is infinite. Hence, for every $n \in A$, we have that

$$a_n > 1 \iff 2a_n > 1 + a_n \iff \frac{a_n}{1+a_n} > \frac{1}{2}.$$

So

$$\sum_{n=1}^{\infty} \frac{a_n}{1+a_n} \geq \sum_{n \in A} \frac{a_n}{1+a_n} > \sum_{n \in A} \frac{1}{2} = \infty.$$

Case 2. Assume that there exists an $n_0 \in \mathbb{N}$ such that $a_n \leq 1$ for every $n \geq n_0$. Then, for every $n \geq n_0$, we have

$$\frac{a_n}{1+a_n} \geq \frac{1}{2}a_n.$$

Hence,

$$\sum_{n=1}^{\infty} \frac{a_n}{1+a_n} \geq \sum_{n=n_0}^{\infty} \frac{a_n}{1+a_n} \geq \sum_{n=n_0}^{\infty} \frac{1}{2} a_n = \frac{1}{2} \sum_{n=n_0}^{\infty} a_n = \infty.$$

(II). Fix $n, k \in \mathbb{N}$. As $(a_n)_{n=1}^{\infty}$ is a sequence of positive terms, we have that the sequence of partial sums $(s_n)_{n=1}^{\infty}$ is strictly increasing. Therefore, for every integer $1 \leq i \leq k$, we have

$$\frac{a_{n+i}}{s_{n+i}} \geq \frac{a_{n+i}}{s_{n+k}}.$$

Hence,

$$\frac{a_{n+1}}{s_{n+1}} + \cdots + \frac{a_{n+k}}{s_{n+k}} \geq \frac{a_{n+1}}{s_{n+k}} + \cdots + \frac{a_{n+k}}{s_{n+k}} = \frac{1}{s_{n+k}} (a_{n+1} + \cdots + a_{n+k})$$

$$= \frac{1}{s_{n+k}} (s_{n+k} - s_n) = 1 - \frac{s_n}{s_{n+k}}.$$

Fix any $n \in \mathbb{N}$. For every integer $m > n$, we know that

$$\frac{a_{n+1}}{s_{n+1}} + \cdots + \frac{a_m}{s_m} \geq 1 - \frac{s_n}{s_m}.$$

Therefore, since $\lim_{m \to \infty} \frac{s_n}{s_m} = 0$, there exists an integer $m_0 > n$ such that

$$\sum_{k=n+1}^{m} \frac{a_k}{s_k} \geq \frac{1}{2},$$

for every $m \geq m_0$. So, by Cauchy's Criterion for series, we obtain that $\sum_{n=1}^{\infty} \frac{a_n}{s_n}$ diverges to ∞.

(III). Fix $n \in \mathbb{N}$. Then,

$$\frac{a_n}{s_n^2} = \frac{s_n - s_{n-1}}{s_n^2} = \frac{s_n - s_{n-1}}{s_{n-1} s_n} \cdot \frac{s_{n-1}}{s_n} = \left(\frac{1}{s_{n-1}} - \frac{1}{s_n} \right) \frac{s_{n-1}}{s_n}.$$

Since $0 < s_{n-1} < s_n$, we know that $0 < \frac{s_{n-1}}{s_n} < 1$ and $\frac{1}{s_{n-1}} > \frac{1}{s_n}$. In particular, $0 < \frac{s_{n-1}}{s_n} < 1$ and $\frac{1}{s_{n-1}} - \frac{1}{s_n} > 0$. So,

$$\frac{a_n}{s_n^2} < \frac{1}{s_{n-1}} - \frac{1}{s_n}.$$

Now take any $n, m \in \mathbb{N}$ with $m > n$. Then, we know that

$$\sum_{k=n+1}^{m} \frac{a_n}{s_n^2} < \sum_{k=n+1}^{m} \left(\frac{1}{s_{n-1}} - \frac{1}{s_n} \right).$$

As the latter sum is telescopic it yields $\displaystyle\sum_{k=n+1}^{m}\left(\frac{1}{s_{n-1}}-\frac{1}{s_n}\right)=\frac{1}{s_n}-\frac{1}{s_m}.$
Hence, as $\frac{1}{s_m}>0$,

$$\sum_{k=n+1}^{m}\frac{a_n}{s_n^2}<\frac{1}{s_n}-\frac{1}{s_m}<\frac{1}{s_n}.$$

Since $\displaystyle\lim_{n\to\infty}\frac{1}{s_n}=0$, we have by Cauchy's Criterion for series that $\displaystyle\sum_{n=1}^{\infty}\frac{a_n}{s_n^2}$ converges.

(IV). Let us consider first the series $\displaystyle\sum_{n=1}^{\infty}\frac{a_n}{1+na_n}$. Assume that $a_n=\frac{1}{n}$, then

$$\sum_{n=1}^{\infty}\frac{a_n}{1+na_n}=\sum_{n=1}^{\infty}\frac{\frac{1}{n}}{1+n\cdot\frac{1}{n}}=\sum_{n=1}^{\infty}\frac{1}{2}\cdot\frac{1}{n}=\frac{1}{2}\sum_{n=1}^{\infty}\frac{1}{n}$$

which is divergent (harmonic series). Now assume that $a_n=\frac{1}{2^n}$, then

$$\sum_{n=1}^{\infty}\frac{a_n}{1+na_n}=\sum_{n=1}^{\infty}\frac{\frac{1}{2^n}}{1+n\cdot\frac{1}{2^n}}=\sum_{n=1}^{\infty}\frac{\frac{1}{2^n}}{\frac{1}{2^n}(2^n+n)}=\sum_{n=1}^{\infty}\frac{1}{2^n+n}.$$

Since $\frac{1}{2^n+n}\leq\frac{1}{2^n}$ for every $n\in\mathbb{N}$ and $\displaystyle\sum_{n=1}^{\infty}\frac{1}{2^n}$ converges, we have by the

Comparison Test that $\displaystyle\sum_{n=1}^{\infty}\frac{1}{2^n+n}$ converges. Therefore, the series $\displaystyle\sum_{n=1}^{\infty}\frac{a_n}{1+na_n}$ sometimes converges and other times diverges depending on a_n.

Let us now consider the series $\displaystyle\sum_{n=1}^{\infty}\frac{a_n}{1+n^2a_n}$. In this case, since $\frac{1}{a_n}>0$, we obtain that

$$\frac{a_n}{1+n^2a_n}=\frac{a_n}{a_n\left(\frac{1}{a_n}+n^2\right)}=\frac{1}{\frac{1}{a_n}+n^2}\leq\frac{1}{n^2}$$

for every $n\in\mathbb{N}$. As $\displaystyle\sum_{n=1}^{\infty}\frac{1}{n^2}$ converges, it follows that $\displaystyle\sum_{n=1}^{\infty}\frac{a_n}{1+n^2a_n}$ also converges by the Comparison Test.

$\square$

Exercise 7.7. *Find the sum of the following series:*

(I) $1 - \dfrac{1}{2} + \dfrac{1}{3} - \dfrac{1}{4} + \dfrac{1}{5} - \dfrac{1}{6} + \cdots,$

(II) $\dfrac{1}{4} - \dfrac{1}{3} + \dfrac{1}{8} - \dfrac{1}{9} + \dfrac{1}{12} - \dfrac{1}{15} + \dfrac{1}{16} - \dfrac{1}{21} + \dfrac{1}{20} - \cdots,$

(III) $1 + \dfrac{1}{3} + \dfrac{1}{5} + \dfrac{1}{7} - \dfrac{1}{2} + \dfrac{1}{9} + \dfrac{1}{11} + \dfrac{1}{13} + \dfrac{1}{15} - \dfrac{1}{4} + \cdots,$

(IV) $1 + \dfrac{1}{3} - \dfrac{1}{2} - \dfrac{1}{4} - \dfrac{1}{6} + \dfrac{1}{5} + \dfrac{1}{7} - \dfrac{1}{8} - \dfrac{1}{10} - \dfrac{1}{12} + \cdots,$

(V) $1 + \dfrac{1}{2^2} + \dfrac{1}{3^2} + \dfrac{1}{5^2} + \dfrac{1}{6^2} + \dfrac{1}{7^2} + \dfrac{1}{9^2} + \cdots.$

Solution to Exercise 7.7.

(I). Observe that we can write the expression $1 - \frac{1}{2} + \frac{1}{3} - \frac{1}{4} + \frac{1}{5} - \frac{1}{6} + \cdots$ as the series:

$$\sum_{n=1}^{\infty} (-1)^{n+1} \frac{1}{n}.$$

This is known as the alternating harmonic series. The general term of the series is $a_n = (-1)^{n+1} \frac{1}{n}$ which satisfies

$$a_n a_{n+1} = (-1)^{n+1} \frac{1}{n} \cdot (-1)^{n+2} \frac{1}{n+1} = (-1)^{2n+3} \frac{1}{n(n+1)} = -\frac{1}{n(n+1)} < 0,$$

that is, the series is an alternating series. Notice that the sequence $(|a_n|)_{n=1}^{\infty} = \left(\dfrac{1}{n} \right)_{n=1}^{\infty}$ decreases and converges to 0. Therefore, by the Leibniz Criterion, the series converges.

Now recall by Exercise 1.8 part (VI) that

$$\sum_{k=n+1}^{2n} \frac{1}{k} = \sum_{k=1}^{2n} \frac{(-1)^{k+1}}{k}.$$

Since $\sum_{n=1}^{\infty} (-1)^{n+1} \frac{1}{n}$ is convergent, we have that

$$\sum_{n=1}^{\infty} (-1)^{n+1} \frac{1}{n} = \lim_{n \to \infty} \sum_{k=1}^{2n} \frac{(-1)^{k+1}}{k} = \lim_{n \to \infty} \sum_{k=n+1}^{2n} \frac{1}{k}.$$

Observe that

$$\sum_{k=n+1}^{2n} \frac{1}{k} = \frac{1}{n+1} + \frac{1}{n+2} + \cdots + \frac{1}{2n}$$

$$= \frac{1}{n}\left(\frac{1}{1+\frac{1}{n}} + \frac{1}{1+\frac{2}{n}} + \cdots + \frac{1}{1+\frac{n}{n}}\right). \qquad (7.1)$$

Notice that $\frac{1}{n}\left(\frac{1}{1+\frac{1}{n}} + \frac{1}{1+\frac{2}{n}} + \cdots + \frac{1}{1+\frac{n}{n}}\right)$ is the Riemann sum of the function $\frac{1}{1+x}$ defined on the interval $[0,1]$. Hence, by taking limits in (7.1), we have

$$\sum_{n=1}^{\infty} (-1)^{n+1} \frac{1}{n} = \lim_{n\to\infty} \frac{1}{n}\left(\frac{1}{1+\frac{1}{n}} + \frac{1}{1+\frac{2}{n}} + \cdots + \frac{1}{1+\frac{n}{n}}\right)$$

$$= \int_0^1 \frac{1}{1+x}dx$$

$$= \log|1+x|\Big|_0^1 = \log 2 - \log 1 = \log 2.$$

(II). If (s_n) is the sequence of the partial sums of this series, it is enough to study the convergence of the subsequence (s_{2k}). We have that

$$s_{2k} = \frac{1}{4} - \frac{1}{3} + \frac{1}{8} - \frac{1}{9} + \cdots + \frac{1}{4k} - \frac{1}{6k-3}$$

$$= \left(\frac{1}{4} + \frac{1}{8} + \cdots + \frac{1}{4k}\right) - \left(\frac{1}{3} + \frac{1}{9} + \cdots + \frac{1}{3(2k-1)}\right).$$

Denoting by h_k the kth partial sum of the harmonic series, we have

$$\frac{1}{4} + \frac{1}{8} + \cdots + \frac{1}{4k} = \frac{1}{4}\left(1 + \frac{1}{2} + \cdots + \frac{1}{k}\right) = \frac{1}{4}h_k.$$

On the other hand,

$$\frac{1}{3} + \frac{1}{9} + \cdots + \frac{1}{3(2k-1)}$$

$$= \frac{1}{3}\left(1 + \frac{1}{3} + \cdots + \frac{1}{2k-1}\right)$$

$$= \frac{1}{3}\left(1 + \frac{1}{2} + \cdots + \frac{1}{2k}\right) - \frac{1}{3}\left(\frac{1}{2} + \frac{1}{4} + \cdots + \frac{1}{2k}\right)$$

$$= \frac{1}{3}\left(1 + \frac{1}{2} + \cdots + \frac{1}{2k}\right) - \frac{1}{6}\left(1 + \frac{1}{2} + \cdots + \frac{1}{k}\right)$$

$$= \frac{1}{3}h_{2k} - \frac{1}{6}h_k.$$

Therefore, we can put

$$s_{2k} = \frac{1}{4}h_k - \left(\frac{1}{3}h_{2k} - \frac{1}{6}h_k\right) = \frac{5}{12}h_k - \frac{1}{3}h_{2k}.$$

Recall the fact that
$$h_k - \log k \xrightarrow[k]{} \gamma,$$
where γ is the Euler-Mascheroni constant. Then we arrive at
$$s_{2k} = \frac{5}{12}(h_k - \log k) - \frac{1}{3}(h_{2k} - \log(2k)) + \frac{5}{12}\log k - \frac{1}{3}\log(2k)$$
$$= \frac{5}{12}(h_k - \log k) - \frac{1}{3}(h_{2k} - \log(2k)) - \frac{1}{3}\log 2 + \frac{1}{12}\log k$$
$$\xrightarrow[k]{} \frac{1}{12}\gamma - \frac{1}{3}\log 2 + \infty = \infty.$$

We conclude that
$$\frac{1}{4} - \frac{1}{3} + \frac{1}{8} - \frac{1}{9} + \frac{1}{12} - \frac{1}{15} + \frac{1}{16} - \frac{1}{21} + \frac{1}{20} - \cdots$$
diverges to ∞.

(III). Let (s_n) be the sequence of partial sums of the series. It is enough to find the limit of the subsequence (s_{5k}). We have
$$s_{5k} = 1 + \frac{1}{3} + \frac{1}{5} + \frac{1}{7} - \frac{1}{2} + \cdots + \frac{1}{8k-7} + \frac{1}{8k-5} + \frac{1}{8k-3} + \frac{1}{8k-1} - \frac{1}{2k}$$
$$= \left(1 + \frac{1}{3} + \frac{1}{5} + \cdots + \frac{1}{8k-3} + \frac{1}{8k-1}\right) - \left(\frac{1}{2} + \frac{1}{4} + \cdots + \frac{1}{2k}\right).$$

Notice that
$$\frac{1}{2} + \frac{1}{4} + \cdots + \frac{1}{2k} = \frac{1}{2}\left(1 + \frac{1}{2} + \cdots + \frac{1}{k}\right) = \frac{1}{2}h_k,$$
where h_k is as the previous item. Also,
$$1 + \frac{1}{3} + \frac{1}{5} + \cdots + \frac{1}{8k-3} + \frac{1}{8k-1}$$
$$= \left(1 + \frac{1}{2} + \frac{1}{3} + \cdots + \frac{1}{8k}\right) - \left(\frac{1}{2} + \frac{1}{4} + \cdots + \frac{1}{8k}\right)$$
$$= \left(1 + \frac{1}{2} + \frac{1}{3} + \cdots + \frac{1}{8k}\right) - \frac{1}{2}\left(1 + \frac{1}{2} + \frac{1}{3} + \cdots + \frac{1}{4k}\right)$$
$$= h_{8k} - \frac{1}{2}h_{4k}.$$

Thus, we obtain
$$s_{5k} = h_{8k} - \frac{1}{2}h_{4k} - \frac{1}{2}h_k$$
$$= (h_{8k} - \log(8k)) - \frac{1}{2}(h_{4k} - \log(4k)) - \frac{1}{2}(h_k - \log k)$$
$$+ \log(8k) - \frac{1}{2}\log(4k) - \frac{1}{2}\log k$$
$$= (h_{8k} - \log(8k)) - \frac{1}{2}(h_{4k} - \log(4k)) - \frac{1}{2}(h_k - \log k) + 2\log 2.$$

Therefore, we find that

$$\lim_k s_{5k} = \gamma - \frac{1}{2}\gamma - \frac{1}{2}\gamma + 2\log 2 = 2\log 2.$$

This means that the sum of the series is $2\log 2$.

(IV). As always, take s_n as the nth partial sum of the series. For every $k \in \mathbb{N}$, we have

$$s_{5k} = 1 + \frac{1}{3} - \frac{1}{2} - \frac{1}{4} - \frac{1}{6} + \frac{1}{5} + \cdots$$

$$+ \frac{1}{4k-3} + \frac{1}{4k-1} - \frac{1}{6k-4} - \frac{1}{6k-2} - \frac{1}{6k}$$

$$= \left(1 + \frac{1}{3} + \frac{1}{5} + \cdots + \frac{1}{4k-1}\right) - \left(\frac{1}{2} + \frac{1}{4} + \frac{1}{6} + \cdots + \frac{1}{6k}\right).$$

On the one hand, we have

$$\frac{1}{2} + \frac{1}{4} + \frac{1}{6} + \cdots + \frac{1}{6k} = \frac{1}{2}\left(1 + \frac{1}{2} + \frac{1}{3} + \cdots + \frac{1}{3k}\right) = \frac{1}{2}h_{3k}.$$

On the other,

$$1 + \frac{1}{3} + \frac{1}{5} + \cdots + \frac{1}{4k-1}$$

$$= \left(1 + \frac{1}{2} + \frac{1}{3} + \cdots + \frac{1}{4k}\right) - \left(\frac{1}{2} + \frac{1}{4} + \frac{1}{6} + \cdots + \frac{1}{4k}\right)$$

$$= \left(1 + \frac{1}{2} + \frac{1}{3} + \cdots + \frac{1}{4k}\right) - \frac{1}{2}\left(1 + \frac{1}{2} + \frac{1}{3} + \cdots + \frac{1}{2k}\right)$$

$$= h_{4k} - \frac{1}{2}h_{2k}.$$

Therefore we arrive at

$$s_{5k} = h_{4k} - \frac{1}{2}h_{2k} - \frac{1}{2}h_{3k}$$

$$= (h_{4k} - \log(4k)) - \frac{1}{2}(h_{2k} - \log(2k)) - \frac{1}{2}(h_{3k} - \log(3k))$$

$$+ \log(4k) - \frac{1}{2}\log(2k) - \frac{1}{2}\log(3k)$$

$$= (h_{4k} - \log(4k)) - \frac{1}{2}(h_{2k} - \log(2k)) - \frac{1}{2}(h_{3k} - \log(3k))$$

$$+ \frac{3}{2}\log 2 - \frac{1}{2}\log 3.$$

This converges to

$$\gamma - \frac{1}{2}\gamma - \frac{1}{2}\gamma + \frac{3}{2}\log 2 - \frac{1}{2}\log 3 = \frac{3}{2}\log 2 - \frac{1}{2}\log 3.$$

We conclude that this is the sum of the series.

(v). Let s_n be the nth partial sum of the present series. We have

$$s_{3k} = 1 + \frac{1}{2^2} + \frac{1}{3^2} + \frac{1}{5^2} + \frac{1}{6^2} + \frac{1}{7^2}$$

$$+ \cdots + \frac{1}{(4k-3)^2} + \frac{1}{(4k-2)^2} + \frac{1}{(4k-1)^2}$$

$$= \left(1 + \frac{1}{2^2} + \cdots + \frac{1}{(4k)^2}\right) - \left(\frac{1}{4^2} + \frac{1}{8^2} + \cdots + \frac{1}{(8k)^2}\right)$$

$$= \left(1 + \frac{1}{2^2} + \cdots + \frac{1}{(4k)^2}\right) - \frac{1}{16}\left(1 + \frac{1}{2^2} + \cdots + \frac{1}{k^2}\right)$$

$$= \beta_{4k} - \frac{1}{16}\beta_k,$$

where

$$\beta_n = 1 + \frac{1}{2^2} + \cdots + \frac{1}{n^2}.$$

Using the fact that

$$\beta_n \xrightarrow[n]{} \frac{\pi^2}{6},$$

we obtain that (s_{3k}), and therefore (s_n), tends to

$$\frac{\pi^2}{6} - \frac{1}{16}\cdot\frac{\pi^2}{6} = \frac{15}{16}\cdot\frac{\pi^2}{6} = \frac{5}{32}\pi^2.$$

$\square$

Exercise 7.8. *Prove that the Cauchy product of two absolutely convergent series is absolutely convergent.*

Solution to Exercise 7.8.

Let $\sum_{n=0}^{\infty} a_n$ and $\sum_{n=0}^{\infty} b_n$ be two absolutely convergent series, that is, there are $A, B \geq 0$ such that $\sum_{n=0}^{\infty} |a_n| = A$ and $\sum_{n=0}^{\infty} |b_n| = B$. Recall that the Cauchy product of $\sum_{n=0}^{\infty} a_n$ and $\sum_{n=0}^{\infty} b_n$ is a series $\sum_{n=0}^{\infty} c_n$ where

$$c_n = \sum_{k=0}^{n} a_k b_{n-k},$$

for every $n \in \mathbb{N} \cup \{0\}$. Then, for every $m \in \mathbb{N} \cup \{0\}$, we have

$$\sum_{n=0}^{m} |c_n| = \sum_{n=0}^{m} \left| \sum_{k=0}^{n} a_k b_{n-k} \right| \leq \sum_{n=0}^{m} \sum_{k=0}^{n} |a_k b_{n-k}|$$

$$= |a_0 b_0| + (|a_0 b_1| + |a_1 b_0|) + \cdots + (|a_0 b_m| + |a_1 b_{m-1}| + \cdots + |a_m b_0|)$$

$$= \sum_{n=0}^{m} \left(|a_n| \sum_{k=0}^{m-n} |b_k| \right) \leq \sum_{n=0}^{m} (|a_n| B) = B \sum_{n=0}^{m} |a_n| \leq AB.$$

Thus, since the sequence $\left(\sum_{n=0}^{m} |c_n| \right)_{m=0}^{\infty}$ is increasing (as m increases we are adding non-negative terms) and bounded above by AB, we have by the Monotone Convergence Theorem that $\left(\sum_{n=0}^{m} |c_n| \right)_{m=0}^{\infty}$ converges. Moreover, since the limit of $\left(\sum_{n=0}^{m} |c_n| \right)_{m=0}^{\infty}$ is $\sum_{n=0}^{\infty} |c_n|$, we have proven our desired result. $\square$

Exercise 7.9. *Study the convergence of the following series and, when convergent, find their sum.*

(I) $\displaystyle\sum_{n=2}^{\infty} \frac{4n - 1}{(n + 2)(n - 1)^2}.$

(VI) $\displaystyle\sum_{n=2}^{\infty} \frac{1}{(n + 1)^2 - 4}.$

(II) $\displaystyle\sum_{n=1}^{\infty} \frac{1}{n(n + 1)}.$

(VII) $\displaystyle\sum_{n=1}^{\infty} \frac{3n^2 + 7n + 6}{n(n + 1)(n + 2)(n + 3)}.$

(III) $\displaystyle\sum_{n=2}^{\infty} \frac{2n + 3}{n(n - 1)(n + 2)}.$

(VIII) $\displaystyle\sum_{n=1}^{\infty} \frac{1}{(n - 1 + \sqrt{3})(n - 2 + \sqrt{3})(n + \sqrt{3})}.$

(IV) $\displaystyle\sum_{n=2}^{\infty} \frac{1}{n^2 - 1}.$

(IX) $\displaystyle\sum_{n=1}^{\infty} \frac{n^2 + 3n + 1}{n^2 (n + 1)^2}.$

(V) $\displaystyle\sum_{n=1}^{\infty} \frac{1}{4n^2 + 16n + 7}.$

(X) $\displaystyle\sum_{n=1}^{\infty} \frac{n^2 (n + 1)^2}{n!}.$

Solution to Exercise 7.9.

(I). Notice first that we have a series of positive terms since it starts at $n = 2$. Now take $a_n = \frac{4n-1}{(n+2)(n-1)^2}$ and $b_n = \frac{1}{n^2}$. Then,

$$\lim_{n \to \infty} \frac{a_n}{b_n} = \lim_{n \to \infty} \frac{\frac{4n-1}{(n+2)(n-1)^2}}{\frac{1}{n^2}} = \lim_{n \to \infty} \frac{(4n - 1)n^2}{(n + 2)(n - 1)^2} = 4.$$

Thus, since the series $\sum_{n=2}^{\infty} b_n = \sum_{n=2}^{\infty} \frac{1}{n^2}$ converges (p-series with $p = 2 > 1$) and $\lim_{n\to\infty} \frac{a_n}{b_n} \in (0, \infty)$, by the Limit Comparison Test the series converges.

Let us now find the value of the series. In order to do so, let us apply partial fraction decomposition to $\frac{4n-1}{(n+2)(n-1)^2}$ as follows

$$\frac{4n-1}{(n+2)(n-1)^2} = \frac{A}{n+2} + \frac{B}{n-1} + \frac{C}{(n-1)^2}.$$

Then,

$$4n - 1 = A(n-1)^2 + B(n+2)(n-1) + C(n+2).$$

For $n = -2$ we have $A = -1$, and for $n = 1$ we obtain $C = 1$. Finally, since $A = -1$ and $C = 1$, we have for $n = 0$ that $-1 = -1 - 2B + 2$; so $B = 1$. Thus, we can write the series as follows:

$$\sum_{n=2}^{\infty} \frac{4n-1}{(n+2)(n-1)^2} = \sum_{n=2}^{\infty} \left(-\frac{1}{n+2} + \frac{1}{n-1} + \frac{1}{(n-1)^2} \right)$$

$$= \sum_{n=2}^{\infty} \left(-\frac{1}{n+2} + \frac{1}{n-1} + \frac{1}{(n-1)^2} + \frac{1}{n} - \frac{1}{n} + \frac{1}{n+1} - \frac{1}{n+1} \right)$$

$$= \sum_{n=2}^{\infty} \left[\frac{1}{n-1} + \frac{1}{n} + \frac{1}{n+1} - \left(\frac{1}{n} + \frac{1}{n+1} + \frac{1}{n+2} \right) + \frac{1}{(n-1)^2} \right].$$

On the one hand, note that the series $\sum_{n=2}^{\infty} \frac{1}{(n-1)^2}$ is convergent. On the other hand, let us study the series

$$\sum_{n=2}^{\infty} \left[\frac{1}{n-1} + \frac{1}{n} + \frac{1}{n+1} - \left(\frac{1}{n} + \frac{1}{n+1} + \frac{1}{n+2} \right) \right].$$

By taking $c_n = \frac{1}{n-1} + \frac{1}{n} + \frac{1}{n+1}$ we have that $c_{n+1} = \frac{1}{n} + \frac{1}{n+1} + \frac{1}{n+2}$. Thus, this latter series is telescopic and satisfies

$$\sum_{n=2}^{\infty} \frac{4n-1}{(n+2)(n-1)^2} = c_2 - \lim_{n\to\infty} c_n$$

$$= 1 + \frac{1}{2} + \frac{1}{3} - \lim_{n\to\infty} \left(\frac{1}{n-1} + \frac{1}{n} + \frac{1}{n+1} \right)$$

$$= \frac{11}{6}.$$

In particular, it is convergent. Therefore, we can split the series in the following way:

$$\sum_{n=2}^{\infty} \frac{4n-1}{(n+2)(n-1)^2}$$

$$= \sum_{n=2}^{\infty} \left[\frac{1}{n-1} + \frac{1}{n} + \frac{1}{n+1} - \left(\frac{1}{n} + \frac{1}{n+1} + \frac{1}{n+2} \right) \right] + \sum_{n=2}^{\infty} \frac{1}{(n-1)^2}$$

$$= \frac{11}{6} + \frac{\pi^2}{6} = \frac{11 + \pi^2}{6}.$$

(II). First, note that we have a series of positive terms. Now take $a_n = \frac{1}{n(n+1)}$ and $b_n = \frac{1}{n^2}$. Then,

$$\lim_{n \to \infty} \frac{a_n}{b_n} = \lim_{n \to \infty} \frac{\frac{1}{n(n+1)}}{\frac{1}{n^2}} = \lim_{n \to \infty} \frac{n^2}{n(n+1)} = 1.$$

Thus, since the series $\sum_{n=1}^{\infty} b_n = \sum_{n=1}^{\infty} \frac{1}{n^2}$ converges (p-series with $p = 2 > 1$) and $\lim_{n \to \infty} \frac{a_n}{b_n} \in (0, \infty)$, we have by the Limit Comparison Test that the series converges.

Let us find the value of the series by applying partial fraction decomposition to $\frac{1}{n(n+1)}$ as follows

$$\frac{1}{n(n+1)} = \frac{A}{n} + \frac{B}{n+1}.$$

Thus,

$$1 = A(n+1) + Bn.$$

For $n = 0$ we have $A = 1$, and for $n = -1$ we obtain $B = -1$. Therefore, we can write the series as below

$$\sum_{n=1}^{\infty} \frac{1}{n(n+1)} = \sum_{n=1}^{\infty} \left(\frac{1}{n} - \frac{1}{n+1} \right).$$

By taking $c_n = \frac{1}{n}$ we have that $c_{n+1} = \frac{1}{n+1}$, so the series is a telescopic series and satisfies

$$\sum_{n=1}^{\infty} \frac{1}{n(n+1)} = c_1 - \lim_{n \to \infty} c_n = 1 - \lim_{n \to \infty} \frac{1}{n} = 1.$$

(III). Observe that we have a series of positive terms. Now take $a_n = \frac{2n+3}{n(n-1)(n+2)}$ and $b_n = \frac{1}{n^2}$. Then,

$$\lim_{n \to \infty} \frac{a_n}{b_n} = \lim_{n \to \infty} \frac{\frac{2n+3}{n(n-1)(n+2)}}{\frac{1}{n^2}} = \lim_{n \to \infty} \frac{2n^3 + 3n^2}{n(n-1)(n+2)} = 2.$$

Thus, since the series $\sum_{n=2}^{\infty} b_n = \sum_{n=2}^{\infty} \dfrac{1}{n^2}$ converges (p-series with $p = 2 > 1$) and $\lim_{n\to\infty} \frac{a_n}{b_n} \in (0, \infty)$, we have by the Limit Comparison Test that the series converges.

In order to find its value, let us apply partial fraction decomposition to $\frac{2n+3}{n(n+1)(n+2)}$ as follows:

$$\frac{2n+3}{n(n-1)(n+2)} = \frac{A}{n} + \frac{B}{n-1} + \frac{C}{n+2}.$$

Then,

$$2n + 3 = A(n-1)(n+2) + Bn(n+2) + Cn(n-1).$$

For $n = 0$ we have $A = -\frac{3}{2}$, for $n = 1$ we obtain $B = \frac{5}{3}$, and for $n = -2$ we get $C = -\frac{1}{6}$. Therefore, we can write the series in the following way:

$$\sum_{n=2}^{\infty} \frac{2n+3}{n(n-1)(n+2)} = \sum_{n=2}^{\infty} \left(-\frac{3}{2n} + \frac{5}{3(n-1)} - \frac{1}{6(n+2)} \right)$$

$$= \frac{1}{6} \sum_{n=2}^{\infty} \left(-\frac{9}{n} + \frac{10}{n-1} - \frac{1}{n+2} \right)$$

$$= \frac{1}{6} \sum_{n=2}^{\infty} \left(-\frac{10}{n} + \frac{1}{n} + \frac{10}{n-1} - \frac{1}{n+2} + \frac{1}{n+1} - \frac{1}{n+1} \right)$$

$$= \frac{1}{6} \sum_{n=2}^{\infty} \left[\frac{10}{n-1} + \frac{1}{n} + \frac{1}{n+1} - \left(\frac{10}{n} + \frac{1}{n+1} + \frac{1}{n+2} \right) \right].$$

By taking $c_n = \frac{10}{n-1} + \frac{1}{n} + \frac{1}{n+1}$ we have that $c_{n+1} = \frac{10}{n} + \frac{1}{n+1} + \frac{1}{n+2}$, which implies that the series is a telescopic series. Hence,

$$\sum_{n=2}^{\infty} \frac{2n+3}{n(n-1)(n+2)} = \frac{1}{6} \left(c_2 - \lim_{n\to\infty} c_n \right)$$

$$= \frac{1}{6} \left[10 + \frac{1}{2} + \frac{1}{3} - \lim_{n\to\infty} \left(\frac{10}{n-1} + \frac{1}{n} + \frac{1}{n+1} \right) \right]$$

$$= \frac{65}{36}.$$

(IV). First of all, note that we have a series of positive terms and it is well defined since the series starts at $n = 2$. Now take $a_n = \frac{1}{n^2-1}$ and $b_n = \frac{1}{n^2}$. Then,

$$\lim_{n\to\infty} \frac{a_n}{b_n} = \lim_{n\to\infty} \frac{\frac{1}{n^2-1}}{\frac{1}{n^2}} = \lim_{n\to\infty} \frac{n^2}{n^2-1} = 1.$$

As the series $\displaystyle\sum_{n=2}^{\infty} \frac{1}{n^2} = \sum_{n=2}^{\infty} \frac{1}{n^2}$ converges (p-series with $p = 2 > 1$) and $\lim_{n\to\infty} \frac{a_n}{b_n} \in (0, \infty)$, we have by the Limit Comparison Test that the series converges.

In order to find the limit of the series, let us apply partial fraction decomposition to $\frac{1}{n^2-1} = \frac{1}{(n+1)(n-1)}$ as follows:

$$\frac{1}{(n+1)(n-1)} = \frac{A}{n+1} + \frac{B}{n-1}.$$

Then,

$$1 = A(n-1) + B(n+1).$$

For $n = -1$ we have $A = -\frac{1}{2}$, and for $n = 1$ we obtain $B = \frac{1}{2}$. Hence, the series can be written in the following way:

$$\sum_{n=2}^{\infty} \frac{1}{n^2 - 1} = \sum_{n=2}^{\infty} \left(\frac{1}{2(n-1)} - \frac{1}{2(n+1)} \right)$$

$$= \frac{1}{2} \sum_{n=2}^{\infty} \left(\frac{1}{n-1} - \frac{1}{n+1} \right)$$

$$= \frac{1}{2} \sum_{n=2}^{\infty} \left(\frac{1}{n-1} + \frac{1}{n} - \frac{1}{n} - \frac{1}{n+1} \right)$$

$$= \frac{1}{2} \sum_{n=2}^{\infty} \left[\frac{1}{n-1} + \frac{1}{n} - \left(\frac{1}{n} + \frac{1}{n+1} \right) \right].$$

By taking $c_n = \frac{1}{n-1} + \frac{1}{n}$, we have $c_{n+1} = \frac{1}{n} + \frac{1}{n+1}$, which shows that the series is a telescopic series. Hence,

$$\sum_{n=2}^{\infty} \frac{1}{n^2 - 1} = \frac{1}{2} \left(c_2 - \lim_{n\to\infty} c_n \right)$$

$$= \frac{1}{2} \left[1 + \frac{1}{2} - \lim_{n\to\infty} \left(\frac{1}{n-1} + \frac{1}{n} \right) \right] = \frac{1}{2} \left(1 + \frac{1}{2} \right) = \frac{3}{4}.$$

(v). First, note that we have a series of positive terms, and now take $a_n = \frac{1}{4n^2 + 16n + 7}$ and $b_n = \frac{1}{n^2}$. Then,

$$\lim_{n\to\infty} \frac{a_n}{b_n} = \lim_{n\to\infty} \frac{\frac{1}{4n^2+16n+7}}{\frac{1}{n^2}} = \lim_{n\to\infty} \frac{n^2}{4n^2 + 16n + 7} = \frac{1}{4}.$$

Hence, since $\displaystyle\sum_{n=1}^{\infty} b_n = \sum_{n=1}^{\infty} \frac{1}{n^2}$ converges (*p*-series with $p = 2 > 1$) and $\lim_{n\to\infty} \frac{a_n}{b_n} \in (0,\infty)$, we have by the Limit Comparison Test that the series converges.

To find the value of the series, apply partial fraction decomposition to $\frac{1}{4n^2+16n+7} = \frac{1}{(2n+1)(2n+7)}$ (use the second degree equation formula if necessary) as follows

$$\frac{1}{4n^2 + 16n + 7} = \frac{A}{2n + 1} + \frac{B}{2n + 7}.$$

Then,

$$1 = A(2n + 7) + B(2n + 1).$$

For $n = -\frac{1}{2}$ we have $A = \frac{1}{6}$, and for $n = -\frac{7}{2}$ we obtain $B = -\frac{1}{6}$. Therefore, the series can be written as

$$\sum_{n=1}^{\infty} \frac{1}{4n^2 + 16n + 7} = \sum_{n=1}^{\infty} \left(-\frac{1}{6(2n + 1)} - \frac{1}{2(n + 1)} \right)$$

$$= \frac{1}{6} \sum_{n=1}^{\infty} \left(\frac{1}{2n + 1} - \frac{1}{2n + 7} \right)$$

$$= \frac{1}{6} \sum_{n=1}^{\infty} \left(\frac{1}{2n + 1} + \frac{1}{2n + 3} - \frac{1}{2n + 3} + \frac{1}{2n + 5} - \frac{1}{2n + 5} - \frac{1}{2n + 7} \right)$$

$$= \frac{1}{6} \sum_{n=1}^{\infty} \left[\frac{1}{2n + 1} + \frac{1}{2n + 3} + \frac{1}{2n + 5} - \left(\frac{1}{2n + 3} + \frac{1}{2n + 5} + \frac{1}{2n + 7} \right) \right].$$

If we take $c_n = \frac{1}{2n+1} + \frac{1}{2n+3} + \frac{1}{2n+5}$, we have $c_{n+1} = \frac{1}{2n+3} + \frac{1}{2n+5} + \frac{1}{2n+7}$, that is, the series is a telescopic series. Thus,

$$\sum_{n=1}^{\infty} \frac{1}{4n^2 + 16n + 7} = \frac{1}{6} \left(c_1 - \lim_{n\to\infty} c_n \right)$$

$$= \frac{1}{6} \left[\frac{1}{3} + \frac{1}{5} + \frac{1}{7} - \lim_{n\to\infty} \left(\frac{1}{2n + 1} + \frac{1}{2n + 3} + \frac{1}{2n + 5} \right) \right]$$

$$= \frac{71}{630}.$$

(VI). Notice that it is a series of positive terms since it starts at $n = 2$. Take $a_n = \frac{1}{(n+1)^2-4}$ and $b_n = \frac{1}{n^2}$. Then,

$$\lim_{n\to\infty} \frac{a_n}{b_n} = \lim_{n\to\infty} \frac{\frac{1}{(n+1)^2-4}}{\frac{1}{n^2}} = \lim_{n\to\infty} \frac{n^2}{(n+1)^2 - 4} = 1.$$

Therefore, as $\sum_{n=2}^{\infty} b_n = \sum_{n=2}^{\infty} \frac{1}{n^2}$ is convergent (p-series with $p = 2 > 1$) and $\lim_{n\to\infty} \frac{a_n}{b_n} \in (0, \infty)$, we have that the series converges by the Limit Comparison Test.

In order to find the limit of the series, let us apply partial fraction decomposition to $\frac{1}{(n+1)^2-4} = \frac{1}{n^2+2n+1-4} = \frac{1}{n^2+2n-3} = \frac{1}{(n-1)(n+3)}$:

$$\frac{1}{(n-1)(n+3)} = \frac{A}{n-1} + \frac{B}{n+3}.$$

Then,

$$1 = A(n+3) + B(n-1).$$

For $n = 1$ we have $A = \frac{1}{4}$, and for $n = -3$ we obtain $B = -\frac{1}{4}$. Thus, we can write the series as

$$\sum_{n=2}^{\infty} \frac{1}{(n+1)^2-4} = \sum_{n=2}^{\infty} \left(\frac{1}{4(n-1)} - \frac{1}{4(n+3)} \right)$$

$$= \frac{1}{4} \sum_{n=2}^{\infty} \left(\frac{1}{n-1} - \frac{1}{n} + \frac{1}{n} - \frac{1}{n+1} + \frac{1}{n+1} - \frac{1}{n+2} + \frac{1}{n+2} - \frac{1}{n+3} \right)$$

$$= \frac{1}{4} \sum_{n=2}^{\infty} \left[\frac{1}{n-1} + \frac{1}{n} + \frac{1}{n+1} + \frac{1}{n+2} - \left(\frac{1}{n} + \frac{1}{n+1} + \frac{1}{n+2} + \frac{1}{n+3} \right) \right].$$

By taking $c_n = \frac{1}{n-1} + \frac{1}{n} + \frac{1}{n+1} + \frac{1}{n+2}$ we have that $c_{n+1} = \frac{1}{n} + \frac{1}{n+1} + \frac{1}{n+2} + \frac{1}{n+3}$, which shows that the series is a telescopic series. Hence,

$$\sum_{n=2}^{\infty} \frac{1}{(n+1)^2-4} = \frac{1}{4} \left(c_2 - \lim_{n\to\infty} c_n \right)$$

$$= \frac{1}{4} \left[1 + \frac{1}{2} + \frac{1}{3} + \frac{1}{4} - \lim_{n\to\infty} \left(\frac{1}{n-1} + \frac{1}{n} + \frac{1}{n+1} + \frac{1}{n+2} \right) \right]$$

$$= \frac{25}{48}.$$

(VII). Observe first that we have a series of positive terms. In order to study the convergence of the series, take $a_n = \frac{3n^2+7n+6}{n(n+1)(n+2)(n+3)}$ and $b_n = \frac{1}{n^2}$. Then,

$$\lim_{n\to\infty} \frac{a_n}{b_n} = \lim_{n\to\infty} \frac{\frac{3n^2+7n+6}{n(n+1)(n+2)(n+3)}}{\frac{1}{n^2}} = \lim_{n\to\infty} \frac{3n^4 + 7n^3 + 6n^2}{n(n+1)(n+2)(n+3)} = 3.$$

Hence, since $\sum_{n=1}^{\infty} b_n = \sum_{n=1}^{\infty} \frac{1}{n^2}$ is convergent (p-series with $p = 2 > 1$) and $\lim_{n\to\infty} \frac{a_n}{b_n} \in (0, \infty)$, we have by the Limit Comparison Test that the series is convergent.

To find its value, let us decompose $\frac{3n^2+7n+6}{n(n+1)(n+2)(n+3)}$ into partial fractions as follows

$$\frac{3n^2 + 7n + 6}{n(n + 1)(n + 2)(n + 3)} = \frac{A}{n} + \frac{B}{n+1} + \frac{C}{n+2} + \frac{D}{n+3}.$$

Then,

$$3n^2 + 7n + 6 = A(n + 1)(n + 2)(n + 3) + Bn(n + 2)(n + 3)$$
$$+ Cn(n + 1)(n + 3) + Dn(n + 1)(n + 2).$$

For $n = 0$ we obtain $A = 1$. For $n = -1$ we have $B = -1$. For $n = -2$, $C = 2$. And for $n = -3$ we have $D = -2$. Thus, we can write the series as

$$\sum_{n=1}^{\infty} \frac{3n^2 + 7n + 6}{n(n + 1)(n + 2)(n + 3)} = \sum_{n=1}^{\infty} \left(\frac{1}{n} - \frac{1}{n+1} + \frac{2}{n+2} - \frac{2}{n+3} \right)$$
$$= \sum_{n=1}^{\infty} \left[\frac{1}{n} + \frac{2}{n+2} - \left(\frac{1}{n+1} - \frac{2}{n+3} \right) \right].$$

If we take $c_n = \frac{1}{n} + \frac{2}{n+2}$, we have that $c_{n+1} = \frac{1}{n+1} + \frac{2}{n+3}$, which implies that the series is in fact a telescopic series. So,

$$\sum_{n=1}^{\infty} \frac{3n^2 + 7n + 6}{n(n + 1)(n + 2)(n + 3)} = c_1 + \lim_{n\to\infty} c_n = \frac{5}{3} + \lim_{n\to\infty} \left(\frac{1}{n} + \frac{2}{n+2} \right) = \frac{5}{3}.$$

(VIII). Let $a_n = \frac{1}{(n-1+\sqrt{3})(n-2+\sqrt{3})(n+\sqrt{3})}$, then

$$\frac{a_{n+1}}{a_n} = \frac{\frac{1}{(n+\sqrt{3})(n-1+\sqrt{3})(n+1+\sqrt{3})}}{\frac{1}{(n-1+\sqrt{3})(n-2+\sqrt{3})(n+\sqrt{3})}} = \frac{(n - 1 + \sqrt{3})(n - 2 + \sqrt{3})(n + \sqrt{3})}{(n + \sqrt{3})(n - 1 + \sqrt{3})(n + 1 + \sqrt{3})}.$$
$$= \frac{n - 2 + \sqrt{3}}{n + 1 + \sqrt{3}},$$

that is, the series is a hypergeometric series since $\frac{a_{n+1}}{a_n} = \frac{\alpha n + \beta}{\alpha n + \gamma}$, where $\alpha = 1 > 0$, $\beta = -2 + \sqrt{3}$ and $\gamma = 1 + \sqrt{3}$. Recall that a hypergeometric series is convergent if $\gamma > \alpha + \beta$. In this case, we have that $\alpha + \beta = 1 + (-2 + \sqrt{3}) = -1 + \sqrt{3} < 1 + \sqrt{3} = \gamma$, so the series converges. Moreover, the sum of the series is $\frac{\gamma a_1}{\gamma - \alpha - \beta}$, so

$$\sum_{n=1}^{\infty} \frac{1}{(n-1+\sqrt{3})(n-2+\sqrt{3})(n+\sqrt{3})} = \frac{(1+\sqrt{3}) \cdot \frac{1}{\sqrt{3}(-1+\sqrt{3})(1+\sqrt{3})}}{1+\sqrt{3}-1+2-\sqrt{3}}$$

$$= \frac{1}{2(3-\sqrt{3})}.$$

(IX). Notice that we have a series of positive terms and let $a_n = \frac{n^2+3n+1}{n^2(n+1)^2}$ and $b_n = \frac{1}{n^2}$. Then,

$$\lim_{n\to\infty} \frac{a_n}{b_n} = \lim_{n\to\infty} \frac{\frac{n^2+3n+1}{n^2(n+1)^2}}{\frac{1}{n^2}}$$

$$= \lim_{n\to\infty} \frac{(n^2+3n+1)n^2}{n^2(n+1)^2}$$

$$= \lim_{n\to\infty} \frac{n^2+3n+1}{(n+1)^2} = 1 < \infty.$$

Thus, since $\sum_{n=1}^{\infty} \frac{1}{n^2}$ converges (p-series with $p = 2 > 1$) and $\lim_{n\to\infty} \frac{a_n}{b_n} \in (0, \infty)$, we have by the Limit Comparison Test that the series converges.

In order to find the value of the series, notice that we can write it as follows

$$\sum_{n=1}^{\infty} \frac{n^2+3n+1}{n^2(n+1)^2} = \sum_{n=1}^{\infty} \frac{(n+1)^2+n}{n^2(n+1)^2} = \sum_{n=1}^{\infty} \left(\frac{(n+1)^2}{n^2(n+1)^2} + \frac{n}{n^2(n+1)^2} \right)$$

$$= \sum_{n=1}^{\infty} \left(\frac{1}{n^2} + \frac{1}{n(n+1)^2} \right).$$

Now we will apply partial fraction decomposition to $\frac{1}{n(n+1)^2}$:

$$\frac{1}{n(n+1)^2} = \frac{A}{n} + \frac{B}{n+1} + \frac{C}{(n+1)^2};$$

then

$$1 = A(n+1)^2 + Bn(n+1) + Cn. \tag{7.2}$$

By taking $n = 0$, we obtain $A = 1$. If we take $n = -1$, then $C = -1$. Finally, by substituting $A = 1$ and $C = -1$ into (7.2) we arrive at

$$1 = (B+1)n^2 + (B+1)n + 1,$$

and therefore $B = -1$. So,

$$\sum_{n=1}^{\infty} \frac{n^2 + 3n + 1}{n^2(n+1)^2} = \sum_{n=1}^{\infty} \left(\frac{1}{n^2} + \frac{1}{n} - \frac{1}{n+1} - \frac{1}{(n+1)^2} \right)$$

$$= \sum_{n=1}^{\infty} \left[\frac{1}{n^2} + \left(\frac{1}{n} - \frac{1}{n+1} \right) - \frac{1}{(n+1)^2} \right].$$

Clearly the series $\sum_{n=1}^{\infty} \frac{1}{n^2}$ and $\sum_{n=1}^{\infty} \frac{1}{(n+1)^2}$ are convergent. Let us show that $\sum_{n=1}^{\infty} \left(\frac{1}{n} - \frac{1}{n+1} \right)$ is also convergent. Indeed, by taking $b_n = \frac{1}{n}$ we have that

$$\sum_{n=1}^{\infty} \left(\frac{1}{n} - \frac{1}{n+1} \right) = \sum_{n=1}^{\infty} (b_n - b_{n+1}),$$

that is, a telescopic series. Thus,

$$\sum_{n=1}^{\infty} \left(\frac{1}{n} - \frac{1}{n+1} \right) = b_1 - \lim_{n \to \infty} b_n = 1 - \lim_{n \to \infty} \frac{1}{n} = 1,$$

so convergent. Hence, we can decompose the original series as

$$\sum_{n=1}^{\infty} \frac{n^2 + 3n + 1}{n^2(n+1)^2} = \sum_{n=1}^{\infty} \frac{1}{n^2} + \sum_{n=1}^{\infty} \left(\frac{1}{n} - \frac{1}{n+1} \right) - \sum_{n=1}^{\infty} \frac{1}{(n+1)^2}$$

$$= \frac{\pi^2}{6} + 1 - \left(\frac{\pi^2}{6} - 1 \right) = 2.$$

(X). Note that we have a series of positive terms and let $a_n = \frac{n^2(n+1)^2}{n!}$. Then,

$$\lim_{n \to \infty} \frac{a_{n+1}}{a_n} = \lim_{n \to \infty} \frac{\frac{(n+1)^2(n+2)^2}{(n+1)!}}{\frac{n^2(n+1)^2}{n!}}$$

$$= \lim_{n \to \infty} \frac{(n+1)^2(n+2)^2 n!}{n^2(n+1)^2(n+1)!} = \lim_{n \to \infty} \frac{(n+2)^2}{n^2(n+1)} = 0 < 1.$$

Thus, by the Ratio Test, the series converges.

To find the value of the series, observe that we can write it in the following way:

$$\sum_{n=1}^{\infty} \frac{n^2(n+1)^2}{n!} = \sum_{n=1}^{\infty} \frac{n(n+1)^2}{(n-1)!} = \sum_{n=1}^{\infty} \frac{n^3 + 2n^2 + n}{(n-1)!}$$

$$= \sum_{n=1}^{\infty} \frac{(n-1)(n-2)(n-3) + 8n^2 - 10n + 6}{(n-1)!}$$

$$= \sum_{n=1}^{\infty} \left(\frac{(n-1)(n-2)(n-3)}{(n-1)!} + \frac{8n^2 - 10n + 6}{(n-1)!} \right)$$

$$= \sum_{n=1}^{\infty} \left(\frac{(n-1)(n-2)(n-3)}{(n-1)!} + \frac{8(n-1)(n-2) + 14n - 10}{(n-1)!} \right)$$

$$= \sum_{n=1}^{\infty} \left(\frac{(n-1)(n-2)(n-3)}{(n-1)!} + \frac{8(n-1)(n-2)}{(n-1)!} + \frac{14n - 10}{(n-1)!} \right)$$

$$= \sum_{n=1}^{\infty} \left(\frac{(n-1)(n-2)(n-3)}{(n-1)!} + \frac{8(n-1)(n-2)}{(n-1)!} + \frac{14(n-1) + 4}{(n-1)!} \right)$$

$$= \sum_{n=1}^{\infty} \left(\frac{(n-1)(n-2)(n-3)}{(n-1)!} + \frac{8(n-1)(n-2)}{(n-1)!} + \frac{14(n-1)}{(n-1)!} + \frac{4}{(n-1)!} \right).$$

Now consider the series $\displaystyle\sum_{n=1}^{\infty} \frac{(n-1)(n-2)(n-3)}{(n-1)!}$, $\displaystyle\sum_{n=1}^{\infty} \frac{8(n-1)(n-2)}{(n-1)!}$ and $\displaystyle\sum_{n=1}^{\infty} \frac{14(n-1)}{(n-1)!}$, and observe that they satisfy the following equalities:

$$\sum_{n=1}^{\infty} \frac{(n-1)(n-2)(n-3)}{(n-1)!} = \sum_{n=4}^{\infty} \frac{1}{(n-4)!},$$

$$\sum_{n=1}^{\infty} \frac{8(n-1)(n-2)}{(n-1)!} = 8 \sum_{n=3}^{\infty} \frac{1}{(n-3)!}$$

and

$$\sum_{n=1}^{\infty} \frac{14(n-1)}{(n-1)!} = 14 \sum_{n=2}^{\infty} \frac{1}{(n-2)!}.$$

Clearly the previous series are all convergent as well as $\sum\limits_{n=1}^{\infty} \dfrac{4}{(n-1)!}$. Hence, the original series can be decomposed into the following combination of series:

$$\sum_{n=1}^{\infty} \frac{n^2(n+1)^2}{n!} = \sum_{n=4}^{\infty} \frac{1}{(n-4)!} + 8\sum_{n=3}^{\infty} \frac{1}{(n-3)!}$$

$$+ 14\sum_{n=2}^{\infty} \frac{1}{(n-2)!} + 4\sum_{n=1}^{\infty} \frac{1}{(n-1)!}$$

$$= e + 8e + 14e + 4e = 27e.$$

$\square$

Exercise 7.10. *Study the convergence of the following series and, when convergent, find their sum.*

(I) $\displaystyle\sum_{n=1}^{\infty} \frac{3^n(n-3)}{n!}.$

(II) $\displaystyle\sum_{n=1}^{\infty} \frac{n^3-n+1}{n!3^n}.$

(III) $\displaystyle\sum_{n=3}^{\infty} \frac{3n^2+8n+6}{(n+2)!}.$

(IV) $\displaystyle\sum_{n=1}^{\infty} \frac{n-1}{n!(n+2)}.$

(V) $\displaystyle\sum_{n=1}^{\infty} \frac{n^3-1}{n!}.$

(VI) $\displaystyle\sum_{n=1}^{\infty} \frac{n^2+1}{(n+1)!}.$

(VII) $\displaystyle\sum_{n=2}^{\infty} \frac{n^2+5n+7}{(n+2)!}.$

(VIII) $\displaystyle\sum_{n=1}^{\infty} (-1)^n \frac{2n+1}{n(n+1)}.$

(IX) $\displaystyle\sum_{n=1}^{\infty} \frac{n(n+1)}{2^n}.$

(X) $\displaystyle\sum_{n=1}^{\infty} \frac{n^2}{3^n}.$

(XI) $\displaystyle\sum_{n=1}^{\infty} (n+1)x^n, \text{ where } x \in \mathbb{R}.$

(XII) $\displaystyle\sum_{n=1}^{\infty} (-1)^n \frac{n^2-n}{3^n}.$

(XIII) $\displaystyle\sum_{n=1}^{\infty} \frac{1}{n\sqrt{n+1}+(n+1)\sqrt{n}}.$

(XIV) $\displaystyle\sum_{n=1}^{\infty} \left(\log\left(\frac{n}{n+1}\right)^n - \frac{1}{2n} + 1 \right).$

Solution to Exercise 7.10.

(I). Note that we have a series of positive terms and let $a_n = \frac{3^n(n-3)}{n!}$. Then,

$$\lim_{n\to\infty} \frac{a_{n+1}}{a_n} = \lim_{n\to\infty} \frac{\frac{3^{n+1}(n-2)}{(n+1)!}}{\frac{3^n(n-3)}{n!}}$$

$$= \lim_{n\to\infty} \frac{n!3^{n+1}(n-2)}{(n+1)!3^n(n-3)} = 3\lim_{n\to\infty} \frac{n-2}{(n+1)(n-3)} = 0 < 1.$$

Hence, by the Ratio Test, the series converges.

To find the value of the series, observe that

$$\sum_{n=1}^{\infty} \frac{3^n(n-3)}{n!} = \sum_{n=1}^{\infty} \frac{n3^n - 3^{n+1}}{n!}$$

$$= \sum_{n=1}^{\infty} \left(\frac{n3^n}{n!} - \frac{3^{n+1}}{n!} \right)$$

$$= \sum_{n=1}^{\infty} \left(\frac{3^n}{(n-1)!} - \frac{3^{n+1}}{n!} \right).$$

By taking $b_n = \frac{3^n}{(n-1)!}$, note that we have

$$\sum_{n=1}^{\infty} \frac{3^n(n-3)}{n!} = \sum_{n=1}^{\infty} (b_n - b_{n+1}),$$

that is, a telescopic series. So,

$$\sum_{n=1}^{\infty} \frac{3^n(n-3)}{n!} = b_1 - \lim_{n \to \infty} b_n = 3 - \lim_{n \to \infty} \frac{3^n}{(n-1)!}.$$

To find the limit $\lim\limits_{n \to \infty} \frac{3^n}{(n-1)!}$, note that

$$\lim_{n \to \infty} \frac{\frac{3^{n+1}}{n!}}{\frac{3^n}{(n-1)!}} = \lim_{n \to \infty} \frac{\frac{3^{n+1}}{n!}}{\frac{3^n}{(n-1)!}} = \lim_{n \to \infty} \frac{3^{n+1}(n-1)!}{3^n n!} = 3 \cdot \lim_{n \to \infty} \frac{1}{n} = 0 < 1.$$

Thus, by the Ratio Test for sequences, we have that $\lim\limits_{n \to \infty} \frac{3^n}{(n-1)!} = 0$. So,

$$\sum_{n=1}^{\infty} \frac{3^n(n-3)}{n!} = 3.$$

(II). Clearly we have a series of positive terms and let $a_n = \frac{n^3 - n + 1}{n!3^n}$. Then,

$$\lim_{n \to \infty} \frac{a_{n+1}}{a_n} = \lim_{n \to \infty} \frac{\frac{(n+1)^3 - (n+1) + 1}{(n+1)!3^{n+1}}}{\frac{n^3 - n + 1}{n!3^n}}$$

$$= \lim_{n \to \infty} \frac{((n+1)^3 - (n+1) + 1)n!3^n}{(n^3 - n + 1)(n+1)!3^{n+1}}$$

$$= \frac{1}{3} \lim_{n \to \infty} \frac{(n+1)^3 - (n+1) + 1}{(n^3 - n + 1)(n+1)} = 0 < 1.$$

Therefore, by the Ratio Test, the series converges.

To find the value of the series, note that we can write it as follows:

$$\sum_{n=1}^{\infty} \frac{n^3 - n + 1}{n!3^n} = \sum_{n=1}^{\infty} \left(\frac{n^3 - n}{n!3^n} + \frac{1}{n!3^n} \right)$$

$$= \sum_{n=1}^{\infty} \left(\frac{n^2 - 1}{(n-1)!3^n} + \frac{1}{n!3^n} \right).$$

Since $\sum_{n=1}^{\infty} \frac{n^3 - n + 1}{n!3^n}$ converges and $\frac{n^2-1}{(n-1)!3^n} \geq 0$ and $\frac{1}{n!3^n} \geq 0$ for every $n \in \mathbb{N}$,

we can split the series $\sum_{n=1}^{\infty} \frac{n^3 - n + 1}{n!3^n}$ into $\sum_{n=1}^{\infty} \frac{n^2 - 1}{(n-1)!3^n}$ and $\sum_{n=1}^{\infty} \frac{1}{n!3^n}$ since

the latter two series will also be convergent $\Big($otherwise at least one of them

would diverge to ∞ contradicting the convergence of $\sum_{n=1}^{\infty} \frac{n^3 - n + 1}{n!3^n}\Big).$

Hence,

$$\sum_{n=1}^{\infty} \frac{n^3 - n + 1}{n!3^n} = \sum_{n=1}^{\infty} \frac{n^2 - 1}{(n-1)!3^n} + \sum_{n=1}^{\infty} \frac{1}{n!3^n}$$

$$= \sum_{n=2}^{\infty} \frac{(n+1)(n-1)}{(n-1)!3^n} + \sum_{n=1}^{\infty} \frac{1}{n!3^n}$$

$$= \sum_{n=2}^{\infty} \frac{n+1}{(n-2)!3^n} + \sum_{n=1}^{\infty} \frac{1}{n!3^n}$$

$$= \sum_{n=2}^{\infty} \frac{n-2+3}{(n-2)!3^n} + \sum_{n=1}^{\infty} \frac{1}{n!3^n}$$

$$= \sum_{n=2}^{\infty} \left(\frac{n-2}{(n-2)!3^n} + \frac{3}{(n-2)!3^n} \right) + \sum_{n=1}^{\infty} \frac{1}{n!3^n}.$$

Using the same ideas as before, notice that we can split the series $\sum_{n=2}^{\infty} \left(\frac{n-2}{(n-2)!3^n} + \frac{3}{(n-2)!3^n} \right)$ into $\sum_{n=2}^{\infty} \frac{n-2}{(n-2)!3^n}$ and $\sum_{n=2}^{\infty} \frac{3}{(n-2)!3^n}$.

Thus,

$$\sum_{n=1}^{\infty} \frac{n^3 - n + 1}{n!3^n} = \sum_{n=2}^{\infty} \frac{n-2}{(n-2)!3^n} + \sum_{n=2}^{\infty} \frac{3}{(n-2)!3^n} + \sum_{n=1}^{\infty} \frac{1}{n!3^n}$$

$$= \sum_{n=3}^{\infty} \frac{1}{(n-3)!3^n} + \sum_{n=2}^{\infty} \frac{3}{(n-2)!3^n} + \sum_{n=1}^{\infty} \frac{1}{n!3^n}$$

$$= \sum_{n=3}^{\infty} \left(\frac{1}{3}\right)^n \frac{1}{(n-3)!} + 3\sum_{n=2}^{\infty} \left(\frac{1}{3}\right)^n \frac{1}{(n-2)!} + \sum_{n=1}^{\infty} \left(\frac{1}{3}\right)^n \frac{1}{n!}$$

$$= \frac{1}{27}\sum_{n=3}^{\infty} \left(\frac{1}{3}\right)^{n-3} \frac{1}{(n-3)!} + \frac{1}{3}\sum_{n=2}^{\infty} \left(\frac{1}{3}\right)^n \frac{1}{(n-2)!} + \sum_{n=1}^{\infty} \left(\frac{1}{3}\right)^n \frac{1}{n!}.$$

Recall that the power series $\sum_{n=0}^{\infty} \frac{x^n}{n!}$ converges when $|x| < 1$ and, in which case, it represents the function e^x. Therefore, in our case, notice that $\sum_{n=3}^{\infty} \left(\frac{1}{3}\right)^{n-3} \frac{1}{(n-3)!}$ and $\sum_{n=2}^{\infty} \left(\frac{1}{3}\right)^n \frac{1}{(n-2)!}$ are series of the type $\sum_{n=0}^{\infty} \frac{x^n}{n!}$ with $x = \frac{1}{3}$, and $\sum_{n=1}^{\infty} \left(\frac{1}{3}\right)^n \frac{1}{n!}$ is of the form $\sum_{n=0}^{\infty} \frac{x^n}{n!}$ with $x = \frac{1}{3}$ but subtracting the first element of the series (when $n = 0$). Thus,

$$\sum_{n=1}^{\infty} \frac{n^3 - n + 1}{n!3^n} = \frac{1}{27}e^{1/3} + \frac{1}{3}e^{1/3} + e^{1/3} - 1 = \frac{37}{27}e^{1/3} - 1.$$

(III). Note that we have a series of positive terms and consider $a_n = \frac{3n^2+8n+6}{(n+2)!}$. Then,

$$\lim_{n\to\infty} \frac{a_{n+1}}{a_n} = \lim_{n\to\infty} \frac{\frac{3(n+1)^2+8(n+1)+6}{(n+3)!}}{\frac{3n^2+8n+6}{(n+2)!}}$$

$$= \lim_{n\to\infty} \frac{(3(n+1)^2 + 8(n+1) + 6)(n+2)!}{(3n^2 + 8n + 6)(n+3)!}$$

$$= \lim_{n\to\infty} \frac{3(n+1)^2 + 8(n+1) + 6}{(3n^2 + 8n + 6)(n+3)} = 0 < 1.$$

Hence, by the Ratio Test, the series converges.

To calculate the limit, notice that we can split the series as follows:

$$\sum_{n=3}^{\infty} \frac{3n^2 + 8n + 6}{(n+2)!} = \sum_{n=3}^{\infty} \frac{3(n+2)(n+1) - n}{(n+2)!}$$

$$= \sum_{n=3}^{\infty} \left(\frac{3(n+2)(n+1)}{(n+2)!} - \frac{n}{(n+2)!} \right)$$

$$= \sum_{n=3}^{\infty} \left(\frac{3}{n!} - \frac{n+2-2}{(n+2)!} \right)$$

$$= \sum_{n=3}^{\infty} \left(\frac{3}{n!} - \frac{n+2}{(n+2)!} + \frac{2}{(n+2)!} \right)$$

$$= \sum_{n=3}^{\infty} \left(\frac{3}{n!} - \frac{1}{(n+1)!} + \frac{2}{(n+2)!} \right).$$

Since the series $\sum_{n=3}^{\infty} \frac{1}{n!}$, $\sum_{n=3}^{\infty} \frac{1}{(n+1)!}$ and $\sum_{n=3}^{\infty} \frac{1}{(n+2)!}$ are convergent, we can split the series into the following combination of series:

$$\sum_{n=3}^{\infty} \frac{3n^2 + 8n + 6}{(n+2)!} = 3 \sum_{n=3}^{\infty} \frac{1}{n!} - \sum_{n=3}^{\infty} \frac{1}{(n+1)!} + 2 \sum_{n=3}^{\infty} \frac{1}{(n+2)!}$$

$$= 3 \left(e - \frac{5}{2} \right) - \left(e - \frac{8}{3} \right) + 2 \left(e - \frac{65}{24} \right)$$

$$= 4e - \frac{41}{4}.$$

(IV). Observe that we have a series of positive terms (except for $n = 1$ in which case we have 0) and let $a_n = \frac{n-1}{n!(n+2)}$. Then,

$$\lim_{n \to \infty} \frac{a_{n+1}}{a_n} = \lim_{n \to \infty} \frac{\frac{n}{(n+1)!(n+3)}}{\frac{n-1}{n!(n+2)}}$$

$$= \lim_{n \to \infty} \frac{n(n+2)n!}{(n-1)(n+3)(n+1)!}$$

$$= \lim_{n \to \infty} \frac{n(n+2)}{(n-1)(n+3)(n+1)}$$

$$= 0 < 1.$$

Thus, by the Ratio Test, the series converges. In order to find the value of the sum of this series, note that we can write the series in the following way:

$$\sum_{n=1}^{\infty} \frac{n-1}{n!(n+2)} = \sum_{n=1}^{\infty} \frac{n+2-3}{n!(n+2)}$$

$$= \sum_{n=1}^{\infty} \left(\frac{n+2}{n!(n+2)} - \frac{3}{n!(n+2)} \right)$$

$$= \sum_{n=1}^{\infty} \left(\frac{1}{n!} - \frac{3}{n!(n+2)} \right)$$

$$= \sum_{n=1}^{\infty} \left(\frac{1}{n!} - 3 \frac{n+1}{(n+2)(n+1)n!} \right)$$

$$= \sum_{n=1}^{\infty} \left(\frac{1}{n!} - 3 \frac{n}{(n+2)!} - 3 \frac{1}{(n+2)!} \right)$$

$$= \sum_{n=1}^{\infty} \left(\frac{1}{n!} - 3 \frac{n+2-2}{(n+2)!} - 3 \frac{1}{(n+2)!} \right)$$

$$= \sum_{n=1}^{\infty} \left(\frac{1}{n!} - 3 \frac{n+2}{(n+2)!} + 6 \frac{1}{(n+2)!} - 3 \frac{1}{(n+2)!} \right)$$

$$= \sum_{n=1}^{\infty} \left(\frac{1}{n!} - 3 \frac{1}{(n+1)!} + 6 \frac{1}{(n+2)!} - 3 \frac{1}{(n+2)!} \right)$$

$$= \sum_{n=1}^{\infty} \left(\frac{1}{n!} - 3 \frac{1}{(n+1)!} + 3 \frac{1}{(n+2)!} \right).$$

Clearly, the series $\displaystyle\sum_{n=1}^{\infty} \frac{1}{n!}$, $\displaystyle\sum_{n=1}^{\infty} \frac{1}{(n+1)!}$ and $\displaystyle\sum_{n=1}^{\infty} \frac{1}{(n+2)!}$ are convergent. Thus, we can split the series in the following way:

$$\sum_{n=1}^{\infty} \frac{n-1}{n!(n+2)} = \sum_{n=1}^{\infty} \frac{1}{n!} - 3 \sum_{n=1}^{\infty} \frac{1}{(n+1)!} + 3 \sum_{n=1}^{\infty} \frac{1}{(n+2)!}$$

$$= e - 1 - 3(e-2) + 3 \left(e - \frac{5}{2} \right)$$

$$= e - \frac{5}{2}.$$

(v). We have a series of positive terms (except for $n = 1$ in which case we have 0) and consider $a_n = \frac{n^3-1}{n!}$. Then,

$$\lim_{n\to\infty} \frac{a_{n+1}}{a_n} = \lim_{n\to\infty} \frac{\frac{(n+1)^3-1}{(n+1)!}}{\frac{n^3-1}{n!}}$$

$$= \lim_{n\to\infty} \frac{((n+1)^3 - 1)n!}{(n^3 - 1)(n+1)!} = \lim_{n\to\infty} \frac{(n+1)^3 - 1}{(n^3 - 1)(n+1)} = 0 < 1.$$

Thus, by the Ratio Test, the series converges.

To find the value of the series, note that we can write the series in the following way:

$$\sum_{n=1}^{\infty} \frac{n^3-1}{n!} = \sum_{n=2}^{\infty} \frac{n^3-1}{n!} = \sum_{n=2}^{\infty} \left(\frac{n^3}{n!} - \frac{1}{n!} \right) = \sum_{n=2}^{\infty} \left(\frac{n^2}{(n-1)!} - \frac{1}{n!} \right)$$

$$= \sum_{n=2}^{\infty} \left(\frac{n^2-1+1}{(n-1)!} - \frac{1}{n!} \right) = \sum_{n=2}^{\infty} \left(\frac{n^2-1}{(n-1)!} + \frac{1}{(n-1)!} - \frac{1}{n!} \right)$$

$$= \sum_{n=2}^{\infty} \left(\frac{(n+1)(n-1)}{(n-1)!} + \frac{1}{(n-1)!} - \frac{1}{n!} \right)$$

$$= \sum_{n=2}^{\infty} \left(\frac{n+1}{(n-2)!} + \frac{1}{(n-1)!} - \frac{1}{n!} \right)$$

$$= \sum_{n=2}^{\infty} \left(\frac{n-2+3}{(n-2)!} + \frac{1}{(n-1)!} - \frac{1}{n!} \right)$$

$$= \sum_{n=2}^{\infty} \left(\frac{n-2}{(n-2)!} + \frac{3}{(n-2)!} + \frac{1}{(n-1)!} - \frac{1}{n!} \right)$$

$$= \frac{7}{2} + \sum_{n=3}^{\infty} \left(\frac{1}{(n-3)!} + \frac{3}{(n-2)!} + \frac{1}{(n-1)!} - \frac{1}{n!} \right).$$

Note that all the series $\sum_{n=3}^{\infty} \frac{1}{(n-3)!}$, $\sum_{n=3}^{\infty} \frac{3}{(n-2)!}$, $\sum_{n=3}^{\infty} \frac{1}{(n-1)!}$ and $\sum_{n=3}^{\infty} \frac{1}{n!}$ are convergent. Thus, we can split the series as follows:

$$\sum_{n=1}^{\infty} \frac{n^3-1}{n!} = \frac{7}{2} + \sum_{n=3}^{\infty} \frac{1}{(n-3)!} + 3\sum_{n=3}^{\infty} \frac{1}{(n-2)!} + \sum_{n=3}^{\infty} \frac{1}{(n-1)!} - \sum_{n=3}^{\infty} \frac{1}{n!}$$

$$= \frac{7}{2} + e + 3(e-1) + e - 2 - \left(e - \frac{5}{2} \right)$$

$$= 4e + 1.$$

(VI). The series is a series of positive terms. Now consider $a_n = \frac{n^2+1}{(n+1)!}$, then

$$\lim_{n\to\infty} \frac{a_{n+1}}{a_n} = \lim_{n\to\infty} \frac{\frac{(n+1)^2+1}{(n+2)!}}{\frac{n^2+1}{(n+1)!}}$$

$$= \lim_{n\to\infty} \frac{((n+1)^2+1)(n+1)!}{(n^2+1)(n+2)!} = \lim_{n\to\infty} \frac{(n+1)^2+1}{(n^2+1)(n+2)} = 0 < 1.$$

Thus, by the Ratio Test, the series converges. Let us calculate the value of its sum. Observe that we can write the series as

$$\sum_{n=1}^{\infty} \frac{n^2+1}{(n+1)!} = \sum_{n=1}^{\infty} \frac{n^2-1+2}{(n+1)!}$$

$$= \sum_{n=1}^{\infty} \left(\frac{n^2-1}{(n+1)!} + \frac{2}{(n+1)!} \right)$$

$$= \sum_{n=1}^{\infty} \left(\frac{(n+1)(n-1)}{(n+1)!} + \frac{2}{(n+1)!} \right)$$

$$= \sum_{n=1}^{\infty} \left(\frac{n-1}{n!} + \frac{2}{(n+1)!} \right)$$

$$= \sum_{n=1}^{\infty} \left(\frac{n}{n!} - \frac{1}{n!} + \frac{2}{(n+1)!} \right)$$

$$= \sum_{n=1}^{\infty} \left(\frac{1}{(n-1)!} - \frac{1}{n!} + \frac{2}{(n+1)!} \right).$$

Note that the series $\sum_{n=1}^{\infty} \frac{1}{(n-1)!}$, $\sum_{n=1}^{\infty} \frac{1}{n!}$ and $\sum_{n=1}^{\infty} \frac{1}{(n+1)!}$ are convergent (the first one being the series representation of the number e). Hence,

$$\sum_{n=1}^{\infty} \frac{n^2+1}{(n+1)!} = \sum_{n=1}^{\infty} \frac{1}{(n-1)!} - \sum_{n=1}^{\infty} \frac{1}{n!} + 2\sum_{n=1}^{\infty} \frac{1}{(n+1)!}$$

$$= e - e + 1 + 2(e-2)$$

$$= 2e - 3.$$

(VII). Note that it is a series of positive terms and consider $a_n = \frac{n^2+5n+7}{(n+2)!}$. Then,

$$\lim_{n\to\infty} \frac{a_{n+1}}{a_n} = \lim_{n\to\infty} \frac{\frac{n^2+5n+7}{(n+2)!}}{\frac{(n+1)^2+5(n+1)+7}{(n+3)!}}$$

$$= \lim_{n\to\infty} \frac{((n+1)^2+5(n+1)+7)}{(n^2+5n+7)(n+3)} = 0 < 1.$$

Hence, by the Ratio Test, the series converges.

Now we will calculate de limit. First, note that we can write the series as

$$\sum_{n=2}^{\infty} \frac{n^2 + 5n + 7}{(n+2)!} = \sum_{n=2}^{\infty} \frac{(n+2)(n+3) + 1}{(n+2)!}$$

$$= \sum_{n=2}^{\infty} \left(\frac{(n+2)(n+3)}{(n+2)!} + \frac{1}{(n+2)!} \right)$$

$$= \sum_{n=2}^{\infty} \left(\frac{n+3}{(n+1)!} + \frac{1}{(n+2)!} \right)$$

$$= \sum_{n=2}^{\infty} \left(\frac{n+1+2}{(n+1)!} + \frac{1}{(n+2)!} \right)$$

$$= \sum_{n=2}^{\infty} \left(\frac{n+1}{(n+1)!} + \frac{2}{(n+1)!} + \frac{1}{(n+2)!} \right)$$

$$= \sum_{n=2}^{\infty} \left(\frac{1}{n!} + \frac{2}{(n+1)!} + \frac{1}{(n+2)!} \right).$$

Note that $\sum_{n=2}^{\infty} \frac{2}{(n+1)!}$ and $\sum_{n=2}^{\infty} \frac{1}{(n+2)!}$ are convergent $\Big($compare them to

$\sum_{n=2}^{\infty} \frac{1}{n!} = e - 2 \Big)$. Thus, we can split the series into the following three series:

$$\sum_{n=2}^{\infty} \frac{n^2 + 5n + 7}{(n+2)!} = \sum_{n=2}^{\infty} \frac{1}{n!} + \sum_{n=2}^{\infty} \frac{2}{(n+1)!} + \sum_{n=2}^{\infty} \frac{1}{(n+2)!}$$

$$= \sum_{n=2}^{\infty} \frac{1}{n!} + 2 \sum_{n=2}^{\infty} \frac{1}{(n+1)!} + \sum_{n=2}^{\infty} \frac{1}{(n+2)!}$$

$$= e - 2 + 2 \left(e - 2 - \frac{1}{2} \right) + e - 2 - \frac{1}{2} - \frac{1}{6}$$

$$= 4e - \frac{29}{3}.$$

(VIII). Notice that it is an alternating series since, by letting $a_n = (-1)^n \frac{2n+1}{n(n+1)}$, we have

$$a_{n+1}a_n = (-1)^{n+1} \frac{2n+3}{(n+1)(n+2)} (-1)^n \frac{2n+1}{n(n+1)} = -\frac{(2n+1)(2n+3)}{n(n+1)^2(n+2)} < 0.$$

Moreover, $|a_n| = \frac{2n+1}{n(n+1)}$. Assume that $|a_{n+1}| > |a_n|$ for some $n \in \mathbb{N}$, then

$$\frac{2n+3}{(n+1)(n+2)} > \frac{2n+1}{n(n+1)} \iff 2n^2 + 3n > 2n^2 + 5n + 2 \iff 2n + 2 < 0,$$

which is absurd. Therefore, the sequence $(|a_n|)_{n=1}^{\infty}$ is decreasing. Furthermore, $\lim_{n \to \infty} \frac{2n+1}{n(n+1)} = 0$. Thus, by the Alternating Series Test, the series is convergent.

We will proceed to find the value of the series. Note that we can write the series in the following way

$$\sum_{n=1}^{\infty}(-1)^n \frac{2n+1}{n(n+1)} = \sum_{n=1}^{\infty}(-1)^n \frac{n+1+n}{n(n+1)}$$

$$= \sum_{n=1}^{\infty}\left((-1)^n \frac{n+1}{n(n+1)} + (-1)^n \frac{n}{n(n+1)}\right)$$

$$= \sum_{n=1}^{\infty}\left((-1)^n \frac{1}{n} + (-1)^n \frac{1}{n+1}\right).$$

Observe that $\displaystyle\sum_{n=1}^{\infty}(-1)^n \frac{1}{n}$ is the alternating harmonic series multiplied by -1

and $\displaystyle\sum_{n=1}^{\infty}(-1)^n \frac{1}{n+1}$ is the alternating harmonic series minus 1 (recall Exercise 7.7 part (I)). Therefore, they are both convergent and we have

$$\sum_{n=1}^{\infty}(-1)^n \frac{2n+1}{n(n+1)} = \sum_{n=1}^{\infty}(-1)^n \frac{1}{n} + \sum_{n=1}^{\infty}(-1)^n \frac{1}{n+1}$$

$$= -\log 2 + \log 2 - 1 = -1.$$

(IX). Observe that it is a series of positive terms and let $a_n = \frac{n(n+1)}{2^n}$.
Then,

$$\lim_{n \to \infty} \frac{a_{n+1}}{a_n} = \lim_{n \to \infty} \frac{\frac{(n+1)(n+2)}{2^{n+1}}}{\frac{n(n+1)}{2^n}}$$

$$= \lim_{n \to \infty} \frac{(n+1)(n+2)2^n}{n(n+1)2^{n+1}} = \frac{1}{2} \lim_{n \to \infty} \frac{n+2}{n} = \frac{1}{2} < 1.$$

Thus, by the Ratio Test, the series is convergent.

Let us calculate the limit. Notice that we can write the series in the following way:

$$\sum_{n=1}^{\infty} \frac{n(n+1)}{2^n} = \sum_{n=1}^{\infty}\left(\frac{n^2}{2^n} + \frac{n}{2^n}\right).$$

By the Comparison Theorem, observe that $\displaystyle\sum_{n=1}^{\infty} \frac{n^2}{2^n}$ and $\displaystyle\sum_{n=1}^{\infty} \frac{n}{2^n}$ are convergent $\left(\text{simply compare them with } \displaystyle\sum_{n=1}^{\infty} \frac{n(n+1)}{2^n}\right)$. Thus,

$$\sum_{n=1}^{\infty} \frac{n(n+1)}{2^n} = \sum_{n=1}^{\infty} \frac{n^2}{2^n} + \sum_{n=1}^{\infty} \frac{n}{2^n}.$$

Let

$$A = \sum_{n=1}^{\infty} \frac{n^2}{2^n} = \frac{1}{2} + \frac{2^2}{2^2} + \cdots + \frac{n^2}{2^n} + \frac{(n+1)^2}{2^{n+1}} + \cdots$$

and take

$$-\frac{1}{2}A = -\frac{1}{2^2} - \frac{2^2}{2^3} - \cdots - \frac{n^2}{2^{n+1}} - \cdots.$$

Since A and $-\frac{1}{2}A$ are convergent, we can add them in the following way:

$$A - \frac{1}{2}A = \left(\frac{1}{2} + \frac{2^2}{2^2} + \cdots + \frac{n^2}{2^n} + \frac{(n+1)^2}{2^{n+1}} + \cdots \right)$$
$$+ \left(-\frac{1}{2^2} - \frac{2^2}{2^3} - \cdots - \frac{n^2}{2^{n+1}} - \cdots \right)$$
$$= \frac{1}{2} + \left(\frac{2^2}{2^2} - \frac{1}{2^2} \right) + \cdots + \left(\frac{(n+1)^2}{2^{n+1}} - \frac{n^2}{2^{n+1}} \right) + \cdots$$
$$= \frac{1}{2} + \frac{3}{2^2} + \cdots + \frac{2n+1}{2^{n+1}} + \cdots$$
$$= \sum_{n=1}^{\infty} \frac{2n-1}{2^n}.$$

Notice that $\displaystyle\sum_{n=1}^{\infty} \frac{2n-1}{2^n}$ is convergent since A is convergent. Moreover,

$$\frac{1}{2}A = \sum_{n=1}^{\infty} \frac{2n-1}{2^n} = \sum_{n=1}^{\infty} \left(\frac{2n}{2^n} - \frac{1}{2^n} \right).$$

Once again, by the Comparison Theorem, we can split the series in the following way:

$$\frac{1}{2}A = 2\sum_{n=1}^{\infty} \frac{n}{2^n} - \sum_{n=1}^{\infty} \frac{1}{2^n}.$$

Let

$$B = \sum_{n=1}^{\infty} \frac{n}{2^n} = \frac{1}{2} + \frac{2}{2^2} + \cdots + \frac{n}{2^n} + \frac{n+1}{2^{n+1}} + \cdots$$

and take

$$-\frac{1}{2}B = -\frac{1}{2^2} - \frac{2}{2^3} - \cdots - \frac{n}{2^{n+1}} - \cdots.$$

Once again, we can do the following:

$$B - \frac{1}{2}B = \left(\frac{1}{2} + \frac{2}{2^2} + \cdots + \frac{n}{2^n} + \frac{n+1}{2^{n+1}} + \cdots\right)$$

$$+ \left(-\frac{1}{2^2} - \frac{2}{2^3} - \cdots - \frac{n}{2^{n+1}} - \cdots\right)$$

$$= \frac{1}{2} + \left(\frac{2}{2^2} - \frac{1}{2^2}\right) + \cdots$$

$$+ \left(\frac{n+1}{2^{n+1}} - \frac{n}{2^{n+1}}\right) + \cdots$$

$$= \frac{1}{2} + \frac{1}{2^2} + \cdots + \frac{1}{2^{n+1}} + \cdots$$

$$= \sum_{n=1}^{\infty} \left(\frac{1}{2}\right)^n.$$

As $\displaystyle\sum_{n=1}^{\infty} \left(\frac{1}{2}\right)^n$ is a geometric series, it yields

$$\sum_{n=1}^{\infty} \left(\frac{1}{2}\right)^n = \frac{1/2}{1 - 1/2} = 1.$$

Thus, $B = 2$, and $A = 2(2 \cdot 2 - 1) = 6$. Hence,

$$\sum_{n=1}^{\infty} \frac{n(n+1)}{2^n} = 6 + 2 = 8.$$

(x). To study the convergence of this series of positive general terms, let us use the Ratio Test with $a_n = \frac{n^2}{3^n}$. Observe that

$$\lim_{n\to\infty} \frac{a_{n+1}}{a_n} = \lim_{n\to\infty} \frac{\frac{(n+1)^2}{3^{n+1}}}{\frac{n^2}{3^n}} = \lim_{n\to\infty} \frac{(n+1)^2 3^n}{n^2 3^{n+1}} = \frac{1}{3}\lim_{n\to\infty} \frac{(n+1)^2}{n^2} = \frac{1}{3} < 1.$$

Hence, by the Ratio Test, the series is convergent.

Now we will calculate the limit. Let

$$S = \sum_{n=1}^{\infty} \frac{n^2}{3^n} = \frac{1}{3} + \frac{2^2}{3^2} + \frac{3^2}{3^3} + \cdots + \frac{n^2}{3^n} + \frac{(n+1)^2}{3^{n+1}} + \cdots$$

and take

$$-\frac{1}{3}S = -\frac{1}{3^2} - \frac{2^2}{3^3} - \frac{3^2}{3^4} - \cdots - \frac{n^2}{3^{n+1}} - \cdots.$$

Since both series S and $-\frac{1}{3}S$ are convergent, we can add in the following way:

$$S - \frac{1}{3}S = \left(\frac{1}{3} + \frac{2^2}{3^2} + \cdots + \frac{n^2}{3^n} + \frac{(n+1)^2}{3^{n+1}} + \cdots\right)$$

$$+ \left(-\frac{1}{3^2} - \frac{2^2}{3^3} - \cdots - \frac{n^2}{3^{n+1}} - \cdots\right)$$

$$= \frac{1}{3} + \left(\frac{2^2}{3^2} - \frac{1}{3^2}\right) + \cdots + \left(\frac{(n+1)^2}{3^{n+1}} - \frac{n^2}{3^{n+1}}\right) + \cdots$$

$$= \frac{1}{3} + \frac{3}{3^2} + \cdots + \frac{(n+1)^2 - n^2}{3^{n+1}} + \cdots$$

$$= \frac{1}{3} + \frac{3}{3^2} + \cdots + \frac{2n+1}{3^{n+1}} + \cdots$$

$$= \sum_{n=1}^{\infty} \frac{2n-1}{3^n},$$

that is,

$$\frac{2}{3}S = \sum_{n=1}^{\infty} \frac{2n-1}{3^n}.$$

Observe that $\displaystyle\sum_{n=1}^{\infty} \frac{2n-1}{3^n}$ is convergent since S is convergent. Let

$$S' = \sum_{n=1}^{\infty} \frac{2n-1}{3^n} = \frac{1}{3} + \frac{3}{3^2} + \cdots + \frac{2n-1}{3^n} + \frac{2n+1}{3^{n+1}} + \cdots$$

and take

$$-\frac{1}{3}S' = \frac{1}{3^2} + \frac{3}{3^3} + \cdots + \frac{2n-1}{3^{n+1}} + \cdots.$$

Hence, once again, we have

$$S' - \frac{1}{3}S' = \left(\frac{1}{3} + \frac{3}{3^2} + \cdots + \frac{2n-1}{3^n} + \frac{2n+1}{3^{n+1}} + \cdots\right)$$

$$+ \left(\frac{1}{3^2} + \frac{3}{3^3} + \cdots + \frac{2n-1}{3^{n+1}} + \cdots\right)$$

$$= \frac{1}{3} + \left(\frac{3}{3^2} - \frac{1}{3^2}\right) + \cdots + \left(\frac{2n+1}{3^{n+1}} - \frac{2n-1}{3^{n+1}}\right) + \cdots$$

$$= \frac{1}{3} + \frac{2}{3^2} + \cdots + \frac{2}{3^{n+1}} + \cdots$$

$$= \frac{1}{3} + \frac{2}{3} \sum_{n=1}^{\infty} \left(\frac{1}{3}\right)^n.$$

As $\displaystyle\sum_{n=1}^{\infty}\left(\frac{1}{3}\right)^{n}$ is a geometric series, we obtain

$$\sum_{n=1}^{\infty}\left(\frac{1}{3}\right)^{n}=\frac{1/3}{1-1/3}=\frac{1}{2}.$$

Thus,

$$\sum_{n=1}^{\infty}\frac{2n-1}{3^n}=\frac{3}{2}\left(\frac{1}{3}+\frac{2}{3}\cdot\frac{1}{2}\right)=1.$$

Hence,

$$\sum_{n=1}^{\infty}\frac{n^2}{3^n}=\frac{3}{2}.$$

(XI). It is obvious that if $x=0$, then the series is convergent and it is equal to 0. So, assume that $x\neq0$. Let $a_n=(n+1)x^n$ and apply the Ratio Test as follows

$$\lim_{n\to\infty}\frac{|a_{n+1}|}{|a_n|}=\lim_{n\to\infty}\frac{(n+2)|x|^{n+1}}{(n+1)|x|^n}=|x|\cdot\lim_{n\to\infty}\frac{n+2}{n+1}=|x|.$$

On the one hand, if $|x|<1$, the series is absolutely convergent (so convergent). On the other hand, if $|x|>1$, the series is not convergent. It remains to study the case of $|x|=1$. It is clear that the series diverges to ∞ when $x=1$. Now, if $x=-1$, let us prove that the sequence of partial sums s_n satisfies

$$s_n:=\begin{cases}k & \text{if } n=2k,\\-(k+1) & \text{if } n=2k-1.\end{cases}$$

For $n=1$: $s_1=-2$. For $n=2$: $s_2=-2+3=1$. For $n=3$: $s_3=1-4=-3$. For $n=4$: $s_4=-3+5=2$. Assume that s_n satisfies the above conditions Induction Hypothesis for some $n\in\mathbb{N}$. Let us show that it is also true for $n+1$. If $n+1=2k$ for some $k\in\mathbb{N}$, then by hypothesis we have $s_n=-(k+1)$, so

$$s_{n+1}=-(k+1)+(-1)^{n+1}(n+2)$$
$$=-(k+1)+(-1)^{2k}(2k+1)=-(k+1)+2k+1=k.$$

If $n+1=2k+1=2(k+1)-1$ for some $k\in\mathbb{N}$, then by hypothesis we have $s_n=k+1$, so

$$s_{n+1}=k+1+(-1)^{n+1}(n+2)$$
$$=k+1+(-1)^{2k+1}(2k+2)=k+1-2k-2=-(k+1).$$

Thus, the sequence of partial sums is divergent. In conclusion, the series is convergent if, and only if, $|x|<1$.

Assume that $|x| < 1$ and let us calculate the precise value of the series. Take the sequence of functions $f_n(x) = (n+1)x^n$. Notice that each $f_n(x)$ is the derivative of the function $g_n(x) = x^{n+1}$. Then, since $\sum_{n=1}^{\infty} g_n(x) = \sum_{n=1}^{\infty} x^{n+1}$ is a power series with interval of convergence $(-1, 1)$, we have

$$\sum_{n=1}^{\infty}(n+1)x^n = \sum_{n=1}^{\infty} f_n(x) = \sum_{n=1}^{\infty} g_n'(x) = \left(\sum_{n=1}^{\infty} g_n(x)\right)'$$

$$= \left(\sum_{n=1}^{\infty} g_n(x)\right)' = \left(\sum_{n=1}^{\infty} x^{n+1}\right)'.$$

As $\sum_{n=1}^{\infty} x^{n+1}$ is a geometric series for every fixed $x \in (-1, 1)$, we arrive at

$$\sum_{n=1}^{\infty}(n+1)x^n = \left(\sum_{n=1}^{\infty} x^{n+1}\right)' = \left(\frac{x^2}{1-x}\right)' = \frac{x^2 - 2x}{(1-x)^2}.$$

(XII). Let us apply the Root Test. Take $a_n = (-1)^n \frac{n^2 - n}{3^n}$ and observe that $|a_n| = \frac{n^2 - n}{3^n}$. Then,

$$\lim_{n \to \infty} \sqrt[n]{|a_n|} = \lim_{n \to \infty} \sqrt[n]{\frac{n^2 - n}{3^n}}$$

$$= \lim_{n \to \infty} \sqrt[n]{\frac{n(n-1)}{3^n}}$$

$$= \lim_{n \to \infty} \frac{1}{3} \sqrt[n]{n} \sqrt[n]{n-1} = \frac{1}{3} < 1.$$

Hence, the series is absolutely convergent and thus, in particular, it is convergent.

Let us calculate the limit of the series. Note that we can write the series as

$$\sum_{n=1}^{\infty}(-1)^n \frac{n^2 - n}{3^n} = \sum_{n=1}^{\infty}\left((-1)^n \frac{n^2}{3^n} - (-1)^n \frac{n}{3^n}\right).$$

Recall that the series $\sum_{n=1}^{\infty} \frac{n^2}{3^n}$ is convergent by Exercise 7.10 part (X); hence, the series $\sum_{n=1}^{\infty}(-1)^n \frac{n^2}{3^n}$ is absolutely convergent (so convergent). Moreover, by the Comparison Theorem, observe that $\sum_{n=1}^{\infty} \frac{n}{3^n}$ is also convergent since $0 < \frac{n}{3^n} \leq \frac{n^2}{3^n}$, thus $\sum_{n=1}^{\infty}(-1)^n \frac{n^2}{3^n}$ is absolutely convergent (so convergent).

Therefore, the following equalities are satisfied:

$$\sum_{n=1}^{\infty}(-1)^n\frac{n^2-n}{3^n} = \sum_{n=1}^{\infty}\left((-1)^n\frac{n^2}{3^n} - (-1)^n\frac{n}{3^n}\right)$$

$$= \sum_{n=1}^{\infty}(-1)^n\frac{n^2}{3^n} - \sum_{n=1}^{\infty}(-1)^n\frac{n}{3^n}.$$

To calculate the value of the series, let us first find the value of $\sum_{n=1}^{\infty}(-1)^n\frac{n^2}{3^n}$.

Let

$$A = \sum_{n=1}^{\infty}(-1)^n\frac{n^2}{3^n} = -\frac{1}{3} + \frac{2^2}{3^2} - \cdots + (-1)^n\frac{n^2}{3^n} + (-1)^{n+1}\frac{(n+1)^2}{3^{n+1}} + \cdots.$$

Then,

$$-\frac{1}{3}A = \frac{1}{3^2} - \frac{2^2}{3^3} + \cdots + (-1)^n\frac{n^2}{3^{n+1}} + \cdots.$$

Hence, since A and $-\frac{1}{3}A$ are convergent, we can subtract them in the following way:

$$A - \left(-\frac{1}{3}A\right) = \left(-\frac{1}{3} + \frac{2^2}{3^2} - \frac{3^2}{3^3} + \cdots + (-1)^n\frac{n^2}{3^n} + (-1)^{n+1}\frac{(n+1)^2}{3^{n+1}} + \cdots\right)$$

$$- \left(\frac{1}{3^2} - \frac{2^2}{3^3} + \frac{3^2}{3^4} - \cdots + (-1)^{n+1}\frac{n^2}{3^{n+1}} + \cdots\right)$$

$$= -\frac{1}{3} + \left(\frac{2^2}{3^2} - \frac{1}{3^2}\right) + \cdots$$

$$\cdots + \left((-1)^{n+1}\frac{(n+1)^2}{3^{n+1}} - (-1)^{n+1}\frac{n^2}{3^{n+1}}\right) + \cdots$$

$$= -\frac{1}{3} + \frac{3}{3^2} + \cdots + (-1)^{n+1}\frac{2n+1}{3^{n+1}} + \cdots$$

$$= \sum_{n=1}^{\infty}(-1)^n\frac{2n-1}{3^n}.$$

Thus,

$$\frac{4}{3}A = \sum_{n=1}^{\infty}(-1)^n\frac{2n-1}{3^n} = 2\sum_{n=1}^{\infty}(-1)^n\frac{n}{3^n} - \sum_{n=1}^{\infty}\left(-\frac{1}{3}\right)^n.$$

Note that in the last equality we are able to split the series into the difference of two series since both of them are convergent (for the second simply apply the Comparison Theorem again). Let

$$B = \sum_{n=1}^{\infty}(-1)^n\frac{n}{3^n} = -\frac{1}{3} + \frac{2}{3^2} \cdots + (-1)^n\frac{n}{3^n} + (-1)^n\frac{n+1}{3^{n+1}} + \cdots.$$

Then

$$-\frac{1}{3}B = \frac{1}{3^2} - \frac{2}{3^3} - \cdots + (-1)^{n+1}\frac{n}{3^{n+1}} + \cdots.$$

Hence, since B and $-\frac{1}{3}B$ are convergent, we can proceed as before:

$$B - \left(-\frac{1}{3}B\right) = \left(-\frac{1}{3} + \frac{2}{3^2} \cdots + (-1)^n\frac{n}{3^n} + (-1)^n\frac{n+1}{3^{n+1}} + \cdots\right)$$

$$- \left(\frac{1}{3^2} - \frac{2}{3^3} - \cdots + (-1)^{n+1}\frac{n}{3^{n+1}} + \cdots\right)$$

$$= -\frac{1}{3} + \left(\frac{2}{3^2} - \frac{1}{3^2}\right) + \cdots$$

$$\cdots + \left((-1)^n\frac{n+1}{3^{n+1}} - (-1)^n\frac{n}{3^{n+1}}\right)$$

$$= -\frac{1}{3} + \frac{1}{3^2} + \cdots + (-1)^{n+1}\frac{1}{3^{n+1}}$$

$$= \sum_{n=1}^{\infty}\left(-\frac{1}{3}\right)^n.$$

Thus,

$$\frac{4}{3}B = \sum_{n=1}^{\infty}\left(-\frac{1}{3}\right)^n.$$

Now let us calculate the value of the series $\displaystyle\sum_{n=1}^{\infty}\left(-\frac{1}{3}\right)^n$ which is a geometric series. So

$$\sum_{n=1}^{\infty}\left(-\frac{1}{3}\right)^n = \frac{-1/3}{1-(-1/3)} = \frac{-1/3}{4/3} = -\frac{1}{4}.$$

Therefore,

$$\frac{4}{3}A = 2B - \left(-\frac{1}{4}\right) = 2\cdot\frac{3}{4}\left(-\frac{1}{4}\right) + \frac{1}{4} = -\frac{3}{8} + \frac{1}{4} = -\frac{1}{8}.$$

Hence,

$$A = \frac{3}{4}\left(-\frac{1}{8}\right) = -\frac{3}{32}.$$

In conclusion, we have that

$$\sum_{n=1}^{\infty}(-1)^n\frac{n^2-n}{3^n} = A - B = -\frac{3}{32} - \left(-\frac{3}{16}\right) = \frac{3}{32}.$$

(XIII). Take $a_n = \dfrac{1}{n\sqrt{n+1}+(n+1)\sqrt{n}}$ the general term of the series and consider $\sum_{n=1}^{\infty}\frac{1}{n^{3/2}}$ which has general term $b_n = \frac{1}{n^{3/2}}$. Notice that $\sum_{n=1}^{\infty}\frac{1}{n^{3/2}}$

is a series of positive general term and it converges since it is a p-series with $p = \frac{3}{2} > 1$. Observe that $\sum_{n=1}^{\infty} \frac{1}{n\sqrt{n+1}+(n+1)\sqrt{n}}$ is also a series of positive general term. Now let us calculate the limit of the sequence $\left(\frac{a_n}{b_n}\right)_{n=1}^{\infty}$:

$$\lim_{n\to\infty} \frac{a_n}{b_n} = \lim_{n\to\infty} \frac{\frac{1}{n\sqrt{n+1}+(n+1)\sqrt{n}}}{\frac{1}{n^{3/2}}}$$

$$= \lim_{n\to\infty} \frac{n^{3/2}}{n\sqrt{n+1}+(n+1)\sqrt{n}}$$

$$= \lim_{n\to\infty} \frac{n^{3/2}}{n^{3/2}\sqrt{1+\frac{1}{n}}+n^{3/2}+n^{1/2}}$$

$$= \frac{1}{2}.$$

By the Limit Comparison Test, as the limit of the sequence $\left(\frac{a_n}{b_n}\right)_{n=1}^{\infty}$ is finite and the series $\sum_{n=1}^{\infty} \frac{1}{n^{3/2}}$ is convergent, we have that $\sum_{n=1}^{\infty} \frac{1}{n\sqrt{n+1}+(n+1)\sqrt{n}}$ converges.

Let us calculate the sum of the series:

$$\sum_{n=1}^{\infty} \frac{1}{n\sqrt{n+1}+(n+1)\sqrt{n}}$$

$$= \sum_{n=1}^{\infty} \frac{n\sqrt{n+1}-(n+1)\sqrt{n}}{\left(n\sqrt{n+1}+(n+1)\sqrt{n}\right)\left(n\sqrt{n+1}-(n+1)\sqrt{n}\right)}$$

$$= \sum_{n=1}^{\infty} \frac{n\sqrt{n+1}-(n+1)\sqrt{n}}{n^2(n+1)-(n+1)^2 n}$$

$$= \sum_{n=1}^{\infty} \frac{n\sqrt{n+1}-(n+1)\sqrt{n}}{n^3+n^2-n^3-2n^2-n}$$

$$= \sum_{n=1}^{\infty} \frac{n\sqrt{n+1}-(n+1)\sqrt{n}}{-n^2-n}$$

$$= \sum_{n=1}^{\infty} \left(\frac{(n+1)\sqrt{n}}{n(n+1)}-\frac{n\sqrt{n+1}}{n(n+1)}\right)$$

$$= \sum_{n=1}^{\infty} \left(\frac{1}{\sqrt{n}}-\frac{1}{\sqrt{n+1}}\right)$$

$$= \lim_{k\to\infty} \sum_{n=1}^{k} \left(\frac{1}{\sqrt{n}}-\frac{1}{\sqrt{n+1}}\right).$$

Observe that $\sum\limits_{n=1}^{k}\left(\dfrac{1}{\sqrt{n}}-\dfrac{1}{\sqrt{n+1}}\right)$ is a telescopic sum whose value is equal to $1-\dfrac{1}{\sqrt{k+1}}$. Therefore,

$$\sum_{n=1}^{\infty}\frac{1}{n\sqrt{n+1}+(n+1)\sqrt{n}}=\lim_{k\to\infty}\sum_{n=1}^{k}\left(\frac{1}{\sqrt{n}}-\frac{1}{\sqrt{n+1}}\right)$$

$$=\lim_{k\to\infty}\left(1-\frac{1}{k+1}\right)=1.$$

(XIV). Let us consider the partial sums of this series:

$$\sum_{n=1}^{k}\left(\log\left(\frac{n}{n+1}\right)^{n}-\frac{1}{2n}+1\right)=\sum_{n=1}^{k}\log\left(\frac{n}{n+1}\right)^{n}-\sum_{n=1}^{k}\frac{1}{2n}+k$$

$$=\log\prod_{n=1}^{k}\left(\frac{n}{n+1}\right)^{n}-\sum_{n=1}^{k}\frac{1}{2n}+k.$$

It can be easily proved by induction that

$$\prod_{n=1}^{k}\left(\frac{n}{n+1}\right)^{n}=\frac{k!}{(k+1)^{k}},$$

and so

$$\sum_{n=1}^{k}\left(\log\left(\frac{n}{n+1}\right)^{n}-\frac{1}{2n}+1\right)=\log\frac{k!}{(k+1)^{k}}-\frac{1}{2}\log k$$

$$+\left(\frac{1}{2}\log k-\sum_{n=1}^{k}\frac{1}{2n}\right)+k$$

$$=\log\frac{k!e^{k}}{(k+1)^{k}\sqrt{k}}-\frac{1}{2}\left(\sum_{n=1}^{k}\frac{1}{n}-\log k\right).$$

On the other hand, using Stirling's Formula, we obtain that, when $k\to\infty$,

$$\frac{k!e^{k}}{(k+1)^{k}\sqrt{k}}\sim\frac{\frac{k^{k}\sqrt{2\pi k}}{e^{k}}e^{k}}{(k+1)^{k}\sqrt{k}}=\frac{\sqrt{2\pi}}{\left(1+\frac{1}{k}\right)^{k}}\longrightarrow\frac{\sqrt{2\pi}}{e}.$$

As a consequence, we have

$$\sum_{n=1}^{\infty}\left(\log\left(\frac{n}{n+1}\right)^n - \frac{1}{2n} + 1\right) = \lim_k \sum_{n=1}^{k}\left(\log\left(\frac{n}{n+1}\right)^n - \frac{1}{2n} + 1\right)$$

$$= \lim_k \log\frac{k!e^k}{(k+1)^k\sqrt{k}} - \frac{1}{2}\lim_k\left(\sum_{n=1}^{k}\frac{1}{n} - \log k\right)$$

$$= \log\frac{\sqrt{2\pi}}{e} - \frac{1}{2}\gamma = \frac{\log 2 + \log\pi - \gamma}{2} - 1,$$

where $\gamma \approx 0.5772156649015$ is the famous Euler-Mascheroni constant. $\square$

Chapter 8

Power Series. Function Sequences and Series

Exercise 8.1. *Consider the sequence (f_n) of functions $f_n : \mathbb{R} \to \mathbb{R}$ given by*

$$f_n(x) = \begin{cases} \frac{1}{n} & \text{if } |x| < n, \\ \frac{n+1-|x|}{n} & \text{if } n \leq |x| < n+1, \\ 0 & \text{if } |x| \geq n+1. \end{cases}$$

Prove that (f_n) converges uniformly on $\mathbb{R}$ to a function $f : \mathbb{R} \to \mathbb{R}$ but

$$\lim_n \int_{-\infty}^{\infty} f_n(x)dx \neq \int_{-\infty}^{\infty} f(x)dx.$$

Solution to Exercise 8.1.
See Figure 8.1 for a representation of the graph of the f_n's.
If $n \leq |x| < n+1$, then $-(n+1) < -|x| \leq -n$, so

$$0 < \frac{n+1-|x|}{n} = f_n(x) \leq \frac{1}{n}.$$

Also, $f_n(x) = \frac{1}{n}$, if $|x| < n$ and $f_n(x) = 0 < \frac{1}{n}$, if $|x| \geq n+1$. In any case $0 \leq f_n(x) \leq \frac{1}{n}$ for all $x \in \mathbb{R}$ and $f_n(n) = \frac{1}{n}$. Then

$$\|f_n\|_{\mathbb{R}} = \sup\{|f_n(x)| : x \in \mathbb{R}\} = \frac{1}{n}$$

for all $n \in \mathbb{N}$. Therefore, if f is the null function on $\mathbb{R}$,

$$\|f_n - f\|_{\mathbb{R}} = \|f_n\|_{\mathbb{R}} = \frac{1}{n} \longrightarrow 0 \quad \text{as } n \to \infty,$$

which means that (f_n) converges to f uniformly on $\mathbb{R}$.
Finally,

$$\int_{-\infty}^{\infty} f(x)dx = \int_{-\infty}^{\infty} 0\,dx = 0$$

DOI: 10.1201/9781003400745-8

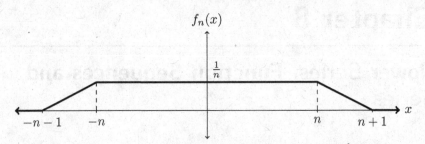

FIGURE 8.1: Representation of the functions $f_n(x)$ corresponding to Exercise 8.1.

whereas

$$\lim_n \int_{-\infty}^{\infty} f_n(x)dx = 2\lim_n \int_0^{\infty} f_n(x)dx$$

$$= 2\lim_n \left[\int_0^n \frac{1}{n}dx + \int_n^{n+1} \frac{n+1-x}{n}dx \right]$$

$$= 2\lim_n \left(1 + \left[\frac{n+1}{n}x - \frac{x^2}{2n} \right]_{x=n}^{x=n+1} \right)$$

$$= 2\lim_n \left(1 + \frac{1}{2n} \right) = 2.$$

$\square$

Exercise 8.2. *Consider the sequence (f_n) of functions $f_n : \mathbb{R} \to \mathbb{R}$ given by*

$$f_n(x) = \begin{cases} a_n \sin(a_n \pi x) & \text{if } 0 \le x \le \frac{1}{a_n}, \\ 0 & \text{if } x > \frac{1}{a_n}, \end{cases}$$

where (a_n) is a sequence of positive real numbers. Prove that

(I) *(f_n) converges uniformly on $\mathbb{R}$ to $f \equiv 0$ if and only if $\lim_n a_n = 0$.*

(II) *If $\lim_n a_n = 0$ then the pointwise limit f of (f_n) satisfies*

$$\lim_n \int_{-\infty}^{\infty} f_n(x)dx \neq \int_{-\infty}^{\infty} f(x)dx$$

despite the fact that there is uniform convergence on $\mathbb{R}$.

(III) *Explain why there is no contradiction in (II) despite the fact that (f_n) converges to f uniformly.*

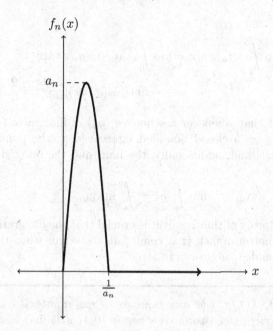

FIGURE 8.2: Representation of the functions $f_n(x)$ corresponding to Exercise 8.2.

Solution to Exercise 8.2.

See Figure 8.2 for a representation of the graph of the f_n's.

(I) For every $n \in \mathbb{N}$ we have that

$$
\begin{aligned}
\|f_n\|_{\mathbb{R}} &= \sup\{|f_n(x)| : x \in \mathbb{R}\} \\
&= \sup\{|a_n||\sin(n\pi x)| : x \in [0, 1/a_n]\} \leq |a_n|.
\end{aligned}
$$

Thus if $\lim_n a_n = 0$ then $\lim_n \|f_n\|_{\mathbb{R}} = 0$, which means that (f_n) converges to 0 uniformly.

On the other hand, if (a_n) does not converge to 0 then there exists $\varepsilon_0 > 0$ and a subsequence (a_{n_k}) of (a_n) such that $a_{n_k} \geq \varepsilon_0$ for every $k \in \mathbb{N}$. Since $|f_{n_k}(\frac{1}{2a_{n_k}})| = |a_{n_k}| = a_{n_k}$, it follows that

$$
\|f_{n_k}\|_{\mathbb{R}} \geq |a_{n_k}| \geq \varepsilon_0
$$

for every $k \in \mathbb{R}$ and therefore the sequence of uniform norms $(\|f_n\|_{\mathbb{R}})$ does not converge to 0, proving that (f_n) does not converge uniformly to $f \equiv 0$.

(II) Assume that (a_n) converges to 0. Then $f \equiv 0$. Clearly

$$
\int_{-\infty}^{\infty} f(x)\,dx = 0
$$

whereas

$$\lim_n \int_{-\infty}^{\infty} f_n(x)dx = \lim_n \int_0^{\frac{1}{a_n}} a_n \sin(a_n \pi x)dx$$

$$= -\frac{1}{\pi} \lim_n \left[\cos(a_n \pi x)\right]_{x=0}^{x=\frac{1}{a_n}} = \frac{2}{\pi}.$$

(III) Observe that whenever a sequence (g_n) of integrable functions converges uniformly on a closed bounded interval $[a, b]$, the pointwise limit is integrable on $[a, b]$ and, additionally, the limit and the integral symbols can be swapped, that is,

$$\lim_n \int_a^b g_n = \int_a^b \lim_n g_n.$$

However, in the proof of that result it is crucial that the integration interval is bounded. As a matter of fact, that result fails to be true when the integration interval is unbounded, as proven in (II).   □

Exercise 8.3. *Let (a_n) be any sequence of real numbers. Assume that (b_n) is an strictly decreasing sequence in $(0, 1]$ with $\lim_n b_n = 0$. Consider the sequence (f_n) of functions $f_n : [0, 1] \to \mathbb{R}$ defined as*

$$f_n(x) = \begin{cases} \frac{a_n}{b_{n+1}}x & \text{if } x \in [0, b_{n+1}], \\ \frac{a_n}{b_{n+1}-b_n}(x - b_n) & \text{if } x \in [b_{n+1}, b_n], \\ 0 & \text{if } x \in [b_n, 1]. \end{cases}$$

(I) *Prove that (f_n) converges pointwise to 0 on $[0, 1]$.*

(II) *Prove that (f_n) converges to 0 uniformly on $[0, 1]$ if and only if $\lim_n a_n = 0$.*

(III) *Provide a choice for (a_n) and (b_n) so that (f_n) does not converge uniformly on $[0, 1]$ but*

$$\lim_n \int_0^1 f_n(x)dx = 0.$$

See Figure 8.3 for a representation of the graph of the f_n's.

Solution to Exercise 8.3.

(I) First, observe that $f_n(0) = 0$ for all $n \in \mathbb{N}$. Now fix $x \in (0, 1]$. Since $\lim_n b_n = 0$, there exists $n_0 \in \mathbb{N}$ with $b_n < x$ for all $n \geq n_0$, that is, $f_n(x) = 0$ for all $n \geq n_0$. Hence $\lim_n f_n(x) = 0$.

(II) It is simple to check that $\|f_n\|_{[0,1]} = |a_n|$ for all $n \in \mathbb{N}$. Clearly (f_n) converges to 0 uniformly on $[0, 1]$ if and only if $\lim_n a_n = 0$.

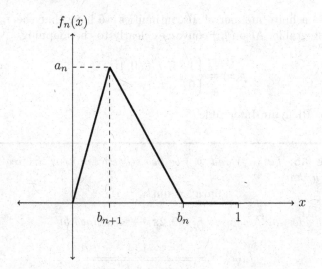

FIGURE 8.3: Representation of the functions $f_n(x)$ corresponding to exercise 8.3.

(III) For any choice of (a_n) and (b_n) as in the statement we have

$$\int_0^1 f_n(x)dx = \frac{a_n b_n}{2}$$

for all $n \in \mathbb{N}$. Then we would have that $\lim_n \int_0^1 f_n(x)dx = 0$ whenever $\lim_n a_n b_n = 0$. A plausible choice for (a_n) and (b_n) would be $a_n = n^r$ and $b_n = \frac{1}{n^s}$ with $0 < r < s$. □

Exercise 8.4. *Provide a sequence (f_n) of Riemann integrable functions $f_n : [0,1] \to \mathbb{R}$ such that (f_n) converges pointwise to a non-Riemann integrable function.*

Solution to Exercise 8.4.

Since $[0,1] \cap \mathbb{Q}$ is countable we can write $[0,1] \cap \mathbb{Q} = \{r_k : k \in \mathbb{N}\}$. Let us define $f_n : [0,1] \to \mathbb{R}$ as

$$f_n(x) = \begin{cases} 1 & \text{if } x \in \{r_1, \ldots, r_n\}, \\ 0 & \text{otherwise.} \end{cases}$$

Since f_n has a finite number of discontinuities we have that the f_n's are all Riemann-Integrable. Also (f_n) converges clearly to the mapping

$$f(x) = \begin{cases} 1 & \text{if } x \in [0,1] \cap \mathbb{Q}, \\ 0 & \text{otherwise,} \end{cases}$$

which is not Riemann-Integrable. □

Exercise 8.5. *Let (a_n) and (b_n) be two sequences of positive real numbers such that*

$$\lim_n a_n = \lim_n b_n = 1.$$

Now consider the functions $f_n : [0, 2\pi] \to \mathbb{R}$ defined by

$$f_n(x) = \frac{1}{\sqrt{a_n \sin^2 x + b_n \cos^2 x}}.$$

(I) *Prove that (f_n) converges uniformly to 1 on $[0, 2\pi]$.*

(II) *Justify that*

$$\lim_n \int_0^{2\pi} \frac{1}{\sqrt{a_n \sin^2 x + b_n \cos^2 x}} dx = 2\pi.$$

Solution to Exercise 8.5.

(I) If $b_n \geq a_n$ and $x \in [0, 2\pi]$, since $0 \leq \cos^2 x \leq 1$, we have

$$\sqrt{a_n} \leq \sqrt{a_n \sin^2 x + b_n \cos^2 x} = \sqrt{a_n + (b_n - a_n) \cos^2 x} \leq \sqrt{b_n}.$$

Similarly, if $a_n \geq b_n$ and $x \in [0, 2\pi]$,

$$\sqrt{b_n} \leq \sqrt{a_n \sin^2 x + b_n \cos^2 x} = \sqrt{a_n + (b_n - a_n) \cos^2 x} \leq \sqrt{a_n}.$$

In other words, if $x \in [0, 2\pi]$,

$$\min\{\sqrt{a_n}, \sqrt{b_n}\} \leq \sqrt{a_n \sin^2 x + b_n \cos^2 x} \leq \max\{\sqrt{a_n}, \sqrt{b_n}\},$$

or, equivalently

$$\frac{1}{\max\{\sqrt{a_n}, \sqrt{b_n}\}} \leq f_n(x) \leq \frac{1}{\min\{\sqrt{a_n}, \sqrt{b_n}\}},$$

for all $x \in [0, 2\pi]$ and $n \in \mathbb{N}$. Moreover,

$$\lambda_n \leq f_n(x) - 1 \leq \mu_n$$

for every $x \in [0, 2\pi]$ and every $n \in \mathbb{N}$, where

$$\lambda_n = \frac{1}{\max\{\sqrt{a_n}, \sqrt{b_n}\}} - 1,$$

$$\mu_n = \frac{1}{\min\{\sqrt{a_n}, \sqrt{b_n}\}} - 1,$$

for every $n \in \mathbb{N}$. Notice that $\lim_n \lambda_n = \lim_n \mu_n = 0$. Hence

$$\|f_n(x) - 1\|_{[0,2\pi]} = \sup\{|f_n(x) - 1| : x \in [0, 2\pi]\} \leq \max\{|\lambda_n|, |\mu_n|\},$$

for all $n \in \mathbb{N}$. Since $\lim_n \max\{|\lambda_n|, |\mu_n|\} = 0$, it follows that

$$\lim_n \|f_n(x) - 1\|_{[0,2\pi]} = 0,$$

and therefore, (f_n) converges uniformly to 1 on $[0, 2\pi]$.

(II) Since (f_n) converges uniformly on the closed, bounded interval $[0, 2\pi]$, we can swap the symbols of limit and integral, obtaining,

$$\lim_n \int_0^{2\pi} \frac{1}{\sqrt{a_n \sin^2 x + b_n \cos^2 x}} dx = \int_0^{2\pi} \lim_n \frac{1}{\sqrt{a_n \sin^2 x + b_n \cos^2 x}} dx$$

$$= \int_0^{2\pi} 1 dx = 2\pi.$$

$\square$

Exercise 8.6. *Let A be a non-empty set and consider the sequence (f_n) of functions $f_n : A \subset \mathbb{R} \to \mathbb{R}$ defined as $f_n(x) = \frac{2nx + (-1)^n x^2}{n}$. Study the pointwise and uniform convergence of (f_n) in the following cases:*

   (I) *A is bounded.*

   (II) *A is unbounded.*

Solution to Exercise 8.6.

The pointwise limit f of (f_n) on $\mathbb{R}$ can be easily calculated:

$$f(x) = \lim_n f_n(x) = \lim_n \left[\frac{2nx + (-1)^n x^2}{n} \right]$$

$$= \lim_n \left[2x + \frac{(-1)^n}{n} x^2 \right] = 2x + x^2 \lim_n \frac{(-1)^n}{n} = 2x.$$

(I) Assume that A is bounded and let $M > 0$ be such that $|x| \leq M$ for all $x \in A$. Then

$$\|f_n - f\|_A = \sup\{|f_n(x) - f(x)| : x \in A\}$$

$$= \sup \left\{ \left| \frac{(-1)^n}{n} x^2 \right| : x \in A \right\} \leq \frac{M^2}{n}.$$

Therefore, $0 \le \lim_n \|f_n - f\|_A \le \lim_n \frac{M^2}{n} = 0$, that is,

$$\lim_n \|f_n - f\|_A = 0.$$

This proves that (f_n) converges to f uniformly on A.

(II) Now assume that A is not bounded. Let $n \in \mathbb{N}$. Then there exists $x_n \in A$ such that $|x_n| > n^2$ (otherwise A would be bounded between $-n^2$ and n^2). The sequence (x_n) of points of A satisfies

$$|f_n(x_n) - f(x_n)| = \left| \frac{(-1)^n}{n} x_n^2 \right| = \frac{x_n^2}{n} > n.$$

Hence

$$\lim_n \|f_n - f\|_A \ge \lim_n |f_n(x_n) - f(x_n)| = \infty$$

and therefore (f_n) does not converge uniformly to f on A.   □

Exercise 8.7. *Calculate the pointwise limit of the following sequences of rational functions and determine whether the convergence is uniform on the given domains.*

(I) $f_n(x) = \dfrac{x^2}{x^2 + n}$, *on* $\mathbb{R}$ *and on* $[a, b]$ *with* $a, b \in \mathbb{R}$ *and* $a < b$.

(II) $f_n(x) = \dfrac{nx}{1 + n^2 x^2}$, *on* $\mathbb{R}$, *on* $[0, \infty)$ *and on* $[a, \infty)$, $a > 0$.

(III) $f_n(x) = \dfrac{nx^2}{1 + nx}$, *on* $[0, \infty)$ *and on* $[a, \infty)$ *with* $a > 0$.

(IV) $f_n(x) = \dfrac{1 - x^n}{1 + x^n}$, *on* $[0, \infty)$, *on* $[0, 1]$ *and on* $[0, a]$ *with* $0 < a < 1$.

Solution to Exercise 8.7.

(I) First, the pointwise limit is the null function $f \equiv 0$ since

$$\lim_n f_n(x) = \lim_n \frac{x^2}{x^2 + n} = 0.$$

However the convergence to $f \equiv 0$ is not uniform on $\mathbb{R}$ since $0 \le f_n(x) < 1$ for all $x \in \mathbb{R}$ and $\lim_{x \to \pm\infty} f_n(x) = 1$ (see Figure 8.4). Therefore, $\|f_n\|_{\mathbb{R}} = 1$ for all $n \in \mathbb{N}$.

At the other end of the scale, the convergence is uniform on intervals of the form $[a, b]$ with $a, b \in \mathbb{R}$ and $a < b$, for in that case

$$\|f_n\|_{[a,b]} = \max\{f_n(a), f_n(b)\}.$$

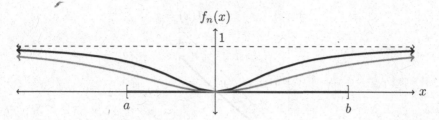

FIGURE 8.4: Representation of $f_n(x) = \frac{x^2}{n+x^2}$ (Exercise 8.7(I)) for two different values of n, namely $n_1 < n_2$. Here f_{n_1} appears in black and f_{n_2} in gray.

for all $n \in \mathbb{N}$. Since we have already seen that there is pointwise convergence to $f \equiv 0$, clearly $\lim_n \|f_n\|_{[a,b]} = 0$.

(II) The sequence converges pointwise to the null function $f \equiv 0$ since

$$\lim_n \frac{nx}{1+n^2x^2} = 0$$

for any choice of $x \in \mathbb{R}$. Notice that f_n is an odd function. To see if the convergence is uniform, having a representation of f_n will be useful (see Figure 8.5). It can be seen that f_n attains its minimum and maximum, respectively at $x = -\frac{1}{n}$ and $x = \frac{1}{n}$. This can be proved using the derivative

$$f_n'(x) = \frac{n(1-n^2x^2)}{(1+n^2x^2)^2}.$$

Clearly f_n' vanishes at $x = \pm\frac{1}{n}$, f_n' is positive on $(-1/n, 1/n)$ and negative on the rest of the real line. Therefore,

$$-1 = f_n(-1/n) \leq f_n(x) \leq f_n(1/n) = 1$$

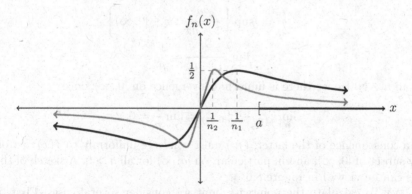

FIGURE 8.5: Representation of $f_n(x) = \frac{nx}{1+n^2x^2}$ (Exercise 8.7(II)) for two different values of n, namely $n_1 < n_2$. Here f_{n_1} appears in black and f_{n_2} in gray.

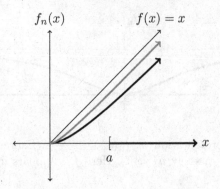

FIGURE 8.6: Representation of $f_n(x) = \frac{nx^2}{1+nx}$ (Exercise 8.7(III)) for two different values of n, namely $n_1 < n_2$. Here f_{n_1} appears in black and f_{n_2} in gray.

for all $x \in \mathbb{R}$, which means that $\|f_n\|_\mathbb{R} = \|f_n\|_{[0,\infty)} = 1$ for all $n \in \mathbb{N}$. This shows that the convergence is not uniform on neither $\mathbb{R}$ nor on $[0,\infty)$.

Now if $a > 0$ and $N \in \mathbb{N}$ satisfies $\frac{1}{N} < a$. Then for every $n \geq N$ we have that $\frac{1}{n} < a$ and therefore f_n is decreasing on $[a,\infty)$. Hence $\|f_n\|_{[a,\infty)} = f_n(a)$ for all $n \geq N$, and since $\lim_n f_n(a) = 0$ we have that the convergence is uniform on $[a,\infty)$.

(III) The pointwise limit $f(x)$ is easy to obtain for any $x \in \mathbb{R}$:

$$f(x) = \lim_n f_n(x) = \lim_n \frac{nx^2}{1+nx} = x.$$

Now $|f_n(x) - f(x)| = \frac{x}{1+nx}$ for every $x \in [0,\infty)$ and since $x \mapsto \frac{x}{1+nx}$ is increasing on $[0,\infty)$, we have that

$$\|f_n - f\|_{[0,\infty)} = \sup\{|f_n(x) - f(x)| : x \in [0,\infty)\}$$

$$= \sup\left\{\frac{x}{1+nx} : x \in [0,\infty)\right\}$$

$$= \lim_{x \to \infty} \frac{x}{1+nx} = \frac{1}{n},$$

for all $n \in \mathbb{N}$. Hence there is uniform convergence on $[0,\infty)$ since

$$\lim_n \|f_n - f\|_{[0,\infty)} = \lim_n \frac{1}{n} = 0.$$

As a consequence of the latter, (f_n) must converge uniformly to $f(x) = x$ on any subset of $[0,\infty)$, and in particular, on $[a,\infty)$ for all $a > 0$. A sketch of the f_n's can be viewed in Figure 8.6.

(IV) To calculate the pointwise limit we consider several cases. First, if $x \in [0,1)$ then $\lim_n x^n = 0$, so

$$\lim_n f_n(x) = \lim_n \frac{1 - x^n}{1 + x^n} = 1.$$

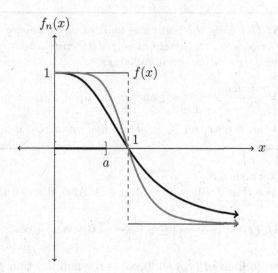

FIGURE 8.7: Representation of $f_n(x) = \frac{1-x^n}{1+x^n}$ (Exercise 8.7(IV)) for two different values of n, namely $n_1 < n_2$. Here f_{n_1} appears in black and f_{n_2} in gray.

Now observe that $f_n(1) = 0$ for all $n \in \mathbb{N}$. Finally, if $x > 1$ then $\lim_n \left(\frac{1}{x}\right)^n = 0$ from which

$$\lim_n f_n(x) = \lim_n \frac{1-x^n}{1+x^n} = \lim_n \frac{\left(\frac{1}{x}\right)^n - 1}{\left(\frac{1}{x}\right)^n + 1} = -1.$$

Therefore, the pointwise limit $f(x)$ of (f_n) on $[0, \infty)$ is

$$f(x) = \begin{cases} 1 & \text{if } x \in [0, 1), \\ 0 & \text{if } x = 1, \\ -1 & \text{if } x \in (1, \infty). \end{cases}$$

There cannot be uniform convergence neither on $[0, 1]$ nor on $[0, \infty)$ since the functions f_n are continuous and f has discontinuity points in both sets.

However, if $a \in [0, 1)$ and $x \in [0, a]$, then

$$|f_n(x) - f(x)| = \left| \frac{1-x^n}{1+x^n} - 1 \right| = \frac{2x^n}{1+x^n} \leq \frac{2a^n}{1+a^n}.$$

The last inequality is due to the fact that $x \mapsto \frac{2x^n}{1+x^n}$ is increasing on $[0, a]$. Therefore,

$$\|f_n - f\|_{[0,a]} \leq \frac{2a^n}{1+a^n}$$

and since $\lim_n a^n = 0$ it follows that $\lim_n \|f_n - f\|_{[0,a]} = 0$, proving that (f_n) converges to 1 uniformly on $[0, a]$ whenever $a \in [0, 1)$.

The functions f_n have been sketched in Figure 8.7.

$\square$

Exercise 8.8. *Calculate the pointwise limit of the following sequences that include trigonometric functions and determine whether the convergence is uniform on the given domains.*

(I) $f_n(x) = \dfrac{\sin nx}{1 + nx}$, *on* $[0, \infty)$ *and on* $[a, \infty)$, $a > 0$.

(II) $f_n(x) = \arctan nx$, *on* $\mathbb{R}$, *in* $[0, \infty)$ *and on* $[a, \infty)$, $a > 0$.

Solution to Exercise 8.8.

(I) First, notice that $f_n(0) = 0$ for all $n \in \mathbb{N}$. Also, if $x > 0$ then

$$|f_n(x)| = \left| \frac{\sin nx}{1 + nx} \right| \leq \frac{1}{1 + nx} \to 0 \quad \text{as } n \to \infty.$$

Hence the pointwise limit of (f_n) on $[0, \infty)$ is the null function $f \equiv 0$.

To study the uniform convergence on $[0, \infty)$ observe that

$$f_n\left(\frac{\pi}{2n}\right) = \frac{1}{1 + \frac{\pi}{2}} = \frac{2}{2 + \pi}.$$

Then $\|f_n - f\|_{[0,\infty)} = \|f_n\|_{[0,\infty)} \geq \frac{2}{2+\pi} > 0$ for every $n \in \mathbb{N}$. The latter implies that $\|f_n - f\|_{[0,\infty)}$ does not converge to zero, so there cannot be uniform convergence on $[0, \infty)$.

In the interval $[a, \infty)$ with $a > 0$ there is uniform convergence since, for $x \geq a$, we have

$$|f_n(x) - f(x)| = |f_n(x)| = \left| \frac{\sin nx}{1 + nx} \right| \leq \frac{1}{1 + nx} \leq \frac{1}{1 + na}.$$

Then

$$\|f_n - f\|_{[a,\infty)} \leq \frac{1}{1 + na} \to 0 \quad \text{as } n \to \infty.$$

The graph of the f_n's can be seen in Figure 8.8 for several choices of n.

(II) In order to calculate the pointwise limit we consider two cases. If $x = 0$ then $f_n(0) = \arctan(0) = 0$ for all $n \in \mathbb{N}$. Now if $x > 0$ then

$$\lim_n f_n(x) = \lim_n \arctan(nx) = \frac{\pi}{2}.$$

Consequently, (f_n) converges pointwise on $[0, \infty)$ to the function

$$f : [0, \infty) \to \mathbb{R} \text{ with } f(x) = \begin{cases} 0 & \text{if } x = 0, \\ \frac{\pi}{2} & \text{if } x > 0. \end{cases}$$

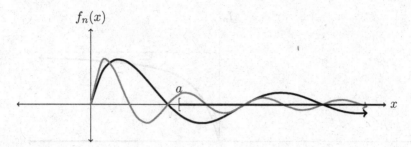

FIGURE 8.8: Representation of $f_n(x) = \frac{\sin nx}{1+nx}$ (Exercise 8.8(I)) for two different values of n, namely $n_1 < n_2$. Here f_{n_1} appears in black and f_{n_2} in gray.

Since f is not continuous at $x = 0$ and the f_n's are all continuous on $[0, \infty)$, there cannot be uniform convergence on $[0, \infty)$. Now notice that $\frac{\pi}{2} - f_n(x) > 0$ for all $x \in \mathbb{R}$ and that

$$\frac{d}{dx}\left[\frac{\pi}{2} - f_n(x)\right] = -\frac{n}{1+n^2 x^2} < 0,$$

so $\frac{\pi}{2} - f_n$ is strictly decreasing. Then, if $a > 0$, we have that

$$\|f_n - \pi/2\|_{[a,\infty)} = \sup\{|f_n(x) - \pi/2| : x \geq a\}$$
$$= \sup\{\pi/2 - f_n(x) : x \geq a\}$$
$$= \frac{\pi}{2} - \arctan(na).$$

Hence

$$\lim_n \|f_n - \pi/2\|_{[a,\infty)} = \frac{\pi}{2} - \lim_n \arctan(na) = 0,$$

which proves that (f_n) is uniformly convergent to $\frac{\pi}{2}$ on $[a, \infty)$.
The graph of the f_n's can be seen in Figure 8.9. $\qquad\square$

Exercise 8.9. *Calculate the pointwise limit of the following sequences of functions involving exponentials and determine whether the convergence is uniform on the given domains.*

(I) $f_n(x) = e^{-nx}$, *on $\mathbb{R}$, on $[0, \infty)$ and on $[a, \infty)$.*

(II) $f_n(x) = xe^{-nx}$, *on $\mathbb{R}$ and on $[0, \infty)$.*

(III) $f_n(x) = x^2 e^{-nx}$, *on $\mathbb{R}$ and on $[0, \infty)$.*

(IV) $f_n(x) = n^2 x^2 e^{-nx}$, *on $\mathbb{R}$, on $[0, \infty)$ and on $[a, \infty)$, $a > 0$.*

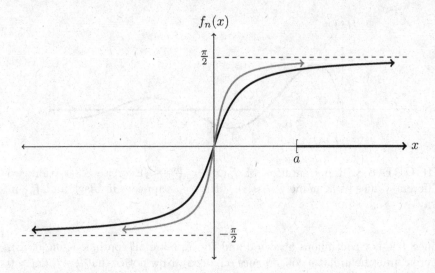

FIGURE 8.9: Representation of $f_n(x) = \arctan nx$ (Exercise 8.8(II)) for two different values of n, namely $n_1 < n_2$. Here f_{n_1} appears in black and f_{n_2} in gray.

Solution to Exercise 8.9.

(I) $f_n(x) = e^{-nx}$. If $x < 0$ then

$$\lim_n f_n(x) = \lim_n e^{-nx} = \infty.$$

Consequently, there is not pointwise convergence on the whole real line. However,

$$\lim_n f_n(0) = \lim_n 1 = 1$$

and, if $x > 0$,

$$\lim_n f_n(0) = \lim_n e^{-nx} = 0,$$

so (f_n) converges pointwise on $[0, \infty)$ to the function $f : [0, \infty) \to \mathbb{R}$ given by

$$f(x) = \begin{cases} 1 & \text{if } x = 0, \\ 0 & \text{if } x > 0. \end{cases}$$

Uniform convergence on $[0, \infty)$ is impossible since the f_n's are continuous but f is not continuous at $x = 0$. Finally, since there is not pointwise convergence for negative x and the convergence on the interval $[0, \infty)$ has already been considered, we can assume that $a > 0$. The functions f_n are strictly decreasing and therefore

$$\|f_n - f\|_{[a,\infty)} = \|f_n\|_{[a,\infty)} = f_n(a) = e^{-na} \to 0 \quad \text{as } n \to \infty.$$

Then (f_n) converges to $f \equiv 0$ uniformly on $[a, \infty)$ for $a > 0$.

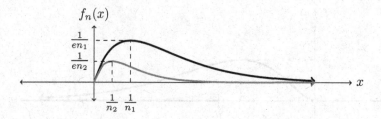

FIGURE 8.10: Representation of $f_n(x) = xe^{-nx}$ (Exercise 8.9(II)) for two different values of n. Here $n_2 > n_1$.

(II) $f_n(x) = xe^{-nx}$. For any negative real number x we have

$$\lim_n f_n(x) = \lim_n xe^{-nx} = -\infty.$$

Then (f_n) does not converge pointwise on the whole real line. However (f_n) converges uniformly to $f \equiv 0$ on $[0, \infty)$. To prove this we will study first the uniform norm of $f_n - f = f_n$ over $[0, \infty)$. Notice that the f_n's are non-negative on $[0, \infty)$. Also $f'_n(x) = (1 - nx)e^{-nx}$ and $f''_n(x) = -n(2 - nx)e^{-nx}$. It is straightforward that the unique root of f'_n, namely, $x = \frac{1}{n}$, satisfies $f''_n(1/n) = -n/e < 0$. Hence f_n must attain its absolute maximum on $[0, \infty)$ at $x = \frac{1}{n}$ and therefore

$$\|f_n - f\|_{[0,\infty)} = \|f_n\|_{[0,\infty)} = f_n\left(\frac{1}{n}\right) = \frac{1}{ne} \to 0 \quad \text{as } n \to 0.$$

Consequently (f_n) converges to $f \equiv 0$ uniformly on $[0, \infty)$.

The functions f_n have been sketched in Figure 8.10.

(III) $f_n(x) = x^2 e^{-nx}$. As in the previous case, if $x < 0$ then

$$\lim_n f_n(x) = \lim_n x^2 e^{-nx} = \infty.$$

Pointwise convergence of (f_n) on the whole real line is therefore impossible. However (f_n) does converge uniformly to $f \equiv 0$ on $[0, \infty)$. To see the latter we proceed an in Exercise 8.9 (II), the f_n's are non-negative on $[0, \infty)$. Also

$$f'_n(x) = (2x - nx^2)e^{-nx}$$

and

$$f''_n(x) = (2 - 4nx + n^2x^2)e^{-nx}.$$

It is easy to check that $x = 0$ and $x = \frac{2}{n}$ are the only roots of f'_n. Also, $f''_n(0) = 2 > 0$ and $f''_n(2/n) = -e^{-1} < 0$. Hence f_n has a minimum at $x = 0$ and a maximum at $x = \frac{2}{n}$. Clearly f_n must attain its absolute maximum on $[0, \infty)$ at $x = \frac{2}{n}$ and therefore

$$\|f_n - f\|_{[0,\infty)} = \|f_n\|_{[0,\infty)} = f_n\left(\frac{2}{n}\right) = \frac{4}{n^2e^2} \to 0 \quad \text{as } n \to 0.$$

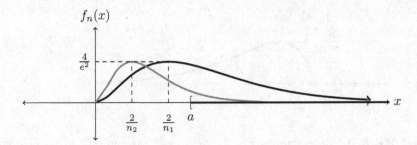

FIGURE 8.11: Representation of $f_n(x) = n^2 x^2 e^{-nx}$ (Exercise 8.9(IV)) for two different values of n. Here $n_2 > n_1$.

Consequently (f_n) converges to $f \equiv 0$ uniformly on $[0, \infty)$.

(IV) $f_n(x) = n^2 x^2 e^{-nx}$. As in the previous cases, (f_n) does not converge pointwise for $x < 0$. On the contrary, $f_n(0) = 0$ for all $n \in \mathbb{N}$. If $x > 0$ then

$$\lim_n f_n(x) = \lim_n n^2 x^2 e^{-nx} = x^2 \lim_n \frac{n^2}{e^{nx}} = 0.$$

Then (f_n) converges pointwise to $f \equiv 0$ on $[0, \infty)$. However the convergence is not uniform on $[0, \infty)$. Indeed, the f_n's are non-negative and they attain their absolute maximum on $[0, \infty)$ at $x = \frac{2}{n}$. To see this just observe that the f_n is nothing but a constant, namely n^2, multiplied by the functions studied in Exercise 8.9 (III), which attain their absolute maximum at $x = \frac{2}{n}$. Then

$$\|f_n - f\|_{[0,\infty)} = \|f_n\|_{[0,\infty)} = f_n\left(\frac{2}{n}\right) = \frac{4}{e^2} > 0 \quad \text{for all } n \in \mathbb{N}.$$

Hence (f_n) does not converge uniformly to $f \equiv 0$ on $[0, \infty)$.

On the other hand, (f_n) converges uniformly to $f \equiv 0$ on $[a, \infty)$ if $a > 0$. To prove the latter notice that f_n is strictly decreasing on $\left(\frac{2}{n}, \infty\right)$. Also, since $\lim_n \frac{2}{n} = 0$, there exists $N \in \mathbb{N}$ such that $\frac{2}{n} < a$ for all $n \geq N$. Thus f_n is decreasing on $[a, \infty)$, so

$$\|f_n - f\|_{[a,\infty)} = \|f_n\|_{[a,\infty)} = f_n(a) = \frac{n^2}{e^{na}} \to 0 \quad \text{as } n \to \infty.$$

The functions f_n have been represented in Figure 8.11. $\qquad \square$

Exercise 8.10. *Let (f_n) be a sequence of continuous functions on a set A converging uniformly on A to a function f. If (x_n) is a sequence in A, $c \in A$ and $\lim_n x_n = c$, prove that $\lim_n f_n(x_n) = f(c)$.*

Solution to Exercise 8.10.

Let $\varepsilon > 0$. Since (f_n) converges to f uniformly on A there exists $N_1 \in \mathbb{N}$ such that

$$|f_n(x) - f(x)| < \frac{\varepsilon}{2}, \quad \text{for all } x \in A \text{ and every } n \geq N_1.$$

Next observe that f must be continuous at $x = c$. Then there exists $\delta > 0$ such that

$$|f(x) - f(c)| < \frac{\varepsilon}{2}, \quad \text{for all } x \in A \text{ such that } |x - c| < \delta.$$

Since (x_n) converges to c, there exists $N_2 \in \mathbb{N}$ so that

$$|x_n - c| < \delta, \quad \text{for every } n \geq N_2.$$

In particular,

$$|f(x_n) - f(c)| < \frac{\varepsilon}{2}, \quad \text{for all } n \geq N_2.$$

If $N = \max\{N_1, N_2\}$, combining the previous inequalities we would have

$$|f_n(x_n) - f(c)| = |[f_n(x_n) - f(x_n)] + [f(x_n) - f(c)]|$$
$$\leq |f_n(x_n) - f(x_n)| + |f(x_n) - f(c)| < \frac{\varepsilon}{2} + \frac{\varepsilon}{2} = \varepsilon,$$

for all $n \geq N$. Thus, $\lim\limits_{n} f_n(x_n) = f(c)$. $\qquad\square$

Exercise 8.11. *Consider the sequences (f_n) and (g_n) given by*

$$f_n(x) = x\left(1 + \frac{1}{n}\right), \qquad x \in \mathbb{R},$$

$$g_n(x) = \begin{cases} \frac{1}{n}, & \text{if } x = 0 \text{ or } x \notin \mathbb{Q}, \\ q + \frac{1}{n}, & \text{if } x = \frac{p}{q}, \ p, q \in \mathbb{Z}, \ q > 0, \ \gcd\{p, q\} = 1. \end{cases}$$

Let $h_n(x) = f_n(x) g_n(x)$.

(I) *Prove that (f_n) and (g_n) converge uniformly on every bounded interval.*

(II) *Prove that (h_n) does not converge uniformly on any bounded interval.*

Solution to Exercise 8.11.

(I) It will be enough to consider closed bounded intervals such as $[a, b]$ with $a, b \in \mathbb{R}$ and $a < b$. The sequence converges pointwise to $f(x) = x$ on $\mathbb{R}$ since

$$\lim_n f_n(x) = \lim_n x\left(1 + \frac{1}{n}\right) = x \lim_n \left(1 + \frac{1}{n}\right) = x,$$

for all $x \in \mathbb{R}$. There is uniform convergence on $[a, b]$ because the uniform norms of $f_n - f$ on $[a, b]$ converge to zero. Indeed,

$$\|f_n - f\|_{[a,b]} = \sup\{|f_n(x) - f(x)| : x \in [a, b]\}$$

$$= \sup\left\{\frac{|x|}{n} : x \in [a, b]\right\}$$

$$= \frac{\max\{|a|, |b|\}}{n} \longrightarrow 0 \quad \text{as } n \to \infty.$$

The functions g_n can be written as $g_n(x) = g(x) + \frac{1}{n}$ for $x \in \mathbb{R}$ where

$$g(x) = \begin{cases} 0 & \text{if } x = 0 \text{ or } x \notin \mathbb{Q}, \\ q & \text{if } x = \frac{p}{q}, \ p, q \in \mathbb{Z}, \ q > 0, \ \gcd\{p, q\} = 1. \end{cases}$$

Then it is straightforward that (g_n) converges pointwise to g on $\mathbb{R}$. Also

$$\|g_n - g\|_{\mathbb{R}} = \frac{1}{n} \longrightarrow 0 \quad \text{as } n \to 0.$$

Hence (g_n) converges uniformly to g on the whole real line, and in particular, on every bounded interval.

(II) However, the sequence $(f_n g_n)$ does not converge uniformly on any bounded interval. It will be enough to consider open intervals (a, b) with $a, b \in \mathbb{R}$ and $0 < a < b$. We say that a positive integer q is a plausible denominator for (a, b) if there exists $p \in \mathbb{Z}$ such that $\gcd\{p, q\} = 1$ and $\frac{p}{q} \in (a, b)$. In that case we say that p is a plausible numerator for (a, b) corresponding to q. For each plausible denominator q there is only a finite number of plausible numerators p since $aq < p < nb$. Therefore, if there were only a finite number of plausible denominators for (a, b), then $(a, b) \cap \mathbb{Q}$ would be finite. Then the set $D(a, b)$ of plausible denominators for (a, b) is unbounded. Thus, n^2 is not an upper bound for $D(a, b)$ for any $n \in \mathbb{N}$, so there is $q_n \in D(a, n)$ such that $q_n > n^2$ for all $n \in \mathbb{N}$. If p_n is a plausible numerator for (a, b) corresponding to q_n for each $n \in \mathbb{N}$, then $a < \frac{p_n}{q_n} < b$, from which $p_n > aq_n > an^2$. Then,

$$\left|(f_n g_n)\left(\frac{p_n}{q_n}\right) - (fg)\left(\frac{p_n}{q_n}\right)\right| = \left|\frac{p_n}{q_n}\left(1 + \frac{1}{n}\right)\left(q_n + \frac{1}{n}\right) - p_n\right|$$

$$= \frac{1}{n} + \frac{1}{n^2} + \frac{p_n}{n} > \frac{1}{n} + \frac{1}{n^2} + an.$$

The latter shows that

$$\|f_n g_n - fg\|_{[a,b]} \geq \frac{1}{n} + \frac{1}{n^2} + an \longrightarrow \infty \quad \text{as } n \to \infty.$$

Therefore $(f_n g_n)$ does not converge uniformly to fg on (a, b).   □

Exercise 8.12. *Study the uniform convergence of the following series of functions on the specified domains.*

(I) $\displaystyle\sum_{n=1}^{\infty} \frac{x^n}{n!}$ *on* $[-1, 1]$ *and on* $(0, \infty)$.

(II) $\displaystyle\sum_{n=1}^{\infty} (1 - x)x^n$ *on* $[0, 1]$.

(III) $\displaystyle\sum_{n=1}^{\infty} \frac{x}{((n - 1)x + 1)(nx + 1)}$ *on* $(0, \infty)$.

(IV) $\displaystyle\sum_{n=1}^{\infty} \frac{nx}{(1 + x)(1 + 2x) \cdots (1 + nx)}$ *on* $[0, 1]$ *and on* $[1, \infty)$.

(V) $\displaystyle\sum_{n=1}^{\infty} \frac{nx}{1 + n^5 x^2}$ *on* $\mathbb{R}$.

(VI) $\displaystyle\sum_{n=1}^{\infty} \frac{n^2}{\sqrt{n!}} (x^n + x^{-n})$ *on* $[1/3, 2]$.

Solution to Exercise 8.12.

(I) Notice that the series of functions is, in fact, a power series. So, we begin by calculating its radius of convergence using the Ratio Test. Let $a_n = \dfrac{1}{n!}$. Then:

$$\lim_{n \to \infty} \frac{a_{n+1}}{a_n} = \lim_{n \to \infty} \frac{\frac{1}{(n + 1)!}}{\frac{1}{n!}} = \lim_{n \to \infty} \frac{n!}{(n + 1)!} = \lim_{n \to \infty} \frac{1}{n + 1} = 0.$$

Hence, the radius of convergence is $R = \infty$.

By the Cauchy-Hadamard Theorem, we know that $\displaystyle\sum_{n=1}^{\infty} \frac{x^n}{n!}$ converges uniformly on any $[-r, r]$ with $0 < r < R$. Hence, since $R = \infty$, for $r = 1$ we have that $\displaystyle\sum_{n=1}^{\infty} \frac{x^n}{n!}$ converges uniformly on $[-1, 1]$.

Not let us study the case of the interval $(0, \infty)$. Assume, by means of contradiction, that $\displaystyle\sum_{n=1}^{\infty} \frac{x^n}{n!}$ is uniformly convergent on $(0, \infty)$. Then, by Cauchy's Uniform Convergence Criterion for series of functions, we have that for each

$\varepsilon > 0$ there exists an $n_0 \in \mathbb{N}$ such that $\left| \sum\limits_{n=j}^{k} \dfrac{x^n}{n!} \right| < \varepsilon$ for every $j, k \geq n_0$ and

$x \in (0, \infty)$. However, by fixing any $k \geq n_0$ and taking $x \geq \sqrt[k+1]{\varepsilon(k+1)!} > 0$, it yields that

$$\left| \sum_{n=k+1}^{k+1} \frac{x^n}{n!} \right| = \frac{x^{k+1}}{(k+1)!} \geq \varepsilon,$$

a contradiction.

In conclusion, the power series $\sum\limits_{n=1}^{\infty} \dfrac{x^n}{n!}$ is uniformly convergent on $[-1, 1]$ but not on $(0, \infty)$.

(II) Let us show that $\sum\limits_{n=1}^{\infty} (1-x)x^n$ is not uniformly convergent on $[0, 1]$. Assume otherwise. Then, by Cauchy's Uniform Convergence Criterion for series of functions, for any $\varepsilon > 0$, there exists $n_0 \in \mathbb{N}$ such that $\left| \sum\limits_{n=j}^{k} (1-x)x^n \right| < \varepsilon$ for any $k \geq j \geq n_0$ and $x \in [0, 1]$. However, for all $k \geq j \geq n_0$, we have that

$$\left| \sum_{n=j}^{k} (1-x)x^n \right| = |x^j - x^{j+1} + x^{j+1} - x^{j+2} + \cdots + x^k - x^{k+1}|$$

$$= |x^j - x^{k+1}| < \varepsilon.$$

As a consequence, by Cauchy's Uniform Convergence Criterion for series of functions, it yields that the sequence of functions $(x^n)_{n=1}^{\infty}$ converges uniformly on $[0, 1]$, which is not true.

(III) Assume that the series of functions $\sum\limits_{n=1}^{\infty} \dfrac{x}{((n-1)x+1)(nx+1)}$ converges uniformly on $(0, \infty)$. Then, by Cauchy's Uniform Convergence Criterion for series of functions, for any $\varepsilon > 0$, there exists $n_0 \in \mathbb{N}$ such that $\left| \sum\limits_{n=j}^{k} \dfrac{x}{((n-1)x+1)(nx+1)} \right| < \varepsilon$ for any $j, k \geq n_0$ and $x \in (0, \infty)$. Now, for any positive integers $k \geq j > 1$ and $x > 0$, we obtain that

$$\left| \sum_{n=j}^{k} \frac{x}{((n-1)x+1)(nx+1)} \right| = \sum_{n=j}^{k} \frac{x}{((n-1)x+1)(nx+1)}$$

$$\geq \frac{x}{((k-1)x+1)(kx+1)}$$

$$= \frac{x}{(k-1)kx^2 + (2k-1)x + 1}.$$

But, it we take the limit when x goes to 0 in the latter inequality, since $k > 1$, we arrive at

$$\lim_{x \to 0} \left| \sum_{n=j}^{k} \frac{x}{((n-1)x+1)(nx+1)} \right|$$

$$\geq \lim_{x \to 0} \left| \frac{x}{(k-1)kx^2 + (2k-1)x + 1} \right| = \infty,$$

which is impossible.

(IV) Let us analyze first the uniform convergence on $[1, \infty)$. We will try to find a uniform bound for $\left| \dfrac{nx}{(1+x)(1+2x)\cdots(1+nx)} \right|$ so that we can apply the Weierstrass M-test. For every $x \in [1, \infty)$ and $n \in \mathbb{N}$, we have

$$\left| \frac{nx}{(1+x)(1+2x)\cdots(1+nx)} \right| = \frac{nx}{(1+x)(1+2x)\cdots(1+nx)}$$

$$\leq \frac{nx}{x(2x)\cdots(nx)} = \frac{nx}{n!x^n} = \frac{1}{(n-1)!x^{n-1}}$$

$$\leq \frac{1}{(n-1)!}.$$

Now, since $\displaystyle\sum_{n=1}^{\infty} \frac{1}{n!}$ converges (in fact, to the number e), we have, by the Weierstrass M-test, that $\displaystyle\sum_{n=1}^{\infty} \frac{nx}{(1+x)(1+2x)\cdots(1+nx)}$ is uniformly convergent on $[1, \infty)$.

(V) Let us find a uniform bound for $\left| \dfrac{nx}{1+n^5x^2} \right|$ in order to apply the Weierstrass M-test. For all $x \in \mathbb{R}$ and $n \in \mathbb{N}$, we have

$$\left| \frac{nx}{1+n^5x^2} \right| = \frac{n|x|}{1+n^5x^2} = \frac{1}{2n^{3/2}} \cdot \frac{2n^{5/2}|x|}{1+n^5x^2}.$$

Now recall that the inequality $\dfrac{2|ab|}{a^2+b^2} \leq 1$ holds for all $a, b \in \mathbb{R}$ with $a \neq 0 \neq b$. Indeed, since $0 \leq (|a|-|b|)^2 = a^2 + b^2 - 2|ab|$, we obtain $2|ab| \leq a^2 + b^2$ and, as a consequence, the inequality is proved. Then, by taking $a = 1$ and $b = n^{5/2}|x|$, we have that

$$\left| \frac{nx}{1+n^5x^2} \right| = \frac{1}{2n^{3/2}} \cdot \frac{2n^{5/2}|x|}{1+n^5x^2} \leq \frac{1}{2n^{3/2}}.$$

Finally, since the numerical series $\displaystyle\sum_{n=1}^{\infty} \frac{1}{n^{3/2}}$ is a p-series with $p = \dfrac{3}{2} > 1$,

we have that $\displaystyle\sum_{n=1}^{\infty} \frac{1}{2n^{3/2}}$ converges. So, by Weierstrass M-test, we have that

$\displaystyle\sum_{n=1}^{\infty} \frac{nx}{1 + n^5 x^2}$ is uniformly convergent on $\mathbb{R}$.

(VI) Let us find a uniform bound for $\left| \dfrac{n^2}{\sqrt{n!}}(x^n + x^{-n}) \right|$ so that we can apply the Weierstrass M-test. Observe that for all $n \in \mathbb{N}$ and $x \in [1/3, 2]$, we have the following:

$$\left| \frac{n^2}{\sqrt{n!}}(x^n + x^{-n}) \right| = \frac{n^2}{\sqrt{n!}}|x^n + x^{-n}| \leq \frac{n^2}{\sqrt{n!}}(|x|^n + |x|^{-n})$$

$$\leq \frac{n^2}{\sqrt{n!}}(2^n + (1/3)^{-n}) = \frac{n^2}{\sqrt{n!}}(2^n + 3^n) \leq \frac{2n^2 3^n}{\sqrt{n!}}.$$

Let us show that $\displaystyle\sum_{n=1}^{\infty} \frac{n^2 3^n}{\sqrt{n!}}$ is convergent. In order to do so, we will apply the Ratio Test with $a_n = \dfrac{n^2 3^n}{\sqrt{n!}}$:

$$\lim_{n \to \infty} \frac{a_{n+1}}{a_n} = \lim_{n \to \infty} \frac{\dfrac{(n+1)^2 3^{n+1}}{\sqrt{(n+1)!}}}{\dfrac{n^2 3^n}{\sqrt{n!}}} = \lim_{n \to \infty} \frac{(n+1)^2 3^{n+1} \sqrt{n!}}{n^2 3^n \sqrt{(n+1)!}}$$

$$= \lim_{n \to \infty} \frac{3(n+1)^2}{n^2 \sqrt{n+1}} = \lim_{n \to \infty} \frac{3(n+1)^{3/2}}{n^2} = 0 < 1.$$

So, by the Ratio Test, $\displaystyle\sum_{n=1}^{\infty} \frac{n^2 3^n}{\sqrt{n!}}$ converges. Hence, by the Weierstrass M-test,

the series of functions $\displaystyle\sum_{n=1}^{\infty} \frac{n^2}{\sqrt{n!}}(x^n + x^{-n})$ uniformly converges on $[1/3, 2]$. $\quad\square$

Exercise 8.13. *For each $n \in \mathbb{N}$ consider the function $f_n : [0, 1] \to \mathbb{R}$ given by*

$$f_n(x) = \begin{cases} 0 & \text{if } x < \frac{1}{n+1}, \\ \sin^2 \frac{\pi}{x} & \text{if } \frac{1}{n+1} \leq x \leq \frac{1}{n}, \\ 0 & \text{if } \frac{1}{n} < x, \end{cases}$$

for all $x \in [0, 1]$.

(I) *Show that (f_n) converges to a continuous function, but not uniformly on $\mathbb{R}$.*

(II) *Prove that the series $\sum_{k=1}^{\infty} f_k$ converges absolutely for all $x \in \mathbb{R}$, but not uniformly on $\mathbb{R}$.*

Solution to Exercise 8.13.

(I) The sequence (f_n) converges pointwise to the null function $f \equiv 0$ on $[0, 1]$. It is obvious that $f_n(0) = 0$ for all $n \in \mathbb{N}$. Now if $x \in (0, 1]$ then there exists $N \in \mathbb{N}$ such that $\frac{1}{n} < x$ for all $n \geq N$. Hence $f_n(x) = 0$ for all $n \geq N$, so $\lim_n f_n(x) = 0$.

However, the convergence is not uniform. Indeed, letting $x_n = \frac{1}{n+\frac{1}{2}}$, we have that $\frac{1}{n+1} \leq x_n \leq \frac{1}{n}$ and

$$f_n(x_n) = \sin^2 \left(n + \frac{1}{2} \right) \pi = 1.$$

Hence, for all $n \in \mathbb{N}$ we have

$$\|f_n - f\|_{[0,1]} = \|f_n\|_{[0,1]} \geq 1,$$

from which uniform convergence on $[0, 1]$ is not possible.

(II) Since the f_n's are non-negative, pointwise convergence here is equivalent to absolute convergence. Observe that the sets of points where the f_n's do not vanish are pairwise disjoint. This idea allows us to prove easily that

$$\sum_{k=1}^{\infty} f_k(x) = \sin^2 \frac{\pi}{x},$$

for all $x \in [0, 1]$. This proves the existence of pointwise convergence. However there is not uniform convergence on $[0, 1]$. To see this, consider the sequence S_n of partial sums of $\sum_{k=1}^{\infty} f_k$. Then

$$S_n(x) = \begin{cases} \sin^2 \frac{\pi}{x} & \text{if } x \geq \frac{1}{n+1}, \\ 0 & \text{if } x < \frac{1}{n+1}, \end{cases}$$

for all $x \in [0, 1]$. We have

$$\left| \sum_{k=1}^{\infty} f_k(x) - S_n(x) \right| = \left| \sin^2 \frac{\pi}{x} - S_n(x) \right| = \begin{cases} 0 & \text{if } x \geq \frac{1}{n+1}, \\ \sin^2 \frac{\pi}{x} & \text{if } x < \frac{1}{n+1}, \end{cases}$$

for all $x \in [0,1]$. Now letting $x_n = \frac{1}{n+\frac{3}{2}} < \frac{1}{n+1}$, it follows that

$$\left\| \sum_{k=1}^{\infty} f_k - S_n \right\|_{[0,1]} \geq \left| \sum_{k=1}^{\infty} f_k(x_n) - S_n(x_n) \right|$$

$$= \sin^2\left(n + \frac{3}{2}\right)\pi = 1.$$

Hence there cannot be uniform convergence. $\qquad\square$

Exercise 8.14. *If* $\displaystyle\sum_{k=1}^{\infty} a_k$ *is a convergent series of positive numbers, prove that the series*

$$\sum_{k=1}^{\infty} (-1)^k \left(a_k x^2 + \frac{1}{k} \right)$$

converges uniformly on all bounded intervals, but it does not converge absolutely for any x.

Solution to Exercise 8.14.

Put $f_n(x) = (-1)^n \left(a_n x^2 + \frac{1}{n} \right)$ for all $n \in \mathbb{N}$ and every $x \in \mathbb{R}$. The series $\displaystyle\sum_{k=1}^{\infty} (-1)^k a_k$ converges absolutely, say to α, whereas $\displaystyle\sum_{k=1}^{\infty} \frac{(-1)^k}{k}$ converges conditionally to, say, β. Then $\displaystyle\sum_{k=1}^{\infty} f_k(x)$ converges pointwise to $\alpha x^2 + \beta$ on the whole real line. Let us see next that the convergence is uniform over all bounded intervals I. It will be enough to consider the case where $I = [a,b]$ with $a, b \in \mathbb{R}$ and $a < b$. Notice that $x^2 \leq \max\{a^2, b^2\}$ for all $x \in [a,b]$. Since $\displaystyle\sum_{k=1}^{\infty} (-1)^k a_k$ and $\displaystyle\sum_{k=1}^{\infty} (-1)^k \frac{1}{k}$ converge to α and β respectively, for all $\varepsilon > 0$ there exist $N_1, N_2 \in \mathbb{N}$ such that

$$\left| \sum_{k=1}^{n} (-1)^k a_k - \alpha \right| < \frac{\varepsilon}{2\max\{a^2, b^2\}} \quad \text{if } n \geq N_1,$$

$$\left| \sum_{k=1}^{n} \frac{(-1)^k}{k} - \beta \right| < \frac{\varepsilon}{2} \quad \text{if } n \geq N_2.$$

If $N = \max\{N_1, N_2\}$ and $n \geq N$ then

$$\left| \sum_{k=1}^{n} f_n(x) - (\alpha x^2 + \beta) \right| = \left| \left[\sum_{k=1}^{n} (-1)^k a_k - \alpha \right] x^2 + \left[\sum_{k=1}^{n} \frac{(-1)^k}{k} - \beta \right] \right|$$

$$\leq \left| \sum_{k=1}^{n} (-1)^k a_k - \alpha \right| x^2 + \left| \sum_{k=1}^{n} \frac{(-1)^k}{k} - \beta \right|$$

$$< \frac{\varepsilon}{2 \max\{a^2, b^2\}} x^2 + \frac{\varepsilon}{2} \leq \frac{\varepsilon}{2} + \frac{\varepsilon}{2} = \varepsilon,$$

for all $x \in [a, b]$. Hence $\sum_{k=1}^{\infty} f_k(x)$ converges uniformly to $\alpha x^2 + \beta$ on $[a, b]$.

At the other end of the scale, if $\sum_{k=1}^{\infty} |f_k(x)|$ were convergent for some $x \in \mathbb{R}$, from

$$\sum_{k=1}^{\infty} |f_k(x)| = \sum_{k=1}^{\infty} \left(a_k x^2 + \frac{1}{k} \right) = \left[\sum_{k=1}^{\infty} a_k \right] x^2 + \sum_{k=1}^{\infty} \frac{1}{k}$$

it would follow that the harmonic series $\sum_{k=1}^{\infty} \frac{1}{k}$ is convergent and that is not true. □

Exercise 8.15. *Determine the points where the function*

$$f(x) = \sum_{n=1}^{\infty} \frac{|x|}{n^2 + x^2}$$

is differentiable.

Solution to Exercise 8.15.

First, observe that

$$\left| \frac{1}{n^2 + x^2} \right| \leq \frac{1}{n^2}$$

for every $x \in \mathbb{R}$ and every $n \in \mathbb{N}$ and that $\sum_{n=1}^{\infty} \frac{1}{n^2}$ is a convergent numerical series. Hence, by Weierstrass M-test $\sum_{n=1}^{\infty} \frac{1}{n^2 + x^2}$ converges uniformly to a function $g : \mathbb{R} \to \mathbb{R}$. Next we prove that g is differentiable on $\mathbb{R}$. Let $g_n(x) = \frac{1}{n^2 + x^2}$ for all $x \in \mathbb{R}$ and every $n \in \mathbb{N}$. If we fix $n \in \mathbb{N}$, then

$$g_n'(x) = \frac{-2x}{(n^2 + x^2)^2},$$

for $x \in \mathbb{R}$. Since the $|g_n'|$'s are even, we only need to search for the local extrema of $|g_n'|$ on $(-\infty, 0]$, where $|g_n'| = g_n'$. On the other hand

$$g_n''(x) = \frac{-2n^2 + 2x^2}{(n^2 + x^2)^3} = 2\frac{(x-n)(x+n)}{(n^2 + x^2)^3},$$

for all $x \in \mathbb{R}$. Clearly, $g_n''(x) < 0$ for $x \in (-\infty, n)$ and $g_n''(x) > 0$ for $x \in (-n, 0]$, so g_n' has a maximum at $x = -n$ and it must attain its absolute maximum at $x = -n$. Hence

$$|g_n'(x)| \le |g_n'(\pm n)| = \frac{2}{n^3},$$

for all $x \in \mathbb{R}$. Since $\sum_{n=1}^{\infty} \frac{2}{n^3}$ is convergent, by Weierstrass M-test, $\sum_{n=1}^{\infty} g_n'(x)$ is uniformly convergent on $\mathbb{R}$. Therefore, g is differentiable on $\mathbb{R}$. So far we have seen that $f(x) = |x|g(x)$ for all $x \in \mathbb{R}$, where g is differentiable. It is straightforward that f is differentiable at any $x \ne 0$. However f is not differentiable at $x = 0$. Indeed,

$$\lim_{x \to 0^+} \frac{f(x) - f(0)}{x - 0} = \lim_{x \to 0^+} \frac{|x|g(x)}{x} = \lim_{x \to 0^+} g(x) = g(0),$$

$$\lim_{x \to 0^-} \frac{f(x) - f(0)}{x - 0} = \lim_{x \to 0^-} \frac{|x|g(x)}{x} = -\lim_{x \to 0^-} g(x) = -g(0),$$

and since $g(0) = \sum_{n=1}^{\infty} \frac{1}{n^2} > 0$, f is not differentiable at 0. $\qquad\square$

Exercise 8.16. *Let $A \subset \mathbb{R}$ be any non-empty set and (f_n) a sequence of bounded functions defined on A that converges uniformly to f on A. Prove that f is bounded and that the f_n's are uniformly bounded, that is, there exists $M > 0$ such that*

$$|f_n(x)| \le M, \quad \text{for all } x \in I.$$

Solution to Exercise 8.16.

If we put $\varepsilon = 1$ in the definition of uniform convergence there must be $N \in \mathbb{N}$ such that

$$|f_n(x) - f(x)| < 1,$$

for all $x \in A$ and every $n \ge N$. Now f_N is bounded on A, so there is $M_N > 0$ such that

$$|f_N(x)| \le M_N,$$

for all $x \in A$. Hence

$$|f(x)| = |[f(x) - f_N(x)] + f_N(x)| \le |f_N(x) - f(x)| + |f_N(x)| < 1 + M_N,$$

for all $x \in A$. This proves that $|f|$ is bounded by $1 + M_N$. Next, if $n \geq N$ then

$$|f_n(x)| = |[f_n(x) - f(x)] + f(x)| \leq |f_n(x) - f(x)| + |f(x)| < 2 + M_N,$$

for all $x \in A$. Since $f_1, \ldots, f_{N-1}$ are bounded on A, there are $M_1, \ldots, M_{N-1} > 0$ such that

$$|f_k(x)| \leq M_k,$$

for all $x \in A$ and $k \in \{1, \ldots, N-1\}$. Finally, if

$$M = \max\{M_1, \ldots, M_{N-1}, 2 + M_N\},$$

it is clear that

$$|f_n(x)| \leq M,$$

for all $n \in \mathbb{N}$ and every $x \in A$. $\qquad \square$

Exercise 8.17. *Let (f_n) be a sequence of continuous functions on a closed, bounded interval $[a, b]$ and suppose (f_n) converges to f uniformly on $[a, b]$. Justify whether or not we have that*

$$\lim_{n \to \infty} \int_a^{b - \frac{b-a}{n}} f_n = \int_a^b f.$$

Solution to Exercise 8.17.

Since f is continuous, it must be bounded on $[a, b]$. Then (f_n) is uniformly bounded on $[a, b]$, that is, there exists $M > 0$ such that

$$|f_n(x)| \leq M,$$

for all $x \in [a, b]$ and every $n \in \mathbb{N}$. We can use the uniform boundedness of the f_n to prove that $\lim_n \int_{b - \frac{b-a}{n}}^b f_n = 0$. Indeed,

$$\left| \int_{b - \frac{b-a}{n}}^b f_n \right| \leq \int_{b - \frac{b-a}{n}}^b |f_n| \leq \int_{b - \frac{b-a}{n}}^b M = \frac{M(b-a)}{n} \longrightarrow 0 \quad \text{as } n \to \infty.$$

Taking into consideration that the existence of uniform convergence guaranties that $\lim_n \int_a^b f_n = \int_a^b f$, we arrive at

$$\lim_n \int_a^{b - \frac{b-a}{n}} f_n = \lim_n \left[\int_a^b f_n - \int_{b - \frac{b-a}{n}}^b f_n \right]$$

$$= \lim_n \int_a^b f_n - \lim_n \int_{b - \frac{b-a}{n}}^b f_n = \int_a^b f.$$

$\qquad \square$

Exercise 8.18. *Determine the interval of convergence of the following power series:*

(I) $\displaystyle\sum_{n=0}^{\infty}\left(\frac{n!}{3\cdot 5\cdots(2n+1)}\right)^{2}x^{n}$,   (VIII) $\displaystyle\sum_{n=0}^{\infty}\sqrt{n}x^{n}$,

(II) $\displaystyle\sum_{n=0}^{\infty}\binom{2n}{n}x^{n}$,   (IX) $\displaystyle\sum_{n=0}^{\infty}x^{n!}$,

(III) $\displaystyle\sum_{n=0}^{\infty}n(\sqrt[n]{2}-1)x^{n}$,   (X) $\displaystyle\sum_{n=1}^{\infty}n^{-\sqrt{n}}x^{n}$,

(IV) $\displaystyle\sum_{n=1}^{\infty}\frac{2^{n}}{n^{2}}x^{n}$,   (XI) $\displaystyle\sum_{n=1}^{\infty}\frac{3^{n}}{\sqrt{n}}x^{2n+1}$,

(V) $\displaystyle\sum_{n=0}^{\infty}\frac{2^{n}}{n!}x^{n}$,   (XII) $\displaystyle\sum_{n=1}^{\infty}n!\left(\frac{x}{n}\right)^{n}$,

(VI) $\displaystyle\sum_{n=1}^{\infty}\frac{3^{n}}{n4^{n}}x^{n}$,   (XIII) $\displaystyle\sum_{n=1}^{\infty}\frac{\log n}{n}x^{n}$,

(VII) $\displaystyle\sum_{n=1}^{\infty}\frac{(-1)^{n}}{n^{2}4^{n}}x^{n}$,   (XIV) $\displaystyle\sum_{n=0}^{\infty}x^{n}\tan\frac{a}{2^{n}}$, *where* $a\in\mathbb{R}$.

Solution to Exercise 8.18.

(I) We begin by finding the radius of convergence R of the power series. Let $a_{n}=\left(\dfrac{n!}{3\cdot 5\cdots(2n+1)}\right)^{2}$ and apply the Ratio Test:

$$\lim_{n\to\infty}\frac{a_{n+1}}{a_{n}}=\lim_{n\to\infty}\frac{\left(\dfrac{(n+1)!}{3\cdot 5\cdots(2n+1)(2n+3)}\right)^{2}}{\left(\dfrac{n!}{3\cdot 5\cdots(2n+1)}\right)^{2}}$$

$$=\lim_{n\to\infty}\frac{((n+1)!)^{2}(3\cdot 5\cdots(2n+1))^{2}}{(n!)^{2}(3\cdot 5\cdots(2n+1)(2n+3))^{2}}$$

$$=\lim_{n\to\infty}\frac{(n+1)^{2}}{(2n+3)^{2}}=\lim_{n\to\infty}\frac{n^{2}+2n+1}{4n^{2}+12n+9}=\frac{1}{4}.$$

Thus, the radius of convergence is $R=\dfrac{1}{1/4}=4$, which implies that the series converges on the interval $(-4,4)$.

Now we must study the convergence at the endpoints of the interval $(-4, 4)$. We will study both of them at the same time by showing that

$$\sum_{n=0}^{\infty} \left| \left(\frac{n!}{3 \cdot 5 \cdots (2n+1)} \right)^2 (-4)^n \right| = \sum_{n=0}^{\infty} \left(\frac{n!}{3 \cdot 5 \cdots (2n+1)} \right)^2 4^n$$

converges. In order to do so, consider $b_n = \left(\dfrac{n!}{3 \cdot 5 \cdots (2n+1)} \right)^2 4^n$ and $c_n = 4^n$, and let us apply the Limit Comparison Test:

$$\lim_{n \to \infty} \frac{b_n}{c_n} = \lim_{n \to \infty} \frac{\left(\dfrac{n!}{3 \cdot 5 \cdots (2n+1)} \right)^2 4^n}{4^n}$$

$$= \lim_{n \to \infty} \left(\frac{n!}{3 \cdot 5 \cdots (2n+1)} \right)^2 = \lim_{n \to \infty} a_n.$$

Observe that we have already proven that $(a_n)_{n=1}^{\infty}$ converges by the Ratio Test. Now, as $\displaystyle\sum_{n=0}^{\infty} \frac{1}{4^n}$ converges since it is a geometric series with ratio $|r| = \dfrac{1}{4} < 1$, we have by the Limit Comparison Test that the power series also converges at -4 and 4. Therefore, the interval of convergence is $[-4, 4]$.

(II) First, we will find the radius of convergence R. Let $a_n = \dbinom{2n}{n}$ and use the Ratio Test:

$$\lim_{n \to \infty} \frac{a_{n+1}}{a_n} = \lim_{n \to \infty} \frac{\dbinom{2n+2}{n+1}}{\dbinom{2n}{n}} = \lim_{n \to \infty} \frac{\dfrac{(2n+2)!}{(n+1)!}}{\dfrac{(2n)!}{n!}} = \lim_{n \to \infty} \frac{(2n+2)!n!}{(2n)!(n+1)!}$$

$$= \lim_{n \to \infty} \frac{(2n+2)(2n+1)}{n+1} = \lim_{n \to \infty} \frac{4n^2 + 5n + 1}{n+1} = \infty.$$

Hence, the radius of convergence is $R = 0$, which implies that the power series only converges at 0. So, the interval of convergence is $\{0\}$.

(III) We commence by obtaining the radius of convergence R of the power series. To do so, let $a_n = n(\sqrt[n]{2} - 1)$ and apply the Root Test:

$$\lim_{n \to \infty} \sqrt[n]{n(\sqrt[n]{2} - 1)} = \lim_{n \to \infty} e^{\log(n(\sqrt[n]{2}-1))^{1/n}} = \lim_{n \to \infty} e^{\frac{1}{n} \log(n(\sqrt[n]{2}-1))}.$$

Independently, we will analyze the limit $\lim_{n \to \infty} n(\sqrt[n]{2} - 1)$. Take $f(x) = x(\sqrt[x]{2} - 1)$, then

$$\lim_{n \to \infty} n(\sqrt[n]{2} - 1) = \lim_{x \to \infty} f(x) = \lim_{n \to \infty} \log x(\sqrt[x]{2} - 1) = \lim_{x \to \infty} \frac{2^{1/x} - 1}{1/x}.$$

Now applying L'Hopital's Rule yields

$$\lim_{n \to \infty} n(\sqrt[n]{2} - 1) = \lim_{x \to \infty} \frac{\log 2 \cdot 2^{1/x}\left(-\frac{1}{x^2}\right)}{-\frac{1}{x^2}} = \log 2 \cdot \lim_{x \to \infty} 2^{1/x} = \log 2.$$

Hence,

$$\lim_{n \to \infty} \sqrt[n]{n(\sqrt[n]{2} - 1)} = e^{\lim_{n \to \infty} \frac{1}{n} \cdot \log(\log 2)} = 1.$$

Thus, the radius of convergence is $R = 1$.

We have proven so far that the power series converges on the interval $(-1, 1)$. Now, since

$$\lim_{n \to \infty} n(\sqrt[n]{2} - 1) = \log 2 \neq 0.$$

the series does not converge for $x = \pm 1$.

(IV) We begin by calculating the radius of convergence R of the power series. Let $a_n = \dfrac{2^n}{n^2}$ and apply the Ratio Test:

$$\lim_{n \to \infty} \frac{a_{n+1}}{a_n} = \lim_{n \to \infty} \frac{\frac{2^{n+1}}{(n+1)^2}}{\frac{2^n}{n}} = \lim_{n \to \infty} \frac{2^{n+1} n^2}{2^n (n+1)^2} = \lim_{n \to \infty} \frac{2n^2}{n^2 + 2n + 1} = 2.$$

Therefore, the radius of convergence is $R = \dfrac{1}{2}$.

So far we have shown that the power series converges on the interval $(-1/2, 1/2)$. Now let us study the endpoints of the latter interval. Assuming that $x = \dfrac{1}{2}$, we obtain

$$\sum_{n=0}^{\infty} \frac{2^n}{n^2} \left(\frac{1}{2}\right)^n = \sum_{n=0}^{\infty} \frac{2^n}{n^2 2^n} = \sum_{n=0}^{\infty} \frac{1}{n^2},$$

which is a p-series with $p = 2 > 1$, so convergent. As an immediate consequence, we get that $\sum_{n=0}^{\infty} \dfrac{2^n}{n^2} \left(-\dfrac{1}{2}\right)^n$ is absolutely convergent. Thus, the interval of convergence is $\left[-\dfrac{1}{2}, \dfrac{1}{2}\right]$.

(V) Let us find the radius of convergence R of the power series. Consider $a_n = \dfrac{2^n}{n!}$ and use the Ratio Test:

$$\lim_{n \to \infty} \frac{a_{n+1}}{a_n} = \lim_{n \to \infty} \frac{\frac{2^{n+1}}{(n+1)!}}{\frac{2^n}{n!}} = \lim_{n \to \infty} \frac{2^{n+1} n!}{2^n (n+1)!}$$

$$= \lim_{n \to \infty} \frac{2n}{n+1} = 2.$$

So, the radius of convergence is $R = \dfrac{1}{2}$.

It remains to study the convergence of the power series at $-1/2$ and $1/2$. Taking $x = \dfrac{1}{2}$, we have

$$\sum_{n=0}^{\infty} \frac{2^n}{n!} \left(\frac{1}{2}\right)^n = \sum_{n=0}^{\infty} \frac{2^n}{n!2^n} = \sum_{n=0}^{\infty} \frac{1}{n!}$$

which converges to the number e. And, in particular, we also obtain that $\sum_{n=0}^{\infty} \dfrac{2^n}{n!} \left(-\dfrac{1}{2}\right)^n$ is absolutely convergent. Hence, the interval of convergence is $\left[-\dfrac{1}{2}, \dfrac{1}{2}\right]$.

(VI) We start with finding the radius of convergence R of the power series. Let $a_n = \dfrac{3^n}{n4^n}$ and apply the Ratio Test:

$$\lim_{n\to\infty} \frac{a_{n+1}}{a_n} = \lim_{n\to\infty} \frac{\dfrac{3^{n+1}}{(n+1)4^{n+1}}}{\dfrac{3^n}{n4^n}} = \lim_{n\to\infty} \frac{3^{n+1}n4^n}{3^n(n+1)4^{n+1}}$$

$$= \lim_{n\to\infty} \frac{3n}{4n+4} = \frac{3}{4}.$$

Thus, the radius of convergence is $R = \dfrac{1}{3/4} = \dfrac{4}{3}$.

We know that the power series converges on the interval $(-4/3, 4/3)$ and it remains to analyze the endpoints. Assume first that $x = \dfrac{4}{3}$, then

$$\sum_{n=1}^{\infty} \frac{3^n}{n4^n} \left(\frac{4}{3}\right)^n = \sum_{n=1}^{\infty} \frac{3^n 4^n}{n4^n 3^n} = \sum_{n=1}^{\infty} \frac{1}{n},$$

which is the harmonic series, so divergent. Now take $x = -\dfrac{4}{3}$, then

$$\sum_{n=1}^{\infty} \frac{3^n}{n4^n} \left(-\frac{4}{3}\right)^n = \sum_{n=1}^{\infty} \frac{(-1)^n 3^n 4^n}{n4^n 3^n} = \sum_{n=1}^{\infty} \frac{(-1)^n}{n},$$

which is the alternating harmonic series, so convergent. Therefore, the interval of convergence is $\left[-\dfrac{4}{3}, \dfrac{4}{3}\right)$.

(VII) Let us begin by analyzing the radius of convergence R of the power series. Consider $a_n = \dfrac{(-1)^n}{n^2 4^n}$ and use the Ratio Test:

$$\lim_{n\to\infty} \frac{|a_{n+1}|}{|a_n|} = \lim_{n\to\infty} \frac{\frac{1}{(n+1)^2 4^{n+1}}}{\frac{1}{n^2 4^n}} = \lim_{n\to\infty} \frac{n^2 4^n}{(n+1)^2 4^{n+1}}$$

$$= \lim_{n\to\infty} \frac{n^2}{4n^2 + 8n + 4} = \frac{1}{4}.$$

Hence, the radius of convergence is $R = \dfrac{1}{1/4} = 4$.

Currently we know that the power series converges at least on the interval $(-4, 4)$. Now let us study the convergence at the endpoint $x = 4$. In particular, we will show that the series (when $x = 4$) is absolutely convergent:

$$\sum_{n=1}^{\infty} \left| \frac{(-1)^n}{n^2 4^n} 4^n \right| = \sum_{n=1}^{\infty} \frac{1}{n^2},$$

which is a p-series with $p = 2 > 1$, so absolutely convergent. Analogously, when $x = -4$, we have that the series is absolutely convergent. Hence, the interval of convergence of the power series is $[-4, 4]$.

(VIII) We begin by studying the radius of convergence of the power series. Let $a_n = \sqrt{n}$ and apply the Ratio Test:

$$\lim_{n\to\infty} \frac{a_{n+1}}{a_n} = \lim_{n\to\infty} \frac{\sqrt{n+1}}{\sqrt{n}} = 1.$$

So, the radius of convergence is $R = 1$.

We have proven that the power series converges at least on the interval $(-1, 1)$. It is clear that the power series diverges when $x = 1$ since $\lim_{n\to\infty} \sqrt{n} = \infty$. But also, when $x = -1$, observe that the sequence

$$\lim_{n\to\infty} |\sqrt{n}(-1)^n| = \lim_{n\to\infty} \sqrt{n} = \infty,$$

which implies that the power series does not converge at $x = -1$. Thus, the interval of convergence of the power series is $(-1, 1)$.

(IX) Notice that the coefficients a_k of this power series are

$$a_k = \begin{cases} 1 & \text{if } k = n! \text{ for some } n \in \mathbb{N}. \\ 0 & \text{otherwise.} \end{cases}$$

In order to calculate the radius of convergence R we apply directly the Cauchy-Hadamard theorem, according to which R is given by

$$R = \frac{1}{\limsup_{k\to\infty} \sqrt[k]{|a_k|}}.$$

Since a_k can only be 0 or 1 (whenever $k = n!$ for some $n \in \mathbb{N}$), it is clear that

$$\limsup_{k \to \infty} |a_k|^{\frac{1}{k}} = \lim_{n \to \infty} |a_{n!}|^{\frac{1}{n!}} = \lim_{n \to \infty} |1|^{\frac{1}{n!}} = 1.$$

Hence $R = 1$.

It remains to study the convergence at $x = 1$ and $x = -1$. The series $\sum_{n=1}^{\infty} a_n$ is not convergent since the subsequence $(a_{n!})$ of (a_n) converges to 1, so (a_n) does not converge to 0. As a consequence of the latter, $\sum_{n=1}^{\infty} (-1)^n a_n$ is not convergent either. Then the power series is not convergent for $x = \pm 1$. Therefore, the interval of convergence is $(-1, 1)$.

(x) We start by analyzing the radius of convergence R of the power series. Let $a_n = n^{-\sqrt{n}}$ and apply the Cauchy-Hadamard theorem:

$$R = \frac{1}{\lim_{n \to \infty} \sqrt[n]{a_n}} = \lim_{n \to \infty} n^{\frac{\sqrt{n}}{n}} = e^{\lim_{n \to \infty} \frac{\log n}{\sqrt{n}}} = e^0 = 1.$$

The last limit is straightforward. It can be justified in any case using the L'Hôpital's rule as follows:

$$\lim_{n \to \infty} \frac{\log n}{\sqrt{n}} = \lim_{x \to \infty} \frac{\log x}{\sqrt{x}} = \lim_{x \to \infty} \frac{\frac{1}{x}}{\frac{1}{2\sqrt{x}}} = \lim_{x \to \infty} \frac{1}{2\sqrt{x}} = 0.$$

Thus, the radius of convergence is $R = 1$.

Now we have to study the convergence at the endpoints of the interval $(-1, 1)$. In particular, we will show that $\sum_{n=1}^{\infty} n^{-\sqrt{n}} (-1)^n$ converges absolutely, which also implies that the power series converges at $x = 1$. Let $b_n = (-1)^n n^{-\sqrt{n}}$ and $c_n = \frac{1}{n^2}$. Then,

$$\lim_{n \to \infty} \frac{|b_n|}{c_n} = \lim_{n \to \infty} \frac{\frac{1}{n^{\sqrt{n}}}}{\frac{1}{n^2}} = \lim_{n \to \infty} \frac{n^2}{n^{\sqrt{n}}} = 0 < 1.$$

So, by the Limit Comparison Test, the series $\sum_{n=1}^{\infty} n^{-\sqrt{n}} (-1)^n$ absolutely converges. Hence, the interval of convergence is $[-1, 1]$.

(XI) We commence finding the radius of convergence R of the power series. Let us apply the Ratio Test in the following way:

$$\lim_{n\to\infty} \left| \frac{\dfrac{3^{n+1}}{\sqrt{n+1}}x^{2n+3}}{\dfrac{3^n}{\sqrt{n}}x^{2n+1}} \right| = \lim_{n\to\infty} |x|^2 \cdot \frac{3^{n+1}\sqrt{n}|x|^{2n+1}}{3^n\sqrt{n+1}|x|^{2n+1}}$$

$$= |x|^2 \lim_{n\to\infty} \frac{3\sqrt{n}}{\sqrt{n+1}} = 3|x|^2.$$

The latter limit is less than 1 if, and only if, $|x| < \dfrac{1}{\sqrt{3}}$; and it is greater than 1 if, and only if, $|x| > \dfrac{1}{\sqrt{3}}$. Hence, by the Ratio Test, if $x \in (-1/\sqrt{3}, 1/\sqrt{3})$, then the power series converges. But also, if $x \in (-\infty, -1/\sqrt{3}) \cup (1/\sqrt{3}, \infty)$, then the power series diverges.

It remains to study the convergence at the endpoints of the interval $(-1/\sqrt{3}, 1/\sqrt{3})$. In particular, we will prove that $\displaystyle\sum_{n=1}^{\infty} \frac{3^n}{\sqrt{n}}\left(-\frac{1}{3}\right)^{2n+1}$ absolutely converges, which implies that $\displaystyle\sum_{n=1}^{\infty} \frac{3^n}{\sqrt{n}}\left(\frac{1}{3}\right)^{2n+1}$ also converges. Notice that

$$\sum_{n=1}^{\infty} \left| \frac{3^n}{\sqrt{n}}\left(-\frac{1}{3}\right)^{2n+1} \right| = \sum_{n=1}^{\infty} \frac{3^n}{3^{2n+1}\sqrt{n}}$$

$$= \sum_{n=1}^{\infty} \frac{1}{3^{n+1}\sqrt{n}}$$

$$\leq \sum_{n=1}^{\infty} \frac{1}{3^{n+1}}$$

$$= \frac{1}{3}\sum_{n=1}^{\infty} \left(\frac{1}{3}\right)^n.$$

Now, since $\displaystyle\sum_{n=1}^{\infty} \left(\frac{1}{3}\right)^n$ is a geometric series with ratio $|r| = \dfrac{1}{3} < 1$, we have by the Direct Comparison Test that $\displaystyle\sum_{n=1}^{\infty} \frac{3^n}{\sqrt{n}}\left(-\frac{1}{3}\right)^{2n+1}$ absolutely converges.

Hence, the interval of convergence is $\left[-\dfrac{1}{\sqrt{3}}, \dfrac{1}{\sqrt{3}}\right]$.

(XII) We begin calculating the radius of convergence R of the power series. First of all, note that

$$\sum_{n=1}^{\infty} n! \left(\frac{x}{n}\right)^n = \sum_{n=1}^{\infty} \frac{n!}{n^n} x^n.$$

Let $a_n = \dfrac{n!}{n^n}$ and use the Root Test:

$$\lim_{n\to\infty} \sqrt[n]{a_n} = \lim_{n\to\infty} \sqrt[n]{\frac{n!}{n^n}} = \lim_{n\to\infty} \frac{\sqrt[n]{n!}}{n} = \lim_{n\to\infty} \frac{n/e}{n} = \frac{1}{e},$$

where we have applied Stirling's Weak Formula. Thus, the radius of convergence is $R = \dfrac{1}{1/e} = e$.

Now we must study the convergence of the power series at the endpoints of the interval $(-e, e)$. First, let us analyze the endpoint $x = e$. Observe that

$$\lim_{n\to\infty} \frac{n! e^n}{n^n} = \lim_{n\to\infty} \sqrt{2\pi n} = \infty,$$

where we have applied Stirling's Formula. Then, since the limit of the sequence of general terms does not converge to 0, it yields by the Divergence Test that the power series diverges at $x = e$. Finally, we will study the endpoint $x = -e$. Note that, if the power series converges at $x = -e$, then the sequence $\left(\dfrac{n!(-e)^n}{n^n}\right)_{n=1}^{\infty}$ must converge to 0. In particular, any subsequence of $\left(\dfrac{n!(-e)^n}{n^n}\right)_{n=1}^{\infty}$ must converge to 0. However,

$$\lim_{n\to\infty} \frac{(2n)!(-e)^{2n}}{(2n)^{2n}} = \lim_{n\to\infty} \frac{(2n)! e^{2n}}{(2n)^{2n}} = \lim_{n\to\infty} \sqrt{4\pi n} = \infty,$$

a contradiction. Thus, the interval of convergence is $(-e, e)$.

(XIII) Let us begin by finding the radius of convergence R of the power series. Consider $a_n = \dfrac{\log n}{n}$ and use the Ratio Test:

$$\lim_{n\to\infty} \frac{a_{n+1}}{a_n} = \lim_{n\to\infty} \frac{\dfrac{\log(n+1)}{n+1}}{\dfrac{\log n}{n}} = \lim_{n\to\infty} \frac{n \log(n+1)}{(n+1) \log n} = \lim_{n\to\infty} \frac{\log(n+1)}{\log n} = 1.$$

The last limit can be easily justified using the L'Hôpital's rule:

$$\lim_{n\to\infty} \frac{\log(n+1)}{\log n} = \lim_{x\to\infty} \frac{\log(x+1)}{\log x} = \lim_{x\to\infty} \frac{\frac{1}{x+1}}{\frac{1}{x}} = \lim_{x\to\infty} \frac{x}{x+1} = 1.$$

Then, the radius of convergence is $R = 1$.

It remains to analyze the convergence of the power series at the endpoints of the interval $(-1, 1)$. For $x = 1$ we have,

$$a_n = \frac{\log n}{n} > \frac{1}{n}$$

for every $n \geq 3$. Since the harmonic series $\sum_{n=1}^{\infty}$ diverges to $+\infty$, the power series at $x = 1$, that is, $\sum_{n=1}^{\infty} a_n$, diverges too. Now let us study the endpoint $x = -1$. Consider the function $f(x) = \dfrac{\log x}{x}$. It is clear that $f(n) = \dfrac{\log n}{n}$ for every $n \in \mathbb{N}$ and

$$\lim_{n \to \infty} \frac{\log n}{n} = \lim_{x \to \infty} \frac{\log x}{x} = \lim_{n \to \infty} \frac{1/x}{1} = \lim_{x \to \infty} \frac{1}{x} = 0,$$

where we have applied L'Hopital's Rule. Moreover, for any $x > 0$,

$$f'(x) = \frac{1 - \log x}{x^2} < 0 \iff 1 - \log x < 0 \iff \log x > 1 \iff x > e.$$

This proves that the sequence $(f(n))$ is essentially decreasing. Hence, by the Alternating Series Test, the power series converges at $x = -1$. We conclude that the interval of convergence is $[-1, 1)$.

(XIV) Observe that the power series clearly converges when $a = 0$ since

$$\sum_{n=0}^{\infty} x^n \tan \frac{0}{2^n} = 0.$$

Thus, if $a = 0$, then the interval of convergence is $\mathbb{R}$.

Next assume that $a \neq 0$. We begin calculating the radius of convergence R of the power series. Let $a_n = \tan \dfrac{a}{2^n}$ and apply the Ratio Test:

$$\lim_{n \to \infty} \frac{|a_{n+1}|}{|a_n|} = \lim_{n \to \infty} \left| \frac{\tan \dfrac{a}{2^{n+1}}}{\tan \dfrac{a}{2^n}} \right| = \lim_{n \to \infty} \left| \frac{\dfrac{\sin \dfrac{a}{2^{n+1}}}{\cos \dfrac{a}{2^{n+1}}}}{\dfrac{\sin \dfrac{a}{2^n}}{\cos \dfrac{a}{2^n}}} \right|$$

$$= \lim_{n \to \infty} \left| \frac{\sin \dfrac{a}{2^{n+1}} \cos \dfrac{a}{2^n}}{\sin \dfrac{a}{2^n} \cos \dfrac{a}{2^{n+1}}} \right| = \lim_{n \to \infty} \left| \frac{\sin \dfrac{a}{2^{n+1}}}{\sin \dfrac{a}{2^n}} \right| \left| \frac{\cos \dfrac{a}{2^n}}{\cos \dfrac{a}{2^{n+1}}} \right|$$

$$= \lim_{n \to \infty} \frac{\dfrac{1}{2} \left| \cos \dfrac{a}{2^n} \right|}{\left| \cos \dfrac{a}{2^{n+1}} \right|} = \frac{1}{2},$$

where we have used the fact that $\sin \dfrac{a}{2^n} \sim \dfrac{a}{2^n}$ and $\lim\limits_{n \to \infty} \cos \dfrac{a}{2^{n+1}} =$ $\lim\limits_{n \to \infty} \cos \dfrac{a}{2^n} = 1$. Hence, the radius of convergence is $R = \dfrac{1}{1/2} = 2$.

It remains to study the endpoints of the interval $(-2, 2)$. For $x = 2$, we have

$$\lim_{n \to \infty} 2^n \sin \frac{a}{2^n} \cos \frac{a}{2^n} = \lim_{n \to \infty} 2^n \cdot \frac{a}{2^n} \cos \frac{a}{2^n} = \lim_{n \to \infty} a \cos \frac{a}{2^n} = a \neq 0.$$

Thus, the power series does not converge at $x = 2$. Finally, let us analyze the endpoint $x = -2$. Note that, if the power series at $x = -2$ were convergent, then, the sequence $\left((-2)^n \sin \dfrac{a}{2^n} \cos \dfrac{a}{2^n} \right)_{n=1}^{\infty}$ would converge to 0. Moreover, any subsequence of $\left((-2)^n \sin \dfrac{a}{2^n} \cos \dfrac{a}{2^n} \right)_{n=1}^{\infty}$ would also converge to 0. However,

$$\lim_{n \to \infty} (-2)^{2n} \sin \frac{a}{2^{2n}} \cos \frac{a}{2^{2n}} = \lim_{n \to \infty} (-1)^{2n} 2^{2n} \cdot \frac{a}{2^{2n}} \cos \frac{a}{2^{2n}}$$

$$= \lim_{n \to \infty} a \cos \frac{a}{2^{2n}} = a \neq 0,$$

a contradiction. Therefore, the interval of convergence is $(-2, 2)$. $\qquad \square$

Exercise 8.19. *Consider* $f(x) = \int_0^x \sqrt{8 - t^3} \, dt$, *for* $x \in (-\infty, 2]$. *Find the Maclaurin series of* f, *calculating its radius of convergence. Find* $f^{(10)}(0)$.

Solution to Exercise 8.19.

Recall that, for any $n \in \mathbb{N}$, we have

$$(1 + x)^\alpha = \sum_{k=0}^{n} \binom{\alpha}{k} x^k + o(x^n)$$

with $x > -1$, $\alpha \neq 0$ and

$$\binom{\alpha}{k} = \begin{cases} \dfrac{\alpha \cdot (\alpha - 1) \cdots (\alpha - k + 1)}{k!} & \text{if } k > 0, \\ 1 & \text{if } k = 0. \end{cases}$$

Then the Maclaurin series of $(1 + x)^\alpha$ is given by

$$\sum_{k=0}^{\infty} \binom{\alpha}{k} x^k.$$

On the other hand, according to the fundamental theorem of calculus,

$$f'(x) = \sqrt{8 - x^3} = 2\sqrt{2} \left[1 - \frac{x^3}{8} \right]^{\frac{1}{2}}.$$

Combining the previous facts, the Maclaurin series of f' would be given by

$$2\sqrt{2}\left[\sum_{k=0}^{\infty}\binom{\frac{1}{2}}{k}\frac{(-1)^k}{8^k}x^{3k}\right].$$

Consequently, the Maclaurin series of f would be

$$2\sqrt{2}\left[\sum_{k=0}^{\infty}\binom{\frac{1}{2}}{k}\frac{(-1)^k}{(3k+1)8^k}x^{3k+1}\right].$$

According to Cauchy-Hadamard formula, the radius of convergence R of the previous series would be given by

$$R = \frac{1}{\lim_k \sqrt[3k+1]{\left|\frac{(-1)^k 2\sqrt{2}\binom{1/2}{k}}{(3k+1)8^k}\right|}} = \frac{2}{\sqrt[3]{\lim_k \sqrt[k]{\left|\binom{1/2}{k}\right|}}}.$$

On the other hand,

$$\lim_k \sqrt[k]{\left|\binom{1/2}{k}\right|} = \lim_k \frac{\left|\binom{1/2}{k+1}\right|}{\left|\binom{1/2}{k}\right|} = \lim_k \frac{|1/2 - k|}{k+1} = 1.$$

Therefore, $R = 2$. We conclude that

$$f(x) = 2\sqrt{2}\left[\sum_{k=0}^{\infty}\binom{\frac{1}{2}}{k}\frac{(-1)^k}{(3k+1)8^k}x^{3k+1}\right]$$

if $|x| < 2$. Observe that

$$\frac{f^{(n)}(0)}{n!} = \begin{cases} 2\sqrt{2}\binom{\frac{1}{2}}{k}\frac{(-1)^k}{(3k+1)8^k} & \text{if } n = 3k+1, \\ 0 & \text{if } n = 3k \text{ or } n = 3k+2. \end{cases}$$

Hence

$$f^{(10)}(0) = f^{(3\cdot 3+1)} = (10!)2\sqrt{2}\binom{1/2}{3}\frac{(-1)^3}{10\cdot 8^3} = -\frac{7\cdot 5\cdot 3^4}{2^5}\sqrt{2}.$$

□

Exercise 8.20. *Find the interval of convergence I of the series*
$$\sum_{n=1}^{\infty}\frac{x^{4n-1}}{4n-1}$$
and prove that for all $x \in I$ we have

$$\sum_{n=1}^{\infty}\frac{x^{4n-1}}{4n-1} = \frac{1}{4}\log\frac{1+x}{1-x} - \frac{1}{2}\arctan x.$$

Solution to Exercise 8.20.

According to Cauchy-Hadamard formula, the radius of convergence is given by

$$R = \frac{1}{\lim_n \sqrt[4n-1]{4n-1}} = \frac{1}{\lim_n \sqrt[n]{n}} = 1.$$

The series is not convergent at the endpoints of $[-1,1]$. Indeed, for $x = -1$ and $x = 1$ we would have the sequences $-\sum_{n=1}^{\infty} \frac{1}{4n-1}$ and $\sum_{n=1}^{\infty} \frac{1}{4n-1}$ respectively, which are clearly divergent. The interval of convergence is thus $(-1,1)$, which means that the function

$$f(x) = \sum_{n=1}^{\infty} \frac{x^{4n-1}}{4n-1}$$

is well-defined on $(-1,1)$. In particular, $f(0) = 0$.

On the other hand, for $x \in (-1,1)$,

$$f'(x) = \sum_{n=1}^{\infty} x^{4n-2} = x^2 \left[\sum_{n=0}^{\infty} (x^4)^n \right] = \frac{x^2}{1 - x^4}.$$

Observe that in the last equality we just need to put $r = x^4$ in the well-known formula for the sum of a geometric series of common ratio $r \in (-1,1)$, given by

$$\sum_{n=0}^{\infty} r^n = \frac{1}{1 - r}.$$

Then f is nothing but a primitive of $\frac{x^2}{1-x^4}$. To integrate that function we have find the partial fraction decomposition of

$$\frac{x^2}{1 - x^4} = \frac{-x^2}{(x-1)(x+1)(x^2+1)},$$

namely

$$\frac{-x^2}{(x-1)(x+1)(x^2+1)} = \frac{A}{x-1} + \frac{B}{x+1} + \frac{Cx+D}{x^2+1}$$
$$= \frac{A(x+1)(x^2+1) + B(x-1)(x^2+1) + (Cx+D)(x^2-1)}{(x-1)(x+1)(x^2+1)},$$

from which

$$-x^2 = A(x+1)(x^2+1) + B(x-1)(x^2+1) + (Cx+D)(x^2-1).$$

Next, letting $x = \pm 1$, $x = 0$ and $x = 2$ in the previous equality we arrive at the system,

$$\begin{array}{rcrcrcrcl}
4A & & & & & & & = & -1, \\
& & 4B & & & & & = & 1, \\
A & - & B & & & - & D & = & 0, \\
15A & + & 5B & + & 6C & + & 3D & = & -4.
\end{array}$$

From the first two equations $A = -\frac{1}{4}$ and $B = \frac{1}{4}$. Plugging A and B in the third equation, $D = -\frac{1}{2}$. Using the last equation

$$C = \frac{-4 - 15A - 5B - 3D}{3} = \frac{1}{3}\left(-4 + \frac{15}{4} - \frac{5}{4} + \frac{3}{2}\right) = 0.$$

Hence

$$\int \frac{x^2}{1 - x^4}\,dx = -\frac{1}{4}\int \frac{dx}{x-1} + \frac{1}{4}\int \frac{dx}{x+1} - \frac{1}{2}\int \frac{dx}{1+x^2}$$

$$= -\frac{1}{4}\log|x-1| + \frac{1}{4}\log|x+1| - \frac{1}{2}\arctan(x) + C$$

$$= \frac{1}{4}\log\left|\frac{x+1}{x-1}\right| - \frac{1}{2}\arctan(x) + C,$$

from which, for $x \in (-1, 1)$,

$$f(x) = \frac{1}{4}\log\frac{x+1}{1-x} - \frac{1}{2}\arctan(x) + C.$$

Taking into account that $f(0) = 0$, it follows that $C = 0$, concluding the proof. $\qquad\square$

Exercise 8.21. *Find the unique power series* $f(x) = \displaystyle\sum_{n=0}^{\infty} a_n x^n$ *with positive radius of convergence such that* $f'' + f = 0$, $f(0) = 1$, $f'(0) = 0$. *Can you identify the function* f?

Solution to Exercise 8.21.

Notice that $a_0 = f(0) = 1$ and $a_1 = f'(0) = 0$. On the other hand

$$0 = f''(x) + f(x) = \sum_{n=2}^{\infty} n(n-1)a_n x^{n-2} + \sum_{n=0}^{\infty} a_n x^n$$

$$= \sum_{n=0}^{\infty} (n+2)(n+1)a_{n+2} x^n + \sum_{n=0}^{\infty} a_n x^n$$

$$= \sum_{n=0}^{\infty} \left[(n+2)(n+1)a_{n+2} + a_n\right] x^n.$$

It is obvious that the coefficients a_n must satisfy the equality

$$(n+2)(n+1)a_{n+2} + a_n = 0,$$

for all natural numbers n including 0. We will prove, by induction, that $a_{2k+1} = 0$ and $a_{2k} = \frac{(-1)^k}{(2k)!}$ for every natural number k including 0. The result holds for $k = 0$ since, as seen above, $a_0 = 1$ and $a_1 = 0$. Now assume that $a_{2k+1} = 0$ and $a_{2k} = \frac{(-1)^k}{(2k)!}$ for a given $k \in \mathbb{N}$. Then

$$0 = (2k+3)(2k+2)a_{2k+3} + a_{2k+1} = (2k+3)(2k+2)a_{2k+3},$$

so $a_{2k+3} = 0$. Also

$$0 = (2k+2)(2k+1)a_{2k+2} + a_{2k} = (2k+2)(2k+1)a_{2k+2} + \frac{(-1)^k}{(2k)!},$$

from which

$$a_{2k+2} = \frac{(-1)^{k+1}}{(2k+2)!},$$

concluding the induction.

To finish, we have arrived at the conclusion that

$$f(x) = \sum_{k=0}^{\infty} \frac{(-1)^k}{(2k)!} x^{2k}$$

which is the Maclaurin series expansion of $\cos x$.   □

Exercise 8.22. *Given the power series* $f(x) = \displaystyle\sum_{n=1}^{\infty} \frac{x^{3n}}{n(3n-1)}$*:*

(I) *Find the interval of convergence.*

(II) *Obtain* $f(x)$ *explicitly for any* x *on the open interval of convergence.*

(III) *Calculate* $\displaystyle\sum_{n=1}^{\infty} \frac{(-1)^n}{n(3n-1)}$*.*

Solution to Exercise 8.22.

(I) Let $a_n = \frac{1}{n(3n-1)}$. Then, the radius of convergence is

$$R = \frac{1}{\lim_n \frac{a_{n+1}}{a_n}} = \frac{1}{\lim_n \frac{n(3n-1)}{(n+1)(3n+2)}} = 1.$$

If we plug $x = \pm 1$ in the power series we obtain the numerical series $\sum_{n=1}^{\infty} \frac{(-1)^n}{n(3n-1)}$ and $\sum_{n=1}^{\infty} \frac{1}{n(3n-1)}$. Since both are clearly absolutely convergent, the interval of convergence of f is $[-1, 1]$.

(II) Observe that $f(x) = xg(x)$, where $g(x) = \sum_{n=1}^{\infty} \frac{x^{3n-1}}{n(3n-1)}$ for all $x \in (-1, 1)$. Then, if $x \in (-1, 1) \setminus \{0\}$,

$$g'(x) = \sum_{n=1}^{\infty} \frac{x^{3n-2}}{n} = \frac{h(x)}{x^2},$$

where $h(x) = \sum_{n=1}^{\infty} \frac{x^{3n}}{n}$. Let us calculate h first. We have

$$h'(x) = 3 \sum_{n=1}^{\infty} x^{3n-1} = 3x^2 \sum_{n=0}^{\infty} \left(x^3\right)^n = \frac{3x^2}{1 - x^3},$$

in the last equality we have used the formula $\sum_{n=0}^{\infty} r^n = \frac{1}{1 - r}$ whenever $r \in (-1, 1)$ with $r = x^3$. By elementary integration and taking into consideration that $h(0) = 0$ we end up with

$$h(x) = -\log(1 - x^3)$$

for all $x \in (-1, 1)$. Hence

$$g'(x) = -\frac{\log(1 - x^3)}{x^2}$$

for all $x \in (-1, 1) \setminus \{0\}$. If we put $u = \log(1 - x^3)$ and $dv = -\frac{1}{x^2}$, then $du = \frac{3x^2}{x^3 - 1} dx$ and $v = \frac{1}{x}$, so integrating by parts,

$$g(x) = \frac{\log(1 - x^3)}{x} - 3 \int \frac{x}{x^3 - 1} dx.$$

To calculate the last integral we need to find the partial fraction decomposition of $\frac{x}{x^3 - 1}$, which we do next:

$$\frac{x}{x^3 - 1} = \frac{1}{(x - 1)(x^2 + x + 1)} = \frac{A}{x - 1} + \frac{Bx + C}{x^2 + x + 1}$$

$$= \frac{A(x^2 + x + 1) + (Bx + C)(x - 1)}{x^3 - 1}$$

$$= \frac{(A + B)x^2 + (A - B + C)x + A - C}{x^3 - 1},$$

from which $(A + B)x^2 + (A - B + C)x + A - C = x$. We obtain the system,

$$\begin{array}{ccccccc} A & + & B & & & = & 0, \\ A & - & B & + & C & = & 1, \\ A & & & - & C & = & 0. \end{array}$$

Summing the three equations we get $3A = 1$, so $A = \frac{1}{3}$. Using that $A = \frac{1}{3}$ in the first and third equations, $B = -\frac{1}{3}$ and $C = \frac{1}{3}$ respectively. Then

$$3 \int \frac{x}{x^3 - 1} dx = \int \frac{dx}{x - 1} - \int \frac{x - 1}{x^2 + x + 1} dx$$

$$= \log |x - 1| - \frac{1}{2} \int \frac{2x + 1}{x^2 + x + 1} dx + \frac{3}{2} \int \frac{dx}{x^2 + x + 1}$$

$$= \log |x - 1| - \log \sqrt{x^2 + x + 1} + \frac{3}{2} \int \frac{dx}{x^2 + x + 1}$$

$$= \log \frac{|x - 1|}{\sqrt{x^2 + x + 1}} + \frac{3}{2} \int \frac{dx}{\frac{3}{4} + \left(x + \frac{1}{2}\right)^2}$$

$$= \log \frac{|x - 1|}{\sqrt{x^2 + x + 1}} + 2 \int \frac{dx}{1 + \left(\frac{2x+1}{\sqrt{3}}\right)^2}$$

$$= \log \frac{|x - 1|}{\sqrt{x^2 + x + 1}} + \sqrt{3} \arctan \left(\frac{2x + 1}{\sqrt{3}}\right) + C.$$

Therefore,

$$g(x) = \frac{\log(1 - x^3)}{x} - \log \frac{1 - x}{\sqrt{x^2 + x + 1}} - \sqrt{3} \arctan \left(\frac{2x + 1}{\sqrt{3}}\right) - C$$

for $x \in (-1, 1) \setminus \{0\}$. To calculate C we just need to take into consideration that

$$0 = g(0) = \lim_{x \to 0} g(x) = \lim_{x \to 0} \frac{\log(1 - x^3)}{x} - \sqrt{3} \arctan \left(\frac{1}{\sqrt{3}}\right) - C$$

$$= \lim_{x \to 0} \frac{x^3}{x} - \frac{\sqrt{3}\pi}{6} - C = -\frac{\sqrt{3}\pi}{6} - C,$$

so $C = -\frac{\sqrt{3}\pi}{6}$, and

$$f(x) = \log(1 - x^3) - x \log \frac{1 - x}{\sqrt{x^2 + x + 1}} - \sqrt{3}x \arctan \left(\frac{2x + 1}{\sqrt{3}}\right) + \frac{\sqrt{3}\pi}{6} x,$$

for $x \in (-1, 1)$.

(III) Notice that

$$\left| \frac{x^{3n}}{n(3n - 1)} \right| \leq \frac{1}{n(3n - 1)},$$

for all $x \in [-1, 1]$ and that $\sum_{n=1}^{\infty} \frac{1}{n(3n-1)}$ is convergent. Then, by Weierstrass M-test, the series $\sum_{n=1}^{\infty} \frac{x^{3n}}{n(3n - 1)}$ is uniformly convergent in $[-1, 1]$. Thus f is continuous in $[-1,1]$, so using the expression for $f(x)$ obtained above, we get

$$f(-1) = \lim_{x \to -1^-} f(x) = \log 2 + \log 2 + \sqrt{3}\arctan\left(\frac{-1}{\sqrt{3}}\right) - \frac{\sqrt{3}\pi}{6}$$

$$= \log 4 - \frac{\sqrt{3}\pi}{3}.$$

$\square$

Bibliography

[1] L. Abellanas and A. Galindo, *Métodos de cálculo*, Schaum, McGraw-Hill Interamericana de España, Madrid, 1989.

[2] L. Abellanas and M. R. Spiegel, *Fórmulas y tablas de matemática aplicada*, Schaum, McGraw-Hill Interamericana de España, Madrid, 1991.

[3] T. A. Apostol, *Análisis Matemático*, 2nd ed., Reverté, Barcelona, 1991.

[4] R. G. Bartle and D. R. Sherbert, *Introducción al análisis matemático de una variable*, Limusa, México, 1990.

[5] L. Bernal González, J. L. Gámez Merino, G. A. Muñoz Fernández, V. M. Sánchez de los Reyes, and J. B. Seoane Sepúlveda, *Análisis de variable real*, Madrid, Paraninfo, 2023.

[6] L. Bernal González, J. L. Gámez Merino, G. A. Muñoz Fernández, and J. B. Seoane Sepúlveda, *Real Analysis: An Undergraduate Textbook for Mathematicians, Applied Scientists, and Engineers*, CRC Press, Boca Raton, FL (in press), 2024.

[7] A. J. Durán, *Historia, con personajes, de los conceptos del cálculo*, Alianza, Madrid, 1996.

[8] I. Grattan-Guinness, *Del cálculo a la teoría de conjuntos, 1630—1910: Una introducción histórica*, Alianza, Madrid, 1984.

[9] M. de Guzmán, *El rincón de la pizarra: ensayos de visualización en análisis matemático*, Pirámide, Madrid, 1996.

[10] M. de Guzmán and B. Rubio, *Problemas, conceptos y métodos del análisis matemático*, Vol. I, Pirámide, Madrid, 1993.

[11] P. D. Lax, S. Z. Burstein, and A. Lax, *Calculus with applications and computing*, Vol. I, Courant Institute of Mathematical Sciences, Universidad de Nueva York, New York, 1972. Notas basadas en un curso impartido en la Universidad de Nueva York.

[12] F. Le Lionnais, *Les nombres remarquables*, Herman, París, 1983.

[13] E. Maor, *The story of a number*, Princeton University Press, Princeton University, Princeton, NJ, 1994.

[14] K. A. Ross, *Elementary analysis: the theory of calculus*, Undergraduate Texts in Mathematics, Springer-Verlag, New York, Heidelberg, 1980.

[15] B. Rubio, *Números y convergencia*, B. Rubio, Madrid, 2006.

[16] ———, *Funciones de variable real*, B. Rubio, Madrid, 2006.

[17] W. Rudin, *Principios de análisis matemático*, Madrid, McGraw-Hill, 1976.

[18] M. Spivak, *Cálculo infinitesimal*, 2nd ed., Barcelona, Reverté, 1994.

[19] V. A. Zorich, *Mathematical analysis*, Universitext, vol. I, Springer-Verlag, Berlin, 2004. Translated from the 4th Russian edition of 2002 by.

Printed in the United States
by Baker & Taylor Publisher Services